KU-793-396
UNIVERSITY LIBRARY LIVERPOOL

MATERIALS FOR CIVIL AND HIGHWAY ENGINEERS

UNIVERSITY LIBRARY LIVERPOOL

MATERIALS FOR CIVIL AND HIGHWAY ENGINEERS

Kenneth N. Derucher

Professor of Civil Engineering
Division of Civil Engineering
Stevens Institute of Technology
Hoboken, NJ

Conrad P. Heins

Professor of Civil Engineering
Institute for Physical Science and Technology
University of Maryland
College Park, MD

Prentice-Hall, Inc., Englewood Cliffs, New Jersey 07632

Library of Congress Cataloging in Publication Data

DERUCHER, KENNETH N *1949-*
Materials for civil and highway engineers.

Includes bibliographies and index.
1. Materials. 2. Civil engineering--Equipment and supplies. I. Heins, Conrad P., joint author II. Title.
TA403.2.D47 620.1'1 80-18313
ISBN 0-13-560490-7

Editorial/production supervision
and interior design by Karen Skrable
Cover design by Edsal Enterprises
Manufacturing buyer: Anthony Caruso

© 1981 by Prentice-Hall, Inc., Englewood Cliffs, N.J. 07632

All rights reserved. No part of this book may be reproduced in any form or by any means without permission in writing from the publisher.

Printed in the United States of America

10 9 8 7 6 5 4 3 2 1

Prentice-Hall International, Inc., *London*
Prentice-Hall of Australia Pty. Limited, *Sydney*
Prentice-Hall of Canada, Ltd., *Toronto*
Prentice-Hall of India Private Limited, *New Delhi*
Prentice-Hall of Japan, Inc., *Tokyo*
Prentice-Hall of Southeast Asia Pte. Ltd., *Singapore*
Whitehall Books Limited, *Wellington*, *New Zealand*

DEDICATED TO

Barbara Frick Derucher

Richard D. Walker

William W. Payne

Oscar E. Manz

Presley A. Wedding

CONTENTS

800724

PREFACE

There have been many books on engineering materials written for Material Scientists, but none written for Civil Engineers. Unfortunately, Civil Engineers have had to rely on these books for lack of better or more appropriate texts.

Materials for Civil and Highway Engineers was designed specifically for Civil Engineers. Unlike other texts, *Materials for Civil and Highway Engineers* covers all materials used in civil engineering. The text discusses the engineering performance of metals, cements, concrete, timber, asphalt, and plastics.

Materials for Civil and Highway Engineers is intended for use by second- or third-year students taking introductory courses in civil engineering materials. The text is meant for use at a four-year institute, but it would be equally acceptable at a two-year school. *Materials for Civil and Highway Engineers* complements the laboratories that are a major part of civil engineering materials courses.

Chapters 1 and 2 cover metals: the metallic state and ferrous and nonferrous metals. Chapter 3 discusses mineral aggregates. Chapter 4 deals with, in addition to portland cement, many types of cement. Chapters 5, 6, and 7 are concerned with concrete: its general strength, its failures, and the design of concrete mixes. Chapter 8 is devoted to timber: classifications, physical characteristics, mechanical properties, decay, durability, and preservation. Chapter 9 covers asphalt and asphaltic cements such as petroleum asphalts, liquid asphalts, and road tars; and their specifications and mix design procedures.

Finally, the appendix provides ASTM standards for various tests.

As principal author, I would like to thank Conrad Heins for obtaining some of the material for Chapter 2.

Kenneth N. Derucher

MATERIALS FOR CIVIL AND HIGHWAY ENGINEERS

THE METALLIC STATE 1

To understand the macroscopic properties of metals and alloys in any detail, it is necessary to have some knowledge of the metallic state at the atomic level. All of the large-scale effects metals exhibit under stress may be accurately explained by the interactions occurring microscopically. Indeed, it is knowledge of the physics involved between individual atoms and groups of atoms that allow predictions of large-scale behavior to be made. The degree to which one can correlate actual test results with known atomic structure will mark the extent to which one can make correct decisions in the selection of metals for a given design. Such ability is only possible through an understanding of the manner in which metallic properties derive from the nature of the metallic state. To facilitate this understanding, we will begin with a discussion of the crystalline structure of metals.

THE CRYSTALLINE NATURE OF METALS

The Metallic Bond

In a pure metal or alloy, each atom is independent of its neighbors; that is, there are no definable molecules. The valence electrons of each atom exist in an unconfined state, free to move randomly among the other atoms at very high velocity. This freedom of movement is due to the nearly equal energy levels existing on the valence shells of the atoms in the lattice. Thus, electrons may move from one atom to the next without disrupting the electronic equilibrium of the structure. The aggregation of all free valence electrons is termed the *electron cloud,* a singular property of the metallic state. As we shall see, the freedom of electrons from individual atoms is responsible for the unique elastic behavior of metals.

When the metal atoms give up their valence electrons to the

electron cloud in metallic bonding, they become positive ions, held in position through their attraction to the electron cloud and their mutual repulsion. For the structure to exist in equilibrium, these forces must balance in such a way that the resultant is zero. At some specific atomic spacing a level of lowest energy is reached for which the internal forces cancel each other. The structure will seek to return to this equilibrium position if displaced, giving rise to the elastic behavior of metals in tension and compression. Figure 1-1 is a graph of force versus distance for two ions. The points above the horizontal axis correspond to repulsive force, those below represent the action of attraction, and the point where

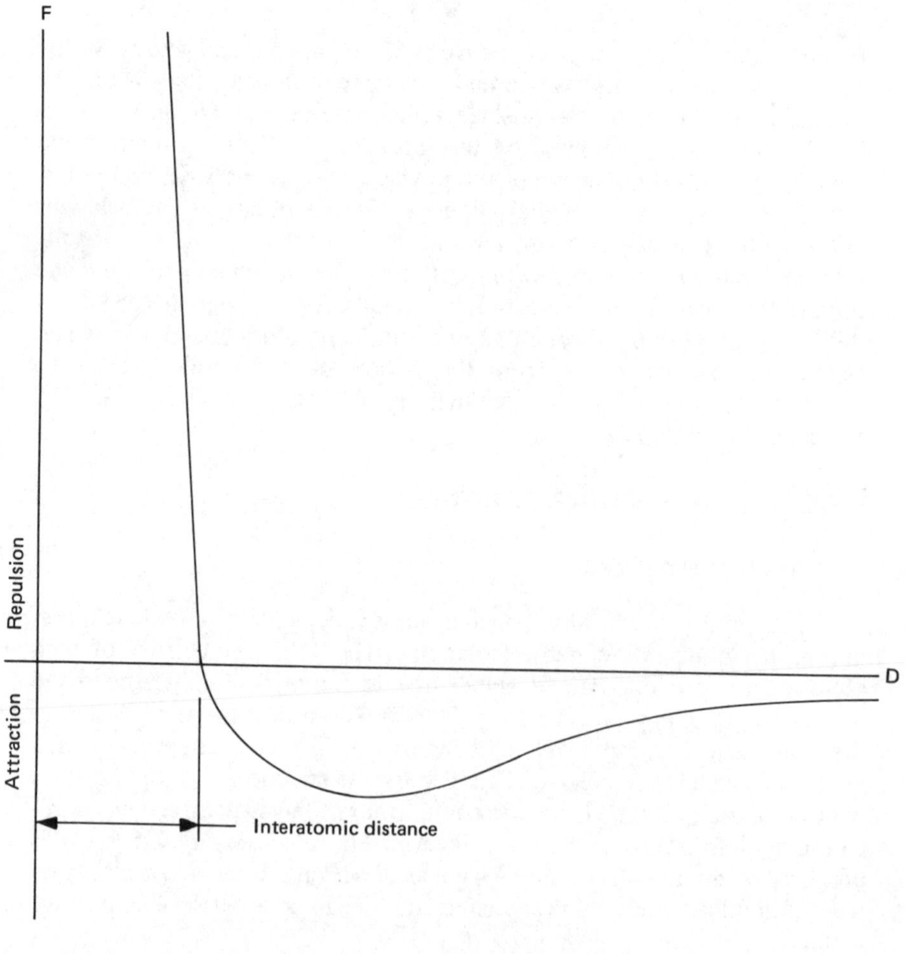

Fig. 1-1 Force versus Distance

the curve crosses the horizontal axis is the equilibrium spacing or mean interatomic distance. If the structure is placed in tension (i.e., an attempt is made to increase the atomic spacing), the attractive force predominates, resisting the deformation and restoring the equilibrium position after load is removed. Similarly, when compressive forces are applied, the ions are pushed closer together and the repulsive force due to their like charges resists the deformation, again restoring the lattice to its equilibrium spacing when load is removed.

For equilibrium to exist throughout the structure (in the absence of any external or internal forces), the interatomic distance must be the same for all atoms in the lattice. Each metal or alloy is associated with a unique interatomic distance which allows individual metals or alloys to be identified through their characteristic x-ray diffraction pattern. This pattern is a result of the crystal structure and atomic spacing, thus possessing the same uniqueness as the interatomic distance of an individual metal.

Unit Cells

The absence of any bonding requirement beyond a uniform spacing results in a very closely packed structure resembling spheres stacked in a box. The majority of metals have as lattice structure one of the following three types, illustrated in Figure 1-2: body-centered cubic (BCC), face-centered cubic (FCC), or hexagonal close-packed (HCP). These geometrical figures, called *unit cells,* represent the smallest configuration of atoms that preserve the characteristic arrangement of the lattice. Table 1-1 lists some of the important metallic elements with their interatomic spacing and unit cell configuration in order of increasing interatomic distance.

Allotropic Behavior

Inspection of Table 1-1 will reveal several elements that have more than one crystalline configuration. The different crystal formations characteristic of an element are called its allotropic forms or *allotropes*. The property itself is termed *polymorphism* and occurs due to the almost identical interatomic distances existing in the allotropes of elements that exhibit this behavior. Examination of the interatomic spacing corresponding to the allotropes of the polymorphous elements cobalt, iron, and titanium listed in Table 1-1 will reveal this similarity.

The interatomic spacing, and hence the crystal structure of an allotrope, is determined by the energy level of the lattice. Iron, for example, has a FCC structure with an interatomic distance of 2.58

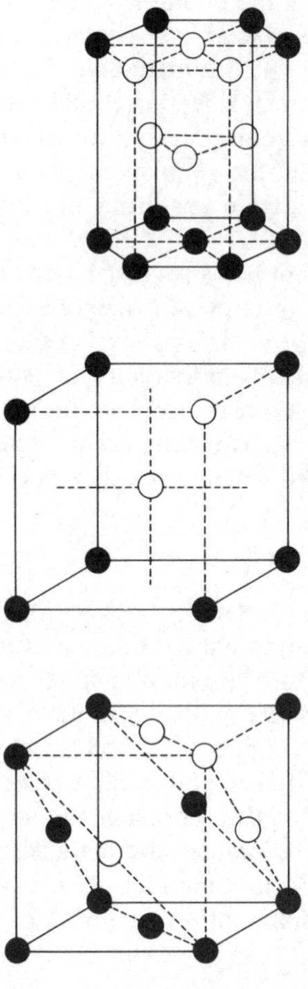

Face-centered cubic (FCC)

Body-centered cubic (BCC)

Hexagonal close-packed (HCP)

Fig. 1-2 Lattice Structure

angstroms (Å) at temperatures above 3038°F (1670°C). This is a higher-energy state than that of the BCC structure. It occurs at lower temperatures with the closer spacing of 2.48 Å.

Grain Formation and Growth

The majority of metals used in industry are polycrystalline; they are composed of a great many small crystals, called *grains*, each of which is made up of metallic ions arranged in the particular space lattice characteristic of that metal. Grain size is determined by the rate of cooling from the liquid state. As the temperature of the liquid metal falls

TABLE 1-1 Interatomic Spacing

Element	*Unit Cell*	*Interatomic Spacing (Å)*
Beryllium	HCP	2.23
Iron	BCC	2.48
	FCC	2.58
Nickel	FCC	2.49
Chromium	BCC	2.50
Cobalt	HCP	2.506
	FCC	2.511
Copper	FCC	2.56
Vanadium	BCC	2.63
Zinc	BCC	2.66
Molybdenum	BCC	2.73
Tungsten	BCC	2.74
Platinum	FCC	2.78
Aluminum	FCC	2.86
Tantalum	BCC	2.86
Gold	FCC	2.88
Titanium	HCP	2.89
	BCC	2.89
Silver	FCC	2.89
Magnesium	HCP	3.20
Lead	FCC	3.50

below its melting point, a phase change begins to occur; low-energy atoms precipitate out of the liquid state first, forming the nuclei upon which further growth takes place. The rate at which nuclei precipitate and growth progresses is a direct function of the cooling rate of the melt. If cooling is slow, few nuclei precipitate and growth is slow. This ten to produce a matrix of large, well-developed crystals. If cooling is rap as when the metal is quenched (immersed in a medium such as s water, or brine at much lower temperature), many nuclei form alm instantaneously and the structure produced is very fine grained.

Regardless of the rate of cooling, individual grains form in dendritic pattern illustrated in Figure 1-3, with growth occurring pre inantly at the ends of the dendrites. Eventually, the growth individual dendrite progresses to the point that other dendrites r its expansion. When this happens, growth continues on the existin until all internal liquid is frozen. This is a random process, dete by the charge positions nuclei occupy when they precipitate. T final shape of a grain is irregular and bears little resemblance to i ordered internal structure.

The space separating individual grains, typically one or t wide, is a region of higher energy (less order) than the grains th

Fig. 1-3 Density Pattern

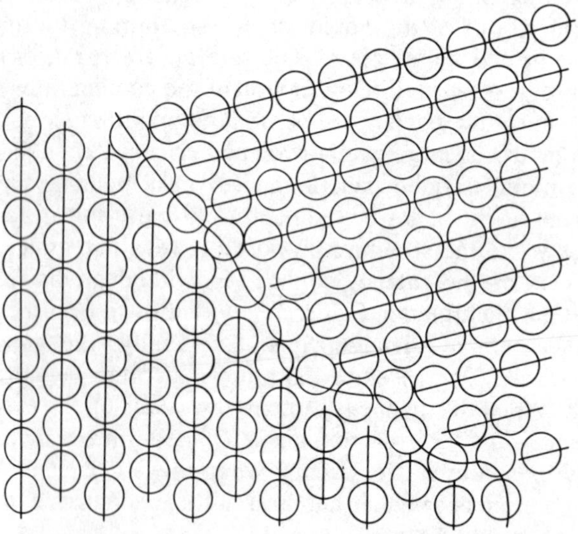

Fig. 1-4 Grain Boundary

It is a transition zone from the ordered crystalline direction of one grain to the differing orientation of adjacent grains. We shall see that grain boundaries play a significant role in influencing the characteristic behavior of a metal when stresses exceed its elastic limit. A diagram of a grain boundary is shown in Figure 1-4.

Alloying

There are certain additional elements, predominantly other metals, which will combine with a pure metal in such a way that the resulting solid solution, called an *alloy,* is also metallic. This is possible only if the alloying element lies within specific boundaries of size and valence. The size factor divides alloys into two distinct groups: substitutional and interstitial alloys. For simplicity, we consider only binary alloys, those composed of two constituent elements, in this discussion. Alloys composed of three or more elements are very common. However, the complexity of dealing with them increases rapidly as their number of constituent elements increases. We limit the scope of our examination accordingly.

In the *substitutional* alloy, the alloying element replaces the primary element in random positions in its lattice. To accomplish this without destroying the structure, the alloying element must have an atomic radii within 15 percent of that of the primary element. If the radial difference is greater than this, the energy required to place the alloying element in position is more than the structure can accommodate and still retain its crystalline form.

As the name implies, an *interstitial* alloying element fits into the spaces or interstices of the primary element's lattice. There is a definite geometrical restriction on the size of the interstitial atom; to fit into the lattice spaces, the alloying element's radius must be less than approximately six-tenths that of the primary element. This corresponds to the maximum-size sphere that will fit into the spaces existing in a close-packed lattice of larger spheres. The transition metals are the main group of elements which act as primary elements in interstitial alloys. There are only four elements small enough to meet the size requirement for interstitial alloying to occur when the primary elements are of this group. They are hydrogen, boron, carbon, and nitrogen and are termed *metalloid elements*.

Table 1-2 is a list of the metallic elements in a form known as the *electromotive series*. As one moves down the series, elements contain increasingly more complete valence shells and thus tend to lose electrons less easily than do the elements lying above them.

For two metallic elements to form a substitutional alloy over a wide range of composition, they must lie near each other in the series; that is, they must be of similar valence. If their valences differ by a sufficient amount, they will tend to exchange electrons and form a compound rather than a metallic bond when their relative proportions approach the necessary ratio. This is also true of interstitial alloys where the metalloid elements tend to form ionic compounds with nontransition metals which are relatively much more electronegative.

Alloys containing more than two elements are often both substitutional and interstitial. An important class of alloys in this category is the nickel steels; nickel and carbon are substitutional and interstitial elements, respectively, in iron, the primary element.

TABLE 1-2 Metallic Elements

Lithium	Cobalt
Potassium	Nickel
Sodium	Tin
Barium	Lead
Calcium	Antimony
Magnesium	Bismuth
Aluminum	Copper
Manganese	Mercury
Zinc	Silver
Chromium	Platinum
Iron	Gold

Phase Diagram

The *phase diagram* is a very useful tool in illustrating the characteristics of an alloy in relation to its temperature and composition. Also called an *equilibrium diagram,* this is a graph of the phases existing in an alloy for any combination of temperature and composition. Figure 1-5 is a typical phase diagram for substitutional binary alloys. In the diagram, relative percentages of one of the two constituent elements are plotted on the horizontal axis with the vertical scale corresponding to temperature. Any two distinct phases are separated by a region in which both exist simultaneously. The curves separating the different regions are the specific points of temperature and composition where one phase or combination of phases can exist in equilibrium with another. These are the loci of saturated equilibrium states.

Three different phases are shown in Figure 1-5: α, β, and L, where α and β are solid phases and L is a liquid phase. Also shown are the

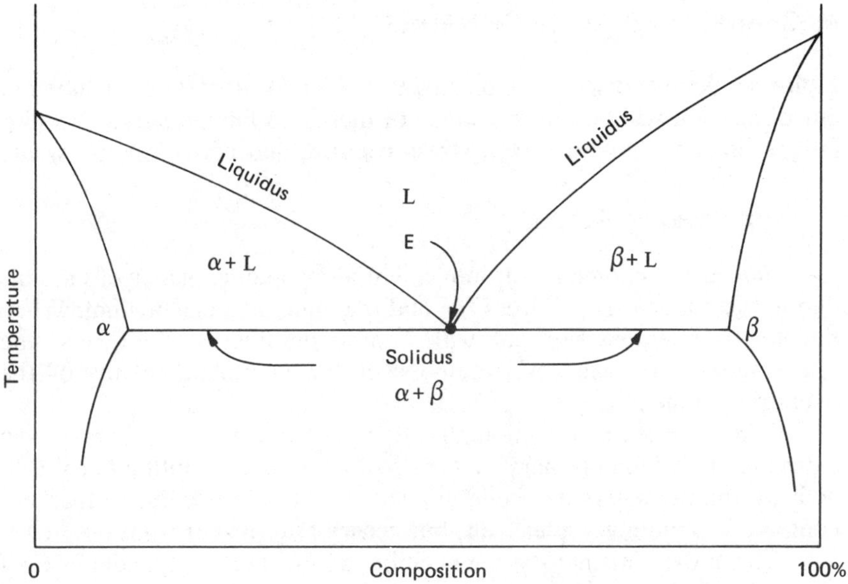

Fig. 1-5 Phase Diagram of Binary Alloy

three two-phase regions separating the single-phase regions, $\alpha + \beta$, $\alpha + L$, and $\beta + L$. The liquidus and the solidus are the boundaries between the liquid and solid phases, respectively, and the mixed regions. They converge to point *E,* the intersection of three regions, which describes the singular temperature and composition at which three phases may exist simultaneously. Such a point, which has a single liquid phase above two solid phases, is called a *eutectic;* one with a single solid phase above two other solid phases is termed a *eutectoid*. Many binary alloys exhibit phase relationships similar to Figure 1-5, with the sizes of the regions and position of the eutectic varying substantially from one alloy to the next. There are also binary alloys with a completely different form from that of Figure 1-5, and eutectoids. An example of an alloy with a phase diagram similar to Figure 1-5 is the copper–tin system. One differing markedly from this is the aluminum–nickel binary system.

For alloys with three or more elements, graphic illustration of phase relationships becomes increasingly difficult. The ternary or three-element alloys, for example, require a three-dimensional graph for complete description. Two-dimensional plots of the system may be made by taking a horizontal cross section of the three-dimensional graph, resulting in the phase relationships of the three constituents for varying compositions at a constant temperature.

MECHANICAL BEHAVIOR OF METALS

In this section we discuss the mechanical behavior of metals. Included in our discussion will be the resistance of metals to failure caused by slip, fatigue, impact, flow or creep, stress rupture, and corrosion and wear.

Slip (Inelastic Action)

Slip is a phenomenon of movement along gliding planes of crystals due to the application of stress beyond the material's elastic limit. When a material undergoes slip, one will observe slip lines or slip bends (fine lines) across the faces of a number of the crystalline grains of the overstressed metal.

Slip is an inelastic action that does not occur to any appreciable extent in any of the ordinary materials of construction until a fairly well defined limiting stress has been applied; and if slip occurs, it does not continue indefinitely, under load, but ceases after a short time.

The most satisfactory explanation of the fact that under a load above the elastic strength of a metal slip soon ceases, and of the observed increase of elastic strength after such overload, was furnished by the slip interference theory of Jeffries and Archer. As slip takes place, the crystalline grains of metal are fragmented into thin plates and the surfaces of the sliding plates tend to "dig into" each other, causing increasing resistance as slip proceeds. In addition to this, adjacent grains, in general, have slip planes lying in different directions, and as slip proceeds, the slip in one grain tends to hinder the slip in the adjacent grain.

Moreover, in metals composed of two or more different kinds of crystalline grains, the stronger grains act as "keys," tending to stop or to hinder slip in weaker grains. Slip interference theory further holds that a metal may contain definite, very small grain sizes which will be more effective as brakes on slip than larger, more widely scattered grains, and also more effective than extremely small grains, grains so small that slip can proceed right by them without much deviation from its direct path. In the case of steel, such an "optimum" size of grain may be produced by heat treatment.

Fatigue

For structural parts (axles, machine parts, shafts, etc.) under infrequent variation of load, the outstanding danger of failure is failure by inelastic distortion (slip).

Examination under the microscope and the scanning electron microscope of the behavior of the crystalline constituents of metals under repeated stress has shown that failure of crystals is caused by a succession of shear slips on parallel planes of vast strength. The failure of the piece as a whole is due to the successive failures of individual crystals and the development of cracks. Failure very seldom occurs at crystalline boundaries.

Fatigue was once thought of as a phenomenon of the cohesion between crystalline constituents of the metal subjected to many repetitions or reversals under comparatively low stresses. However, the failure of a material under repeated stress is a process of gradual or progressive fracture of the crystals themselves. Thus, the term "fatigue" is improperly used and the appropriate term should be "progressive fracture."

In a general way, the difference in the behavior of material under repeated loads and under a single load is shown by the tendency toward gradual cracking of the material under repeated loads. Under a single application of load the material of a structure either withstands the load or fails. Under a repeated load, the load may be applied many thousands of times and the material may withstand the load for a while, and then fail by the gradual spread of cracks in the material. Under repeated loads, local strains (which would be of little importance in a single load) may form a nucleus for damage which gradually spreads until the whole member fails. Slip in crystals is likely to occur at points where high localized stresses are set up. These stresses depend upon the distribution of stress between crystals, which in turn depends upon the homogeneity of the structure. Thus, flaws and cracks in the internal structure tend to cause internal stresses, which were formed by cooling, heat treatment, or mechanical working of the metal. These stresses would not normally affect the static structural strength but tend to produce slip in crystals when the metal is subjected to repeated stresses.

Impact

Impact is the resistance of a material to failure due to brittleness under service conditions in a structure. In selecting materials for members in a structure that must resist impact, two factors must be considered:

1. Total stress allowable.
2. Total strain allowable.

Resistance to impact is a function of both of these factors and both must be taken into consideration in design.

For materials that must withstand heavy accidental impact without actual rupture, toughness of the material is of primary concern. Toughness of a material may be measured by the area under the stress–strain curve of the material. According to Moore, a striking illustration of the resistance of materials to rupture under impact is furnished by comparing the action of oak with that of cast iron. Under static load cast iron is about three times as strong as oak, but the strain that oak will stand before rupture is about nine times the strain that cast iron will stand. The area under the stress–strain diagram for oak is about three times that for cast iron, and under impact loads oak requires about three times as much energy for fracture as cast iron.

The ability of a material to resist impact without a permanent distortion is measured by the area under the stress–strain diagram up to the elastic limit: in other words, the area under the straight-line portion of the stress–strain curve. If elastic resistance to impact is desired, a material with a high elastic strength or a low modulus of elasticity should be used.

Flow or Creep

Creep is the very slow flow of a material at elevated temperatures under sustained stress. The rate of flow of a given metal depends on the magnitude of the stress and the temperature. Lead and zinc, for example, will creep under normal temperatures at relatively low stresses. However, most metals require a relatively high temperature before they will begin to flow under low stresses. Creep may continue for an indefinite amount of time under a sustained load tending to distort the material and eventually cause failure of the material by rupture. Unlike repeated loads (fatigue), creep affects the entire body of the material under stress instead of producing a localized rupture.

Stress Rupture

With the development of jet propulsion and aircraft gas turbines, greater attention has been placed on metals to withstand stresses at high temperature. In the aircraft industry this temperature may be as high as 982°C (1800°F). Various laboratory tests have been developed and proven reliable in indicating the performance of the metal at high temperatures under stress. These laboratory tests eliminate time-consuming engine tests. In determining the material's life under continued high temperature and loading, stress necessary to cause rupture at a given time and temperature is the property of greatest importance. This property is sometimes referred to as *stress rupture*.

In jet engines and aircraft gas turbines operations, severe local overheating and overstressing may occur in a very short time. Thus, designs must provide for short-term strengths of alloys at temperatures somewhat higher than at expected operating levels as well as sufficient stress rupture strengths at the anticipated operating temperature levels.

Corrosion and Wear

Each year millions and millions of dollars worth of damage to iron and steel structures is caused by corrosion. Further, millions of dollars are spent to replace worn-out parts to machine structures. Corrosion and wear damage take place gradually over a period of time. They seldom cause sudden or dramatic structural disasters, as parts damaged by corrosion or wear can be replaced or repaired before failure occurs.

Most metals associated with construction materials first come in contact with water which contains dissolved oxygen or with moist air and enter into solution readily. The rate of solution is usually retarded by a film of hydrogen forming on the metal or by coating the metal with a protective coating. However, oxygen will combine with the hydrogen and over a period of time will strip it away from the metal, and thus further corrosion results.

Five classifications of corrosion for metals exist:

1. Atmospheric.
2. Water immersion.
3. Soil.
4. Chemicals other than water.
5. Electrolytic.

In atmospheric corrosion a large excess of oxygen is available and the rate of corrosion is largely determined by the quantity of moisture in the air and the length of time in contact with the iron.

When metals are immersed in water, the amount of oxygen dissolved in the water is an important factor. If the water did not contain any dissolved oxygen, the metal would not corrode. If the water is acidic, the corrosion rate is increased, whereas water that is alkaline has very little corrosion activity unless the solution is highly concentrated.

In soil corrosion and in corrosion by chemicals other than water, the most important item is the ingredient coming in contact with the iron or steel.

Corrosion by electrolysis due to stray currents from power circuits may be disastrous, but in nearly all cases it can be prevented by suitable electrical precautions.

The most common protective coating against corrosion for iron and steel is paint. The paint coating is usually mechanically weak and it cracks and wears out. Thus, for it to do a satisfactory job, the paint must be renewed every 2 or 3 years. When painting the structure, it should first be cleaned and the rust removed.

If the structure is to be immersed in water or if it comes in contact with water, paint provides little protection. Thus, the portion that is in contact with water might require a coating of asphalt or coal tar to protect it.

Another exellent method of preventing corrosion is to encase the iron or steel in concrete. Although concrete is porous, it will provide adequate protection for years. However, if the concrete becomes cracked, it loses most of its protecting ability and should be replaced if possible, or patched.

Metals under stress, especially those beyond thin elastic strength, corrode more rapidly than do unstressed metals.

In nearly all cases the failure of materials by mechanical wear under abrasion occurs gradually, the progress of wear is evident, and the failure is not a definitely defined event but one whose occurrence is a matter of judgment on the part of the user of the material. Failure by wear rarely leads to disaster, and usually involves repair or replacement of a part.

TESTING AND EVALUATION OF METALS

Tests are conducted on materials of construction in order to determine their quality and their suitability for specific uses in machines and structures. It is necessary for the producer, consumer, and the general public to have tests for the determination of quantitative properties of materials such that the material may be properly selected, specified, and designed. Tests are further needed to duplicate materials and to check upon the uniformity of different shipments.

Testing and evaluation of the many various metals and alloys requires hundreds of tests and specifications. The ASTM specifications dealing with metals and their alloys comprise 10 volumes. ASTM, Part 4, which covers various tests for steel products, is a primary reference for civil and highway engineers.

The most common mechanical tests (for metals and metallic products) are listed in ASTM A370 as follows:

1. ASTM E8: Tension Testing of Metallic Materials.
2. ASTM E10: Test for Brinell Hardness of Metallic Materials.
3. ASTM E18: Test for Rockwell Hardness and Rockwell Superficial Hardness of Metallic Materials.
4. ASTM E23: Notched-Bar Impact Testing of Metallic Materials.

ASTM E8: Tension Testing of Metallic Materials

Most commercial specifications for metals have requirements for physical properties as determined by the tensile strength test. The tension test is one of the most important tests for determining the structural and mechanical properties of metals. These properties include the ultimate strength (tensile), ductility (elongation and reduction of area), modulus of elasticity, yield point, offset yield strength, proportional limit, and the elastic and inelastic range, among others.

The *tensile test* is performed by gripping the opposite ends of a test specimen called a "coupon" and pulling it apart. Various standard ASTM coupons are shown in Figure 1-6. Standard test specimens are obtained from sheared, blanked, sawed, trepanned, or oxygen-cut materials which are being tested. In all cases, special attention should be taken to ensure the removal by machining all distorted cold-worked or heat-affected areas. Test specimens should be machined such that they have a reduced cross section at the midpoint to localize the zone of fracture. In addition, the specimen should be gaged—marked such that the percent elongation may be determined.

When loading a metal specimen, the rate of loading is usually unimportant but should not exceed 690 MPa/min. Most modern testing machines are equipped with electronic strain gages; otherwise, extensometers or foil gages (in conjunction with strain indicators and switching and balancing units) may be employed. When the coupon is fastened in the machine and the strain gage is fastened, the test begins. As the specimen is pulled apart, a load–strain (stress–strain) diagram may be plotted (automatically if electronic strain gages are used) as shown in Figure 1-7. From this diagram the various physical and/or mechanical properties may be obtained.

In a tensile test it is customary to give the strength of a material in terms of unit stress or internal force per unit of area. Also, the point at which yielding starts is expressed as unit stress.

Further, in a tensile test, strain is measured as unit strain. *Unit strain* in any direction is the deformation per unit of length in that direction. If electronic strain gages are employed in a tensile test, the plot on the stress-strain diagram is a function of the force per unit area versus the unit strain.

However, in some cases the unit strain cannot be obtained directly (as in the use of an extensometer), so the total strain or deformation is measured. *Deformation* in any direction is the total change in the dimension of a member in that direction. When the loading is such that the unit strain is constant over the length of a member, it may be

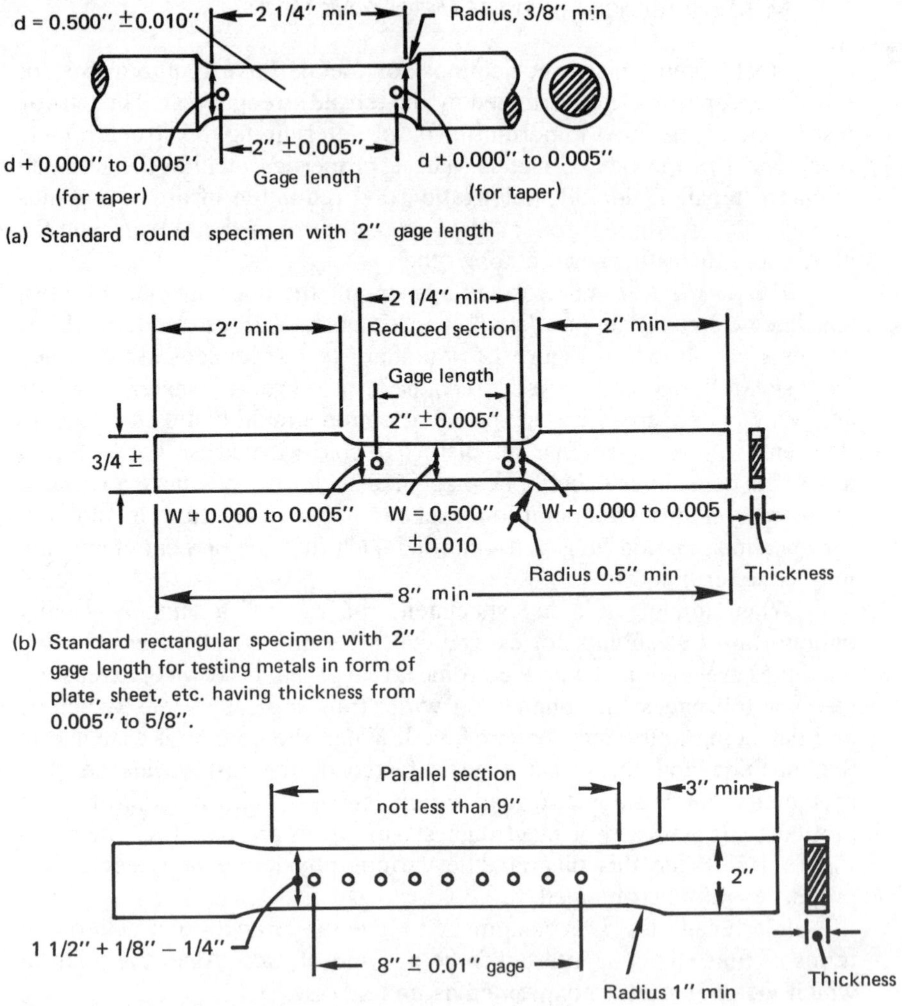

(a) Standard round specimen with 2″ gage length

(b) Standard rectangular specimen with 2″ gage length for testing metals in form of plate, sheet, etc. having thickness from 0.005″ to 5/8″.

Fig. 1-6 Various Coupons (Courtesy of ASTM)

computed by dividing the deformation by the original length of the member (gage length). Thus, unit strain equals total strain divided by the gage length.

Tensile Strength The *tensile strength* of metals is the maximum axial load (ultimate load) observed in a tension test divided by the original cross-sectional area. The strength increases and reaches a maximum in mild steel, after extensive elongation and necking. As indicated, it is

Yield Point At the termination of the linear portion of the stress–strain curve, some materials (such as a low-carbon steel) develop a yield point. The *yield point* is the first load at which there is a marked increase in strain without an increase in stress. This behavior may be a consequence of inertia due to the effects of the testing machine and the deformation characteristics of the test specimen. The yield point is sometimes taken as the proportional limit and elastic limit, which is an incorrect practice. Most metals do not have a yield point, and thus an offset method is utilized. In our example and viewing Figure 1-7, the yield point would correspond to 41,000 psi.

Offset Yield Strength The *offset yield strength* is defined as the stress corresponding to a permanent deformation, usually 0.01 or 0.20 percent (0.000) or 0.002 in./in.). The offset method is usually used with materials that have a definite straight-line portion to their stress–strain curve. One measures the corresponding offset percentage on the stress–strain curve and projects upward a straight line parallel with the straight-line portion of the stress–strain curve. Where the line intersects the stress–strain curve, the value is read off as the offset yield strength. In Figure 1-7 this value would be 42,000 psi.

In situations where the stress–strain curve does not exhibit a straight-line portion, the secant modulus or tangent modulus method may be used.

Proportional Limit The *proportional limit* is the greatest stress that a material is capable of without deviating from the law of proportionality of stress to strain (Hooke's law). *Hooke's law* is defined as

$$f = E\epsilon \tag{1.5}$$

where f = unit stress

ϵ = unit strain

E = Young's modulus of elasticity

Metals are elastic within the proportional limit and thus the proportional limit has significance in the elastic stability of columns and shells. In our example, as shown by Figure 1-7, the proportional limit is 33,000 psi.

Elastic Limit and Inelastic Limit The *elastic limit* is the largest unit stress that can be developed without a permanent set remaining after the load is removed. In most cases the elastic limit is difficult to determine, and many materials do not have a well-defined proportional limit, or even have one at all; thus, the offset yield strength is used to measure the beginning of plastic deformation (inelastic limit).

Modulus of Rigidity The *modulus of rigidity* or the shearing modulus of elasticity, as it is sometimes called, is defined as

$$G = \frac{\nu}{\gamma} \tag{1.6}$$

where G = modulus of rigidity

ν = unit shearing stress

γ = unit shearing strain

The modulus of rigidity may also be rewritten and related to the modulus of elasticity by

$$G = \frac{E}{2(1 + \mu)} \tag{1.7}$$

where μ is a constant known as *Poisson's ratio* and for structural steel equals 0.3.

Modulus of Toughness The *toughness* of a material is the ability of a material to absorb large amounts of energy. The *modulus of toughness* can be related to the area under the entire stress–strain curve, and it depends on both strength and ductility. Because of the difficulty of determining toughness analytically, often toughness is measured by the energy required to fracture a specimen, usually notched and sometimes at low temperatures, in an impact test such as the Charpy or the Izod.

Modulus of Resilience The *resilience* of a material is that property of an elastic body by which energy can be stored up in the body by loads applied to it and given up in recovering its original shape when the loads are removed. Thus, the *modulus of resilience* is equal to the area under the straight-line portion of the stress–strain curve (a triangle).

ASTM E10: Brinell Hardness of Metallic Materials

The *hardness* of metals is usually determined by measuring the resistance to penetration of a ball, cone, or pyramid. The results of a hardness test can be directly related to the tensile strength of a material.

Thus, it provides a quick check on the tensile strength of a material without going through a time-consuming tensile test.

The *Brinell hardness method* is based upon determining the resistance offered to indentation by a hard ball of specific diameter that is subjected to a given pressure. The pressure used in testing steel is usually 6600 lb (3000 kg) and the diameter of the ball is 0.4 in. (10 mm). When softer metals are utilized, a pressure of 1100 lb (500 kg) is used. The Brinnell hardness number can be computed by the following formula:

$$\mathrm{BH} = \frac{2P}{\pi D(D - \sqrt{D^2 - d^2})} \tag{1.8}$$

where P = pressure, kg

D = diameter of the steel ball, mm

d = average diameter of indentation, mm

As is obvious from the equation, the smaller the indentation, the greater the Brinell hardness number.

ASTM E18: Test for Rockwell Hardness

ASTM E18 specifies the test for the determination of the Rockwell hardness and Rockwell superficial hardness of metallic materials. The *Rockwell hardness method* employs either a ball or a diamond cone in a precision testing instrument that is designed to measure depth of penetration accurately. Two superimposed impressions are made, one with a load of 22 lb (10 kg) and the second with a load of 220 lb (100 kg). The depth to which the major load drives the ball or cone below that depth to which the minor load has previously driven it is a measure of the hardness. This method is usually designated as *Rockwell B,* as the ball is used here. For the harder steels, greater accuracy is obtained by use of a diamond cone applied under a major load of 330 lb (150 kg) and a minor load of 22 lb (10 kg) and is designated as *Rockwell C*.

Rockwell superficial hardness machines are used for the testing of very thin shells or thin surface layers of materials. In this case the minor load is 6.6 lb (3 kg) and the major load varies from 33 to 99 lb (15 to 45 kg).

In all cases the Rockwell hardness number is read directly from the scales on the machine. The relationship of the Brinell and Rockwell hardness numbers to tensile strength is shown in Table 1-3.

TABLE 1-3 Relationship of Brinell and Rockwell Hardness Numbers to Tensile Strength

Brinell Indentation Diameter (mm)	Brinell Hardness Number		Rockwell Hardness Number		Rockwell Superficial Hardness Number, Superficial Diamond Penetrator			(Approximate) Tensile Strength (MPa)
	Standard Ball	Tungsten Carbide Ball	B Scale	C Scale	15–N Scale	30–N Scale	45–N Scale	
2.50		601		57.3	89.0	75.1	63.5	2262
2.60		555		54.7	87.8	72.7	60.6	2055
2.70		514		52.1	86.5	70.3	47.6	1890
2.80		477		49.5	85.3	68.2	54.5	1738
2.90		444		47.1	84.0	65.8	51.5	1586
3.00	415	415		44.5	82.8	63.5	48.4	1462
3.10	388	388		41.8	81.4	61.1	45.3	1331
3.20	363	363		39.1	80.0	58.7	42.0	1220
3.30	341	341		36.6	78.6	56.4	39.1	1131
3.40	321	321		34.3	77.3	54.3	36.4	1055
3.50	302	302		32.1	76.1	52.2	33.8	1007
3.60	285	285		29.9	75.0	50.3	31.2	952

3.70	269	269		27.6	73.7	48.3	28.5	897
3.80	255	255		25.4	72.5	46.2	26.0	855
3.90	241	241	100.0	22.8	70.9	43.9	22.8	800
4.00	229	229	98.2	20.5	69.7	41.9	20.1	766
4.10	217	217	96.4					710
4.20	207	207	94.6					682
4.30	197	197	92.8					648
4.40	187	187	90.7					621
4.50	179	179	89.0					607
4.60	170	170	86.8					579
4.70	163	163	85.0					566
4.80	156	156	82.9					552
4.90	149	149	80.8					503
5.00	143	143	78.7					490
5.10	137	137	76.4					462
5.20	131	131	74.0					448
5.30	126	126	72.0					434
5.40	121	121	69.0					414
5.50	116	116	67.6					400
5.60	111	111	65.7					386

ASTM E23: Notched-Bar Impact Testing of Metallic Materials

Impact tests are performed primarily for two reasons:

1. To determine the ability of the material to resist impact under service conditions.
2. To determine the quality of the metal from a metallurgical standpoint.

As previously indicated, two types of impact tests are usually performed, the Charpy and the Izod tests. Both apply a dynamic load by use of a pendulum that has enough kinetic energy to rupture a specimen in its path.

PROBLEMS

1.1. Explain the crystalline nature of metals.

1.2. What is meant by allotropic behavior?

1.3. Why is grain formation and growth important in metals?

1.4. Discuss the phenomenon of slip.

1.5. Why is "fatigue" an improper term as it applies to metals?

1.6. Explain the process of flow or creep.

1.7. What are the five classifications of corrosion? Discuss.

1.8. Explain by use of your own diagram the tension testing of steel. Show all physical and mechanical properties and explain each.

1.9. Explain the two most common hardness tests.

1.10. Using the library and ASTM standards, explain the Charpy impact test.

REFERENCES

BRICK, R. M., GORDEN, R. B., and PHILLIPS, A., *The Structure and Properties of Alloys,* McGraw-Hill Book Company, New York, 1964.

CORDON, W.A., *Properties, Evaluation, and Control of Engineering Materials,* McGraw-Hill Book Company, New York, 1979.

COTTRELL, A. H., *An Introduction to Metallurgy,* Edward Arnold (Publishers) Ltd., London, 1967.

DAVIS, H.E., TROXELL, G. E., and WISKOCIL, C. T., *The Testing and Inspection of Engineering Materials,* 3rd ed., McGraw-Hill Book Company, New York, 1964.

DIETER, G. E., *Mechanical Metallurgy,* McGraw-Hill Book Company, New York, 1961.

DOAN, G. E., and MAHLA, E. M., *The Principles of Physical Metallurgy,* McGraw-Hill Book Company, New York, 1941.

GILLET, H.W., *The Behavior of Engineering Metals,* John Wiley & Sons, Inc., New York, 1951.

JASTRZEBSKI, Z. D., *The Nature and Properties of Engineering Materials,* John Wiley & Sons, Inc., New York, 1959.

JEFFRIES, Z. and ARCHER, R. S., *The Science of Metals,* McGraw-Hill Book Company, New York, 1924.

MOORE, H. F., *Materials of Engineering,* 7th ed. McGraw-Hill Book Company, New York, 1947.

SEITZ, F. B., *The Physics of Metals,* McGraw-Hill Book Company, New York, 1943.

VAN VLACK, L. H., *Elements of Material Science,* 2nd ed., Addison-Wesley Publishing Company, Inc., Reading, Mass., 1964.

FERROUS AND NONFERROUS METALS 2

In general, metals can be classified into two major groups: ferrous and nonferrous. A *ferrous metal* is one in which the principal element is iron, as in cast iron, wrought iron, and steel. A *nonferrous metal* is one in which the principal element is not iron, as in copper, tin, lead, nickel, aluminum, and refractory metals.

In this chapter we discuss both ferrous and nonferrous metals. Special attention will be placed on the production techniques of steel.

GENERAL

Sources of Metals

In general, over 45 metals of industrial importance are found within the earth's crust. With the exception of aluminum, iron, magnesium, and titanium, which occur in appreciable percentages within the earth's crust, all other metals comprise less than 1 percent of the earth's crust. Thus, most metals occur in the form of ore, in which the metal has to be extracted. An ore is usually referred to as a *mineral,* which is a chemical compound or mechanical mixture. The material associated with the ore which has no commercial use is referred to as *gangue*. Basically, six classifications of ore exist:

1. Native metals.
2. Oxides.
3. Sulfides.
4. Carbonates.
5. Chlorides.
6. Silicates.

The native metals consist of copper and precious metals. Oxides

are the most important ore source, in that iron, aluminum, and copper can be extracted from them. Sulfides include ores of copper, lead, zinc, and nickel. Carbonates include ores of iron, copper, and zinc. The chlorides include ores of magnesium, and the silicates include ores of copper, zinc, and beryllium.

Production of Metals

Four operations are required for the production of most metals:

1. Mining the ore.
2. Preparing the ore.
3. Extracting the metal from the ore.
4. Refining the metal.

In the mining operation, both the methods of open-pit borrowing and underground mining are utilized. The most famous examples of an open-pit operation are the iron mines in the Mesabi Range in Minnesota. An example of underground mines are the copper mines in upper Michigan. Most underground mines are of the room-and-pillar type for working horizontal veins or of the stepping type for working vertical veins by cutting a series of steps.

In the preparation process the ore is crushed and large quantities of gangue are removed by a heavy-media-separation method. In some cases, the preparation of the ore may involve roasting or calcining. In roasting, the ore of sulfide is heated to remove the sulfur and in calcining the carbonate ores are heated to remove carbon dioxide and water.

The extraction of the metal from the ore is accomplished through a chemical process. These chemical processes reduce the compounds, such as oxides, by releasing the oxygen from chemical combination and thus freeing the metal.

Basically, three types of processes of extraction are used:

1. Pyrometallurgy.
2. Electrometallurgy.
3. Hydrometallurgy.

In the *pyrometallurgy* process (generally referred to as *smelting*) the ore is heated in a furnace producing a molten solution, from which the metal can be obtained by chemical separation. The blast furnace or reverberatory furnace is used in this process.

In the *electrometallurgy* process metals are obtained from ores by electrical processes utilizing an electric furnace or an electrolytic process.

Hydrometallurgy or *leaching* involves subjecting the ore to an aqueous solution from which the metal is dissolved and recovered.

As a result of the extraction process, the metals will contain impurities, which must be removed by a refining process. If the metal was extracted by the pyrometallurgy process, the most common method of refining is by oxidizing the impurities in a furnace (steel from pig iron). However, other methods are utilized, such as liquation (tin), distillation (zinc), electrolysis (copper), and the addition of a chemical reagent (manganese to molten steel).

FERROUS METALS

Ferrous metals comprise three general classes of materials of construction:

1. Cast iron.
2. Wrought iron.
3. Steel.

All of these classes are produced by the reduction of iron ores to pig iron and the subsequent treatment of the pig iron to various metallugical processes. Both cast iron and wrought iron have fallen in production with the advent of steel, as steel tends to exhibit better engineering properties than do cast and wrought iron. The application of steel and steel alloys is so widespread it has been estimated that there are over a million uses. In construction, steel has three principal uses:

1. Structural steel.
2. Reinforcing steel.
3. Forms and pans.

Classification of Iron and Steel

Iron products may be grouped under six headings:

1. Pig iron.
2. Cast iron.
3. Malleable cast iron.
4. Wrought iron.
5. Ingot iron.
6. Steel.

Pig iron is obtained by reducing the iron ore in a blast furnace.

This is accomplished by charging alternate layers of iron ore, coke, and limestone in a continuously operating *blast furnace*. Blasts of hot air are forced up through the charge to accelerate the combustion of coke while at the same time raising the temperature sufficiently to reduce the iron ore to molten iron. The limestone is a flux which unites with impurities in the iron ore to form slag. The blast furnace accomplishes three functions:

1. Reduction of iron ore.
2. Absorption of carbon.
3. Separation of impurities.

The amount of carbon present in pig iron is usually greater than 2.5 percent but less than 4.5 percent. The iron may be cast into bars, referred to as *pigs*.

Cast iron is remelted pig iron after being cast into pigs or about to be cast in final form. It does not differ from pig iron in composition and it is not in a malleable form.

Malleable cast iron is cast iron that has undergone special annealing treatment after casting and has been made malleable or semimalleable.

Wrought iron is a form of iron that contains slag, is initially malleable but normally possesses little to no carbon, and will harden quickly when rapidly cooled.

Ingot iron is a form of iron (or a low-carbon steel) that has been cast from a molten condition.

Steel is an iron–carbon alloy which is cast from a molten mass whose composition is such that it is malleable in some temperature range. Carbon steel is steel that has a carbon content of less than 2 percent and generally of less than 1.5 percent; its properties are dependent on the amount of carbon it contains. Alloy steels are steels in which the properties are due to elements other than carbon.

MANUFACTURE OF STEEL

As previously stated, the first process in the manufacture of steel is the reduction of iron ore to pig iron by use of a blast furnace. This is followed by the removal of impurities, and four principal methods are used to refine the pig iron and scrap metal:

1. Open-hearth furnace.
2. Bessemer furnace.
3. Electric furnace.
4. Basic oxygen furnace.

In this section we discuss in detail the blast furnace, the open-hearth furnace, the Bessemer converter, the electric furnace, and the basic oxygen furnace. Most of the discussion is taken directly from the *Steel Products Manual* of the American Iron and Steel Institute.

Blast Furnace

Blast furnaces (Figure 2-1) are so named because of the continuous blast of air required to bring about the necessary heat and chemical reactions in the raw materials in the stack. As much as 4.5 tons of air may be needed to make 1 ton of pig iron.

Most blast furnaces, and all of the more modern installations, are served by *turboblowers*. These enormously powerful steam engines are designed to receive steam at 700 psi (492,100 kg/m^2) at a temperature of 750°F (399°C).

The blowers force air from the atmosphere through piping into *stoves,* which are the most prominent auxiliaries serving a blast furnace. Each furnace must have at least two stoves and may have three.

Essentially, a stove consists of two parts: a *combustion chamber,* which is a vertical passageway wherein cleaned blast furnace gas is burned, and *brick checkerwork,* which contains many small passageways for heating the air as it passes over the hot masonry.

The design of the stoves varies in complexity, but it is important to remember that each stove serves two alternating functions. A stove will receive cold air from the blowers and pass it on through the heated brick checkerwork to the blast furnace; when the atmospheric air from the blower has cooled the brick checkerwork, another stove will take over the function and hot air from the top of the blast furnace will burn in the combustion chamber to reheat the brick checkerwork in the furnace stove.

Modern stoves for large furnaces are somewhat like farm silos in appearance (cylindrical in shape with a domed top). They may be up to 28 ft (8.5 m) in diameter and about 120 ft (36 m) high from the bottom to the top of the dome. Depending upon the type of brick checkerwork used, the stoves may contain upwards of 250,000 ft^2 (22,500 in.^{2}) of heating surface.

It is common for steel columns to support steel grids inside the stove, and these grids bear the brick checkerwork. Insulation between the brick and the steel shells prevents distortion of the stove.

Special valves control the flow of air from the blower to the heated stove, which is referred to as being "on blast." The stove or stoves in the process of having their checkerwork heated are referred to as being

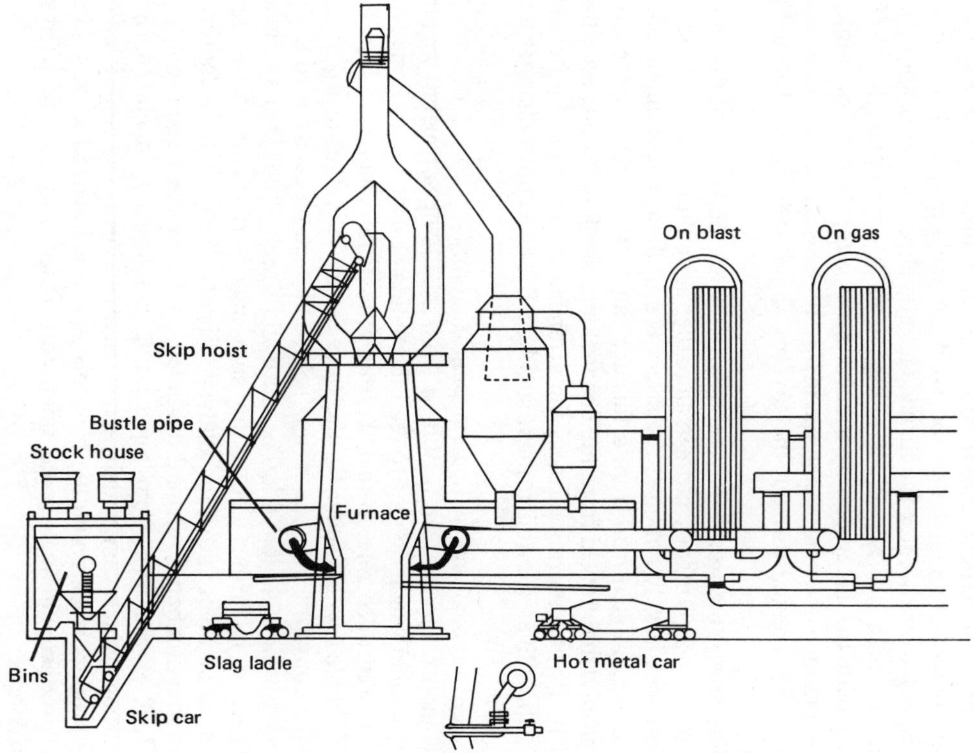

Fig. 2-1 Blast Furnace (Courtesy of AISI)

"on gas." In a three-stove arrangement serving one blast furnace, each stove is on gas twice as long as it is on blast.

The air from an on-blast stove speeds toward the furnace at a temperature that ranges from 1400 to 2100°F (760 to 1150°C). Air enters a bustle pipe made of heavy steel plate which completely encircles the stack of the blast furnace. From this bustle pipe smaller pipes extend at an angle downward and enter the stack of the furnace. These entry pipes, called *tuyeres,* enter the furnace at a level considerably above the hearth, in which lies a pool of molten iron at approximately 2700°F (1482°C) with the layer of impurities, called *slag,* floating on top. In modern practice it is not uncommon for fuels, such as oil or a coal sledge, to be injected through the tuyeres along with the air for higher output.

The term "pig iron" may seem obscure to modern city dwellers, but its origin was perfectly clear in an earlier agricultural age when small blast furnaces largely confined their services to their own communities. In those days molten iron from a furnace ran down a channel prepared in the ground and flowed from the channel into small holes dug on either side of it. To neighboring farmers the arrangement resembled a familiar sight in their lives (newborn pigs, sucklings). The central channel was known as a "sow."

Today's typical pig casting machines feature twin conveyor lines of shallow molds. Molten iron is poured from a ladle or hot metal car into a short channel which divides to feed the molds passing on the conveyor lines at a controlled rate so that there is little or no spilling. Iron in the molds is cooled with water. By the time the mold is overturned at the end of the conveyor line, the iron is solid and each piece is called a *pig*. It probably weighs 40 lb (18 kg) or more and simply drops into a railway car. There it is further cooled by spraying. Samples are taken and transported to a laboratory for analysis. Finally, the carload of cool iron is taken to a numbered pile and unloaded until customers require it.

Most of the iron from pig casting machines is made to rigid specifications which vary widely according to end use. The generic name for this product is *merchant pig iron,* because it is intended to be sold as an end product, although some small percentage of it may be used in steel making.

The customers for merchant pig iron are foundrymen who cast pipe for water and waste, for oil burner parts, and for automobile engine blocks, to name a few of a hundred uses. Among others are the dutch ovens used in many household kitchens, intricately shaped pipe fittings, heavy bases of machine tools, and major elements in a magnet developed at the Department of Energy's Argonne Laboratories.

A merchant pig iron producer may have several hundred piles of different grades of iron, each manufactured to meet the specifications of a given customer. A merchant pig iron producer may blend a dozen or more grades of iron ore to arrive at a desired chemical composition. Thus, the different products may be considered as iron alloys.

Most merchant pig iron furnaces are considerably smaller than the machines producing iron for steelmaking purposes. Even so, it is worthy of note that while, for example, 30,000 ft^3 (900 m^3) of solids weighing just under 1000 tons is being processed in a furnace stack, the chemical composition of the pigs is being controlled within fractions of 1 percent relative to carbon, silicon, sulfur, phosphorus, and manganese.

Foundries buy and melt scrap to supplement merchant pig iron in making their products. Some of them also buy imported iron. In a recent year about 20 percent of the available supply of merchant pig iron in the United States was of foreign origin. But domestic pig iron remains the standard for quality control in making castings because foundrymen are dependent on the chemistry of the iron they buy from various producers in the United States.

Serious changes in chemistry occur when too much poorly graded scrap and iron are consumed. The precise specifications to which American merchant iron is made provide a standard by which foundrymen can control the quality of their product.

Open-Hearth Furnace

Open-hearth furnaces (Figure 2-2) are so named because limestone, scrap steel, and molten iron are charged into a shallow steelmaking area called a "hearth" and are exposed, or "open," to the sweep of flames that emanate from opposite ends of the furnace alternately. The first open hearths for steelmaking were built in the United States during the late 1800s, and they have been developed to a high degree of sophistication over the years since. In the beginning, as now, one of their most attractive features concerns the use of steel scrap.

Theoretically, an open-hearth furnace can operate using either blast furnace iron alone or steel scrap alone, but most are designed to operate using both in approximately equal proportions. These proportions vary according to such economic factors as the price of scrap or the availability of molten iron. For more than half a century open-hearth furnaces were the outstanding high-tonnage producers of steel for America's expanding industry and are still major factors in the steel industry. A furnace that will produce a fairly typical 350 tons (318,000 kg) of steel in 5 to 8 hours may be about 90 ft (27 m) long and 30 ft (9 m) wide. It is very likely to be one of several furnaces contained in a single building referred to as an *open-hearth shop*. These very large furnaces are installed close together

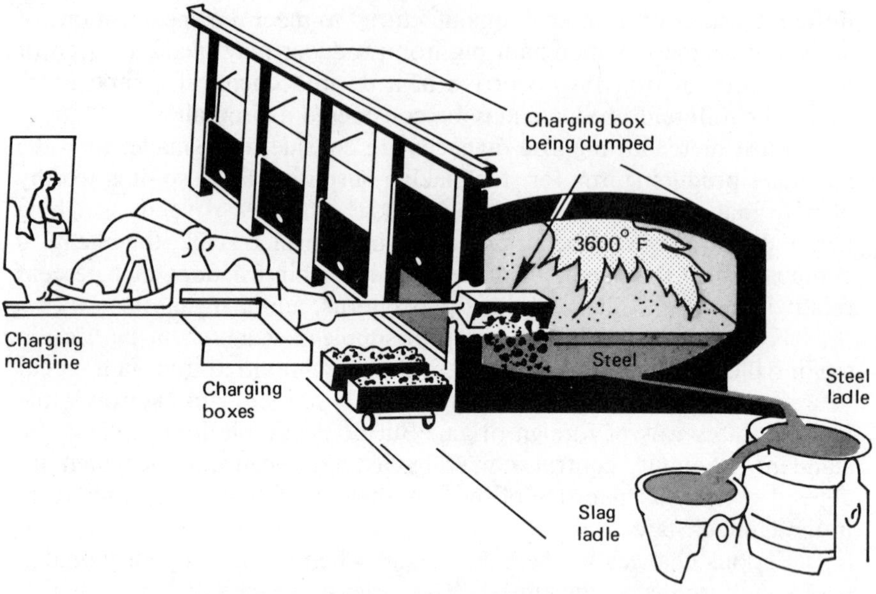

Fig. 2-2 Open-Hearth Furnace (Courtesy of AISI)

so that they may be serviced by the same units of highly specialized auxiliary equipment.

Because open-hearth shops are so large, it is difficult to obtain meaningful photographs. The arrangement can be likened to that of a split-level house with the entrance on the top floor and with a bottom level exit and heating system.

The charging side of the furnace is on the upper level and faces a charging floor full of highly specialized equipment. The most notable piece of equipment is a charging machine, which is electrically operated on a very broad gage track. Between the door side of the furnace and the charging machine is another track system. Special cars containing boxes of raw materials are run into the open-hearth shop on this track and are parked in position near the furnace doors. The charging machines are equipped to pick up the boxes of raw materials from their buggies and to thrust them in through the temporarily opened doors of the furnace, empty them by overturning them and then put them back on to their buggies. The empty charging boxes are then hauled away and refilled in preparation for the next charge.

In the rafters above the tracks for raw material, buggies, and charging machines are enormously strong overhead cranes, also running on tracks extending the full length of the shop. This crane system is used in many ways, but primarily to carry ladles of molten iron from a blast furnace or a mixer to the open hearths. Special troughs are wheeled into place in front of a furnace door and the huge ladles of molten iron are tilted so that a stream of the white-hot metal pours from the lip of the ladle through the trough and into the furnace. These paragraphs are devoted primarily to the layout of the shop; the actual processes of operation will be described later.

The only other major elements on the charging-floor side of the open-hearth furnace are the control areas, which contain panels permitting operators to control most of the processes at a safe distance from the heat and moving equipment.

The dish-shaped hearth of an open-hearth steelmaking furnace is quite shallow. The walls containing the charging doors are vertical and each unit is covered by an arched refractory roof. On the side of the furnace opposite the charging floor is a *taphole,* which is so arranged that molten steel can rush by gravity through a spout into a large ladle on the pouring floor or pit side of the open-hearth shop, which is at a considerably lower level than the charging floor.

A very considerable portion of an open-hearth facility is not visible at all. Brick *checker chambers* are located at both ends of each furnace below the level of the charging floor. The bricks in these chambers are arranged to leave a great number of passages through which the hot waste gases from the furnace pass and heat the brickwork prior to going through the gas cleaning equipment. Later, the flow of gas is reversed, and the atmospheric air supporting combustion in the furnace passes through the heated bricks and is heated on its way to the hearth.

Today, practically all open-hearth furnaces have been converted to the use of oxygen. The gas is fed into the open hearth through the roof by means of retractable lances. The use of gaseous oxygen in open hearth increases temperatures and thereby speeds up the melting process.

Each open-hearth shop is supervised by a melter foreman. This man is in charge of all furnaces in the shop and their crews. He operates the controls, directs any repairs to the furnace in operation, and reports any variations from standard procedure. There are many variables in open-hearth steelmaking practice, the most significant being the ratio of pig iron to scrap. For the purpose of this book, the rather common "fifty-fifty" practice is chosen for description. In this practice the limits of molten pig iron are usually 45 to 55 percent and the charge may include relatively small amounts of solid pig iron or iron scrap generated with the plant. Limestone and some ore agglomerates are also charged.

Assuming that the raw steel has been tapped from an open-hearth furnace, the first step in making a new heat of steel is to examine the interior of the furnace carefully to determine if any damage has been done to the hearth or roof. If so, a special patching gun may be wheeled into position and the damage patched by literally shooting a refractory material into the holes or cracks.

As the furnace is being prepared, buggies containing boxes of raw materials are rolled into position next to the furnace. The long-armed charging machine picks up boxes of limestone and fairly light steel scrap. One by one these boxes are thrust through the furnace doors and dumped. The flame of burning fuel oil, tar, or gases shoots from one end of the furnace across the solid materials in the hearth, partially melting them. (Most of the combustion gases are collected at the other end of the furnace and used to heat the checker brick.) Molten iron is then poured into the furnace.

All of this is much more complex than it sounds, because timing is extremely important. It is necessary to have the solid materials at such a temperature and degree of oxidation that, on the one hand, the molten pig iron will not be chilled by the scrap and, on the other hand, the oxidation of the metalloids of the pig iron will not be delayed by insufficient oxygen support from the oxidized scrap. The present-day usage of oxygen roof lances has greatly aided this critical phase of charging.

The chemical reactions for making steel in an open-hearth furnace relate to the removal of carbon, manganese, phosphorus, sulfur, and silicon from the metallic bath. First, silicon and manganese are oxidized and become part of the slag, which is a layer of molten limestone and other materials floating on top of the molten steel. Next, the oxidation of carbon speeds up, creating carbon monoxide gas, which causes agitation as it leaves the metal. Eventually, phosphorus and sulfur are transferred to the slag.

The agitation of the metal bath caused by carbon monoxide is called the *ore boil*. The more violent turbulence caused by calcination of limestone is called the *lime boil*. After both "boils" have subsided, the refining phase of open-hearth steelmaking begins, with the aim of lowering the phosphorus and sulfur contents to meet end-product specifications, controlling the carbon content, and generally bringing the composition and grade of the steel to predetermined levels.

The manufacture of steel in open-hearth furnaces takes much more time per ton than it does by the basic oxygen process, but it affords very good control of chemical analysis and is capable, in some instances, of making very large batches—sometimes 600 tons (540,000 kg) of steel of a given analysis in a single heat.

When a heat of steel is ready to be tapped, a member of the furnace crew, working from the tapping side, breaks out the clay plug and refractory that has closed the taphole since before the furnace was charged. The operation may involve the use of a special explosive charge and is undertaken following the best safety practice.

The highest level of the taphole is located at the lowest part of the hearth and slopes down to a spout. Most of the steel surges out of the furnace into a large ladle before any of the slag floating on top of the molten bath appears. The late appearance of slag permits alloy and recarburized and deoxidized materials to be added to the steel in the ladle. The ladle is so placed beneath the spout that the stream of molten steel is given a swirling motion which tends to mix and make more homogeneous the metal and the additions.

The capacity of the ladle is matched to the amount of steel in the furnace. When the slag comes through the taphole, a layer of it is permitted to lie on top of the molten metal as a covering while the remainder overflows through a special notch into a *slag thimble,* a small vessel placed next to the steel ladle.

Some very large open-hearth furnaces are provided with two tapholes, therefore requiring two ladles on the pit-side floor.

Bessemer Process

In the Bessemer process (Figure 2-3) the impurities in pig iron are removed by oxidation, by finely divided air currents blowing through a bath of molten iron contained in a vessel called a *converter.*

The Bessemer converter consists of a heavy steel pear-shaped shell with refractory brick. It is supported by two trunnions upon which it can rotate 180°. The upper portion can be either concentric or eccentric. The bottom of the converter is pierced with a large number of small holes, called tuyeres, through which the air blast is forced by means of a blower from the windbox up through the molten metal, oxidizing the impurities.

Converters generally have a capacity of 1 to 40 tons, with the average around 25 tons. For a small converter of 15-ton capacity, the clear-mouth opening is 2 to 2.5 ft (0.6 to 0.75 m) in diameter with the inside cylindrical diameter at the bottom opening 8 ft (2.4 m) and the height of the converter approximately 15 ft (4.5 m).

The lining of the converter is usually 12 to 15 in. (0.3 to 0.33 m) thick and is made of a refractory material of strongly acidic character, with silica being the primary constituent. This lining lasts for several months before it has to be replaced.

The bottom is lined with 24 to 30 in. (0.6 to 0.75 m) of damp siliceous material bound together with clay in which the tuyere bricks

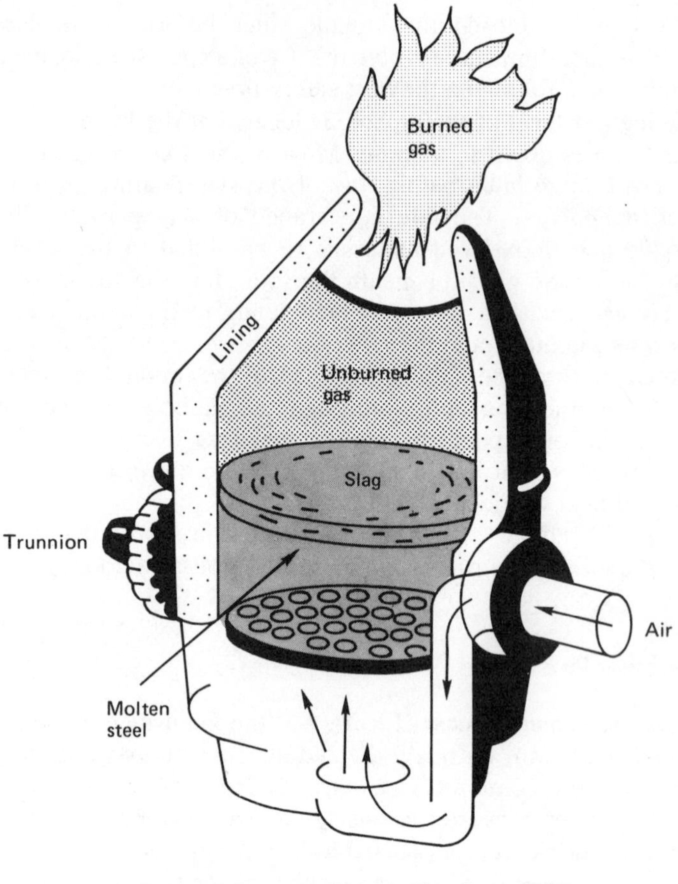

Fig. 2-3 Bessemer Process (Courtesy of AISI)

are set. Each tuyere brick is about 30 in. long and has about 10 blast holes, each approximately 0.25 in. (0.6 cm) in diameter. Tuyere bricks have a useful life of about 1 month.

In the operation of the Bessemer process, the converter is tilted to a horizontal position to receive a charge. The blast is turned on after charging and before righting, to prevent the metal from entering the tuyere.

As soon as the blow is on, the silicon and manganese begin to burn to form oxides and thus are reduced to traces before the oxidation of the carbon becomes appreciable. As soon as the carbon is burned out, the converter is turned down, the blasts are shut off, and the recarburizer is added.

This method of making steel is of little importance today because of the advent of better production facilities.

Electric Furnace

Electric arc furnaces (Figure 2-4) have a long history of producing alloy, stainless, tool, and other specialty steels. More recently, operators have also learned to make larger heats of carbon steels in these furnaces. Therefore, the electric steelmaking process is presently becoming a high-tonnage producer.

Electric arc furnaces are shallow steel cylinders lined with refractory brick. In the first half of the twentieth century most electric furnaces were loaded, or changed, with scrap iron and steel through a door in the side of the cylinder. However, today, the entire roof of an electric furnace is mounted on cantilevered steel beams so that it can be lifted and swung to one side. Thus, electric furnaces are now charged in one operation from buckets or other containers brought in by overhead cranes.

The roof of an electric furnace is pierced so that three carbon or graphite electrodes can be lowered into the furnace, and these electrodes give the electric arc furnace its name, because, in operation, the current arcs from one electrode to the metallic charge and then from the charge to the next electrode. This provides intense heat.

Opposite one of the doors is a tapping spout through which the molten steel is poured into a ladle. The entire furnace is mounted on

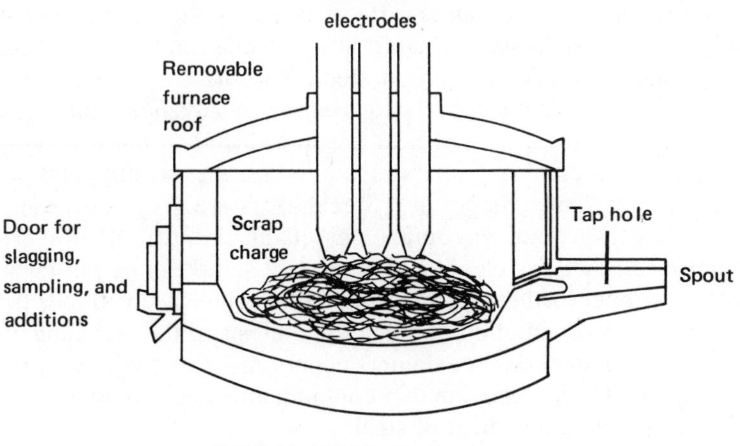

Fig. 2-4 Electric Arc Furnace

"rockers" so that it can be tilted to permit the molten steel to emerge through the spout.

From this description and the accompanying diagram it can be seen that the charge for electric furnace steelmaking is entirely solid. The principal metallic charge is steel scrap. When a complex specialty steel is to be made, the scrap is carefully selected so that no unwanted elements are present. Pig iron is charged into some electric furnaces and, more recently, prereduced iron ore, which may contain up to 98 percent of the element iron. Such ore, under these conditions, is handled as a primary metallic source.

One of the major factors in making electric furnaces economically competitive as high-tonnage producers of high-carbon steel is the increased size of the units. Most experienced operators remember when a furnace capable of producing 50 tons of steel was considered large. Today, some furnaces average 300 tons per heat.

A relatively few years ago new electric arc furnaces customarily operated on 25,000 to 35,000 kilovolt-amperes compared with up to 80,000 kVA for today's units. Electric furnace steelmakers figure their electricity costs on the basis of kilowatt-hours per ton of raw steel. As power input and furnace size increase, the time required to produce a heat of steel decreases. Therefore, the kilowatt-hours per ton also decrease. As indicated, this economy has thrust electric arc furnaces into the high-tonnage production picture, particularly in areas where molten iron is not readily available.

The charge to most electric furnaces is largely scrap with small amounts of burned lime and mill scale. The scrap is segregated (separated) into stockpiles of identified grades, a procedure that is necessary for several reasons. Primary among these is the matter of economics. It would be wasteful to use valuable alloy steel scrap as a material for making common grades of carbon steel. The most economical way of operating an electric furnace is to assure, in advance, that only the elements desired for a special heat are introduced to the furnace. This means that the preliminary operations in collecting, sorting, and preparing scrap are extremely important. Steelmakers are very concerned with the recycling aspect of environmental quality control. However, this does not mean that any classification of steel scrap can be put back into a furnace without further treatment, and this is the case with electric arc furnaces. The fragmentation of old automobiles and the recycling of "tin cans" and other steel-base containers cannot make up too large a portion of the charge. These items contain contaminants such as copper and tin which might spoil a new heat of steel.

Once the sorted scrap has been placed into the hearth of the furnace, ferroalloys and other elements that are not easily oxidized may be charged in the furnace prior to melting down. If the metallic charge is too low in carbon, additions of coke or of scrap electrodes are included with the scrap. In all instances, the amount of carbon charged into the electric arc furnace is higher than is required to produce the desired chemical analysis in the finished steel.

The top of the furnace is now swung into position, the electrodes are lowered through the roof, and the electric power is turned on. The heat resulting from the arcing between electrodes and metal begins to melt the solid scrap. The electrodes, in some cases 2 ft thick and 24 ft long, turn white-hot to a considerable distance above the arc. As the charge melts, they may rise and fall vertically, a process called *searching*.

At a given point, iron ore is added to reduce the carbon content and cause a "bubbling" or boiling action. This is one of the most important factors in the production of high-quality steel.

Limestone and flux are charged on top of the molten bath. Through a chemical interaction, impurities in the steel rise in the molten slag, which floats on top of the melt. When the desired product is a carbon steel, a *single-slag* practice is used. This means that the furnace is tilted slightly and much of the slag is raked off through the charge door. In the *double-slag* method, an oxidizing slag is first formed, then raked off, and an additional slag is formed for reducing purposes.

The direct use of oxygen gas is extremely important in modern practice. It is of great value in the rapid removal of carbon from the bath.

In the double-slag method, the original slag is removed from the surface of the bath by cutting off the power to the electrodes, back-tilting the furnace slightly, and then pouring the slag out through the charge door into a slag thimble. The original slag must be removed thoroughly to prevent delay in making up the second slag. This might cause the reversion of some elements from slag to metal. Once the first slag is removed, additional lime and iron ore can be added.

The steel should not be held under the second slag any longer than is absolutely necessary. As soon as the results of the last analysis are reported, additions are made to the bath to adjust the carbon and alloying element content. When all of the additions are in solution, ferrosilicon may be added and aluminum thrust through the slag into the bath to control the grain size of the steel.

When the chemical composition of the steel meets specifications, the furnace (roof, electrodes, and all) is tilted forward so that molten metal may pour out the taphole through the spout. The slag comes after the steel and serves as an insulating blanket during tapping into a ladle.

Basic Oxygen Process

The basic oxygen steelmaking process (Figure 2-5) uses as its principal raw material molten pig iron from a blast furnace. The other source of metal is scrap. Lime, rather than limestone, is the fluxing agent. As the name implies, heat is provided by the use of oxygen.

Although the basic oxygen steel production in the United States was first reported by the American Iron and Steel Institute in 1954, the modern technique was pioneered in Austria starting shortly after World War II, and the concept of using oxygen in a pneumatic steelmaking process originated with Sir Henry Bessemer in the mid-1800s, at the time when he was developing the process now bearing his name. Bessemer recognized that blowing pure oxygen into his converter would be advantageous, but the technology for the bulk production of oxygen had not yet been conceived.

The first European basic oxygen furnaces were too small to be of commercial value in this country. Much of the developmental work in adapting the processes to big furnaces, containing over 300 tons of molten metal, was done in the United States.

The basic oxygen furnace is a steel shell lined with refractory materials. The body of the furnace is cylindrical, the bottom is slightly cupped, and the top is shaped like a truncated cone with its open base set on top of the cylinder. The narrow part of the truncated cone is open to receive raw materials and a jet of oxygen and also to permit dirty waste gases from the steelmaking process to escape into extensive air treatment facilities. The entire furnace is supported on horizontal trunnions so that it can be tilted.

Usually, these furnaces are installed in pairs so that one of them can be making steel while the other is being filled with raw materials. A basic oxygen furnace (BOF) may produce batches of over 300 tons in 45 minutes as against 5 to 8 hours for the older open-hearth process. Most grades of steel can now be made in BOFs, although this was not true in their early history. Even today, many foreign BOFs concentrate on making only a few uncomplicated high-tonnage steels, whereas in the United States it is the practice to produce a wide range of compositions.

The first step for making a heat of steel in a BOF is to tilt the furnace and charge it with steel scrap. This is done by swinging the furnace on its trunnions through an arc so that the open top can be reached by a charger which runs on rails at an appropriate level. The charger looks like a combination small railroad car and dump truck. The boxlike part of the charger is filled with carefully graded steel scrap and is carried by the railcar to a position where it can dump its contents into the tilted furnace. Pistons, or jointed arms, then lift and tilt the box so

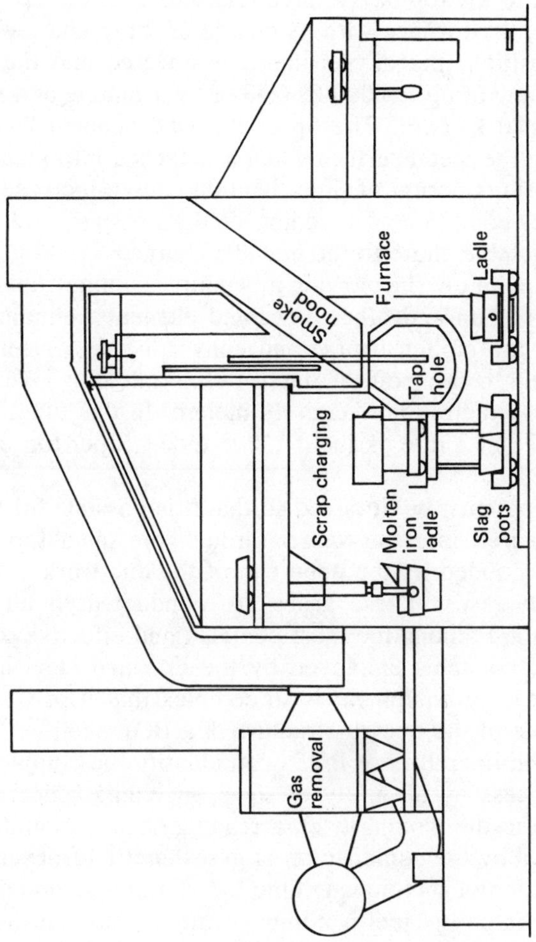

Fig. 2-5 Basic Oxygen Steelmaking Process

that the material slides out the open end into the preheated furnace. The furnace is rotated forward to distribute the scrap over the furnace bottom.

Immediately following the scrap charge, an overhead crane presents a ladle of molten iron from a blast furnace or from a holding device called a *mixer*. The iron, accounting for 65 to 80 percent of the charge, is also poured into the top of the tilted furnace.

As soon as the furnace with its charge of scrap and molten iron is in a vertical position, the oxygen lance is lowered and the oxygen is turned on to a flow of up to 6000 ft^3 (180 m^3) per minute at a pressure of up to 160 psi (480 kg/cm^2). The tip of the water-cooled lance may be about 6 ft above the metal surface when it is locked into place.

In a very short period of time, ignition causes increased heat and provides correct conditions for adding lime, fluorspar, and sometimes scale via a retractable chute to the metallic charge.

From that point on, the blowing procedure is uninterrupted. Oxygen combines with carbon and other unwanted elements, eliminating those impurities from the molten charge and converting it to steel. The lime and fluorspar help to carry off the impurities as a floating layer of slag on top of the metal which is now entirely molten. In this function, lime is usually consumed at a rate of about 150 lb (680 kg) per ton of raw steel produced.

A BOF shop must be designed so that it is possible for the oxygen lance to be lowered into the vessel through the open top while that aperture is also hooded by the open end of the ductwork which carries away the waste gases. These gases are conducted to air treatment facilities, which are among the most complex and effective of the many pollution control systems employed by the iron and steel industry. In fact, the gas cleaning equipment is so complex that it may account for one-third or more of the overall structure of a BOF shop.

Experienced operators of BOFs can identify the completion of the steelmaking process by a variety of signs, including a decrease in the flame, a change in the wound level, a reading of the amount of oxygen used as indicated by the composition of gases emitted from the furnace, and a consideration of the blowing time for iron of a given composition.

When the batch of steel is complete, the clamps on the lance are released and the lance is retracted through the hood. The furnace is then rotated back toward the charging floor until the slag floating on top of the metal is even with the lip at the top of the furnace. Special equipment is then used to determine the temperature of the bath, which may be 3000°F (1649°C) for many of the more common types of steel. Once a correct temperature reading has been obtained, it may be necessary to test for carbon content. When both temperature and carbon content are acceptable, the furnace is rotated away from the charging platform past the vertical position and onto the opposite tilt. In this position the taphole, set into the cone-shaped portion of the furnace top, is aimed at a ladle which receives the molten steel. The slag, which floats on top of the steel, is caused to stay above the taphole by the progressive tilt of the furnace. Alloys are added to the ladle of steel, often by chutes extended from above the teeming floor.

The ladle into which the metallic contents of the furnace have been teemed is usually mounted on a railcar which is removed to a position where an overhead crane can lift it. The overhead crane carries the ladle of molten steel to a position where it can be poured into ingot molds or into a strand casting machine for solidification.

Meanwhile, the molten slag remaining in the furnace is emptied into a receptacle which is carried away for disposal. Except for possible minor refractory lining repair, the BOF is then ready to be recharged immediately. At approximately the same time, the other furnace in a pair should be ready to begin the steelmaking process.

STRUCTURE OF IRON AND STEEL

Carbon Steel

Carbon steel is an alloy of iron and carbon. The carbon atoms actually replace or enter into solution among the lattice structure of the iron atoms and limit the slip planes in the lattice structure. The amount of carbon within the lattice determines the properties of the steel.

Alloys containing less than 0.008 percent carbon are classed as irons. Steel is an iron–carbon alloy in which the carbon content is less than 20 percent. These steel products, including structural steel and reinforcing steel, can be rolled and molded into a shape. However, as the carbon content goes above 2.0 percent, the material becomes increasingly hard and brittle. Thus, cast iron has a carbon content above 2.0 percent. *High-strength steels* are alloys containing less than 0.8 percent carbon (the eutectoid composition) and are sometimes referred to as *hypoeutectoid steels*. *Structural steels* are alloys containing less than 2.0 percent but more than 0.8 percent carbon and are referred to as *hypereutectoid steels*. Wrought iron is a combination of iron and slag.

Phase Diagrams

A typical phase diagram of iron–carbon is shown in Figure 2-6. This equilibrium diagram is plotted for amounts of carbon up to 5 percent, which is sufficient for practical purposes. The diagram at times may be extended up to 6.67 percent of carbon, as this corresponds to 100 percent cementite. *Cementite,* a compound of iron and carbon, is a

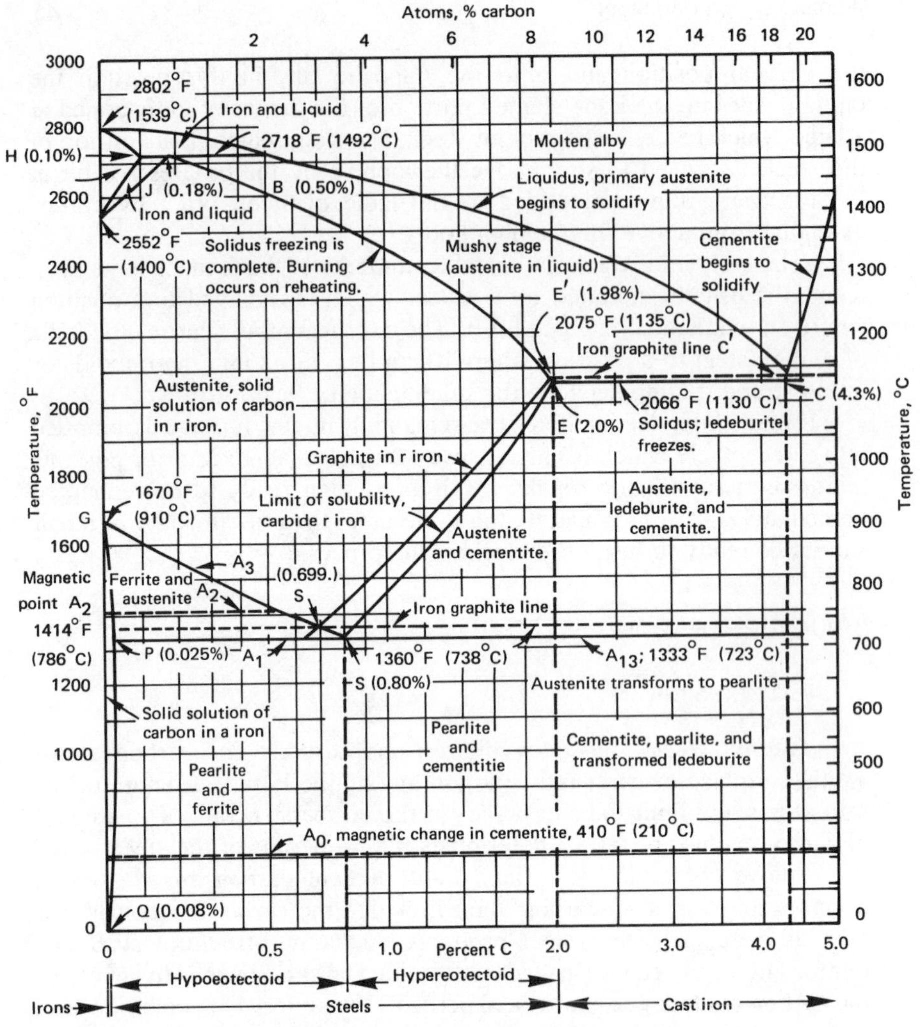

Fig. 2-6 Phase Diagram of Iron–Carbon

high-carbon steel that exists in a stable phase as iron carbide (Fe_3C). Cementite contains 6.67 percent carbon and 93.33 percent iron and is very hard and brittle. Other terms of importance shown on the phase diagram of Figure 2-6 are austenite, eutectic, eutectoid, ferrite, graphite, and pearlite. These terms are used to define the various stages or phases of iron–carbon alloys. *Austenite* is gamma iron with carbon in solution. The *eutectic* of the carbon–steel alloy is the combination that melts at the lowest temperature. Thus, according to Figure 2-6, the eutectic at 4.3 percent carbon melts at 2060°F (1130°C) and will contain a solid solution of austenite and a solid cementite. Below the 1330°F (723°C) mark, the solution changes to cementite and pearlite. The *eutectoid* in an equilibrium diagram for a solid solution is the point at which the solution on cooling is converted to a mixture of solids. Pearlite is a eutectoid and changes to a solid at 1330°F (723°C). *Ferrite* is iron that has not combined with carbon in pig iron as steel. This fact allows the steel to be cold-worked. *Graphite* (black lead) occurs as small flakes of carbon that become mixed with the steel. *Pearlite* is a lamellar aggregate of ferrite and cementite often occurring in carbon steels and in cast irons.

Properties of Cementite, Ferrite, and Pearlite

Cementite is a hard and very brittle material with a Brinell hardness of 650 and a diamond hardness of 760. Ferrite is a soft and ductile material with a Brinell hardness of 90 and a diamond hardness of 170. It has a 40 percent elongation in 2-in. (5.08-cm) specimen. Pearlite is harder and less ductile than ferrite but is softer and less brittle than cementite. Pearlite has a Brinell hardness of 275 and a diamond hardness of 300. It has an elongation of 15 percent in a 2-in. (5.08-cm) specimen.

Impurities in Steel

The principal impurities in steel are silicon, phosphorus, sulfur, and manganese.

The amount of silicon in structural steel is less than 1 percent and forms a solid solution with iron. This small amount of silicon increases both the ultimate strength and the elastic limit of steel with no appreciable change in its ductility. Silicon may further prevent the solution of carbon in iron.

The phosphorus in steel is in the form of iron phosphide (Fe_3P). For low-grade structural steel the amount of phosphorus is about 0.1 percent and decreasing to 0.05 percent for high-grade structural steel. Tool steel is approximately 0.02 percent phosphorus.

Sulfur in steel combines with the iron to form iron sulfate (FeS). This compound has a low melting point and segregation may take place.

Manganese has an affinity for sulfur and combines with such as well as with other impurities to form slag. In other words, manganese acts like a cleanser. Manganese is used to harden steels.

Structural Steel

Table 2-1 illustrates the uses of structural steels. It becomes obvious that the steel must have strength, toughness, and, above all, durability. The requirements for structural steel are presented in Table 2-2. In most cases the maximum percent of carbon is less than 0.27, but most structural steels average 0.2 percent.

TABLE 2-1 Structural Steel[a]

ASTM Designation	*Product*	*Use*
A36	Carbon-steel shapes, plates, and bars	Welded, riveted, and bolted construction; bridges, buildings, towers, and general structural purposes
A53	Welded or seamless pipe, black or galvanized	Welded, riveted, and bolted construction; primary use in buildings, particularly columns and truss members
A242	High-strength, low-alloy shapes, plates, and bars	Welded, riveted, and bolted construction; bridges, buildings, and general structural purposes; atmospheric-corrosion resistance about four times that of carbon steel; a weathering steel
A245	Carbon-steel sheets, cold- or hot-rolled	Cold-formed structural members for buildings, especially standardized buildings; welded, cold-riveted, bolted, and metal-screw construction
A374	High-strength, low-alloy, cold-rolled sheets and strip	Cold-formed structural members for buildings, especially standardized buildings; welded, cold-riveted, bolted, and metal-screw construction
A440	High-strength shapes, plates, and bars	Riveted or bolted construction; bridges, buildings, towers, and other structures; atmospheric-corrosion resistance double that of carbon steel
A441	High-strength, low-alloy manganese–vanadium steel shapes, plates, and bars	Welded, riveted, or bolted construction but intended primarily for welded construction; bridges, buildings, and other structures; atmospheric-corrosion resistance double that of carbon steel
A446	Zinc-coated (galvanized) sheets in coils or cut lengths	Cold-formed structural members for buildings, especially standardized buildings; welded, cold-riveted, bolted, and metal-screw construction

A500	Cold-formed welded or seamless tubing in round, square, rectangular, or special shapes	Welded, riveted, or bolted construction; bridges, buildings, and general structural purposes
A501	Hot-formed welded or seamless tubing in round, square, rectangular, or special shapes	Welded, riveted, or bolted construction; bridges, buildings, and general structural purposes
A514	Quenched and tempered plates of high yield strength	Intended primarily for welded bridges and other structures; welding technique must not affect properties of the plate, especially in heat-affected zone
A529	Carbon-steel plates and bars to ½ in. thick	Buildings, especially standardized buildings; welded, riveted, or bolted construction
A570	Hot-rolled carbon-steel sheets and strip in coils or cut lengths	Cold-formed structural members for buildings, especially standardized buildings; welded, cold-riveted, bolted, and metal-screw construction
A572	High-strength, low-alloy columbian–vanadium steel shapes, plates, sheet piling, and bars	Welded, riveted, or bolted construction of buildings in all grades; welded bridges in grades 42, 45, and 50 only
A588	High-strength, low-alloy steel shapes, plates, and bars	Intended primarily for welded bridges and buildings; atmospheric-corrosion resistance about four times that of carbon steel; a weathering steel
A606	High-strength, low-alloy hot- and cold-rolled sheet and strip	Intended for structural and miscellaneous purposes where savings in weight or added durability are important

[a]After E. H. Gaylord, Jr., and C. N. Gaylord, *Design of Steel Structures,* McGraw-Hill Book Company, New York, 1972.

Table 2-2 ASTM Structural Steel Requirements

Product:	*Shapes*[a]	*Plates*					*Bars*			
Thickness [in. (mm)]:	*All*	*To ¾ (19), incl.*	*Over ¾ to 1½ (19 to 38), incl.*	*Over 1½ to 2½ (38 to 64), incl.*	*Over 2½ to 4 (64 to 102), incl.*	*Over 4 (102)*	*To¾ (19), incl.*	*Over ¾ to 1½ (19 to 38), incl.*	*Over 1½ to 4 (102), incl.*	*Over 4 (102)*
Carbon, maximum (%)	0.26	0.25	0.25	0.26	0.27	0.29	0.26	0.27	0.28	0.29
Manganese, (%)	—	—	0.80–1.20	0.80–1.20	0.85–0.20	0.85–0.20	—	0.60–0.90	0.60–0.90	0.60–0.90
Phosphorus, max. (%)	0.04	0.04	0.04	0.04	0.04	0.04	0.04	0.04	0.04	0.04
Sulfur, max. (%)	0.05	0.05	0.05	0.05	0.05	0.05	0.05	0.05	0.05	0.05
Silicon (%)	—	—	—	0.15–0.30	0.15–0.30	0.15–0.30	—	—	—	—
Copper when copper steel is specified, min. (%)	0.20	0.20	0.20	0.20	0.20	0.20	0.20	0.20	0.20	0.20

ASTM Designation	Yield (ksi)	Minimum Tensile Properties of Structural Steels Strength (ksi)	Elongation (8 in. Unless Noted) (%)
Carbon steels			
A36	36	58–80	20
A529	42	60–85	19
High strength steels			
A242, A440, A441			
To ¾ in. thick	50	70	18
Over ¾ in. to 1½ in.	46	67	19
Over 1½. in. to 4 in.	42	63	16
A572			
Grade 42, to 4 in. incl.	42	60	20
Grade 45, to 1½ in. incl.	45	60	19
Grade 50, to 1½ in. incl.	50	65	18
Grade 55, to 1½ in. incl.	55	70	17
Grade 60, to 1 in. incl.	60	75	16
Grade 65, to ½ in. incl.	65	80	15
A588			
To 4 in. thick	50	70	19–21[b]
Over 4 in. to 5 in.	46	67	19–21[b]
Over 5 in. to 8 in.	42	63	19–21[b]
Quenched and tempered steels			
A514			
To 2½ in. thick	100	115–135	18[b]
Over 2½ in. to 4 in.	90	105–135	17[b]

[a]Manganese content of 0.85–1.35% and silicon content of 0.15–0.30% is required for shapes over 426 lb/ft.
[b]In 2 in.

Reinforcing Steel

As will be explained in Chapter 5, concrete exhibits great compressive strength but little tensile or flexural strength. Thus, deformed bars of structural steel are embedded in the concrete to take up the tensile or fluxural forces. These deformed bars have been developed in such a way as to force the concrete between the deformations such that failure in shear will occur before slippage. Table 2-3 lists the ASTM standard-size reinforcing bars as to these weights and dimensions. ASTM has also standardized reinforcing steel according to its yield point, as shown in Table 2-4.

HEAT TREATMENTS

Hardening or Quenching

Whenever a solid solution, such as steel, decomposes due to a falling temperature into the eutectoid, the decomposition may be more or less completed, depending on the cooling rate. This process is utilized in the hardening of steel. If the steel is cooled slowly, the changes just discussed will take place; however, if the steel is cooled too quickly, decomposition into the eutectoid will be prevented and a structure called *martensite* is produced rather than steel. Martensite is a hard structure with little ductility (a necessary property in steels).

The successful hardening of steel may be achieved by the application of three general principles:

1. Steel should always be annealed before hardening, to remove forging or cooling stains.
2. Heating for hardening should be slow.
3. Steel should be quinched on a rising, not on a falling temperature.

Quenching media vary, but are basically of three types:

1. Brine for maximum hardness.
2. Water for rapid cooling of the common steels.
3. Oils (light, medium, or heavy) for use with common steel parts of irregular shapes or for alloy steels.

All hardened steel is in a state of strain, and steel pieces with sharp angles or grooves sometimes crack immediately after hardening. For this

TABLE 2-3 ASTM Standard-Size Reinforcing Bars[a]

		Nominal Dimensions[b]			Deformation Requirements (mm)		
Bar Designation Number[c]	Nominal Weight (kg/m)	Diameter (mm)	Cross-Sectional Area (cm^2)	Perimeter (mm)	Maximum Average Spacing	Minimum Average Height	Maximum Gap (Chord 12½% of Nominal Perimeter)
3	0.560	9.52	0.71	29.9	6.7	0.38	3.5
4	0.994	12.70	1.29	39.9	8.9	0.51	4.9
5	1.552	15.88	2.00	49.9	11.1	0.71	6.1
6	2.235	19.05	2.84	59.8	13.3	0.96	7.3
7	3.042	22.22	3.87	69.8	15.5	1.11	8.5
8	3.973	25.40	5.10	79.8	17.8	1.27	9.7
9	5.059	28.65	6.45	90.00	20.1	1.42	10.9
10	6.403	32.26	8.19	101.4	22.6	1.62	11.4
11	7.906	35.81	10.06	112.5	25.1	1.80	13.6
14[d]	11.384	43.00	14.52	135.1	30.1	2.16	16.5
18[d]	20.238	57.33	25.81	180.1	40.1	2.59	21.9

[a]Based on ASTM 6615, 6616, 6617.

[b]The nominal dimensions of a deformed bar are equivalent to those of a plain round bar having the same weight per foot as the deformed bar.

[c]Bar numbers are based on the number of eighths of an inch included in the nominal diameter of the bars.

[d]Available in billet steel only.

TABLE 2-4 Kinds and Grades of Reinforcing Bars

Type of Steel and ASTM Specification Number	*Grade Designation*	*Size Numbers*	*Yield Point Minimum (MPa)*	*Tensile Strength Minimum (MPa)*	*Elongation in 203-mm Minimum (%)*	*Diameter Bend Test Pin*[a]
Billet steel A615	40	3	276	483	11	4*d*
		4, 5			12	4*d*
		6			12	5*d*
		7			11	5*d*
		8			10	5*d*
		9			9	5*d*
		10			8	5*d*
		11			7	5*d*
		14, 18			—	None
	60	3, 4, 5	415	621	9	4*d*
		6			9	5*d*
		7, 8			8	6*d*
		9, 10, 11			7	8*d*
		14, 18			7	None
Rail Steel A616	50	3	345	550	6	6*d*
		4, 5, 6			7	6*d*
		7			6	6*d*

		8			5	6*d*
		9, 10, 11[b]			5	8*d*
	60	3	415	620	6	6*d*
		4, 5, 6			6	6*d*
		7			5	6*d*
		8			4.5	6*d*
		9, 10, 11[b]			4.5	8*d*
Axle steel A617	40	3	275	480	11	4*d*
		4, 5			12	4*d*
		6			12	5*d*
		7			11	5*d*
		8			10	5*d*
		9			9	5*d*
		10			8	5*d*
		11			7	5*d*
	60	3, 4, 5			8	4*d*
		6			8	5*d*
		7			7	6*d*
		8	415	620	8	6*d*
		9, 10, 11			7	8*d*

[a]*d*, diameter of specimen.

[b]Number 11, 90° bend; all others 180°.

reason, tempering must follow the quenching operation as soon as possible.

Tempering

Tempering of steel is defined as the process of reheating a hardened steel to a definite temperature below the critical temperature, holding it at that point for a time, and cooling it usually by quenching for the purpose of obtaining toughness and ductility in the steel.

Annealing

Annealing is basically the opposite objective of hardening. *Annealing* is the process of heating a metal above the critical temperature range, holding it at that temperature for the proper period of time, and then slowly cooling. During the cooling process, pearlite, ferrite, and/or cementite form. The objectives of annealing are:

1. To refine the grain.
2. To soften the steel to meet definite specifications.
3. To remove internal stresses caused by quenching, forging, and cold working.
4. To change ductility, toughness, electrical, and magnetic properties.
5. To remove gases.

PHYSICAL PROPERTIES OF STEELS

In general, and as previously mentioned, three principal factors influence the strength, ductility, and elastic properties of steel:

1. The carbon content.
2. The percentages of silicon, sulfur, phosphorus, manganese, and other alloying elements.
3. The heat treatment and mechanical working.

Carbon

The various properties of different grades of steel are due more to variations in the carbon content of the steel than to any other single factor. Carbon acts as both a hardener and a strengthener, but at the same time it reduces the ductility.

Silicon, Sulfur, Phosphorus, and Manganese

The effect of silicon on strength and ductility in ordinary proportions (less than 0.2 percent) is very slight. If the silicon content is increased to 0.3 or 0.4 percent, the elastic limit and ultimate strength of the steel are raised without reducing the ductility. This is a procedure used for steel castings.

Sulfur within ordinary limits (0.02 to 0.10 percent) has no appreciable effect upon the strength or ductility of steels. It does, however, have a very injurious effect upon the properties of the hot metal, lessening its malleability and weldability, thus causing difficulty in rolling, called "red-shortness."

Phosphorus is the most undesirable impurity found in steels. It is detrimental to toughness and shock-resistance properties, and often detrimental to ductility under static load.

Manganese improves the strength of plain carbon steels. If the manganese content is less than 0.3 percent, the steel will be impregnated with oxides that are injurious to the steel. With a manganese content of between 0.3 and 1.0 percent, the beneficial effect depends upon the amount of carbon content. As the manganese content rises above 1.5 percent, the metal becomes brittle and worthless.

Effect of Heat Treatment upon Physical Properties

The effects of various heat treatments upon the mechanical properties of wrought or rolled carbon steels of various compositions are shown in Table 2-5.

Tensile Strength

The tensile strength and properties of various carbon steels as given by the ASTM are shown in Table 2-6. The modulus of elasticity of all grades varies from 28,000,000 to 30,000,000 psi (1.9×10^{11} to 2.1×10^{11} N/m^2).

Structural Steel

The chemical compositions for structural steel used in bridges and buildings as given by the ASTM are listed in Table 2-7. Notice that the carbon content is not specified. The reason for this is that the manufacturer is permitted to vary the carbon percentage to meet the tensile strength requirements of Table 2-6.

TABLE 2-5 Mechanical Properties of Heat-treated Wrought or Rolled Carbon Steels[a]

Composition (%)							
Carbon	Manganese	Heat Treatment	Yield Point (psi)	Tensile Strength (psi)	Elongation in 2 in. (%)	Reduction of Area (%)	Brinell Hardness
0.14	0.45	Hot-rolled	45,000	59,500	37.5	67.0	112
		Annealed	31,000	54,500	39.5	67.0	107
		Quenched in water	—	90,000	21.0	67.0	170
		Quenched in oil	56,500	71,500	34.0	75.5	134
0.32	0.5	Hot-rolled	49,500	75,500	30.0	51.9	144
		Annealed	41,000	70,000	30.5	51.9	131
		Quenched in water	—	135,000	8.0	16.9	255
		Quenched in oil	67,500	101,000	23.5	62.3	207
		Quenched in oil, tempered at 650°F	61,500	84,000	30.0	71.4	163
0.46	0.40	Hot-rolled	52,500	86,500	22.5	30.7	160
		Annealed	48,000	79,500	28.5	46.2	153
		Quenched in water	—	220,000	1.0	0.0	600
		Quenched in oil	87,500	126,500	20.5	51.9	225
		Quenched in oil, tempered at 560°F	81,500	111,500	24.0	57.2	—
		Quenched in oil, tempered at 650°F	73,000	98,000	25.5	59.8	192

0.57	0.65	Hot-rolled	57,000	106,500	19.0	27.4	220
		Annealed	50,000	95,000	25.0	40.3	183
		Quenched in water	—	215,000	0.0	0.0	578
		Quenched in oil	105,000	152,000	16.5	40.3	311
		Quenched in oil, tempered at 460°F	97,500	145,000	16.0	46.2	293
		Quenched in oil, tempered at 650°F	79,500	113,000	24.0	62.3	228
0.71	0.67	Hot-rolled	66,000	128,000	15.0	20.5	240
		Annealed	46,500	111,500	16.5	24.0	217
		Quenched in oil	100,000	184,500	1.5	0.0	364
		Quenched in oil, tempered at 460°F	115,500	177,000	10.0	34.0	340
		Quenched in oil, tempered at 560°F	106,000	148,500	17.0	43.0	311
		Quenched in oil, tempered at 650°F	91,000	125,500	19.5	57.2	269

[a]Data from the American Society for Steel Treating.

TABLE 2-6 Tensile Properties of Various Steels

Kind and Use of Steel	*Tensile Strength (psi)*	*Yield Point, Min. (psi)*	*Elongation Min. in 8 in. (%)*	*Elongation Min. in 2 in. (%)*	*Reduction of Area, Min. (%)*
Structural steel for bridges and buildings					
Structural	60,000–72,000	33,000	21	22	
Rivet	52,000–62,000	28,000	24		
Structural steel for ships					
Structural	58,000–71,000	32,000	21	22	
Rivet	55,000–65,000	30,000	23		
Carbon steel forgings for locomotives and cars					
Annealed or normalized	75,000 min.	37,500		20	33
Normalized quenched, and tempered	115,000 min.	75,000		16	35
Carbon steel bolting material, heat-treated					
Grade BO	100,000 min.	75,000		16	45
Boiler and firebox steel					

Flange	55,000–65,000	0.5 u.t.s.[a]	1,500,000/u.t.s.	1,750,000/u.t.s.
Firebox, grade A	55,000–65,000	0.5 u.t.s.	1,550,000/u.t.s.	1,750,000/u.t.s
Firebox, grade B	48,000–58,000	0.5 u.t.s.	1,550,000/u.t.s.	1,750,000/u.t.s.
Boiler-rivet steel				
Grade A	45,000–55,000	0.5 u.t.s.	1,500,000/u.t.s.	
Grade B	58,000–68,000	0.5 u.t.s.	1,500,000/u.t.s.	
Billet-steel-concrete reinforcing bars				
Plain bars				
Structural grade	55,000–70,000	33,000	1,400,000/u.t.s.	
Intermediate grade	70,000–90,000	40,000	1,300,000/u.t.s.	
Hard grade	80,000 min.	50,000	1,100,000/u.t.s.	
Deformed bars				
Structural grade	55,000–75,000	33,000	1,200,000/u.t.s.	
Intermediate grade	70,000–90,000	40,000	1,100,000/u.t.s.	
Hard grade	80,000 min.	50,000	1,000,000/u.t.s.	
Concrete reinforcing bars from rerolled steel rails				
Plain bars	80,000 min.	50,000	1,000,100/u.t.s.	
Deformed bars	80,000 min.	50,000	1,000,000/u.t.s.	

[a]u.t.s., ultimate tensile strength.

TABLE 2-7 Chemical Composition of Structural Steel for Bridges and Buildings, ASTM Specifications

Element	*Ladle Analysis, Max. (%)*	*Check Analysis, Max. (%)*
Open-Hearth and Electric-Furnace Structural and Structural-Rivet Steel		
Phosphorus		
Acid process	0.06	0.075
Basic process	0.04	0.05
Sulfur	0.05	0.063
Copper (when specified)	0.020	0.18
Acid Bessemer Structural Steel[a]		
Phosphorus	0.11	0.138

[a]Not permitted in bridges or in building members subject to dynamic loads.

Torsional Shear

The strength of steel in torsional shear is shown in Table 2-8. The modulus of elasticity for shear is about 12,000 psi (8.2×10^7 N/m^2).

ALLOY STEELS

Alloy steels are steels that owe their distinctive properties to elements other than carbon. Common alloys include chromium, nickel, manganese, molybdenum, silicon, copper, vanadium, and tungsten.

These alloys can be classified into two groups: those which combine with the carbon to form carbides, such as nickel, silicon, and copper,

TABLE 2-8 Strength of Steel in Torsional Shear

Class of Steel	*Computed Extreme Fiber Stress (psi)*	*Shearing Modulus of Elasticity (psi)*
Mild Bessemer	64,200	11,320,000
Medium Bessemer	68,300	11,570,000
Hard Bessemer	74,000	11,700,000
Cold-rolled	79,900	11,950,000

and those which do not combine with carbon to form carbides, such as manganese, chromium, tungsten, molybdenum, and vanadium.

Alloys are added to steel for three principal reasons:

1. To increase hardness.
2. To increase the strength.
3. To add special properties, such as
   a. Toughness.
   b. Improved magnetic and electrical properties.
   c. Corrosion resistance.
   d. Machinability.

Chromium

Chromium is primarily a hardening agent and is generally added to steel in amounts of 0.70 to 1.20 percent, with a variation in carbon content of 0.17 to 0.55 percent. Its value is due principally to its property of combining intense hardness after quenching with very high strength and elastic limit. Thus, it is well suited to withstand abrasion, cutting, or shock. It does lack ductility, but this is unimportant in view of its high elastic limit. Chromium steels corrode less rapidly than do carbon steels.

Nickel–Chromium

Nickel–chromium steels, when properly heat-treated, have a very high tensile strength and elastic limit, with considerable toughness and ductility. The nickel content is usually 3.5 percent, with a carbon content ranging from 0.15 to 0.50 percent.

One very important property of nickel–chromium steels is that by adding aluminum, cobalt, copper, manganese, silicon, silver, or tungsten, stainless steel results.

Manganese

As previously indicated, manganese is present in all steels as a result of the manufacturing process. When the manganese is 1.0 percent or greater in solution with steel, it is considered an alloy. Manganese will add hardness to steel if used within the proper range.

Molybdenum

Molybdenum provides strength and hardness in steel. It inhibits grain growth on heating as a result of its slow solubility of austenite.

When in solution in the austenite, it decreases the cooling rate and, therefore, increases the depth of hardening.

Silicon

Silicon is added to carbon steel for the purpose of deoxidizing. For this reason, silicon may be added in amounts of up to 0.25 percent. Silicon does not form carbides but does dissolve in the ferrite up to about 15 percent. Silicon decreases hysteresis and eddy-current losses, and thus is valuable for electrical machinery.

Vanadium

Vanadium is a powerful element for alloying in steel. It forms stable carbides and improves the hardenability of steels. Vanadium promotes a fine-grained structure and promotes hardness at high temperatures. The amount of vanadium present is 0.10 to 0.30 percent when used.

Copper

Copper increases the yield strength, tensile strength, and hardness of steel. However, ductility may be decreased by about 2 percent. The most important use of copper is to increase the resistance of steel to atmospheric corrosion.

Tungsten

Tungsten increases the strength, hardness, and toughness of steel. After moderately rapid cooling from high temperatures, tungsten steel exhibits remarkable hardness, which is still retained upon heating to temperatures considerably above the ordinary tempering heats of carbon steels. It is this property of tungsten that makes it a valuable alloy, in conjunction with chromium or manganese, for the production of high-speed tool steel.

NONFERROUS METALS

In this section, various nonferrous metals will be listed with only a brief statement; specific details will be omitted.

Basically, three groups of nonferrous metals exist. The first group, those of greatest industrial importance, are comprised of aluminum, copper, lead, magnesium, nickel, tin, and zinc. The second group includes antimony, bismuth, cadmium, mercury, and titanium. The third and final group, important in that they are used to form alloy steels, includes chromium, cobalt, molybdenum, tungsten, and vanadium.

The nonferrous alloys of the greatest importance are alloys of copper with tin (bronzes), alloys of copper with zinc (brasses), and alloys of aluminum, magnesium, nickel, and titanium.

PROBLEMS

2.1. Explain the purpose and use of the blast furnace.

2.2. Discuss four manufacturing processes of steel.

2.3. Why is carbon so important in steel?

2.4. Define austenite, cementite, eutectic, eutectoid, ferrite, graphite, and pearlite.

2.5. Discuss the various heat-treatment processes.

2.6. What is tempering?

2.7. What factors influence the strength, ductility, and elastic properties of steel?

2.8. Discuss the importance of silicon, sulfur, phosphorus, and manganese in steel.

2.9. List and discuss the various alloy steels.

2.10. Discuss briefly the use of nonferrous metals that are alloyed with steel.

REFERENCES

American Iron and Steel Institute, *The Making of Steel,* Washington, D.C., 1970.

American Society for Metals, *Metals Handbook,* Cleveland, Ohio, 1948.

American Society for Testing and Materials, "Specifications for Structural Steel," *ASTM A-36, C1020.*

BULLENS, D. K., *Steel and Its Heat Treatment,* 5th ed., 3 vols., John Wiley & Sons, Inc., New York, 1948. See especially vol. 1, *Principles,* and vol. 2, *Tools, Processes, and Control.*

DIGGES, T. G., "Effect of Carbon on the Hardening of High Purity Iron Carbon Alloys," *Trans. Am. Soc. Met., 25* (1938).

GAYLORD, E. H., JR., and GAYLORD, C. N., *Design of Steel Structures,* McGraw-Hill Book Company, New York, 1972.

Lipson, H., and Parker, A. M. B., "The Structure of Martensite," *J. Iron Steel Inst. (Lond.), 149* (1944).

Sisco, F. T., *Modern Metallurgy for Engineers,* 2nd ed., Pitman Publishing Corp., New York, 1948.

Teichert, E. J., *Ferrous Metallurgy,* 2nd ed., McGraw-Hill Book Company, New York, 1944.

MINERAL AGGREGATES 3

Aggregates play a very important role in the design and construction of highway and airport pavements, as the underlying material upon which the pavement rests. They are also important as an ingredient in rigid (concrete) and flexible (asphalt) pavements or structures.

Aggregates are the most important factor in the cost of pavement construction, accounting for more than 30 percent of the total cost. Aggregates make up approximately 65 to 85 percent of a concrete structure and 92 to 96 percent of an asphalt structure.

AGGREGATE TYPES AND PROCESSING

Aggregate is a combination of sand, gravel, crushed stone, slag, or other material of mineral composition, used in combination with a binding medium to form such materials as bituminous and portland cement concrete, macadam, mastic, mortar, and plaster, or alone, as in railroad ballast, filter beds, and various manufacturing processes such as fluxing.

Aggregates may be further classified as natural or manufactured. *Natural aggregates* are taken from natural deposits without change in their nature during production, with the exception of crushing, sizing, grading, or washing. In this group, crushed stone, gravel, and sand are the most common, although pumice, shells, iron ore, and limerock may also be included. *Manufactured aggregates* include blast furnace slag, clay, shale, and lightweight aggregates.

A further classification would be to divide the aggregate into two types: fine and coarse. According to ASTM C125 (Concrete and Concrete Aggregates), *fine aggregate* is defined as aggregate passing a ⅜-in. (9.5-mm) sieve and almost entirely passing a No. 4 (4.75-mm) sieve and predominantly retained on the No. 200 (75-μm) sieve or that portion of an aggregate passing the No. 4 (4.75-mm) sieve and retained on the No. 200 (75-μm) sieve. *Coarse aggregate* is defined as aggregate predominantly retained on the No. 4 (4.75-mm) sieve or that portion of an aggregate retained on the No. 4 (4.75-mm) sieve. These definitions are for concrete aggregates; for bituminous concrete mixtures the dividing line between fine and coarse aggregate is the No. 8 (9.5-mm) or the No. 10 (11.8-mm) sieve.

Processing

The main fundamental rule of good aggregate processing is to obtain aggregates of the highest quality at the least cost. Each process is completed with these objectives in mind. The processes include, but are not limited to, excavation, transportation, washing, crushing, and sizing. Processing begins with excavation and quarrying of the material and ends upon being stockpiled or delivered to the site.

In the excavation process the overburden is removed (if applicable), as its presence in aggregate in the form of silt or clay cannot be tolerated. The removal of the overburden is carried out through the use of power shovels, draglines, or scrapers. Overburden removal is usually considered only if there is a depth of 50 ft (15.24 m) or more. If the overburden is light, it will wash out in the processing of the aggregate.

After the aggregate is excavated, it is transported by rail, truck, or conveyor belt to the processing plant.

At the processing plant, unacceptable (deleterious) materials are removed. A deleterious material is a material that may prove harmful to the final product for which the aggregate is to be used. One method of removing deleterious materials (clay, mud, leaves, etc.) is to wash the raw material. Sometimes conveyor belts are used to haul the aggregate through flumes which are flushed with water.

The next process is to reduce the size of the stone or gravel. In this process many types of crushers are used. The oldest is the jaw crusher, which consists of a fixed jaw and a reciprocating jaw, which are suited for hard rocks of all types. Newer crushers have a higher capacity than the jaw crusher, but this is the only disadvantage of the jaw crusher. The usual practice is to reduce the size of the rock at a ratio of 1 : 6 or less.

For sizing, vibratory sieves are used for the coarse materials and hydraulic classification devices for fine material. The screens vary of course, in design, capacity, and efficiency. In the screening process about 70 percent of the material will pass through the screen, so that the goals of high efficiency and capacity are met. In most cases some removal of oversize particles, called scapping, will take place.

Particles

The screened aggregate particles may be rounded or angular. Gravel consists of naturally rounded particles resulting from disintegration and abrasion of rock or processing of weakly bound conglomerate. Sand consists of rock particles that have been disintegrated naturally; the grains are generally angular but have been subjected to weathering. Sand is fine aggregate resulting from natural disintegration and abrasion of rock or processing of completely friable sandstone. Crushed stone is a product of the artificial crushing of rocks, boulders, or large cobblestones, substantially all faces of which result from the crushing operation. Stone sand is a finely crushed rock corresponding to sand in size. Gravel and crushed stone are considered to be coarse aggregate.

Aggregate particles vary considerably in texture. Gravel particles have a very smooth texture, whereas crushed stone has a rough surface texture. Aggregate particles have considerable variations in porosity. Usually, crushed stone and gravel have low porosity.

Gradation and Aggregate Blending

The gradual gradation in size from coarse to fine is a key property of aggregates. Aggregate gradation affects the workability of portland cement concrete mixes and the stability and durability of bituminous concrete mixes, as well as the stability, drainage, and frost resistance of base courses. Therefore, aggregates should be tailored to their proposed usage: in concrete or bituminous or as base courses. For specific examples, refer to succeeding chapters.

Aggregates may be dense, gap-graded, uniform, well-graded, or open-graded. The terms "dense" and "well-graded" are essentially the same, as are "gap," "uniform," and "open-graded." Figure 3-1 illustrates the five types of gradations. Other methods of expressing size distribution have been developed. One such method is the Fuller–Thompson method. An empirical formula can be used to determine the gradation:

$$P = 100\left(\frac{d}{D}\right)^{N} \tag{3.1}$$

where P = percent finer than the sieve
d = sieve size in question
D = maximum size aggregate to be used (top size)
N = coefficient of adjustment, which adjusts the curve for fineness or coarseness

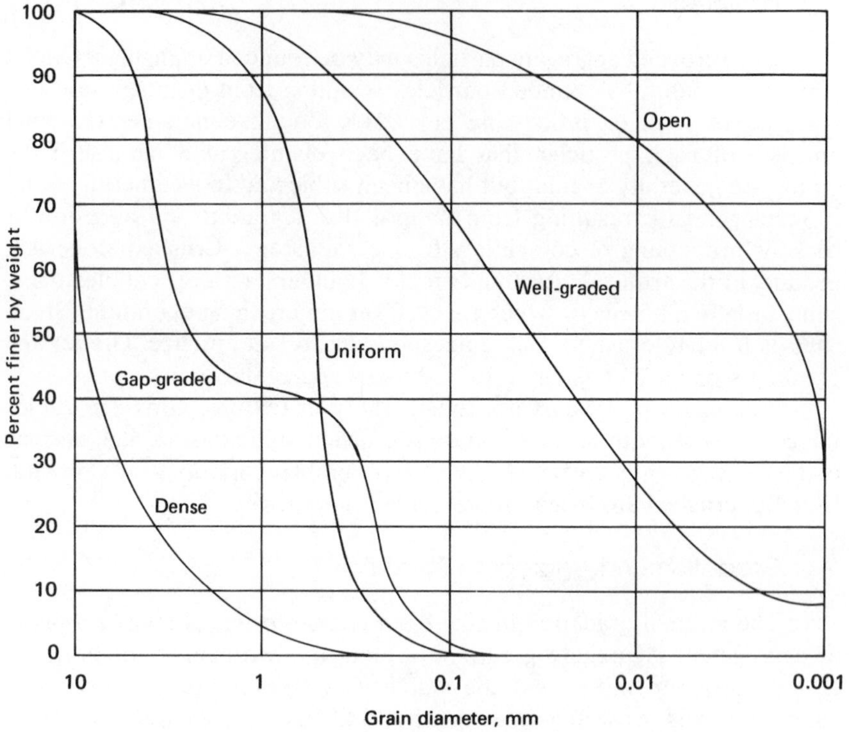

Fig. 3-1 Five Types of Gradation

Figure 3-2 shows a typical plot utilizing this formula at values of N equal to 1.3, 0.5, and 0.7. The maximum size of the aggregate is 1 in. (2.54 cm). Notice that a fine gradation is represented by $N = 0.3$ and a coarse-graded material is represented by $N = 0.7$. Therefore, a dense material would be $N = 0.5$, as the Fuller–Thompson experiments indicate. Originally, the work of Fuller and Thompson included the cement in a cement–aggregate mixture, but the relationship (equation) has become known as *Fuller's maximum density curve,* regardless of the constituents. The maximum density curve is only an approximation, since the actual gradation may depend on the nature of the material. However, if employed properly, it can be a valuable tool as a point of beginning in designing aggregate blends for maximum density.

Other research to determine maximum density of aggregates has been conducted. One theory states that if aggregates were screened into three sizes, coarse, medium, and fine, the combination with the greatest density would be a mixture of approximately 2 parts coarse aggregate

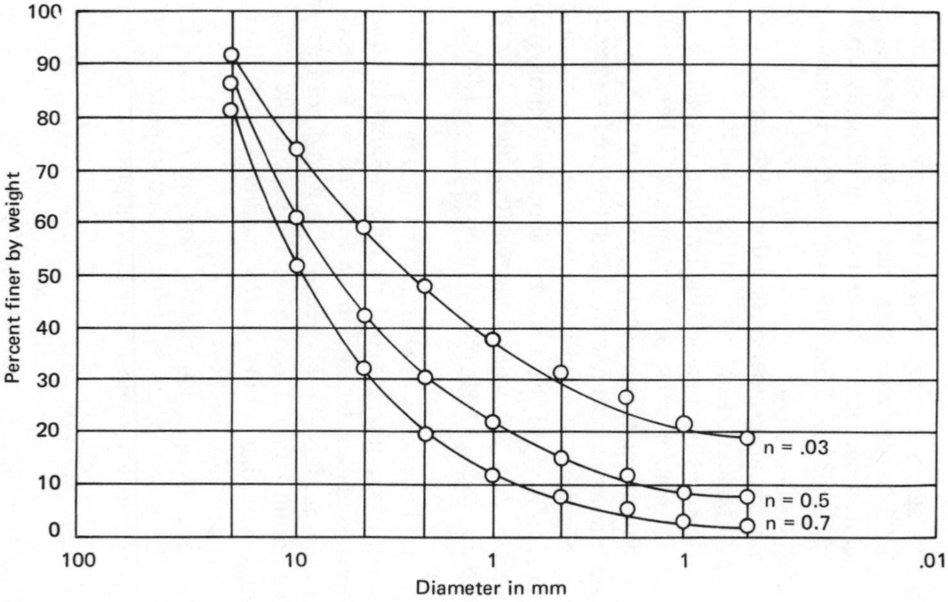

Fig. 3-2 Fuller–Thompson Curves

and 1 part fine aggregate, with no medium fraction. Further, research showed that high densities could be obtained by using equal portions of each size.

Manufactured Aggregates

Manufactured aggregates are man-made aggregates. One such aggregate, air-cooled blast-furnace slag, has become one of the most common manufactured aggregates to be used in highway construction. ASTM C125 (Concrete and Concrete Aggregates) defines *air-cooled blast-furnace slag* as the material reulting from solidification of molten blast-furnace slag under atmospheric conditions; subsequent cooling may be accelerated by application of water to the solidified surface. *Blast-furnace slag* is defined as the nonmetallic product, consisting essentially of silicates and aluminosilicates of calcium and other bases, that is developed in a molten condition simultaneously with iron in a blast furnace. It is not uncommon to find this manufactured aggregate used in industrial areas, especially if steel mills are present, as in the states of Alabama, Indiana, Maryland, New York, Ohio, and Pennsylvania. Slag is lighter in weight than are natural aggregates. Slags have a specific gravity between 2.0 and 2.5, whereas natural aggregates have a specific gravity between 2.3 and 3.2.

Although the definition of manufactured aggregate given above is sufficient for most uses, others further classify manufactured aggregates as those man-made aggregates that have resulted as a direct product rather than as a by-product. In this sense, slag would not be a manufactured aggregate but an artificial aggregate. Thus, an artificial aggregate is a man-made aggregate that results as a by-product from the manufacturing of some other product. Lightweight aggregates are considered artificial aggregates. Lightweight aggregates may be cinders, clay, shale, shells, or slag. These type of aggregates are used to produce lightweight concrete in a structure where dead weight is important.

AGGREGATE AS A BASE-COURSE MATERIAL

The importance of aggregates in a base course lying between the compacted subgrade and the portland cement concrete slab or the bituminous concrete slab cannot be overstated. An improper design of a base course can lead to structural failure of the slab. Base courses may also include several types of undercourses, such as subbases, filter beds, and leveling courses. Base courses serve a variety of purposes depending upon construction practices and the environment. They serve to provide structural capacity to bituminous concrete slabs, drainage for portland cement concrete slabs, and low susceptibility to frost. As previously indicated, gradation is the key factor in the success of aggregates as a base course.

Gradation

The gradation of the aggregate can affect structural capacity, drainage, and frost susceptibility. Thus, the control over gradation is a principal concern for most engineers. Related to this control is the hardness of the aggregate as soft, weak, or friable articles undergo aggregate degradation (a process whereby fines are generated by aggregate breakdown during the placement or the use of the material).

According to Krebs and Walker, three general types of aggregate mixtures can be recognized with respect to fines:

1. Aggregates only, no fines.
2. Fines just filling the voids of aggregate fraction.
3. Fines overfilling the voids of aggregate fraction.

In the first case the aggregate derives its strength from grain-to-grain contact of the aggregate particles. In this situation the base-course material would be unstable unless it was confined, but it does provide excellent drainage and is completely non-frost-susceptible.

In the second case, also, the aggregate derives its strength from grain-to-grain contact of the aggregate particles. However, in this situation the base-course material would be stable even if unconfined, because of the inherent cohesive properties of the fine content that fills the voids between the aggregate particles. Further, the drainage is adequate and can be non-frost-susceptible.

In the final case, the strength is derived from the grain-to-grain contact of the fines rather than of the aggregate particles; thus, a strength reduction occurs. The drainage characteristics of this base course would be poor and it would be very frost-susceptible.

In most highway construction techniques the base course falls between types 1 and 2 for best practice. Thus, a base-course material should have sufficient fines to just fill the voids within the aggregate particles, with a gradation curve approaching that of Fuller's maximum density curve.

Particle Strength, Shape, and Texture

To resist the stress of repeated loads and to avoid aggregate degradation, base-course aggregates must exhibit strength and toughness to function for their intended purpose. Open-graded base courses are more susceptible to degradation than is dense-graded material. In addition, the hardness of the aggregate adds strength to the base course. Sandstones and shale generally degrade easily, although their individual particles may be strong.

Table 3-1 gives the average values for the physical properties of the principal types of rocks: igneous, sedimentary, and metamorphic. In each group the specific gravity, hardness, and toughness are given together with other properties. The purpose of the table is to give the reader an indication of which materials would serve best as a base-course material.

Of equal importance to base-course material other than strength is particle shape and toughness. Table 3-1 also showed the toughness of this rock. These two properties are very important to base-course materials. Angular, nearly equidimensional particles rough in texture are extremely preferred for base-course material. The angularity contributes to aggregate interlocking, and a rough surface texture prevents movement of one particle upon another. Rounded particles tend to role over one another as they do not interlock with one another. Smooth-textured aggregate particles allow slippage when they are in contact with each other. Thus, rounded aggregate with a smooth surface texture is the least preferred for base-course usage. Table 3-2 summarizes the engineering properties of various aggregate types.

AGGREGATES FOR PORTLAND CEMENT CONCRETE

Aggregate properties for portland cement concrete are in many cases different from aggregates used for base courses or for use in bituminous concrete. Aggregate gradation becomes a key factor as it controls the workability of the plastic concrete. Further, to this, the aggregates used in portland cement concrete are a blend of fine and coarse aggregate to achieve an economical mix. For coarse aggregate the same definitions apply as previously given. For fine aggregate the same definitions as previously given apply, but we introduce a new term, fineness modulus. The *fineness modulus,* which denotes the relative fineness of the sand, is defined as one one-hundredth of the sum of the cumulative percentages held on the standard sieves in a sieve test of sand. Six sieves are used in the determination, Nos. 4, 8, 16, 30, 50, and 100. The smaller the value

TABLE 3-1 Average Values for Physical Properties of the Principal Type of Rocks[a]

Type of Rock	Bulk Specific Gravity	Absorption (%)[b]	Loss by Abrasion (%)		Hardness[e]	Toughness[f]
			Deval[c]	Los Angeles[d]		
Igneous						
Granite	2.65	0.3	4.3	38	18	9
Syenite	2.74	0.4	4.1	24	18	14
Diorite	2.92	0.3	3.1	—	18	15
Gabbro	2.96	0.3	3.0	18	18	14
Peridotite	3.31	0.3	4.1	—	15	9
Felsite	2.66	0.8	3.8	18	18	17
Basalt	2.86	0.5	3.1	14	17	19
Diabase	2.96	0.3	2.6	18	18	20
Sedimentary						
Limestone	2.66	0.9	5.7	26	14	8
Dolomite	2.70	1.1	5.5	25	14	9
Shale	1.8–2.5	—	—	—	—	—
Sandstone	2.54	1.8	7.0	38	15	11
Chert	2.50	1.6	8.4	26	19	12
Conglomerate	2.68	1.2	10.0	—	16	8
Breccia	2.59	1.8	6.4	—	17	11
Metamorphic						
Gneiss	2.74	0.3	5.9	45	18	9
Schist	2.85	0.4	5.5	38	17	12
Amphibolite	3.02	0.4	3.9	35	16	14
Slate	2.74	0.5	4.7	20	15	18
Quartzite	2.69	0.3	3.3	28	19	16
Marble	2.63	0.2	6.3	47	13	6
Serpentinite	2.62	0.9	6.3	19	15	14

[a]From Bureau of Public Roads manual, *The Identification of Rock Types* (1950).
[b]After immersion in water at atmospheric temperature and pressure.
[c]AASHTO T3 or ASTM D289.
[d]AASHTO T96 or ASTM C131.
[e]Dorry hardness test, *U.S. Dept. Arg. Bull. 949.*
[f]AASTHO T5 or ASTM D3a.

of the fineness modulus, the finer the sand. The fineness modulus for a good sand should range between 2.25 and 3.25.

In portland cement concrete the strength of the aggregate is not as important as it would be if it was used as a base-course material. The aggregate in portland cement concrete acts as a filler so that not as much sand, cement, and water are needed.

The bond between the aggregate and the cementing materials of portland cement concrete is influenced by surface texture. In most cases

TABLE 3-2 Summary of Engineering Properties of Rocks[a]

Type of Rock	*Mechanical Strength*	*Durability*	*Chemical Stability*	*Surface Characteristics*	*Presence of Undesirable Material*	*Crushed Shape*
Igneous						
Granite, syenite, diorite	Good	Good	Good	Good	Possible	Good
Felsite	Good	Good	Questionable	Fair	Possible	Fair
Basalt, diabase, gabbro	Good	Good	Good	Good	Seldom	Fair
Peridotite	Good	Fair	Questionable	Good	Possible	Good
Sedimentary						
Limestone, dolomite	Fair	Fair	Good	Good	Possible	Good
Sandstone	Fair	Fair	Good	Good	Seldom	Good
Chert	Good	Poor	Poor	Fair	Likely	Poor
Conglomerate, breccia	Fair	Fair	Good	Good	Seldom	Fair
Shale	Poor	Poor		Good	Possible	Fair
Metamorphic						
Gneiss, schist	Good	Good	Good	Good	Seldom	Good
Quartzite	Good	Good	Good	Good	Seldom	Fair
Marble	Fair	Good	Good	Good	Possible	Good
Serpentinite	Fair	Fair	Good	Fair	Possible	Fair
Amphibolite	Good	Good	Good	Good	Seldom	Fair
Slate	Good	Good	Good	Fair	Seldom	Poor

[a]From Bureau of Public Roads manual, *The Identification of Rock Types* (1950).

the bond strength is improved by an aggregate with a rough surface texture. However, the degree of texture need not be great, but some texture is desired.

In most portland cement concrete mixtures, the effect of surface texture is minimal. If one was to compare two types of aggregates, a smooth-surfaced gravel and a rough-surfaced crushed stone, to be utilized in two separate concrete mixes of equal cement factor and consistency, the smooth-surfaced gravel mix would require less water. Thus, the strength achieved from using the smooth-surfaced gravel would be the same as the rough-surfaced crushed stone. Thus, in the smooth-surfaced gravel mix the less water resulted in a lower water cement ratio increasing its strength. The mix containing rough-surfaced crushed stone required more water, increasing the water–cement ratio, resulting in a weaker concrete. Thus, the two effects balanced out and the resulted strengths were equal.

Particle shape influences the workability of the concrete mix, but the interlocking characteristic needed for base-course material is not important here. Angular aggregates require more mortar to fill voids and separate aggregate particles for workability, which results in a higher water–cement ratio for a given cement factor and consistency.

Freezing and Thawing

Probably the single most important reason why concrete fails is due to the effects of freezing and thawing. Other reasons are deleterious materials and chemical reactions. The latter two reasons will be discussed in more detail in Chapter 7. The freezing and thawing phenomenon is also discussed in Chapter 7, but some detail will be given here.

The freezing and thawing of concrete is much more important at present than it was several years ago with respect to aggregates. Sources of aggregates of known satisfactory performance are being depleted through one means or another. This fact can only lead to three major possibilities:

1. Production of synthetic aggregate.
2. Beneficiation of unsuitable material.
3. Use of manufactured and waste materials as supplements and replacements for conventional aggregates in construction.

Therefore, the need for an understanding of the influence of aggregates on the resistance of concrete to freezing and thawing is becoming increasingly important. If the aggregate is unsound, three major classifications of distress result:

1. Pitting and popouts.
2. D-line cracking deterioration.
3. Map cracking.

Pitting, the disintegration of weak, friable pieces of aggregates due to frost action, usually occurs by the gradual deterioration of deleterious particles near the surface. *Popouts* are caused by the rapid disruption of harder but saturated pieces of rocks or aggregates. Pitting and popouts themselves do not affect the structure in terms of strength, but they allow water to enter the structure through accesses provided by the pitting and popout and ultimately result in deterioration.

D-line cracking is caused by the change in volume of coarse-aggregate particles breaking the bond with the mortar matrix. It is aided by the entrance of water. Ordinarily, the cracks are filled with a calcareous deposit. D-line cracking appears first along transverse joints, then later along longitudinal joints.

Map cracking is a form of disintergration in which random cracks develop in a well-distributed pattern over an entire surface instead of concentrated along joints or free edges as in D-line cracking. D-line cracking and map cracking can aid structural failure of pavements by causing blowups.

As one views the three major classifications of distress, the question arises as to what materials cause these harmful effects. Several aggregates cause deterioration of concrete, including chert, limestone, and shale. But what is the specific nature of these materials that cause deterioration? Chert is a fine-grained hard rock that will usually pass conventional specifications. However, chert is sufficiently weathered as to be deleterious to concrete exposed to freezing and thawing. Shale and argillaceous limestone cause considerable damage to concrete exposed to severe weathering.

The different responses of aggregates to freezing when saturated depend upon the pore charactristics of the aggregate and the cement paste. Saturated aggregates of low porosity may accommodate pore-water freezing by simple elastic expansion. Saturated aggregates of moderate to high porosity may fail because the particle dimension exceeds a certain critical size or may cause failure in the paste immediately adjacent to the aggregate particles because of aggregate pore-water displacement. As for the shale and argillaceous limestones, these materials have small interconnected voids and are as capable of attaining a high degree of saturation.

The disruption of concrete by aggregates is a result of hydraulic pressures. The hydraulic prssure is a result of the degree of saturation (proportional to total void space filled with water) and the permeability and size of the aggregate particles. Upon freezing, water expands 9 percent, and if the degree of saturation of the aggregate is critical, 91.7 percent water will be expelled into the paste surrounding the aggregate particles, and potentially destructive hydraulic pressure may develop there also. So the properties of paste, its permeability, air content, and porosity are also involved in the problem. Three additional factors, composition, texture, and structure, also play important roles in the freezing and thawing of concrete. As shown by Powers, maximum hydraulic pressure is greatest when the aggregate has a low coefficient of permeability. At a given porosity, the smaller the mean size of the capillaries, the lower the coefficient of permeability and the higher the resistance to flow at a given ratio of efflux. Thus, smallness of capillaries is one characteristic of an unfavorable capillary system. In short, low water capacity and high permeability (i.e., low porosity and coarse texture) is a favorable combination, resulting in a relatively large critical size. Critical size varies from rock to rock (because it is dependent on capillry size), so no value is given here. The limit of water capacity for elastic accommodations is 0.3 percent. It should be noted that a wide range of critical sizes exist at a given strength.

The critical saturation point is 91.7 percent, as shown by Powers. One would immediately try to keep the water below the 91.7 percent level, but it must be remembered that water is not uniformly distributed in an aggregate particle at the time of freezing. For an aggregate that is not uniformly distributed, the critical saturation may be something less. Critical diameter is usually considered 4 μm (0.004 mm). The surface texture, which is directly related to the porosity of the material, is considered important in aggregates as well as the permeability. A rough, porous texture will produce high loading strengths and a favorable critical size. Irregular particles have much more void space than do smooth particles. Irregular particles need more mortar to fill the void space, and they also have a greater demand for water—because more mortar, more sand; more fines, more water. Another aspect to remember is that some aggregates expand when they come in contact with water, which may result in deterioration.

If the situation is such that the aggregate might become critically saturated, the paste is sure to become saturated, and when it does it can quickly become destroyed by freezing, regardless of the kind of cement of which it is made or the extent of curing.

When a specimen of saturated cement paste becomes frozen, it, too, has a critical thickness of about three thousandths of an inch. Pastes are highly porous and have an extremely fine texture. The structure of the paste is such that 72 percent is solid and 28 percent accessible to evaporated water. The cavities in the cement paste are seldom completely saturated at the time of freezing, and therefore hydraulic pressure does not necessarily develop at the time of freezing. Nevertheless, it is possible for the paste to become dilated.

Concrete freezes from the outside to the inside. When freezing starts, the void spaces contain water. If the water is above the critical saturation point of 91.7 percent, the capillaries will freeze and expel water to the surrounding area—probably into the cement paste because of its small capillaries. Since the paste and aggregate cannot accommodate this expansion of 9 percent, the hydraulic pressure is so great that cracking or popouts result. Once an opening is formed in the concrete, water is allowed to enter and deterioration is underway.

The most acceptable explanation of why concrete may fail during freezing and thawing was given by Powers. Concrete contains many air-filled cavities, consisting of entrained air bubbles, accessible pores in the aggregate particles, and thin fissures under the aggregate particles. All empty cavities of the type mentioned, especially air bubbles, are difficult to fill with water. They cannot be filled by capillary action because a liquid cannot flow from a small capillary to fill a larger one. However, pressures caused by freezing water are more than sufficient to free unfrozen water into such spaces. The resistance to movement of water must be the primary source of pressures, for practically all concrete contains enough air-filled space to accommodate the water and its volume of expansion when frozen.

It is therefore apparent that test procedures must be used or developed to determine the soundness of the aggregates used in the making of concrete.

AGGREGATES FOR BITUMINOUS MIXTURES

The influence of aggregates on the properties and performance of bituminous mixes is great. The ideal aggregate for a bituminous mix would have proper gradation and size, be strong and tough, and be angular in particle shape. Other properties would consist of low porosity, surfaces that are free of dirt, rough texture, and hydrophobic nature. Tables 3-1 and 3-2, which illustrate aggregate porosity, abrasion, hardness, toughness, strength, durability, surface characteristics, shape, and so on, also apply to bituminous mixes. The aggregate gradation and size, strength, toughness, and shape are important considerations for stability of the structure. The porosity and the surface characterisitics are important to the aggregate–bitumen interaction. The asphalt cement or its product must adhere to the aggregate and at the same time coat all the aggregate particles. If the aggregate particle has a low porosity and is smooth, the asphalt cement will not adhere to the aggregate. Adhesion becomes an extremely important property during periods when the mix is exposed to water. If the aggregate wets easily, the water will compete with the bitumen for adsorption onto the aggregate surface, and the aggregate will separate from the bitumen, which is known as slippage.

Aggregate Gradation and Size

Depending upon the specific use of the bituminous mix, the size and gradation of the aggregate varies tremendously. A high-quality bituminous mix that is to be used as a pavement for heavy traffic will generally utilize a dense-graded aggregate (a well-graded aggregate from coarse to fine). In this particular case one would not utilize Fuller's maximum density curve because it does not leave sufficient room for the asphalt cement. Therefore, the best procedure would be to open the grading somewhat more than the maximum. This opening of the gradation is achieved by the addition of fines (material less than the No. 200 sieve).

Strength, Toughness, Shape, and Porosity

The aggregate in a bituminous mix, unlike that in portland cement concrete, supplies most of the stability and thus should have a certain amount of strength and toughness; otherwisen loss of stability will result. Open-graded mixes are subject to greater mechanical breakdown than a dense-graded mix. Thus, if the material selected for a bituminous mix is of minimal strength, a denser mix of the same material would be utilized. ASTM C136 (Resistance to Abrasion of Small Size Coarse Aggregate by Use of the Los Angeles Machine) attempts to measure the effective strength and toughness of an aggregate. Table 3-1 shows the abrasive results of various aggregates subjected to the Los Angeles machine. The test shows very little correlation with the field performance of the aggregates. It does, however, have some use in distinguishing aggregates that are unsuitable in surface treatment.

Particle shape is a property of the aggregate that is more important than gradation and size, strength, and toughness when it comes to bituminous mixes. When sounded aggregates are used in an open-mix gradation, very little stability is achieved. Thus, when an open-mix gradation is used, angular aggregate should be used. If rounded aggregate has to be used in a bituminous mix, it should be crushed, thus resulting in a fractured plane.

Porosity of the aggregate strongly affects the economics of a mix. In each mix the aggregate should have a certain amount of porosity Zpmt percent). In general, the higher the porosity, the more asphalt will be absorbed into the aggregate, thus requiring a higher percent asphalt in the mix design. Thus, an aggregate that is porous gives rise to the possibility of selective absorptionm In *selective absorption* the oily portions of the asphalt are selectively absorbed, leaving a hard residue on the surface of the aggregate particle. This process could lead to stripping of the aggregate from the asphalt cement.

AGGREGATE BENEFICIATION

In various parts of the country, sound aggregates are scarce and transportation costs for this good aggregate are expensive. In these areas, the aggregate that is available may have certain deleterious materials which prevent these aggregates from passing specifications, or these aggregates have an adverse field performance record. In a few of these cases the aggregate can be beneficiated so that it becomes a useful economical product.

In the beneficiating of aggregates, several processes have been developed:

1. Washing.
2. Heavy-media separation.
3. Elastic fractionation.
4. Jigging.

Washing

Undesirable aggregates are washed to remove the particle coatings or to change the gradation. Fine content can be removed by exposing the aggregate to streams of water while it moves over screens or in special wash tanksmjawStokes lawjax can also be utilized to remove certain fractions of fine aggregate by differential settlement. For water at 77°F (25°C) and using 2.65 as the specific gravity of sand, Stokes' law is

$$V = 9000 D^2 \tag{3.2}$$

where V = velocity of settlement

D = particle climates

Thus, if sand is introduced near the top of an elongated water tank, various size fractions can be removed from pockets spaced away from the sand–water stream, which is entering horizontally.

Heavy-Media Separation

Heavy-media separation utilizes the principle that the specific gravity of much deleterious material is lighter than the specific gravity of sound aggregate. In this method, referred to as the *sink float method,* the suspension is composed of water and magnetite or ferrosilicon, or a combination of the two. The suspension is maintained at a specific gravity that will allow the deleterious material to float to the top and the sound material to sink.

Elastic Fractionation

Elastic fractionation is a procedure whereby heavy but soft particles can be removed. Aggregates fall on an inclined plate, and their quality is measured by the distance they bounce from the surface. The bouncing stones are collected in three different compartments, placed so as to collect particles in categories of their bouncing characteristics. Poor, soft, or friable particles bounce only a short distance, whereas the harder, sound aggregate particles bounce much farther. The elastic fractionation process removes only those particles having elastic properties that cause them to bounce poorly. It does not remove deleterious particles that have a high modulus of elasticity. Thus, heavy-media separation should be used in conjunction with elastic fractionation to remove all the deleterious material.

Jigging

Jigging is a specific-gravity method of removing light particles such as coal, lignite, or sticks. Upward pulsations created by air tend to hinder settlement of lighter particles, which are removed by shimming devices. The advantage of the process is that either fine or coarse aggregate may be used.

ASTM TEST SPECIFICATIONS FOR AGGREGATES

In the following section we will look at various ASTM test specifications, with special emphasis on the following categories:

1. Tests concerning the general quality of aggregates.
2. Tests concerning deleterious materials in aggregates.
3. Tests used in the design of portland cement concrete and bituminous mix design.

In each case we will look at the purpose of the test, followed by a brief description of the test and the authors' opinions and conclusions concerning the procedure, results, and validity of the test.

Tests Concerning the General Quality of Aggregates

ASTM C136 (Resistance to Abrasion of Small Size Coarse Aggregate by Use of the Los Angeles Machine) The purpose of this specification is to test coarse aggregate smaller than 1.5 in. (3.81 cm.) for resistance to abrasion using the Los Angeles testing machine and to evaluate base-course aggregates for possible degradation.

In this procedure, the test sample is placed in the Los Angeles testing machine after it has been prepared for testing in accordance with this specification. The machine is rotated at a speed of 30 to 33 rpm for 500 revolutions. The material is discharged from the machine and a preliminary separation of the sample is made on a sieve coarser than the No. 12. The finer portion is sieved using No. 12 sieve in a manner conforming to the specification. The material coarser than the No. 12 sieve is washed and oven-dried at 221 to 230°F (105 to 110°C) to constant weight and weighed to the nearest gram. The difference between the original weight and the final weight of the test sample is expressed as a percent of the original weight. The value is reported as a percent of wear.

According to the specifications, backlash or slip in the driving mechanism is very likely to furnish results that are not duplicated by other laboratories—an apparent disadvantage. In 1937, Woolf compared the Los Angeles abrasion results with the service records of coarse aggregates and concluded that the Los Angeles test gives accurate indications of the quality of the material under test and that its use in specifications controlling the acceptance of coarse aggregates is warranted.

However, lack of sufficient data from any one test procedure makes it impossible to suggest limits or specifications for the abrasion resistance on any type of concrete surface. Also, different surfaces may require different abrasion values. For example, a sidewalk would not have to have a high-wearing resistance as compared to one in which heavy roller cars are constantly passing over it. Some materials can pass this test but ultimately are dangerous to the concrete.

Unless someone places limits on wearing-resistance values of material such as limestone, blast-furnace slag, hard quartz, and other materials, the test is only a fair representation of what may happen. It is a good test for base-course aggregates, in that it can give an idea of degradation characteristics.

ASTM C88 (Soundness of Aggregates by Use of Sodium Sulfate or Magnesium Sulfate) The purpose of this specification is to determine the potential resistance of an aggregate to weathering.

In this procedure 5000 g of an aggregate having a known sieve analysis is immersed in a solution of sodium or magnesium sulfate for 16 to 18 hours. Next, it is placed in an oven at 230°F (110°C) and dried to constant weight. The procedure is repeated for the desired period (usually 5 or 10 cycles); the sample is cooled, washed, and dried to constant weight; then sieved, weighed, and recorded as the percent of weight lost.

This method furnishes information helpful in judging the soundness of aggregates subjected to weathering action, particularly when adequate information is not available from service records of the material exposed to actual weathering conditions. Attention is called to the fact that test results by the use of the two salts differ considerably, and care must be exercised in fixing proper limits in any specifications that may include requirements for these tests.

ASTM C666 (Resistance of Concrete to Rapid Freezing and Thawing) The purpose of this specification is to determine how concrete will react under continuous cycles of freezing and thawing and to rank aggregates.

In this procedure two methods are used. Method A involves rapid freezing and thawing in water, and method B involves rapid freezing and thawing in air. Immediately after curing, the specimen is brought to a temperature of 42.5 ± 5°F (5.8 ± 2.8°C), tested for fundamental transverse frequency, weighed, and measured in accordance with ASTM C215 (Fundamental Transverse, Longitudinal, and Torsional Frequencies of Concrete Specimens). The specimen is protected against loss of moisture between the time of removal from curing and the start of the freeze–thaw test.

The freezing and thawing test is started by placing the specimens in the thawed water at the beginning of the thawing phase of the cycle. After each cycle the specimen is tested for the fundamental transverse frequency at a temperature of 425°F (5.8 ± 2.8°C), weighed, and returned to the apparatus. For procedure A, the container is rinsed out, and clean water is added, and the specimen is returned to the freezer. This procedure is continued for 300 cycles or until the relative dynamic modulus of elasticity reaches 60 percent of the initial modulus. If the test is interrupted, the specimen is stored in the frozen condition. For procedure B it is undesirable to store the specimens in the thawed conditions for more than 2 days.

For procedure A, the specimen must have at least ⅛ in. (0/32 cm) of water all around it. In order to do this, it must be kept in a container. If the container is made of metal, the water will freeze from the top down, and as it expands there is no place for the water to go except into the concrete pores. This situation results in scaling and may cause misleading interpretations. If the container is made of rubber or other material that will expand as the water expands, the problem of scaling in the initial stages is of no concern, as it will not occur. With scaling there is an eventual weight loss, and in the early stages, especially in a metal container, the results may be misleading. Another problem may be the way the specimen is cured: whether it is cured in water or cured in air and saturated. Another problem arrives because of the rate of freezing. The Corps of Engineers freeze concrete at a rate of 12 cycles per day; other agencies freeze at a rate of 4 to 6 cycles per day; and still others freeze at a rate of 1 cycle per day. Thus, the thermal properties of the aggregates begin to play an important role in the process. The test should be used to rank vaious aggregates rather than to determine their performance characteristics when used in concrete.

ASTM C215 (Fundamental Transverse, Longitudinal, and Torsional Frequencies of Concrete Specimens) The purpose of this test is to determine the relationship between strength loss and cycles of freezing and thawing.

Test specimens are made in accordance with ASTMC192 (Making and Curing Concrete Test Specimens in the Laboratory). The weight and the average length will be determined. The most important test for portland cement concrete is the transverse frequency. The specimen is placed on supports such that it may vibrate without restrictions in a free transverse mode. The specimen is forced to vibrate at various frequencies. Record the frequency of the test specimen that results in maximum indication having a well-defined peak on the indicator and at which observation of nodal points indicates fundamental transverse vibration as the fundamental transverse frequency. *Young's modulus* is then calculated as follows:

$$\text{durability factor} = \mathrm{DF}_{300} = \frac{PN}{M} = \frac{(\text{relative } E)\,(N \text{ cycles})}{\text{duration of test}} \tag{3.3}$$

In this test it is not necessary to perform the test for 300 cycles of freezing and thawing. It is only necessary to perform the test for 150 cycles and then calculate the durability factor at 50% and the DF_{300} can be calculated. The only problem with this test is to make sure that the specimen is vibrating at its fundamental transverse mode. The test is a good one in that it is nondestructive and gives a relationship between strength loss and the number of cycles of freezing and thawing.

ASTM C597 (Pulse Velocity through Concrete) The major purpose of this specification is to check the uniformity in mass concrete, to indicate characteristic changes in concrete, and in the survey of field structures estimate the severity of deterioration, cracking, or both.

In this procedure a sound wave is transmitted through the concrete mass and the length of time it takes to travel from one end to the other is recorded. Knowing the time and the path length, the velocity can be computed.

According to the ASTM, results obtained from this test should not be considered as a means of measuring strength or as adequate for establishing the compliance of the modulus of elasticity of field concrete. The primary use of the test is to check dams for weak spots. There have been complaints about its lack of precision, and so far it has not proven to be completely satisfactory.

ASTM C671 (Critical Dilation of Concrete Specimens Subjected to Freezing) The purpose of this specification is to determine the test period of frost immunity to concrete specimens measured by the water immersion time required to produce critical dilation when subjected to a prescribed slow-freezing procedure.

In this procedure, the test specimen is molded and cured as prescribed by ASTM C192 (Making and Curing Concrete Test Specimens in the Laboratory). Once the test specimen is prepared and conditioned, the test starts. The test cycle consists of cooling the specimen in water-saturated kerosene from 35 to 15°F(1.67 to −9.44°C) at a rate of 5 ± 1°F (−2.8 ± 0.5°C) per hour followed immediately by returning the specimen to the 35°F (1.67°C) water bath, where the specimen will remain until the next cycle. Normally, one test cycle would be carried out every 2 weeks. The length changes are measured during the cooling process. The test is continued until critical dilation is exceeeded or until the period of interest is over.

Basically, this is one of the most poorly written procedures in the ASTM specifications. If a good aggregate is utilized, the aggregate may never exceed its critical dilation.

ASTM C682 (Evaluation of Frost Resistance of Coarse Aggregates in Air-entrained Concrete by Critical Dilation Procedure) The purpose of this procedure is to evaluate the frost resistance of coarse aggregates in air-entrained concrete.

This procedure is basically the same as that for the preceding specification. The only difference is that the sample is prepared in accordance with ASTM C295 (Petrographic Examination of Aggregates for Concrete). The aggregate is graded in accordance with field use; otherwise, equal portions of the No.4, ⅜-in. ½-in., and 1-in. sieves are used. Further, the aggregate should be used in this test as it is used in the field. Portland cement should meet the specifications of ASTM C150

Portland Cement), and the fine aggregate should meet the specification of ASTM C33 (Concrete Aggregates). The mix proportion should be in accordance with the ACI method of mix design with an air content of 6 percent and a slump of 2.5 ± 0.5 in.

This test is supposed to simulate the field performance. The significance of the results in terms of potential field performance will depend upon the degree to which field conditions can be expected to correlate with those employed in the laboratory. Thus, the field conditions must be assessed. Obviously, one of the main problems is to simulate field conditions. Some problems are: degree of saturation, age of concrete when the first freeze comes, length of freezing season, amount of water available during freezing, curing procedure, and condition of the aggregate when it enters the mixer. All of these conditions must be duplicated and thus there are definite chances for error. Also, the project should be planned well in advance and the conditions known before the test is run.

ASTM C672 (Scaling Resistance of Concrete Surfaces Exposed to Deicing Chemicals) The purpose of this specification is to evaluate the effect of mix design, surface treatment, curing, or other variables of concrete subjected to scaling due to freezing and thawing, and to determine the resistance to scaling of a horizontal concrete surface subjected to freezing and thawing in the presence of deicing chemicals.

In this specification the concrete at the age of 28 days, after proper curing, is covered with approximately 0.25 in. (0.64 cm) of calcium chloride and water solution having a concentration such that each 100 ml of solution contains 4 g of anhydrous calcium chloride. The specimen is then placed in a freezing chamber for 16 to 18 hours. The specimen is then removed and placed in air at 75 ± 3°F (23 ± 1.7°C) with a relative humidity of 45 to 55 percent for 6 to 8 hours. Water is added to the chamber between each cycle to maintain the depth of the solution. The procedure is repeated daily, and at the end of five cycles the surface of the concrete is flushed thoroughly. A visual inspection of the concrete is

TABLE 3-3 Concrete Scaling Ratings

Rating	*Condition of Surface*
0	No scaling
1	Very slight scaling
2	Slight to moderate scaling
3	Moderate scaling
4	Moderate to severe scaling
5	Severe scaling

made with the ratings given in Table 3-3. These ratings are recorded and the test continues.

Generally, this test is performed up to 50 cycles before final evaluation is made. The only problem with this test is that of the rating system. What represents moderate scaling to one person may not be that to another.

ASTM C295 (Petrographic Examination of Aggregates for Concrete) The purpose of this specification is to screen the good from the bad aggregates. It has eight specific purposes:

1. Preliminary determination of quality.
2. Establish properties and probable performance.
3. Correlation of samples with aggregates previously tested and used.
4. Selecting and interpreting other tests.
5. Detection of contamination.
6. Determining effects of processing.
7. Determining physical and chemical properties.
8. Describe and classify constituents.

In this specification the procedure should be carried out by a geologist utilizing x-ray diffraction, differential thermal analysis, electron microscopy, electron diffraction, electron probe, infrared spectroscopy, microscope, and the naked eye.

The main purpose of petrographic examinations is to determine physical and chemical properties of aggregates. The relative abundance of specific types of rocks and minerals is established as well as particle shape, surface texture, pore characteristics, hardness, and potential alkali reactivity. Coatings are identified and described; and the presence of contaminating substances is determined.

If the petrographic examination predicts potential alkali reactivity of the aggregate, it is very helpful, in that the time required for this test is substantially shorter than that of the freeze–thaw test or most other available tests.

This specification also sets down a fundamental principle regarding aggregates: if an unfamiliar source is found, it can be compared to known data.

Petrographic examination is the best method by which deleterious and extraneous substances can be detected and determined quantitatively.

ASTM D1075 (Effects of Water on Cohesion of Compacted Bituminous Mixtures) The purpose of this specification is to measure the loss of cohesion resulting from the action of water on compacted bituminous mixtures containing penetration-grade asphalts. In other words, it evaluates the stripping properties of aggregates.

In this specification a 4-in. (10.16-cm) cylindrical specimen 4 in. (10.16 cm.) high is tested in accordance with ASTM D1074 (Compressive Strength of Bituminous Mixtures). Then the bulk specific gravity of each specimen is determined. Each set of six test specimens is sorted into two groups of three specimens each so that the average bulk specific gravity is the same in each group. Group 1 is tested in accordance with procedure A and group 2 in accordance with procedure B. In test procedure A, the test specimens are brought to the test temperature of 77 ± 1.8°F (25 ± 1°C) by storing them in an air bath maintained at the test temperature for not less than 4 hours, their compressive strength determined in accordance with ASTM D1074. In test procedure B, the test specimen is immersed in water for 4 days at 120 ± 1.8°F (49 ± 1°C). The specimen is transferred to a second water bath at 77 ± 1.8°F (25 ± 1° C) and stored for 2 hours. At that time the compressive strength is determined and the numerical index of resistance of bituminous mixtures to the detrimental effect of water as the percentage of the original strength that is retained after the immersion period is calculated as follows:

$$\text{index of retained strength (\%)} = \frac{S_2}{S_1} \times 100 \tag{3.4}$$

where S_1 = compressive strength of dry specimens (group 1)

S_2 = compressive strength of immersed specimens (group 2)

This test gives an excellent indication of the retained strength but sets no definite values as to what is a good or a bad limit.

Tests Concerning Deleterious Materials in Aggregates

ASTM C235 (Scratch Hardness of Coarse Aggregate Particles) This specification is primarily intended for field use in estimating the quality of a deposit of coarse aggregate. It affords a fast, convenient means of determining the amount of soft particles in aggregates. The test identifies materials that are soft or those that are poorly bonded (i.e., the separate particles in the aggregate piece are easily detached from the mass). The test is not intended to identify other types of deleterious materials in aggregates.

In this procedure the test should be made on a freshly broken surface of the aggregate particle. If the particle contains more than one type of rock and is partly hard and partly soft, it should be classified as soft only if the soft portion is one-third or more of the volume of the particle. A scratch hardness test can be made on the exposed surface of a particle provided that consideration is given to softening of the surface due to weathering. A particle with a thin, soft, weathered surface and a hard core should be classified as soft.

Each particle of aggregate under test is subjected to a scratching motion of a brass rod, using a pressure of 2 ± 0.1 lb (8.9 ± 0.4 N). [The brass rod should be 1/16 in. (0.16 cm) in diameter with a rounded point.] The brass rod should be of suitable hardness so that when filed to a sharp point, it will scratch a copper penny but fail to scratch a nickel. Particles are considered soft if, during the scratching process, a groove is made in them without removing metal from the brass rod, or particles from the rock mass.

Although this test is intended primarily for field use in estimating qualities, it is felt that soft particles can be determined very easily. However, breakage, segregation, and contamination of aggregate can occur during handling and stockpiling. This test should be used as a preliminary test only and the results should not go beyond the preliminary stages. One factor to consider is the pressure applied to the particle. This pressure would seem very hard to obtain, as what might feel like 2 lb (8.9 N) to one person may not to another. The problems of breakage, segregation, and contamination are important, because if the aggregate is considered hard, it will be handled and stockpiled. That is, if the particles pass the test and are considered to be 30 percent soft, then because of handling the aggregate may increase to 35 percent soft particles and thus be unacceptable. Also, if this aggregate was used for portland cement concrete, it may break down during mixing, resulting in the introduction of fines in the mix which may be detrimental to the general quality of the concrete.

ASTM C33 (Concrete Aggregates) The purpose of this specification is to ensure that satisfactory materials are used in concrete. The specification covers both fine and coarse aggregates but does not cover lightweight aggregates.

The specification establishes definitions for fine and coarse aggregate and places restrictions on grading, deleterious substances, and soundness. The specification also establishes methods for testing and sampling.

This specification is good in that it lays down the fundamental rules for fine and coarse aggregates used in concrete as well as the sampling and testing methods to be followed.

ASTM C142 (Clay Lumps and Friable Particles in Aggregates) The purpose of this specification is to measure only particles that might cause unsightly blemishes in concrete surfaces. It is an approximate method for the determination of clay lumps and friable particles in natural aggregates.

Aggregates for this test consist of the material remaining after the completion of ASTM C117 [Materials Finer Than No. 200 (75-μm) Sieve in Mineral Aggregates by Washing]. The aggregate is dried to a constant weight at a temperature of 230°F (105 ± 5°C). Weigh the test sample and spread it into a thin layer on the bottom of the container and examine it for clay lumps and/or friable particles. Particles that can be broken down with the fingers into finely divided particles are classified as friable particles provided that they can be removed by wet sieving. The residue is removed and weighed. The amount of clay lumps and friable particles in fine aggregate or individual sizes of coarse aggregate is computed as follows:

$$P = \frac{W - R}{W} \times 100 \tag{3.5}$$

where P = percent of clay lumps or friable particles

W = weight of test sample passing the layer of sieves but coarser than the No. 16 sieve

R = weight of particles retained on designated sieve

This test is not widely used because of the physical limitations of sorting through all the particles, and also because these particles are merely a symptom of inadequate processing, which can be remedied by improved washing techniques. Also, the techniques of breaking the particles with the fingers vary from one person to another (i.e., the pressure applied by one person is different from that applied by another). Some clay lumps are hard and cannot be broken apart with the fingers. Thus, inaccurate results will be obtained.

ASTM C117 [Materials Finer than No. 200 (75-μm) Sieve in Mineral Aggregates by Washing] The purpose of this test is to determine the amount of material finer than a No. 200 (75-μm) sieve in aggregate by washing. Clay particles and other aggregate particles that are dispersed by the wash water as well as water-soluble materials will also be removed from the aggregate during the test.

A sample of the aggregate is washed in a prescribed manner and the decanted wash water containing suspended and dissolved materials is passed through a No. 200 (75-μm) sieve. The loss in weight resulting from the wash treatment is calculated as weight percent of the original sample and is reported as the percentage of material finer than a No. 200 (75-μm) sieve by washing.

This test provides a measure of fines, including clay and silt, in concrete aggregates. The test has been criticized as not furnishing an indication of harmful clays, which may increase mixing-water requirements and volume-change tendencies of concrete. Experiments show that individual operators should have a 95 percent probability of checking their results within 0.3 percent. The amount of material passing a No. 200 (75-μm) sieve, by washing to the nearest 0.1 percent, is calculated as follows:

$$A = \frac{B - C}{B} \times 100 \tag{3.6}$$

where A = percentage of material finer than a No. 200 (75-μm) sieve, by washing

B = original dry weight of sample, grams

C = dry weight of sample, after washing, grams

Excessive quantities of fines in portland cement concrete detract from the quality of the mix by increasing the mixing requirement. In bituminous mixtures, asphalt demand may increase, although the problem is less serious than that in portland cement concrete. In base-course aggregates, this test can be especially important in determining potential susceptibility to frost action. For this reason, this is an important test.

ASTM C123 (Light Weight Pieces in Aggregates) The purpose of this specification is to determine the approximate percentage of lightweight pieces in aggregates by means of sink-float separation in a heavy liquid of suitable specific gravity.

In this procedure the fine aggregate is allowed to dry and cooled to room temperature after following the procedure prescribed in ASTM D75. The material is sieved using a No. 50 sieve and then brought to saturated-surface dry conditions. It is put into a heavy liquid such as CCl_4 or kerosene and 1,1,2,2,-tetrabromethane. The particles will be separated by the float-sink method provided that the specific gravities are different enough to permit separation. The liquid is poured off into a second container and passed through a skimmer. Care is taken that only the floating pieces are poured off with the liquid and that none of the sand is decanted onto the skimmer. The liquid is returned to the first container, and agitated again, and the decanting process repeated until the liquid is free of friable particles. The pieces are dried and the weight determined. For a coarse aggregate the particles are sieved using a No. 4 sieve and the foregoing process is repeated.

The materials or chemicals used in combination to form the heavy liquid for the separation process are very toxic and must be handled with great care. The test eliminates particles that might produce concrete of low durability when exposed to freezing and thawing. Experiments have shown that test procedures using gravity levels up to about 2.50 result in adequate aggregate. Coal and lignite are separated at a specific gravity of 2.0. Potentially harmful chert is separated at a specific gravity of 2.35.

ASTM C40 (Organic Impurities in Sands for Concrete) This specification covers an approximate determination of the presence of injurious organic compounds in natural sands that are to be used in cement mortar or concrete. The principal value of the test is to furnish a warning that further tests of the sands are necessary before they are approved for use.

The procedure for this specification involves a color test. The sand and a 3 percent solution of sodium hydroxide are mixed vigorously in a graduate and allowed to stand for 24 hours. The color of the liquid is then compared to the color of a solution of postassium dichromate in sulfuric acid. If the solution of the sand and sodium hydroxide is darker than the potassium dichromate, organic impurities are present in the sand.

This is a good quick test, and if impurities are indicated, the mortar strength test should be performed to determine if the impurities are deleterious. Certain types of organic matter, principally tannic acid and its compounds derived from the decay of vegetable matter, interfere with the hardening and strength development of cement. This test detects this type of material but unfortunately also reacts to other organics, such as bits of wood, which might not be harmful to strength. A negative test is conclusive evidence of freedom from harmful organic matter, but a positive test may or may not fortell difficulty. This is an excellent test in that it is a warning of possible dangers.

ASTM C227 [Potential Alkali Reactivity of Cement–Aggregate Combinations (Mortar-Bar Method)] The purpose of this specification is to determine if an aggregate will react with the alkalies in the cement. This test is basically a test to predict the alkali–silica reaction.

The test method consists of molding bars of mortar 1 in. × 1 in. × 12 in. (2.54 cm × 2.54 cm × 30.5 cm) in which the aggregate in question is combined with a cement that is to be used in the field. The proportions should be 1 part cement to 225 parts of graded aggregate by weight. Use enough water to develop a flow of 105 to 120 in accordance with ASTM C109 (Compressive Strength of Hydraulic Cement Mortar), with the exception that the flow table drops ½ in. (1.27 cm) for 10 trips in 6 seconds. After 24 hours in the molds, the lengths of the bars are measured, and they are stored at a constant temperatures of 100°F (37.8°C) in sealed containers containing a small amount of water in the bottom but not in contact with the specimens. Length changes are to be measured after 1, 2, 3, 6, 9, and 12 months and, if necessary, every 6 months after. If expansion is less than 0.04 percent in 6 months, it is considered nonreactive. If the expansion is between 0.04 and 0.07 percent, the aggregate is suspicious. If the expansion is between 0.07 and 1 percent, the aggregate is reactive.

The obvious disadvantage to this test is the length of time involved to perform it. The test would seem to be reliable in rejecting poor aggregate. The limits given are judgments based on reports through research and petrographic examination. The limits are somewhat low for the purpose of conservatism.

ASTM C289 [Potential Reactivity of Aggregates (Chemical Method)] The purpose of this test procedure is to determine the potential reactivity of an aggregate with alkalies in portland cement concrete in a very short time. This is a test for the alkali–silica reaction and is not intended for the alkali–carbonate reaction.

In this procedure the material is ground to the point when it is finer than the No. 50 sieve but coarser than the No. 100 sieve. Twenty-five grams of the material are mixed with 25 ml of a 1 *N* solution of No. OH in a steel vessel about 2 in. (5.08 cm) in diameter and 2½ in. (6.35 cm) high. The vessel is sealed at a temperature of 176°F (80°C) for 24 hours and then the liquid is filtered and tested for alkalinity and dissolved silica.

The results are presented on semilog paper as dissolved silica vs. reduction in alkalinity. If a considerable amount of silica is dissolved and there is no loss of alkalinity, the aggregate is a reactive aggregate. If a considerable amount of silica is dissolved and there is a considerable amount of reduction in alkalinity, we have a potentially reactive material. If there is little dissolved silica and little reduction in alkalinity, the aggregate is of good quality.

This test has the advantage of being quick and accurate, as the results can be run within 24 hours. This test is based upon field experience and should only be used as a screening device. The test will not give satisfactory results with carbonate rocks such as ferrous and magnesium rocks. In these two aggregate types, a reduction in ions results and thus the results are no longer reliable.

Tests Used in the Design of Concrete Mixes (Portland Cement or Bituminous)

ASTM D75 (Sampling Aggregates) The purpose of this test is to sample fine and coarse aggregate for the following purposes:

1. Preliminary investigation of the potential source of supply.
2. Control of the product at the source of supply.
3. Control of the operations at the site of use.
4. Acceptance or rejection of the materials.

In this procedure, sampling plans and acceptance and control tests vary with the type of construction in which the material is used. Samples for preliminary investigation tests are obtained by the party responsible for development of the potential source. The sampler must use every precaution to obtain samples that will show the true nature and condition of the materials they represent. Samples must be inspected and sampling taken from conveyor belts, flowing aggregate stream, or stockpiles. The number of samples taken depends on the variations and the properties measured.

Sampling is as important as testing; thus, the test results depend on the sampling. If the sampler is inexperienced with the techniques involved, the entire test becomes questionable. Thus, in this specification, a person familiar with sampling should perform all sampling procedures.

ASTM C136 (Sieve or Screen Analysis of Fine and Coarse Aggregates) The purpose of this specification is to determine the particle size of fine and coarse aggregates to be used in various tests.

In this procedure, a weighed sample of dry aggregate is separated through a series of sieves or screens of progressively smaller openings for determination of particle-size distribution.

In this specification, the results are dependent upon individual technique. The test is placed in two categories, mechanical sieving and hand sieving. This excellent test determines the gradation of the aggregates, which is so important in mix design procedures.

ASTM C127 (Specific Gravity and Absorption of Coarse Aggregate) The purpose of this specification is to ultimately determine the solid volume of coarse aggregate and the unit volume of the dry rodded aggregate such that a weight–volume characteristic can be determined so that a concrete design mix can be determined. The bulk specific gravity is used to determine the volume occupied by the aggregate.

In this procedure, approximately 5 kg of the aggregate is selected after quartering. After the aggregate is thoroughly washed to remove dust or other coatings from the surface of the particles, the sample is dried to constant weight at a temperature of 212 to 230°F (100 to 110°C) and cooled in air at room temperature for 1 to 3 hours. After cooling, the sample is immersed in water at room temperature for a period of 24 ± 4 hours. Next, the specimen is removed from the water and rolled in a large absorbent cloth towel until all visible films of water are removed. The large particles are wiped by hand. Care is taken not to allow evaporation of water from aggregate pores during the operation of surface drying. The sample is weighed in the saturated-surface-dry condition. Then the sample is weighed in water, making sure that the entrapped air is removed. The sample is dried at 212 to 230°F (100 to 110°C) cooled at room temperature for 1 to 3 hours, and weighed. The bulk and apparent specific gravity and the percent absorption are determined as follows:

$$\text{bulk specific gravity} = \frac{A}{B - C} \tag{3.7}$$

where A = weight of oven-dry specimen in air, grams

B = weight of saturated-surface-dry specimen in air, grams

C = weight of saturated specimen in water, grams

$$\text{bulk specific gravity (saturated-surface-dry)} = \frac{B}{B - C} \tag{3.8}$$

$$\text{apparent specific gravity} = \frac{A}{A - C} \tag{3.9}$$

$$\text{absorption} = \frac{B - A}{A} \times 100 \tag{3.10}$$

The specific gravity of aggregates is important as it is used to determine the aggregate for use in a concrete mix.

ASTM C128 (Specific Gravity and Absorption of Fine Aggregate) The purpose of this specification is to determine the bulk and apparent specific gravity of fine aggregate as well as the absorption.

In this procedure, 500 g of fine aggregate is immersed in a pycnometer which is filled with water to the 90 percent capacity. The pycnometer is rolled, inverted, and agitated to eliminate air bubbles. The temperature is adjusted to 73.4 ± 3°F (23 ± 1.7°C). The total weight of the pycnometer, sample, and water is determined. The fine aggregate is removed, dried to a constant weight at 212 to 230°F (100 to 110°C), cooled at room temperature for 0.5 to 1.5 hours, and weighed. The weight of the pycnometer is determined and the bulk specific gravity, bulk saturated-surface-dry specific gravity, apparent specific gravity, and the absorption are calculated as follows:

$$\text{bulk specific gravity} = \frac{A}{B + 500 - C} \tag{3.11}$$

where A = weight of oven-dry specimen in air, grams

B = weight of pycnometer filled with water, grams

C = weight of pycnometer with specimen and water to calibration mark, grams

$$\text{bulk saturated-surface-dry specific gravity} = \frac{500}{B + 500 - C} \tag{3.12}$$

$$\text{apparent specific gravity} = \frac{A}{B + A - C} \tag{3.13}$$

$$\text{absorption} = \frac{500 - A}{A} \times 100 \tag{3.14}$$

This is an important test because the volume of aggregate is determined for the concrete mix. The results of this test are used in all concrete-mix design procedures, whether portland cement or bituminous.

ASTM C29 (Unit Weight of Aggregate) This method covers the determination of the unit weight of fine, coarse, or mixed aggregate.

In this procedure, the sample is dried to constant weight in an oven at 220 to 230°F (105 to 110°C) and thoroughly mixed. A cylindrical metal bucket is calibrated using water (knowing that water weighs 62.4 lb/ft^3). The measure is filled one-third full and the surface is leveled with the fingers. The layer of aggregate is rodded 25 times with a tamping rod. The strokes are applied evenly over the sample. This procedure is repeated at two-thirds full and at full. The measure is leveled, weighed, and multiplied by the volume of the bucket. This method applies to aggregates of 1.5 in. (3.81 cm) or less. For aggregate over 1.5 in. (3.81 cm), use the jigging method.

The results of this test should check within 1 percent when duplicated. The results of this procedure are important, as they are used in the mix design procedure for both portland cement and bituminous concrete.

PROBLEMS

3.1. What is the difference between a natural aggregate and a manufactured aggregate?

3.2. Aggregates may be classified as fine aggregate or coarse aggregate; explain the difference.

3.3. Explain how aggregates are processed for use as a portland cement concrete ingredient or as a bituminous concrete ingredient.

3.4. How does particle shape affect the use of aggregate in base-course materials? In portland cement concrete? In bituminous concrete?

3.5. Explain the use of Fuller's maximum density curve.

3.6. Why is gradation important in portland cement concrete?

3.7. Which type of aggregates (igneous, sedimentary, or metamorphic) would you expect to be most suitable as a base-course material? Why?

3.8. Review various references on the subject of freezing and thawing and write a short report on how they eventually lead to concrete failure.

3.9. Review several references and explain why aggregate beneficiation is necessary. Include in your report the methods used for aggregate beneficiating.

3.10. Review the ASTM specifications for tests concerning the general quality of aggregates, deleterious materials in aggregates, and the specifications used in the design of portland cement and bituminous concrete mixes, and write a short report on the purposes, procedures, and reasons for the tests.

REFERENCES

ABDUN-NUR, E. A., "Concrete and Concrete Making Materials," *ASTM Spec. Tech. Publ. 169-A,* 1966, pp. 7–17.

ALLEN, C. W., "Influence of Mineral Aggregates on the Strength and Durability of Concrete," Symposium on Mineral Aggregate, *ASTM Spec. Tech. Publ. 83,* 1948.

BATEMAN, J. H., *Materials of Construction,* Pitman Publishing Corp., New York, 1950, pp. 23–74.

FULLER, W. B. and THOMPSON, S. E., "The Laws of Proportioning Concrete," *Trans. Am. Soc. Civil Engrs., 59,* 1907.

Highway Research Board, *Bibliography on Mineral Aggregates,* Washington, D.C., 1949 (Bibliography No. 6).

JACKSON, F. H., "The Durability of Concrete in Service," *Proc. Am. Concr. Inst., 43* (1942), p. 165.

KREBS, R. D. and WALKER, R. D., *Highway Materials,* McGraw-Hill Book Company, New York, 1971.

MINOR, C. E., "Degradation of the Mineral Aggregates," *ASTM Spec. Tech. Publ. 277,* 1960, pp. 109–121.

NORDBERG, B., "Canada's Most Modern Gravel Plant,"*Rock Prod., 55* (Aug. 1952).

PARSONS, W. H., and INSLEY, H., "Observations on Alkali–Aggregate Reaction," *Proc. Am. Concr. Inst., 44* (1948), p. 625.

Pit and Quarry Handbook, 41st ed., Complete Service Publishing Co., Chicago, 1948.

POWERS, T. C., "Basic Considerations Pertaining to Freezing and Thawing Tests," *ASTM Proc., 55* (1955) p. 1132.

STANTON, T. E., "California Experience with the Expansion of Concrete through Reaction between Cement and Aggregate, *J. Am. Concr. Inst., 39* (Jan. 1942).

SWEET, H. S., "Physical and Chemical Tests of Mineral Aggregates and Their Significance," Symposium on Mineral Aggregates, *ASTM Spec. Tech. Publ. 83,* 1948.

VERBECK, V., "Osmotic Studies and Hypothesis Concerning Alkali–Aggregate Reactions," *ASTM Proc., 55* (1955), pp. 1110–1127.

WOOLF, D. O., "Methods for the Determination of Soft Pieces in Aggregates," *Public Roads, 26,* (Apr. 1951), p. 148.

WOOLF, D. O., "Methods for the Determination of Soft Pieces in Aggregate," *Public Roads, 26,* (Apr. 1937).

CEMENTS 4

Cements are materials that exhibit characteristic properties of setting and hardening when mixed to a paste with water. They are a class of products that can be very complex and of somewhat variable composition and constitution.

Cements are divided into two classifications: hydraulic and nonhydraulic. This division is based upon the way in which the cement sets and hardens. The *hydraulic cements* have the ability to set and harden under water. Hydraulic cements include, but are not limited to, the following: hydraulic limes, pozzolan cements, slag cements, natural cements, portland cements, portland–pozzolan cements, portland blast-furnace-slag cements, alumina cements, expansive cements, and a variety of others (white portland cements, colored cements, oil-well cements, regulated cements, waterproofed cements, hydrophobic cements, antibacterial cements, barium and strontium cements).

Nonhydraulic cements do not have the ability to set and harden under water but require air to harden. The main example of a nonhydraulic cement would be a lime.

LIME

Lime, one of the oldest known cementing materials, is readily available and rather inexpensive. Lime is produced by burning limestone (calcium carbonate) with impurities such as magnesia, silica, iron, alkalies, alumina, and sulfur. This burning process takes place in either a vertical or a rotary kiln at a temperature of 1800°F (980°C). Calcium carbonate is decomposed into calcium oxide and carbon dioxide according to the following reaction:

$$CaCO_3 \rightarrow CaO + CO_2 \qquad (4.1)$$

The calcium oxide that is formed is called *quicklime*, which, when in the presence of water, reacts to form calcium hydroxide together with a great evolution of heat:

$$CaO + H_2O \rightarrow Ca(OH)_2 + \text{heat} \tag{4.2}$$

This process is called *slaking* and the product calcium hydroxide is called *slaked lime* or *hydrated lime*. The rate of reaction depends mainly on the purity of the lime. The higher the purity of lime, the greater its reactivity toward water. Commercial quicklime is classified into three groups: quick, medium, and slow slaking.

Depending upon the amount of water added during the slaking process, lime putty or hydrated lime may be formed. Hydrated lime is produced by adding just enough water (one-third of its weight) to quicklime. Lime putty is formed when an overextended amount of water is added to the quicklime.

Both lime putty and hydrated lime are always mixed with mortar sand in proportions of 1 part lime to 3 parts sand by volume, to prevent excessive shrinkage.

The setting of lime mortar is the result of the loss of water either by absorption of water by the block, brick, or whatever, or by evaporation. The hardening is caused by the reaction of carbon dioxide in the air with the hydrated lime as follows:

$$Ca(OH)_2 + CO_2 \rightarrow CaCO_3 + H_2O \tag{4.3}$$

This results in the formation of calcium carbonate crystals, which bind the heterogeneous mixture into a coherent mass. The hardening process is slow and may take several years to develop its full strength. However, it needs the free circulation of air to provide the necessary carbon dioxide to penetrate the innermost portion of the mortar for hardening to take effect.

HYDRAULIC LIMES

Hydraulic limes are made by burning siliceous or argillaceous limestone whose clinker after calcination (in a continuous kiln) contains a sufficient percentage of lime silicate to give hydraulic properties to the product, but which normally contain so much free lime that the mass of clinker will slake on the addition of water.

As the content of alumina and silica in the lime increase, the rate of slaking and heat evolution decrease to a point where no reaction occurs between water and lime. At high temperature, the alumina and silica combine with calcium oxide, calcium silicates, and aluminates, which do not easily combine with water when it is in lump form. Therefore, quicklime is added in the slaking process and the large lumps are broken up into fine powder due to the expansion of the quicklime. The final product consists of lime silicate and about one-fourth hydrated lime. The material in the fine form can now readily combine with water. Hydraulic limes do exhibit hydraulic properties but are not suited to subaqueous construction because they require free access of air during hardening. The air is necessary to ensure carbonation of free calcium hydroxide present in large amounts in calcium carbonate; otherwise, the full strength of the lime cannot be developed.

Hydraulic limes are used for browning plaster coats or for stucco and similar uses.

POZZOLAN CEMENTS

A *pozzolan*, according to ASTM C595 (Blended Hydraulic Cements), is a siliceous or siliceous and aluminous material which, in itself, possesses little or no cementitious value but will, in finely divided form and in the presence of moisture, react chemically with calcium hydroxide at ordinary temperatures to form compounds possessing cementitious properties. Pozzolans are further classified as natural pozzolans or artificial pozzolans. Natural pozzolans are further classified into two groups. The first group is made up of pumicite, obsidian, scoria, tuff, santorin, and trass, which are derived from volcanic rocks. The second group of natural pozzolans contains large quantities of finely dispersed, amorphous silica which react with lime in the presence of water to form hydrated silicates, which accounts for their hydraulic properties.

Artificial pozzolans include fly ash, boiler slag, and by-products from the treatment of bauxite ore.

Pozzolan cements are manufactured by direct grinding of the volcanic rocks or by calcining and grinding clays, shales, and diatomaceous earth. Pozzolan cements include all cementing materials that are made by the incorporation of pozzolans with hydrated lime in which no subsequent calcination is needed.

The physical requirements for pozzolans are given in ASTM C595 (Blended Hydraulic Cements) and are repeated here in Table 4-1. Whenever a pozzolan is used in cement, it must conform to the requirements of ASTM C311 (Sampling and Testing Fly Ash or Natural Pozzo-

TABLE 4-1 Physical Requirements for Pozzolans

Fineness	
Amount retained when wet-sieved on No. 325 (45-μm) sieve, max. (%)	20.0
Pozzolanic activity test	
Pozzolanic strength, min. [psi (MPa)]	800.0 (5.5)

lans for Use as a Mineral Admixture in Portland Cement Concrete) and/ or ASTM C618 (Specifications for Fly Ash and Raw or Calcined Natural Pozzolan for Use as a Mineral Admixture in Portland Cement Concrete).

Thus far, very little use if any has been found for pozzolan cements in the area of structural concrete. Its main use has been where mass is required rather than strength. It has also found limited use when mass concrete is needed with little heat of hydration.

SLAG CEMENTS

Slag cements are hydraulic cements consisting mostly of an intimate and uniform blend of granulated blast-furnace slag and hydrated lime in which the slag constituent is at least 60 percent of the weight of the slag cement. The mixture is often not calcined.

Basically, two types of slag cement exist. The first is designated Type S slag cement by ASTM C595 (Blended Hydraulic Cements). Type S slag cement may be used in combination with portland cement in making concrete and in combination with hydrated lime in making masonry mortar. The second type of slag cement is designated Type SA slag cement by ASTM C595 (Blended Hydraulic Cements). Type SA slag cement, which is air-entrained slag cement, has the same general uses as Type S.

Blast-furnace slags are a nonmetallic product, consisting essentially of silicates and aluminosilicates of calcium and of other bases, that is developed in a molten condition simultaneously with iron in a blast furnace. The blast-furnace slags, which are suitable for use in slag cements, are fusible lime silicates derived as waste products from the operation of blast furnaces in smelting iron from its ore.

When slag cements are manufactured, they go through a variety of operations, such as granulation (which not only renders the slag more hydraulic, but at the same time reduces the harmful sulfides), drying of the slag, preparation of the hydrated lime, proportioning the mix, mixing, and final grinding.

Slag cements are of limited importance in structural concrete, but may find success in projects requiring large masses of concrete masonry where weight and bulk are more important than strength. It may also find use as a masonry cement in that it does not have a staining effect because of its low alkali content.

NATURAL CEMENTS

A natural cement, as defined by ASTM C10 (Natural Cements), is a hydraulic cement produced by calcining a naturally occurring argillaceous limestone at a temperature below the sintering point and then grinding to a fine powder. The amounts of silica, alumina, and iron oxide present are sufficient to combine with all the calcium oxide to form the corresponding calcium silicates and aluminates, which account for the hydraulic properties of natural cements.

There exist two types of natural cements according to ASTM C10 (Natural Cements), Type N and Type NA. Type N natural cement is for use with portland cement in general concrete construction. Type NA natural cement is air-entrained cement and has the same uses as Type N.

Natural cements are made by the calcination of a natural clay limestone, which is made up of clay material (13 to 35 percent), silica (10 to 20 percent), and a balance of alumina and iron oxide. The clay material gives the cement its hydraulic properties.

After calcination and possible slaking to remove the free lime, the clinker is ground into a fine powder known as a natural cement, with the following average composition:

SiO_2	CaO	MgO	Fe_2O_3	Al_2O_3
22–29%	31–57%	1.5–22%	1.5–3.2%	5.2–8.8%

Thus, a quick view of the composition would indicate the possibilities of wide variation among mechanical properties. The physical requirements of natural cements are shown in Table 4-2.

Natural cements should not be used in exposed areas but may be used as a substitute for portland cements in mortars and concrete when the stresses encountered will never be high or in situations in which weight and mass are more essential than strength.

TABLE 4-2 Physical Requirements of Natural Cements

Properties	*Type N*	*Type NA*
Fineness, specific surface (cm^2/g)		
Air permeability apparatus		
Average value, min.	6000	6000
Minimum value, any one sample	5500	5500
Soundness		
Autoclave expansion, of blend of 75 percent natural cement and 25 percent portland cement by weight, max. (%)	0.8	0.8
Time of setting, Vicat test		
Time (min)	30	30
Air entrainment		
No agent used, max. (vol. %)	12	
Using air-entrainment agent (vol. %)	—	19 ± 3
Compressive strength, min. [psi (MPz)]		
Compressive strength of mortar cubes, composed of 1 part natural cement and 1 part standard sand by weight; must be equal to or higher than the values specified for the following ages:		
1 day in moist air, 6 days in water	500 (3.4)	500 (3.4)
1 day in moist air, 27 days in water	1000 (6.9)	1000 (6.9)

PORTLAND CEMENTS

Portland cement is one of the most widely used construction materials and is the most important hydraulic cement. It is used in concrete, mortar, plaster, stucco, and grout. It is used in all types of structural concrete (walls, floors, bridges, tunnels, subways, etc.), whether reinforced or not. It is further used in all types of masonry (foundations, footings, dams, retaining walls, and pavements). When portland cement is mixed with sand and lime, it serves as mortar for laying brick or stone, or as plaster or stucco for interior or exterior walls. When portland cement is mixed with coarse aggregate (aggregate larger than the No. 4 sieve) and fine aggregate (sand) together with enough water to ensure a good consistency, concrete results.

Portland cement is defined, according to ASTM C150, as a hydraulic cement produced by pulverizing clinker consisting essentially of hydraulic calcium silicates, usually containing one or more of the forms of calcium sulfate as an interground addition. The approximate proportions for portland cement are as follows:

Lime (CaO) 60–65%

Silica (SiO_2) 20–25%

Iron oxide and alumina (Fe_2O_3 and Al_2O_3, 7–12%

History of Portland Cement

The name *portland cement* was proposed by Joseph Aspdin in 1824. The name came about because the powdery material, which he patented, set up with water and sand and resembled a natural limestone quarried on the Isle of Portland in England.

The first portland cement manufactured in the United States was produced by David Saylor at Coplay, Pennsylvania, in 1875 by calcining in vertical kilns at a high temperature a mixture of argillaceous limestone rock with a pure limestone. With the increase in demand for quantity and quality, rotary kilns came into production in 1899. The production of portland cement in the United States has increased from 10 million barrels per year in 1900 to 400 million barrels per year in 1970. World production is approximately six times that of the United States.

Raw Materials

The raw materials of portland cement may be classified into three groups: calcareous, argillocalcareous, and argillaceous. Table 4-3 illustrates the materials that make up each group. From the table and a little knowledge of geology, it is evident that the essential constituents of portland cement are lime, silica, and alumina. Lime does not occur in nature but is found in a suitable form in a carbonate. Silica and alumina are found free in nature in the form of clay, shale, or slate.

Limestone (calcium carbonate) contains impurities of magnesia, silica, iron, alkalies, and sulfur. Magnesia in the form of carbonate of magnesia often occurs in limestone and, if present in the amount of 5 percent or more, will make the limestone unsuitable.

TABLE 4-3 Raw Materials in Portland Cements

Calcareous ($CaCO_3 > 75\%$)	*Argillocalcareous* ($CaCO_3 = 40$ to 75%)	*Argillaceous* ($CaCO_3 < 40\%$)
Limestone	Clayey limestone	Slate
Chalk	Clayey chalk	Shale
Shells	Clayey marl	Clay

Silica by itself does not combine with lime in the kiln; thus, small quantities of free silica in the limestone makes the limestone unacceptable. However, if silica is combined with alumina in the limestone, it will combine with lime in the kiln and is acceptable.

Iron in limestone can occur either as an oxide (Fe_2O_3) or as a sulfide (FeS_2). If the iron is in the form of an oxide, it acts as a flux in combining the lime and silica in the kiln. As a sulfide it reacts very strangely and can prove to be quite injurious to the production of portland cement. If the iron in the form of iron sulfide is present in limestone by 4 percent or greater, the limestone should be rejected.

The makeup of limestone is such that it contains alkalies in the form of soda and potash. These alkalies are not harmful and are usually driven off in the kiln.

One of the major roles of alumina in limestone is to combine with silica such that the limestone will combine with the lime in the kiln and thus make the limestone an acceptable product.

Sulfur, the final impurity in limestone, exists in two forms: lime sulfate and iron pyrite. If each is present in amounts of 3 percent or more, the limestone should be rejected.

Chalk is a variety of limestone formed from pelagic, or floating, organisms that are very fine-grained, porous, and friable. It is white or very light-colored and consists almost entirely of calcite. The rock is made up of the calcite shells of microorgansims partially cemented by structureless calcite. The best known chalks are those of the Cretaceous, exposed in cliffs on both sides of the English Channel. The Selma (Cretaceous) chalk of Alabama, Mississippi, and Tennessee and the Niobrara chalk of the same age in Nebraska are well-known deposits in the United States.

Marl is an argillaceous, nonindurated calcium carbonate deposit that is commonly gray or blue-gray. It is somewhat friable, and in some respects resembles chalk, with which it is interbedded in some localities. It is formed in some freshwater lakes, partially by the action of aquatic plants. The clay content of marls varies, and all gradations between small amounts of clay (marly limestone) and large amounts (marly clay) are found.

Slates are clays that have been solidified in a laminated structure and have the property of splitting into thin sheets. They have limited application in the production of portland cement.

Shales are clays that have become hardened by pressure. They have been formed from deposits of sedimentary clay. Shales are preferred over soft clays in the production of portland cement because segregation of the shale and limestone is less likely to occur.

Clays are formed from the debris resulting from the decay of rocks.

Clays take the form of three groups with respect to methods of transportation: residual, sedimentary, and glacial. Clays left where the rocks decayed are *residual*. Clays that have been transported and deposited by stream action are *sedimentary*. *Glacial* clays are those deposited by glacial movement. In any clay the silica content should not be less than 55 to 65 percent, and the amount of alumina and iron oxide combined should be between one-third and one-half the amount of silica.

Manufacture of Portland Cement

Portland cement is the most important hydraulic cement used in construction, for mortars, plasters, grouts, and concrete. The manufacture of portland cement occurs through a series of steps (quarrying, crushing, grinding, mixing, calcining, grinding, addition of retarder, and packing). Portland cement is made by burning an intimate mixture, composed mainly of calcareous, argillocalcareous, or argillaceous materials, at a clinker temperature of 2800°F (1550°C). This partially sintered clinker is then ground to a very fine powder with a very small amount of gypsum (2 to 4 percent) as a retarder. The cement is then packaged into 94-lb (43-kg) bags or into a hypothetical barrel which contains four sacks [376 lb (170 kg)] and is approximately 4 ft^3 (0.1 m^3) loose volume or 1.912 ft^3 (0.05 m^3) solid volume with a specific gravity of 3.15; or it may be bulk-stored. The cement-making process shown in Figure 4-1 is the process as it appears today, that is, as most industries are set up.

Composition of Portland Cement

Eight types of portland cement are recognized by the ASTM under specification ASTM C150:

Type I.	For use when the special properties specified for any other type are not required.
Type IA.	Air-entraining cement for the same uses as Type I, when air entrainment is desired.
Type II.	For general use, more especially when moderate sulfate resistance or moderate heat of hydration is desired.
Type IIA.	Air-entraining cement of the same uses as Type II, where air entrainment is desired.
Type III.	For use when high early strength is desired.
Type IIIA.	Air-entraining cement for the same use as Type III, where air entrainment is desired.
Type IV.	For use when a low heat of hydration is desired.
Type V.	For use when high sulfate resistance is desired.

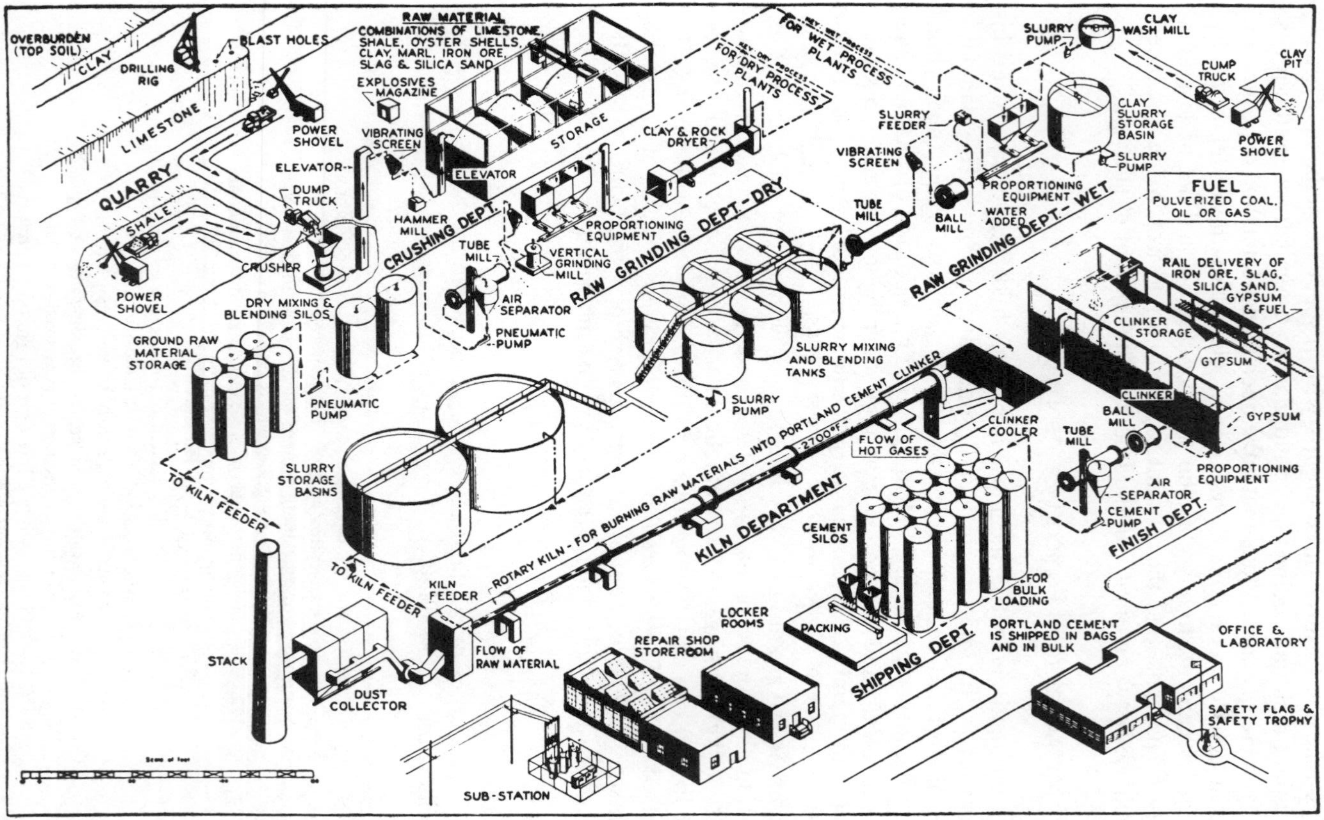

Fig. 4-1 Flowchart of Manufacturing Process of Portland Cement (Courtesy of PCA)

TABLE 4-4 Standard Chemical Requirements

	Cement Type				
	I and IA	*II and IIA*	*III and IIIA*	*IV*	*V*
Silicon dioxide (SiO_2), min. (%)	—	21	—	—	—
Aluminum oxide (Al_2O_3), max. (%)	—	6	—	—	—
Ferric oxide (Fe_2O_3), max. (%)	—	6	—	6.5	—
Magnesium oxide (MgO), max. (%)	6	6	6	6	6
Sulfur trioxide (SO_3), max. (%)	—	—	—	—	—
When ($3CaO \cdot Al_2O_3$) is 8 percent or less	—	—	—	—	—
When ($3CaO \cdot Al_2O_3$) is more than 8 percent	3	3	3.5	2.3	2.3
Loss on ignition, max. (%)	3.5	—	4.5	—	—
Insoluble residue, max. (%)	3	3	3	2.5	3
Tricalcium silicate ($3CaO \cdot SiO_2$), max. (%)	0.75	0.75	0.75	0.75	0.75
Dicalcium silicate ($2CaO \cdot SiO_2$), max. (%)	—	—	—	35	—
Tricalcium aluminate ($3CaO \cdot Al_2O_3$), max. (%)	—	—	—	40	—
Tetracalcium aluminoferrite plus twice the tricalcium aluminate	—	8	15	7	5–20

If a chemical analysis were to be performed on any one of the eight portland cements, the composition would be calcium oxide, silica, alumina, iron oxide, magnesium oxide, sulfure trioxide, and others. The chemical analysis would further reveal that these oxides exist in portland cement as calcium silicates and aluminates, such as tricalcium silicate ($3CaO{\cdot}SiO_2$), dicalcium silicate ($2CaO{\cdot}SiO_2$), tricalcium aluminate ($3CaO{\cdot}Al_2O_3$), and tetracalcium aluminoferrite ($4CaO{\cdot}Al_2O_3{\cdot}Fe_2O_3$). In the nomenclature of the cement industry, these four compounds are written as follows: C_3S, C_2S, C_3A, and C_4AF, respectively. Tables 4-4 and 4-5 show the standard chemical requirements and the optional chemical requirements for portland cement.

TABLE 4-5 Optional Chemical Requirements

	Cement Type					
	I and IA	*II and IIA*	*III and IIIA*	*IV*	*V*	*Remarks*
Tricalcium aluminate ($3CaO{\cdot}Al_2O_3$), max. (%)	—	—	8	—	—	Moderate sulfate resistance
Tricalcium aluminate ($3CaO{\cdot}Al_2O_3$), max. (%)	—	—	5	—	—	High sulfate resistance
Sum of tricalcium silicate and tricalcium aluminate, max. (%)	—	58	—	—	—	Moderate heat of hydration
Alkalies (Na_2O + $0.658K_2O$), max. (%)	0.6	0.6	0.6	0.6	0.6	Low-alkali cement

Properties of the Main Compounds

As indicated, portland cement is made up of four main compounds: tricalcium silicate, dicalcium silicate, tricalcium aluminate, and tetracalcium aluminoferrite. These compounds are present in the clinker in the form of interlocking crystals. Figure 4-2 shows a phase diagram $CaO–SiO_2–Al_2O_3$ ($C_3S–C_2S–C_3A$) in which the field of portland cement lies between. Also shown in the field are aluminous and blast-furnace-slag cements. Table 4-6 lists the characteristics of the major compounds in portland cement.

The most desirable constituent is that of tricalcium silicate (C_3S), because it hardens rapidly and accounts for the high early strength of the cement. When water is added to tricalcium silicate, a rapid reaction occurs as follows:

$$3CaO{\cdot}SiO_2 + H_2O = 2CaO{\cdot}SiO_2{\cdot}xH_2O + Ca(OH)_2 \tag{4.4}$$

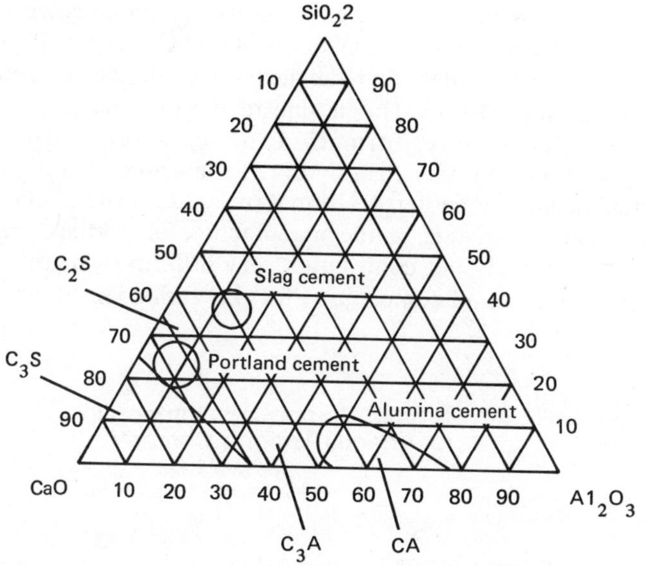

Fig. 4-2 Phase Diagram

TABLE 4-6 Characteristics of the Major Compounds in Portland Cement

Values	*Tricalcium Silicate,* $3CaO{\cdot}SiO_2$ (C_3S)	*Dicalcium Silicate,* $2CaO{\cdot}SiO_2$ (C_2S)	*Tricalcium Aluminate,* $3CaO{\cdot}Al_2O_3$ (C_3A)	*Tetracalcium Aluminoferrite,* $4CaO{\cdot}Al_2O_3{\cdot}Fe_2O_3$ (C_4AF)
Cementing value	Good	Good	Poor	Poor
Rate of reaction	Medium	Slow	Fast	Slow
Amount of Heat liberated	Medium	Small	Large	Small

The results indicate a less basic amorphous hydrated calcium silicate and a crystalline calcium hydroxide. The calcium silicate that is formed is considered the product to which early strength is attributed.

Dicalcium silicate hardens slowly and contributes largely to strength increase at ages beyond 1 week. In the presence of water, dicalcium silicate ($2CaO{\cdot}SiO_2$) hydrates slowly and forms a hydrated calcium silicate ($2CaO{\cdot}SiO_2{\cdot}xH_2O$).

Tricalcium aluminate liberates a large amount of heat during the first few days of hardening. It also contributes slightly to early-strength development. Tricalcium aluminate ($3CaO \cdot Al_2O_3$) hydrates with water to form a hydrated tricalcium aluminate ($3CaO \cdot Al_2O_3 \cdot 6H_2O$). If gypsum is added it acts as a retarder, and the heat of evolution is less and the setting occurs more slowly. This is due to the fact that gypsum, when present, results in the formation of calcium sulfoaluminate ($3CaO \cdot Al_2O_3 \cdot 3CaSO_4$) rather than hydrated tricalcium aluminate.

Tetracalcium aluminoferrite formation reduces the clinkering temperature, thereby assisting in the manufacture of portland cement. It hydrates rather rapidly but contributes very little to strength. Table 4-7 shows typical compound composition for the various portland cements.

TABLE 4-7 Typical Compound Composition for Various Portland Cements

Types of Portland Cement	*Compound Composition (%)*			
	C_3S	C_2S	C_3A	C_4AF
I. Normal	50	24	11	8
II. Moderate	42	33	5	13
III. High early strength	60	13	9	8
IV. Low heat	26	50	5	12
V. Sulfate resisting	40	40	4	9

Types of Portland Cement

As previously mentioned, eight types of portland cement are recognized by ASTM under Specification ASTM C150. The standard five types of portland cement (this excludes the three that are air-entrained) are also recognized by the Canadian Standards Association (CSA) and given specific names. We next investigate each type separately.

ASTM Type I or CSA normal portland cement is a general-purpose cement. It is used when the special properties specified for any other type are not required. It is used where there would be no severe climate changes or severe exposure to sulfate attack from water or soil. Its uses include reinforced-concrete buildings, pavements, sidewalks, bridges, railings, tanks, reservoirs, floors, curbs, culverts, and retaining walls. In general, it is used in nearly all situations calling for portland cement.

ASTM Type II or CSA moderate portland cement is a general-purpose cement to be used when moderate sulfate resistance or moderate heat of hydration is desired. It is used in structures of considerable mass, such as abutments and piers and retaining walls. Its use also minimizes temperature rise when concrete is placed in warm weather.

ASTM Type III or CSA high-early-strength portland cement is used when high early strength is desired, usually less than 1 week. It is usually used when a structure must be put into service as quickly as possible. This cement is made by changing the proportions of raw materials, by finer grinding, and by better burning, such that the dicalcium silicate is less and the tricalcium silicate is greater.

ASTM Type IV or CSA low-heat-of-hydration portland cement is used when a low heat of hydration is required. This type of cement develops strength at a slower rate than does the ASTM Type I. However, it is intended for mass structures such as large gravity dams, where the temperature rise on a continuous pour is great. If the temperature were not minimized, large cracks and/or flows would appear and the structure might prove to be unsound.

ASTM Type V or CSA sulfate-resisting portland cement is used when high sulfate resistance is desired. It is used when concrete is to be exposed to severe sulfate action by soil or water.

The three types of air-entraining cements, Types IA, IIA, and IIIA, as given by ASTM C140, are used in concrete for improved resistance to freezing and thawing action and to action of salt scaling by chemical attack. Typical air-entraining agents include Vinsol resin, Airolon, and Darex AEA.

Table 4-8 compares the strengths of the various types of cements with Type I at five different moist-curing intervals.

TABLE 4-8 Approximate Relative Strengths of Concrete as Affected by Type of Cement

Types of Portland Cement	*Compressive Strength (% of Normal Portland Cement)*				
	1 Day	*3 Days*	*7 Days*	*28 Days*	*3 Months*
I. Normal	100	100	100	100	100
II. Modified	75	80	85	90	100
III. High early strength	190	190	120	110	100
IV. Low heat	55	55	55	75	100
V. Sulfate resisting	65	65	75	85	100

Properties of Portland Cement

Most specifications for portland cement, such as ASTM specifications, place specific chemical composition and physical property requirements on the cement, as shown in Tables 4-9 and 4-10. Next, we will

TABLE 4-9 Optional Chemical Requirements

	Cement Type[a]					
	I and IA	*II and IIA*	*III and IIIA*	*IV*	*V*	*Remarks*
Tricalcium aluminate ($3CaO \cdot Al_2O_3$)[b], max. (%)	—	—	8	—	—	For moderate sulfate resistance
Tricalcium aluminate ($3CaO \cdot Al_2O_3$)[b], max. (%)	—	—	5	—	—	For high sulfate resistance
Sum of tricalcium silicate and tricalcium aluminate[b], max. (%)	—	58[c]	—	—	—	For moderate heat of hydration
Alkalies (Na_2O + $0.658K_2O$), max. (%)	0.6[d]	0.6[d]	0.6[d]	0.6[d]	0.6[d]	Low-alkali cement

[a]Attention is called to the fact that cements conforming to the requirements for all of these types may be carried in stock in some areas. In advance of specifying the use of other than Type I cement, it should be determined whether the proposed type of cement is or can be made available.
[b]The expressing of chemical limitations by means of calculated assumed compounds does not necessarily mean that the oxides are actually or entirely present as such compounds.

When the ratio of percentages of aluminum oxide to ferric oxide is 0.64 or more, the percentages of tricalcium silicate, dicalcium silicate, tricalcium aluminate, and tetracalcium aluminoferrite should be calculated from the chemical analysis as follows:

tricalcium silicate = (4.071 × percent CaO) − C7.600 × percent SiO_2) − (6.718 × percent Al_2O_3) − (1.430 × percent Fe_2O_3) − (2.852 × percent SO_3)

dicalcium silicate = (2.867 × percent SiO_2) − (0.7544 × percent C_3S)

tricalcium aluminate = (2.650 × percent Al_2O_3) − (1.692 × percent Fe_2O_3)

tetracalcium aluminoferrite = 3.043 × percent Fe_2O_3

When the alumina–ferrite oxide ratio is less than 0.64, a calcium aluminoferrite solid solution [expressed as ss(C_4AF + C_2F)] is formed. Contents of this solid solution and of tricalcium silicate should be calculated by the following formulas:

ss(C_4AF + C_2F) = (2.100 × percent Al_2O_3) + (1.702 × percent Fe_2O_3)

tricalcium silicate = (4.071 × percent CaO) − (7.600 × percent SiO_2) − (4.479 × percent Al_2O_3) − (2.859 × percent Fe_2O_3) − (2.852 × percent SO_3)

No tricalcium aluminate will be present in cements of this composition. Dicalcium silicate shall be calculated as previously shown.

In the calculation of C_34, the values of Al_2O_3 and Fe_2O_3 determined to the nearest 0.01 percent should be used. In the calculation of other compounds, the oxides determined to the nearest 0.1 percent shall be used.

All values calculated as described in this note should be reported to the nearest 1 percent.

[c]This limit applies when moderate heat of hydration is required and tests for heat of hydration are not requested.

[d]This limit may be specified when the cement is to be used in concrete with aggregates that may be deleteriously reactive.

TABLE 4-10 Standard Physical Requirements

	Cement Type[a]							
	I	*IA*	*II*	*IIA*	*III*	*IIIA*	*IV*	*V*
Air content of mortar[b], (vol. %)								
Maximum	12	22	12	22	12	22	12	12
Minimum	—	16	—	16	—	16	—	—
Fineness, specific surface (cm^2/g) (alternative methods)[c]								
Turbidimeter test, min.	1600	1600	1600	1600	—	—	1600	1600
Air permeability test, min.	2800	2800	2800	2800	—	—	2800	2800
Autoclave expansion, max. (%)	0.8	0.8	0.8	0.8	0.8	0.8	0.8	0.8
Compressive strength [psi (MPa)], not less than the values shown for the following ages:[d]								
1 Day	—	—	—	—	1800 (12.4)	1450 (10.0)	—	—
3 Days	1800 (12.4)	1950 (10.0)	1500 (10.3) 1000[e] (6.9)[e]	1200 (8.3) 800[e] (5.5)[e]	3500 (24.1)	2800 (19.3)	— —	1200 (15.2)
7 Days	2800 (19.3)	2250 (15.5)	2500 (17.2)	2000 (13.8)	—	—	1000 (6.9)	2200 (15.2)

28 Days	—	—	1700[f] (11.7)[f] —	1350[f] (9.3)[f] —	—	—	2500 (17.2)	3000 (20.7)
Time of setting (alternative methods)[f]								
Gilmore test								
Initial set (min), not less than	60	60	60	60	60	60	60	60
Final set (hr), not more than	10	10	10	10	10	10	10	10
Vicat test						45	45	45
Initial set (min), not less than	45	45	45	45	45			
Final set (hr), not more than	8	8	8	8	8	8	8	8

[a]Attention is called to the fact that cements conforming to the requirements for all of these types may not be carried in stock in some areas. In advance of specifying the use of other than Type I cement, it should be determined whether the proposed type of cement is or can be made available.

[b]Compliance with the requirements of this specification does not necessarily ensure that the desired air content will be obtained in concrete.

[c]Either of the two alternative fineness methods may be used at the option of the testing laboratory. However, in case of dispute, or when the sample fails to meet the requirements of the air-permeability test, the turbidimeter test shall be used, and the requirements in this table for the turbidimeter method shall govern.

[d]The strength at any age shall be higher than the strength at any preceding age.

[e]When the optional heat of hydration on the chemical limit or the sum of the tricalcium silicate and tricalcium aluminate is specified.

[f]The purchaser should specify the type of setting-time test required. In case he does not so specify, or in case of dispute, the requirements of the Vicat test only shall govern.

discuss the fineness, soundness, time of setting, compressive strength, heat of hydration, loss of ignition, and specific gravity of portland cement. Most of these tests are covered by ASTM specifications.

The *fineness* of the cement affects the rate of hydration. The greater the cement fineness, the greater the rate of hydration and hence the greater the strength development during the first 7 days. To measure the fineness of the cement, the Wagner turbidimeter or the Blaine air-permeability apparatus is used.

Soundness of hardened cement paste is a measure of the potential expansion of the several constituent parts or the ability to retain its volume after setting. Lack of soundness (unsound) is attributed to excessive amounts of hard-burned free lime or magnesia. This free lime takes on water and at some later date develops expansive forces. Most specifications call for the use of an autoclave (high-pressure steam boiler) to indicate the soundness or unsoundness of the cement.

Time of setting is measured by the Gilmore and/or Vicat apparatus, which is used to determine the rate at which portland cement hardens: in other words, to determine if a cement paste remains plastic long enough to permit normal placing of the concrete. The length of time that concrete remains plastic is dependent upon the chemical composition, fineness, water content, and temperature.

Compressive strength of portland cement is made by taking the cement specimen and mixing it with a uniform silica sand and water in prescribed proportions and molding the mixture into 2 in. × 2 in. × 2 in. (5.08 cm × 5.08 cm × 5.08 cm) cubes. The cubes are cured and then tested in compression to give an indication of the strength-developing characteristics of the portland cement.

Heat of hydration is the heat generated when cement and water react. The amount of heat generated is dependent chiefly on the chemical composition, fineness of the cement, and the temperature of curing time.

Loss of ignition of portland cement is determined by heating a cement sample of known weight to a full red heat of 1652°F (900°C) until a constant weight is obtained. The weight loss of the sample is then determined. Ignition is the indication of prehydration and carbonation, which may be caused by improper or prolonged storage.

The *specific gravity* of portland cement is generally about 3.15.

PORTLAND–POZZOLAN CEMENTS

Portland–pozzolan cements are hydraulic cements consisting of an intimate and uniform blend of portland cement or portland blast-furnace-slag cement and fine pozzolan produced either by intergrinding portland-cement clinker and pozzolan, by blending portland cement or portland blast-furnace-slag cement and finely divided pozzolan, or a combination of intergrinding and blending, in which the pozzolan constituent is between 15 and 40 percent by weight of the portland-pozzolan cement.

Portland–pozzolan cements consists of four types: Type IP, Type IP-A, Type P, and Type P-A, each with two optional provisions: Type IP portland–pozzolan cement and Type IP-A air-entrained portland–pozzolan cement, both of which are used in general concrete construction. Type P portland–pozzolan cement and Type P-A air-entrained portland–pozzolan cement are used in concrete construction where high strengths at early stages are not required. For Types IP and IP-A cement, moderate sulfate resistance or moderate heat of hydration or both may be specified by adding the suffixes (MS) or (MH) or both to the selected type designation. For Types P and PA, moderate sulfate resistance or low heat of hydration or both may be specified by adding the suffixes (MS) or (LH) or both to the selected type designation.

Portland–pozzolan cements require more water for a given consistency but exhibit greater shrinkage upon drying. Further, they exhibit less strength prior to 28 days of curing, but greater strength after 28 days of curing when compared to normal portland cement concrete.

Portland–pozzolan cements may be used for mass concrete where mass and weight are more important than strength. It also exhibits excellent sulfate resistance and hence is good for seawalls. Portland–pozzolan cements are also used in dam construction because of their low heat of hydration.

PORTLAND BLAST-FURNACE-SLAG CEMENT

Portland blast-furnace-slag cement, according to ASTM C595 (Blended Hydraulic Cements), is a hydraulic cement consisting of an intimate and uniform blend of portland cement and fine, granulated blast-furnace slag produced either by intergrinding portland cement clinker and granulated blast-furnace slag or by blending portland cement and finely ground granulated blast-furnace slag, in which the slag constituent is between 25 and 65 percent of the weight of portland blast-furnace-slag cement.

Portland blast-furnace-slag cement exists in two types, with two optional provisions. Type IS portland blast-furnace-slag cement is for use in general concrete construction. The second type, Type IS-A, air-entrained portland blast-furnace-slag cement, is also used in general concrete construction. Moderate sulfate resistance or moderate heat of hydration or both may be specified by adding the suffixes (MS) or (MH) or both to the selected type designation.

ALUMINA CEMENTS

Alumina cement has a high alumina content because it consists primarily of calcium aluminates. Aluminate cement is sometimes referred to as "high-alumina cement." Alumina cements have a different chemical composition and constitution than does portland cement. However, they have several outstanding properties, such as high early strength (usually setting and hardening to full strength at 48 hours compared to 28 days for portland cement), excellent refractoriness, and good resistance against chemical attacks (hence, resisting disintegrating action of seawater quite well).

They are somewhat limited due to the fact that elevated temperatures can produce permanent strength reductions when moisture is present.

The raw materials used for the manufacture of alumina cement are limestone and bauxite. The two are ground with one another and then placed in a kiln until the mixture melts at 2900°F (1600°C). The clinker is cooled, ground, and gypsum is added and finally packaged.

Alumina cements are used where high early strength is required and moderate temperatures are to be maintained.

EXPANSIVE CEMENTS

An *expansive cement* is a hydraulic cement containing a constituent that, during the process of hydration, setting, and/or hardening, undergoes an expansion (increase in volume) but remains sound and eventually develops into satisfactory strength. An expansive cement is normally used in a situation where shrinkage of the concrete cannot be tolerated; hence, it compensates for the shrinkage that will take place.

There are generally three types of expansive cements: Types M (Soviet), K (Klein), and S (Portland Cement Association). Expansive cements can achieve high strength because of the high-alumina or portland cement components.

SPECIAL PORTLAND CEMENTS

A variety of *special cements* exists (white portland, colored cements, oil-well cements, regulated cements, waterproofed cements, hydrophobic cements, antibacterial cements, barium and strontium cements) but are limited for specific uses and purposes. Each type will be mentioned briefly and the reader should keep in mind that this list is by no means complete.

White portland cement is used for decorative displays. It makes an excellent base when colored aggregates are used. White portland cement is low in iron and manganese, which gives the cement its white look, compared to normal portland cement, which is gray. Hence, by reducing iron and manganese in normal portland cement, a white cement is produced.

Colored cements are made by intergrinding a chemically inert pigment such as metallic oxide in the amount of 3 to 10 percent to portland cement. Colored cements, like white portland cement, are used for decorative purposes. However, they have one disadvantage in that they have a tendency to fade over the years.

Oil-well cements are slow-setting cements which are used to seal deep wells. The cement is made in a slurry and pumped to depths within the well under high temperature and pressure before it is allowed to set. These types of cements are governed by the American Petroleum Institute for each of eight classes.

Regulated cements are rapid-setting and -hardening cements. They are used in the manufacture of blocks, pipes, prestressed and precast concrete, and, of course, for patch work. In strength, they are comparable to portland cement Types I, II, and III.

A *waterproofed cement* is a portland cement interground with a water-repellent material, such as calcium stearate. The purpose is to reduce the water permeability of the concrete.

Hydrophobic cements are similar to waterproofed cements, in that portland cement is interground with a hydrophobic (water-repellent) material. However, the purpose is to prolong the life of the cement during storage or while it is being transported long distances.

An *antibacterial cement* is a portland cement interground with an antibacterial agent with the intention of reducing harmful microorganisms. It is used in food-processing plants to minimize deterioration caused by fermentation.

Barium and strontium cements are portland cements in which the calcium oxide is replaced completely or in part by barium oxide or strontium oxide. Their purpose is to act as a concrete shield in which the barium and strontium absorb x-rays and gamma rays.

PROBLEMS

4.1. Explain the difference between hydraulic and nonhydraulic cements.

4.2. Why is lime important in the manufacturing process of portland cement?

4.3. Discuss the uses of the following: pozzolan cements, slag cements, natural cements, portland cements.

4.4. List eight types of portland cement and explain their uses.

4.5. Explain how the following compounds affect the character of portland cement: (a) tricalcium silicate, (b) dicalcium silicate, (c) tricalcium aluminates, (d) tetracalcium aluminoferrite, and (e) alumina.

4.6. Why are portland–pozzolan cements important?

4.7. What special property of alumina cement makes its use attractive?

4.8. List several special cements and discuss their uses.

4.9. Overall, which types of cements do you view as exceptional? Explain.

4.10. After reviewing outside reference materials, discuss the manufacturing process of portland cements.

REFERENCES

American Society for Testing and Materials, *Book of Standards*, Part 14, 1979.

BAUER, E. E., *Plain Concrete*, 3rd ed., McGraw-Hill Book Company, New York, 1949.

BROWN, L. S., "Tricalcium Aluminate and the Microstructure of Portland Cement Clinker," *Proc. ASTM, 37*, Part II (1937).

DAVIS, R. E., KELLEY, J. W., TROXELL, G. E., and DAVIS, H. E., "Properties of Mortars and Concretes Containing Portland–Pozzolan Cement," *Am. Concr. Inst, 32* (Sept.–Oct. 1935), p. 80.

KREBS, R. D., and WALKER, R. D., *Highway Materials*, McGraw-Hill Book Company, New York, 1971.

LARSON, T. D., *Portland Cement and Asphalt Concretes*, McGraw-Hill Book Company, New York, 1963.

MILLS, A. P., HAYWARD, H. W., and RADER, L. F., *Materials of Construction*, 6th ed., McGraw-Hill Book Company, New York, 1955.

POPOVICS, S., *Concrete-making Materials*, McGraw-Hill Book Company, New York, 1979.

STRENGTH OF CONCRETE 5

Whether used in buildings, bridges, pavements, or any other of its numerous areas of service, concrete must have strength, the ability to resist force. The forces to be resisted may result from applied loads, from the weight of the concrete itself, or, more commonly, from a combination of these. Therefore, the strength of concrete is taken as an important index of its general quality. Hence, tests to determine strength are undoubtedly the most common type made to evaluate the properties of hardened concrete. There are three reasons for this: (1) the strength of concrete, in compression, tension, shear, or a combination of these, has, in most cases, a direct influence on the load-carrying capacity of both plain and reinforced structures; (2) of all the properties of hardened concrete, those concerning strength can usually be determined most easily; and (3) by means of correlations with other more complicated tests, the results of strength tests can be used as a qualitative indication of other important properties of hardened concrete.

The results of tests on hardened concrete are usually not known until it would be very difficult to replace any concrete that is found to be faulty. These tests, however, have a policing effect on those responsible for construction and provide essential information in cases where the concrete forms a vital structural element of any building. The results of tests on hardened concrete, even if they are known late, help to disclose any trends in concrete quality and enable adjustments to be made in the production of future concrete.

COMPRESSIVE STRENGTH

Significance of Compressive Strength

Concrete is used in many ways and is subject to a variety of different loading conditions, and so different types of stress develop.

Very often the dominant stress is compressive in nature, since this material has long been known to exhibit its best strength characteristics when subjected to compressive loading. The compressive strength of concrete is one of its most important and useful properties and one of the most easily determined. The compressive strength of concrete is indicated by the unit stress required to cause failure of a test specimen. Concrete also exhibits tensile and shear strength, in which compressive strength is frequently used as a measure of these properties. The tensile strength of concrete is roughly 10 to 12 percent of the compressive strength, and the flexural strength of plain concrete, as measured by the modulus of rupture, is about 15 to 20 percent of the compressive strength.

In addition to being a significant indicator of load-carrying ability, strength is also indicative of other elements of quality concrete in a direct or indirect manner. In general, strong concrete will be more impermeable, better able to withstand severe exposure, and more resistant to wear. On the other hand, strong concrete may have greater shrinkage and susceptibility to cracking than a weaker material.

Finally, the concrete-making properties of the various ingredients of the mix are usually measured in terms of the compressive strength.

Specimens

Specimens to determine the compressive strength of concrete are generally obtained from four different sources: (1) cylinders made in the laboratory, (2) cylinders made in the field, (3) cores of hardened concrete cut from structures, and (4) portions of beams broken in flexure. Each type of specimen has a specific purpose or purposes.

ASTM C192 (Method of Making and Curing Concrete Compression and Flexure Test Specimens in the Laboratory) describes in detail methods for preparation and examination of the constituent material; proportioning and mixing of concrete; determining the consistency of the mix; and molding, curing, and capping of the specimens. Cylinders made in the laboratory constitute a large portion of the compression specimens. There are three reasons for this: (1) in research, to determine the effect of variations in materials or conditions of manufacture, storage, or testing on the strength and other properties of concrete; (2) as control tests in conjunction with (a) tests on plain or reinforced concrete members or structures or (b) tests to determine other properties of hardened concrete; and (3) to evaluate mix designs for laboratory field use. In the making of such specimens, large variations can be introduced into the results of the compression test if great care is not taken in the manufacture of the test specimen. These variations may be attributed to the character of the cement, conditions of mixing, character and grading of the aggregate, size of the aggregate, size and shape of the specimen, curing and aging, temperature, and moisture content at time of testing.

ASTM C31 (Method of Making and Curing Concrete Compression and Flexure Test Specimens in the Field) describes the detailed method of making standard cylinders in the field. When cylinders are prepared in the field, they should be made from the same concrete used on the job. In addition, the same curing process, or a process as similar as possible, should be used. The purpose of cylinders made in the field may be to check the adequacy of the laboratory mix design, to determine when a structure may be put in service, or to measure and control the quality of the concrete.

Compressive test results of cored hardened concrete usually result in lower compressive strength than anticipated. ASTM C42 (Methods of Obtaining and Testing Drilled Cores and Sawed Beams of Concrete) covers the procedure for securing and testing the cylindrical cores which are most commonly used for determining compressive strength. Cores are only drilled when results of the standard cylinder test are questionable or when investigations are made of old structures.

Finally, ASTM C116 (Test for Compressive Strength of Concrete Using Portions of Beams Broken in Flexure) describes the procedure and apparatus necessary to determine the compressive strength of concrete from broken portions of beams tested in flexure. This test is extremely useful where beam specimens are made to determine the modulus of rupture, as in highway construction, of which one would like an appropriate value of the compressive strength. The method is not meant to be used as a comparison with laboratory cylinder tests. When the method is used and a correlation attempt is made, it becomes necessary to apply a correlation factor.

Making Specimens

The compressive strength of concrete depends primarily on the water–cement ratio. However, other factors, such as character of the cement, conditions of mixing, character and grading of the aggregate, size of the aggregate, size and shape of the specimen, curing and aging, temperature, and moisture content at the time of testing also have a bearing on the compressive strength.

The characterization of the cement for a given water–cement ratio plays an important role in the early compressive strength development of concrete. All portland cements behave more or less similarly, although the gain in strength with age is not always the same. Some cements gain their strength more rapidly at first, whereas others show greater increase at later periods. This applies not only to the five types covered by the

ASTM specifications, but to some extent to different cements within a single group. Tests have shown that the strength for a given water–cement ratio shows the greatest difference among cements at the early ages. For 90 days and later the differences are much less.

The importance of thorough mixing for the development of strength and for uniformity throughout the batch has long been recognized. The earliest studies in concrete showed increases in strengths with continued mixing, but the increase became slight after first a rapid rise. The size of the batch, the type and consistency of the concrete, and the type of mixer are all involved in fixing the period in which gain in strength with time of mixing is significant. The time of mixing is governed by ASTM method C94.

Surface conditions, the size and shape of the particles, and the gradation are the characteristics of the aggregate that are of principal concern to the strength of the concrete. The surface conditions of the aggregate affect the adhesion of the cement paste to the aggregate particles. The presence or absence of adherent dirt or clay, the roughness, and texture affect the adhesion. These characteristics have greater effect on flexural strength than on compressive strength.

The shape of the particles influence the strength of the concrete by affecting the quality and the amount of paste that is required for workability with a given mixture. Also, the bond with the cement paste may be weakened where relatively large surface areas of the flat pieces of aggregate occur, especially when they happen to be combined in planes of shear and tension.

When the water–cement ratio is the same and the mixtures are plastic and workable, considerable changes in grading will affect the strength of the concrete only to a small degree. The principal effect of changing the aggregate grading is to change the amount of cement and water needed to make the mixture workable with the desired water–cement ratio.

In general, as the maximum size of the aggregate is increased, lower water–cement ratios can be used for suitable workability and, therefore, greater strengths are obtained for a given cement content. In the high strength range, over 4500 psi (31 MPa), higher compressive strengths are usually obtained at a given water–cement ratio with smaller maximum sizes of aggregates. Data from compression tests of concrete containing very large aggregates, 4 in. (10.16 cm), are conflicting because of limitations in the size of test specimens.

Compression tests of concrete are ordinarily conducted on cylindrical specimens with height equal to twice the diameter, so that surface rupture, produced upon fracture, will not intersect the end bearings. The ends of the cylinders should be carefully formed to give parallel, smooth surfaces so as to obtain uniform distribution of stress. A uniformly stressed cylinder that has been properly molded will break in the shape of a double cone with vertex in the center of the cylinder.

It is generally accepted that the diameter of the specimen should be at least three times the nominal size of the coarse aggregate. A 6 in. × 12 in. (15.24 cm × 30.48 cm) cylinder is the standard for aggregate smaller than 2 in. (5.08 cm). If the aggregate is too large for the size of mold available, the oversize aggregate may be removed by set screening. If a mold having a diameter less than three times the maximum size of the aggregate is used, the indicated compressive strength will be lowered. However, a large mold may be used; in some cases, molds as large as 36 in. (91.44 cm) in diameter have been used for concrete containing very large aggregates, such as those used in dam construction. Attention must be called to the fact that the size of the cylinder itself affects the observed compressive strength; for example, the strength of a cylinder 36 in. (91.44 cm) in diameter by 72 in. (182.88 cm) high may be only about 82 percent of that of a standard 6 × 12 in. (15.24 cm × 30.48 cm) cylinder. A reduction in the size of the specimen below that of the standard 6 in. × 102 in. (15.24 × 30.48 cm) cylinder will yield a somewhat greater indicated compressive strength.

Unless the specimens are carefully molded, errant and irregular results will be obtained. Generally, a cylinder of poorly compacted concrete will have a lower strength than one that is properly compacted. Thus, it is necessary for the standard methods for making specimens, Methods C31 and C192, to specify procedures for compacting, rodding, or vibrating the concrete in the mold. If the specifications under which the work is being performed do not state the method of consolidation, the choice is determined by the slump. Concrete, with a slump greater than 3 in. (7.62 cm), should be rodded. If the slump is between 1 and 3 in. (2.54 and 7.62 cm), the concrete may be either vibrated or rodded. When the slump is less than 1 in. (2.54 cm), the specimens must be consolidated by vibration. When the concrete is to be rodded, it should be placed in the cylinder in three layers and rodded 25 strokes per layer if the cylinder is 3 to 6 in. (7.62 to 15.24 cm) in diameter, 50 if 8 in. (20.32 cm), 75 if 10 in. (25.40 cm). When vibrating, the mold is filled and vibrated in two layers. Care must be exercised to vibrate only long enough to obtain proper consolidation. Overvibration tends to cause segregation. These methods are specified in order to permit reproducibility of results by different technicians.

Cylinder molds should be of nonabsorbent material and are generally of steel; however, carboard molds are quite often used in the field. ASTM C470 [Specification for Single-Use Molds for Forming 6 × 12 in. (15.24 × 30.48 cm) Concrete Compression Test Cylinders] defines adequate paper molds as well as lightweight sheet-steel molds. Although the cardboard is heavily paraffined, in most cases it absorbs part of the water in the concrete mixture. The use of cardboard molds may lower the observed compressive strength on the average about 3 percent, and reductions as great as 9 percent have been noted.

Capping procedures are standardized under ASTM Method C617 (Capping Cylindrical Concrete Specimens). Any material that is sufficiently strong and can be molded can be used for capping. The most common capping materials are neat portland cement paste, high-strength gypsum plaster, and sulfur compounds. The cap should be as thin as possible and the plane surface of the cap at either end of the specimen should be truly at right angles to the axis of the cylindrical specimen. All surfaces that depart from a plane by more than 0.002 in. (0.005 cm) should be capped and all caps should be checked for this by means of a feeler gage and steel straightedge. Slight irregularities can be made good by scraping if the capping material has not set too hard. Further, according to the ASTM specification, the surface of the capping must not depart by more than 0.5 degree from perpendicularity with the axis of the cylinder, or a cant of 1 in 96.

If neat portland cement is used for capping, it should be mixed about 3 hours before use to minimize any harmful effect due to shrinkage. The capping procedure should be carried out at least 3 days before testing.

Gypsum plaster can, according to the ASTM specification, be used for capping provided that it has a strength of over 5000 psi (31 MPa) when tested as a 2-in. (5.08-cm) cube. Suitable mixtures of sulfur (melted) and granular materials applied about 2 hours prior to testing are also recommended, but care is necessary to avoid overheating the melted mixture to prevent loss of rigidity.

Concrete can gain in strength only as long as moisture is available and used for hydration. The term "curing" is used in reference to the maintenance of a favorable environment for the continuation of the chemical reactions that take place. It is through the early curing process that the internal structure of the concrete is built up to provide strength and water tightness. While simple retaining moisture within the concrete may be sufficient for low to moderate cement contents, mixes that are rich in cement generate considerable heat of hydration, which may expel moisture from the concrete in the period immediately after setting. The standard curing conditions require that the specimen be held at a temperature of 73.4 ± 3°F (23 ± 1.7°C) and in the "moist conditions" until the time of the test (ASTM C192). Any variation from this procedure may produce a specimen having a different strength from that which would be produced under standard conditions.

Cylinders to be used for quality control should be cured according to the standard conditions; however, cylinders made in the field and tested to measure the strength of the concrete in the structure should be cured in the same manner as the structure. Two methods of field curing are presently used: (1) those which interpose a source of water, in the form of ponding, or a wet material to prevent or counteract evaporation; and (2) those which minimize loss of water by interposing an impermeable medium or by other means. A third method of curing that is used in the manufacture of concrete products is the artificial application of heat while the concrete is maintained in a moist condition.

Curing and aging cannot be separated; an increase in age provides for further chemical combinations if the conditions are favorable for continued reaction. Provided that favorable conditions are met, concrete gains in strength with age.

Temperature also plays an important role in the curing process of concrete. The chemical reactions proceed more rapidly at higher temperatures. Tests of specimens sealed against loss of moisture show higher early strength, but lower strengths at later ages as the temperature is increased in the range of 40 to 115°F (4.4 to 26.1°C). The U.S. Bureau of Reclamation has found that for job control, specimens cured at 70°F (21.2°C) lower temperatures at the time of casting and for a few hours thereafter give higher strengths at 1 to 3 months. The rapid stiffening in the first few hours under the higher temperature is apparently detrimental to the later development in strength.

Test Procedure

Once the specimen is made, the method by which it is tested any further affects the strength obtained. Two of the more important influences are the rate of loading and the eccentricity of loading.

The rate of loading has a definite effect on compressive strength, although the effect is usually fairly small over the ranges of speed used in ordinary testing. The results of tests on concrete indicate that the relationship between strength and rate of loading is approximately logarithmic; the more rapid the rate, the higher the indicated strength. A rapid rate of loading may indicate as much as a 20 percent increase in the apparent compressive strength. For this reason, ASTM method C39, which applies also to testing of cores, specifies that the rate of loading for screw-powered machines shall be 0.05 in./1 min (0.11 cm/min) and for hydraulic machines 20 to 50 psi/sec (1.41 to 3.52 $kg/cm^2/sec$).

The effect of eccentric loading is obvious and the alignment of all machines should be checked. Any eccentricity will tend to decrease the strength of the test specimen, the amount of decrease being greater for low-strength than for high-strength concrete. Therefore, to ensure that a concentric and uniformly distributed load is applied to the specimen, a spherically seated bearing block is required on one end, and the specimen should be carefully centered on this bearing block. The object of the block is to overcome the effect of a small lack of parallelism between the head of the machine and the end face of the specimen, giving the specimen as even a distribution of initial load as possible. It is desirable that the spherically seated bearing block be at the upper end of the test specimen. In order that the resultant of the forces applied to the end of the specimen should not be eccentric with the axis of the specimen, it is important that the center of the spherical surface of this block be in the flat face that bears on the specimen and that the specimen itself be carefully centered with respect to the center of this spherical surface. Owing to increased frictional resistance as the load builds up, the spherically seated bearing cannot be relied upon to adjust itself to bending action that may occur during the test.

Significance of Results

The results of a compression test is essentially only comparative as the value of the compressive strength obtained cannot be regarded as equal to the strength of the concrete deposited in the work. The value will only give an indication of the quality of concrete, because of the various factors that affect the mix, which have already been discussed. This lack of knowledge regarding the relationship between the strengths of concrete in a cylinder and in a structure requires the use of a larger factor of safety than would otherwise be necessary.

Compressive strength may be used as a qualitative measure of other properties of hardened concrete. No exact relationship exists between compressive strength and flexural strength, tensile strength, modulus of elasticity, wear resistance, fire resistance, or permeability. Only an approximation can be made of these properties. Nevertheless, this approximation is very useful to the engineer.

Compressive tests further aid in the selection of ingredients that may be used in making concrete. Compressive strength is a measure of the indirect effect of admixtures, which may be beneficial for one purpose but detrimental to another.

TENSILE AND FLEXURAL STRENGTH

Significance of Tensile and Flexural Strength

Flexural tension is most commonly developed in beams and slabs as the result of loads, temperature changes, shrinkage, and, in some cases, moisture changes. The case of simple uniaxial tension is rarely encountered in structures or members and can be obtained in laboratory tests only with care. However, significant principal tension stresses may be associated with multiaxial states of stress in walls, shells, or deep beams.

As concrete that has to withstand tensile stresses is normally reinforced, its tensile strength has not received much attention, although it is of great importance in determining the ability of concrete to resist cracking due to shrinkage on drying and thermal movements. The tensile strength develops more quickly than the compressive strength and is usually about one-tenth the compressive strength at ages up to about 14 days, falling to about 5 percent at later ages. Cracking of concrete is usually a tensile failure, and this alone makes the tensile strength of concrete quite important.

Specimens

It is not easy to perform an axial tension test on concrete. It is difficult to apply the load truly axially and to grip the ends of the specimen without imposing high local stresses; a relatively large specimen must be used if a measurable load is to be applied, and a large number of specimens are required to ensure a reliable average. The beam test for flexural tension and the split-cylinder test are the simplest procedures for obtaining an indication of the tensile strength and are the tests usually performed.

Flexural tension tests may be made in several ways, the most common being ASTM C78 [Test for Flexural Strength of Concrete (Using Simple Beam with Third-Point Loading)]. ASTM C293 [Test for Flexural Strength of Concrete (Using Simple Beam with Center-Point Load)] is also used to a limited extent. The second method is intended for use with small specimens and not as an alternative to the test with third-point loading. The results of the flexural tests are expressed by the formula

$$R = \frac{Mc}{I}$$

where R = modulus of ruptures

M = maximum bending moment

c = one-half the depth of the beam

I = moment of inertia of the cross section

Another procedure for obtaining an indication of tensile strength is given in ASTM C496 (Test for Splitting Tensile Strength of Molded

Concrete Cylinders). In this test a standard cylinder is loaded in compression on its side. Fracture occurs along the plane that includes both lines through which the load is applied. While high compressive stresses occur at the lines where load is applied, the plane on which fracture occurs is subjected largely to a uniform tensile stress. The splitting strength is calculated as follows:

$$T = \frac{2P}{\pi l d}$$

where T = splitting tensile strength

P = maximum applied load

l = length

d = diameter

Making Specimens

ASTM methods C192 and C31 describe the procedures for making flexural test specimens and also cylinders for the splitting tensile strength test in the laboratory and in the field. ASTM method C31 stipulates that the length of the beam should be at least 2 in. (5.08 cm) longer than three times its depth and that its width should be not more than one and one-half times its depth. The minimum depth or width should be at least three times the maximum size of aggregate. A typical specimen used would be 6 in. × 6 in. × 21 in. (15.24 cm × 15.24 cm × 53.34 cm), and is tested under third-point loading on a span of 18 in. (45.72 cm). As the depth of the beam is increased, there is a decrease in the modulus of rupture.

Many of the parameters that affect the compressive strength of concrete also apply to the flexural strength as well.

Test Procedure

Flexural strength measurements are extremely sensitive to all aspects of specimen preparation and testing procedure.

The principal requirements of the supporting and loading blocks of the apparatus for the flexural strength of concrete are as follows: (1) they should be of such shape that they permit use of a definite and known length of span; (2) the areas of contact with the material under test should be such that unduly high stress concentrations (which may cause localized crushing around bearing areas) do not occur; (3) there should be provision for longitudinal adjustment of the position of the supports so that longitudinal restraint will not be developed as loading progresses; (4) there should be provision for some lateral rotational adjustment to accommodate beams that have a slight twist from end to end, so that torsional stresses will not be induced; and (5) the arrangement of parts should be stable under load.

Apparatus for measuring deflection should be so designed that crushing at the supports, settlement of the supports, and deformation of the supporting and loading blocks or parts of the machine do not introduce serious errors into the results. One method of avoiding these sources of errors is to measure deflections with reference to points on the neutral axis above the supports.

Routine flexure tests are usually simple to conduct. Ordinarily, only the modulus of rupture is required; this is determined from the load at rupture and the dimensions of the specimen. When the modulus of elasticity is required, a series of load-deflection observations are made.

The dimensions of the concrete specimen should be measured to the nearest 0.01 in. (0.025 cm). The supporting and loading blocks should be located with a reasonable degree of accuracy (0.2 percent of the span length). The supports and specimens should be placed centrally in the testing machine and checked to see that they are in proper alignment and can function as intended. Deflectometers and strainometers should be located carefully and checked to see that they operate satisfactorily and are set to operate over the range required.

The rate of load application, unless standardized, may cause considerable variation in the results of flexure tests, the variation being as much as 15 percent for the range of rates that may be obtained in the average laboratory. Concrete beams may be loaded rapidly at any desired rate up to 50 percent of the breaking load, after which loads should be applied at a rate such that the extreme fibers are stressed at 150 psi (0.93 MPa) or less per minute.

Beams may be tested under either center-point or third-point loading. Third-point loading invariably gives lower strengths than center-point loading. Tests indicate the following order of decreasing magnitude of the strength obtained: (1) center loading, with moment computed at center; (2) center loading, with moment computed at point of fracture; and (3) third-point loading. Third-point loading probably gives lower strengths because the maximum moment is distributed over a greater length of the beam; since the concrete is not homogeneous, this loading method seeks the weakest section.

Curing affects the tensile strength in much the same manner as it affects the compressive strength. A beam that has been allowed to dry before testing will yield lower flexural strength than one tested in a saturated condition. Consequently, in tests used to determine or control the quality of concrete, uniformity of results will be assured only if the beams are cured in a standard manner and tested when wet.

The temperature of a beam at the time of testing will also affect the results. As the temperature increases, the strength decreases.

ASTM method C78 is also prescribed for tests of beams sawed from hardened concrete. When such beams are used, primarily as a control of concrete quality, they should be turned on their sides before testing and will usually require capping because of the irregularity of the sawed surface (the sides are not plane and parallel).

The test for splitting tensile strength described in method C496 is simple to make. The effectiveness with which the material in the bearing strips is able to conform to the irregularities of the surface of the specimen and distribute the load affects the results. For uniformity, method C496 specifies that ⅛-in. (0.32-cm)-thick plywood should be used for bearing strips. Care must be taken to apply the load through a diametrical plane. The load should be applied such that the stress increases between 100 and 200 psi/min (7.03 and 14.1 kg/cm^2/min).

Significance of Results

In the case of pavement construction, it appears that the flexural strength of concrete is at least as important as the compressive strength. It has been suggested that pavements be designed on the basis of flexural strength. The design procedure would be exactly the same except that the tables would have to be prepared on the basis of flexural strength instead of compressive strength. Many agencies that are involved in pavements make only flexural tests. The results of these tests give an indication of when the concrete has gained sufficient strength that load may be applied or the forms removed.

Splitting tension tests have been used to evaluate the bond-splitting resistance of concrete.

With increased emphasis on the control of cracking in reinforced concrete, appreciation of the importance of tensile strength has increased. However, the three general methods of estimating tensile strength give slightly different results. The splitting-tensile-strength test is the easiest to perform and gives the most uniform results. The tensile strength from the splitting test is about 1½ times greater than that obtained in a direct tension test and about two-thirds of the modulus of rupture.

SHEARING AND TORSIONAL STRENGTH

Significance of Shearing Strength

The shearing strength of concrete is a very important property of the material, as it is the primary determining factor in the compressive strength of short columns. However, under certain conditions the strength of concrete beams also depends upon the shearing strength of the material.

The average strength of concrete in direct shear varies from about 0.5 of the compressive strength for rich mixtures to about 0.8 of the compressive strength for lean mixtures.

Torsion

The application to a concrete specimen of torsion alone produces pure shearing stresses on certain planes. However, the case of pure shear acting on a plane is seldom, if ever, encountered in actual structures. Further, failure under these conditions will occur in tension rather than shear. The strength of concrete subjected to torsion is therefore related to its tensile strength rather than to its shearing strength.

COMBINED STRESSES

Significance of Combined Stresses

Concrete in structures is almost never subjected to a single type of stress. Just as nearly all structural members are acted upon by various combinations of moments, shear, and axial load, the concrete in them is usually subjected to some combination of compressive, tensile, and shearing stresses.

ASTM C801 (Recommended Practice for Determining Mechanical Properties of Hardened Concrete under Triaxial Loads) is useful in determining the strength and deformation characteristics of concrete, such as shear strength at various lateral pressures, angle of shearing resistance, strength in pure shear, deformation modulus, and creep behavior. Prior to the application of this relatively new method, extensive research was conducted on cylinders with combinations of axial tension and lateral compression, and torsion and axial compression.

Results of Tests

There are no universally accepted criteria of failure of concrete; hence, there is no single correct way of discussing the behavior of concrete under combined stresses and there are many ways of presenting the results. ASTM method C801 allows four ways in which the data may be presented:

1. Graphical plots of the following equations:

$$f_1 = f_c' + K(f_3)^a$$

or, for the strength increase beyond the uniaxial strength:

$$f_1 - f_c' = K(f_3)^a$$

where f_1 = largest principal stress

f_3 = smallest principal stress

f_c' = unconfined compressive strength

K, a = empirical coefficients

2. A graphical plot of the stress difference versus axial strain. Stress difference is defined as the maximum principal axial stress minus the minimum principal stress. The value of the minimum principal stress should be indicated on the curve.
3. A graphical plot of axial stress versus axial strain for different confining pressures.
4. Mohr stress circles constructed on an arithmetic plot with shear stresses as ordinates and normal stresses as abscissas. At least three triaxial compression tests, each at a different confining pressure, should be made on the same material to define the envelope to the Mohr stress circle.

FATIGUE STRENGTH

Concrete will, when subjected to repeated load, fail at a load smaller than its static ultimate strength. Early work on fatigue was conducted on low-strength concrete. Thus, in the early classic work of Probst, the strength of the concrete tested was only just over 2000 psi (12.4 MPa). His results nevertheless agree with more recent work conducted on concretes of much higher strength. Probst discovered that there was a critical stress below which concrete, subjected to repeated loading, increased in strength and elasticity. His minimum stress was 5 percent of the ultimate and the maximum or critical stress ranged from 47 to 60 percent of the ultimate. Further tests have shown that the fatigue strength at 10 million cycles is about 55 percent of the static flexural strength. The fatigue strength in axial compression is about 55 percent of the static compressive strength. No data are available to establish the fatigue strength in axial tension, in shear, or under combined stresses. Concrete does not have a fatigue or endurance limit, at least at less than 10 million cycles of load, as most body-centered-cubic metals do. Failure under repeated loads is especially important in pavement design.

Frequent rest periods during a fatigue test may raise the fatigue strength as much as 9 percent higher than had there been no rest periods. The fatigue strength increases as the rest periods are increased in duration to 5 minutes. No additional gain is obtained for longer rest periods.

Tests to determine the fatigue strength of concrete should be made on specimens as large as possible in order to decrease the influence of lack of homogeneity. The cross section of the specimen should be at least three times the maximum nominal size of the aggregate, and even larger dimensions might be desirable in some cases. Because of the large size of specimens used, the large testing machines required are usually capable of applying load at a rate ranging from only a few cycles a day to 500 cycles per minute. Thus, it takes a minimum of 2 weeks to several months to apply as many as 10 million cycles of load. Because of the time involved, the specimens are generally aged and air-dried before being tested in order to prevent gain of strength during the test.

PROBLEMS

5.1. Why are tests to determine the strength of concrete the most common made to evaluate the properties of hardened concrete?

5.2. What is the significance of the compressive strength of concrete?

5.3. How are specimens obtained for the compression test of concrete?

5.4. The compressive strength of concrete depends upon many factors. List them and discuss each in detail.

5.5. In the testing of concrete in compression, what factors can affect the results of the test? Explain.

5.6. What is the significance of flexural strength?

5.7. What is splitting tensile strength?

5.8. What factors during testing affect the final results of the flexure test?

5.9. Explain the significance of the shearing strength; the combined stresses.

5.10. Why is fatigue strength important?

REFERENCES

BERGSTROM, S. G., "Curing Temperature, Age and Strength of Concrete," *Constr. Rev., 27,* No. 3 (July 1954).

"Curing Concrete Specimens," U.S. Bureau of Reclamation, Spec. Rep. 16, 1954.

DAVIS, H. E., TROXELL, G. E., and WISKOCIL, C. T., *The Testing and Inspection of Engineering Materials,* McGraw-Hill Book Company, New York, 1964.

"Effects of Wet-Screening to Remove Large Size Aggregate Particles on the Strength of the Concrete," Corps Eng., Ohio River Div. Lab., Mariemont, Ohio, Jan. 1953.

FORSSBLAD, L., "Investigations of Internal Vibration of Concrete," *Acta Polytech. Scand., Civil Eng. Build. Constr. Ser. No. 29,* Stockholm, (1965).

GOLDBECK, A. T., "Apparatus for Flexural Tests of Concrete Beams," Report of Committee C-9 on Concrete and Concrete Aggregates, Appendix VIII, *Proc. ASTM, 30,* Part 1 (1930), p. 591.

GONNERMAN, H. F., "Effect of End Condition of Cylinder in Compression Tests of Concrete," *Proc. ASTM, 24,* Part II (1924), p. 1036.

KEIL, F., "Cements," *3rd Int. Symp. Chem. Cement,* London, Sept. 1952.

KELLERMAN, W. F., "Effect of Size of Specimen, Size of Aggregate, and Method of Loading upon the Uniformity of Flexural Strength Tests," *Public Roads, 13,* No. 11 (Jan. 1933), p. 177.

KENNEDY, T. B., "A Limited Investigation of Capping Materials for Concrete Test Specimens," *Proc. Am. Concr. Inst., 41* (1944), p. 117.

KESLER, C. E., "Effect of Length to Diameter Ratio on Compressive Strength–An ASTM Cooperative Investigation," *Proc. ASTM, 59* (1959), p. 1216.

KESLER, C. E., "Statistical Relation between Cylinder, Modified Cube, and Beam Strength of Plain Concrete," *Proc. ASTM, 54* (1954), p. 1178.

KESLER, C. E., "Strength of Hardened Concrete," "Significance of Tests and Properties of," in "Concrete and Concrete Making Materials," *ASTM Spec. Tech. Publ. 169A,* 1966.

KLIEGER, P., "Effect of Mixing and Curing Temperature on Concrete Strength," *Proc. Am. Concr. Inst., 54* (1958), p. 1063.

KREBS, R. D., and WALKER, R. D., *Highway Materials,* McGraw-Hill Book Company, New York, 1966.

LEA, F. M., "Modern Developments in Cement in Relation to Concrete Practice," *J. Inst. Civil Eng.* (Feb. 1943).

LECAMUS, B., "Research of Fatigue Strength," *J. Am. Concr. Inst., 63,* No. 1 (Jan. 1966).

MATHER, B., "Effect of Type of Test Specimen on Apparent Compressive Strength of Concrete," *Proc. ASTM, 45* (1945), p. 802.

MCMILLAN, F. R., "Suggested Procedure for Testing Concrete in Which the Aggregate Is More than One-Fouth the Diameter of the Cylinders," *Proc. ASTM, 30,* Part I (1930), p. 521.

"Methods of End Conditions before Capping upon the Compressive Strength of Concrete Test Cylinders," *Proc. ASTM, 41* (1941), p. 1038.

MILLS, A. P., HAYWARD, H. W., and RADER, L. F., *Materials of Construction,* John Wiley & Sons, Inc., New York, 1955.

MURDOCK, J. W., "A Critical Review of Research on Fatigue of Plain Concrete," *Bull. 975,* Eng. Exp. Sta., Univ. Ill., Urbana, Ill., 1965.

NASH, J. P., "Tests of Concrete Road Aggregates," *Proc. ASTM, 17,* Part II (1917), p. 394.

NEVILLE, A. M., *Properties of Concrete,* John Wiley & Sons, Inc., New York, 1963.

NIELSON, K. E. C., "Effect of Various Factors on the Flexural Strength of Concrete Test Beams," *Mag. Concr. Res.,* No. 15 (Mar. 1954), p. 105.

NORDLY, G. M., "Fatigue of Concrete—A Review of Research," *Proc. Am. Concr. Inst., 55* (1959), p. 191.

ORCHARD, D. F., *Concrete Technology,* John Wiley & Sons, Inc., New York, 1973.

PRICE, W. H., "Factors Influencing Concrete Strength," *Proc. Am. Concr. Inst., 47* (1951), p. 417.

PROBST, E., "Further Investigations of Alternating Loads on Concrete," *Cement Concr. Res. 31* (Mar. 1942).

PROBST, E., "The Influence of Rapidly Alternating Loading on Concrete and Reinforced Concrete," *Struct. Eng., 9,* No. 10 (Oct. 1931), and No. 12 (Dec. 1931).

RICHART, F. E., BRANDIZAEG, A., and BROWN, R. L., "A Study of the Failure of Concrete under Combined Compressive Stresses," *Bull. 185,* Eng. Exp. Sta., Univ. Ill., Urbana, Ill., 1928.

SHANK, J. R., "Plastic Flow of Concrete at High Overload," *Proc. Am. Concr. Inst., 45* (1949), p. 493.

SHIDELER, J. J., and MCHENRY, D., "Effect of Speed in Mechanical Testing of Concrete," in "Speed of Testing of Nonmetallic Materials," *ASTM Spec. Tech. Publ. 185,* 1955.

SMITH, F. C., and BROWN, R. Q., "The Shearing Strength of Cement Mortar," *Bull. 106,* Eng. Exp. Sta., Univ. Wash., Seattle, Wash., 1941.

TALBOT, A. N., "Tests of Concrete: I. Shear; II. Bond," *Bull. 8,* Eng. Exp. Sta., Univ. Ill., Urbana, Ill., 1906.

TALBOT, A. A., and RICHART, F. E., "The Strength of Concrete," *Bull. 137,* Eng. Exp. Sta., Univ. Ill., Urbana, Ill., 1923.

TUTHILL, L. H., and DAVIS, H. E., "Over-vibration and Re-vibration of Concrete," *Indian Concr. J., 20* (1946).

WALKER, S., and BLOEM, D. L., "Studies of Flexural Strength of Concrete, Part 2: Effects of Curing and Moisture Distribution," *Proc. Highway Res. Board, 36* (1957), p. 334; "Part 3: Effects of Variations in Testing Procedures," *Proc. ASTM, 57* (1957), p. 1122.

WATERS, T., "The Effect of Allowing Concrete to Dry before It Has Fully Cured," *Mag. Concr. Res., 7,* No. 20 (July 1955).

WATSTEIN, D., "Effect of Straining Rate on the Compressive Strength and Elastic Properties of Concrete," *Proc. Am. Concr. Inst., 49* (1953), p. 729.

WERNER, G., "The Effect of Type of Capping Material on the Compressive Strength of Concrete Cylinders," *Proc. ASTM, 58* (1958), p. 1166.

WERNER, G., "The Effect of Type of Capping Material on the Compressive Strength of Concrete Cylinders," Report of the Bureau of Public Roads presented to the Sixty-first Annual Meeting of the ASTM, June, 1958.

WRIGHT, P. J. F., "Comments on an Indirect Tensile Test on Concrete Cylinders," *Mag. Concr. Res.,* No. 20 (1955), p. 87.

Wright, P. J. F., and Garwood, F., "The Effect of the Method of Test on the Flexural Strength of Concrete," *Mag. Concr. Res.*, No. 15 (Mar. 1954), p. 105.

Wright, P. V. F., "The Design of Concrete Mixes on the Basis of Flexural Strength," *Proc. Symp. Mix Design Quality Control Concr.*, Cement and Concrete Association, London, May 1954.

DESIGN PROCEDURE IN MAKING CONCRETE

6

Concrete is a composite material made up of inert materials of varying sizes which are bound together by a binding medium. Mortar is made up of a mixture of cement, water, air, and fine aggregate. Concrete contains coarse aggregate in addition to cement, water, air, and fine aggregate. The cement, water, and air combine to form a paste that binds the aggregates together. Thus, the strength of the concrete is dependent on the strength of the aggregate–matrix bond. The entire mass of the concrete is deposited or placed in a plastic state and almost immediately begins to develop strength (harden), a process which, under proper airing conditions, may continue for years. Because concrete is initially in a plastic state, it lends itself to all kinds of construction, regardless of size or shape. One drawback is that the concrete in a plastic condition must be placed within forms, and these forms cannot be removed until the concrete has hardened somewhat.

In types of work where concrete is used to counteract compressive stresses, it is an excellent building material. However, if the concrete must counteract tensile stresses, it must be reinforced with steel, as concrete is weak in tension.

CONCRETE MATERIALS

Cement

Usually, portland cement is specified for general concrete construction work and should conform to the standard specifications of the ASTM. From time to time special cements may be required if the project involves unusual requirements. Chapter 4 discussed the various types of portland cement as well as special cement requirements. The reader is directed to Chapter 4 for a description of the specifications established by the ASTM and also to review the importance and uses of special cements.

Water

There is usually very little trouble in obtaining water for use in concrete. Almost any water that is drinkable may be used to make concrete. Drinking water with a noticeable taste or odor should not be used until it is tested for organic impurities. Impurities in mixing water may cause any one or all of the following:

1. Abnormal setting time.
2. Decreased strength.
3. Volume changes.
4. Efflorescence.
5. Corrosion of reinforcement.

Some of the impurities in mixing water that cause these undesirable effects in the final concrete are:

1. Dissolved chemicals.
2. Seawater.
3. Sugar.
4. Algae.

Dissolved chemicals may either accelerate or retard the set and can substantially reduce the concrete strength. Further, such dissolved chemicals can actively attack the cement–sand bond, leading to early disintegration of the concrete.

Seawater containing less than 3 percent salt is generally acceptable for plain concrete but not for reinforced or prestressed concrete. The presence of salt can lead to corrosion of the reinforcing bars and prestressing tendons.

If sugar is present in even small amounts, it can cause rapid setting and reduced concrete strength.

Algae can cause a reduction in the strength of concrete by increasing the amount of air captured in the paste and reducing the bond strength between the paste and the aggregate.

Aggregates

The requirements for fine and coarse aggregate for concrete construction were described in Chapter 3.

PRINCIPAL REQUIREMENTS FOR CONCRETE

In the design of concrete mixes, three principal requirements for concrete are of importance:

1. Quality.
2. Workability.
3. Economy.

Quality

The *quality* of concrete is measured by its strength and durability. Hardened concrete must have sufficient strength to resist the stresses from loads as well as the stresses created by its own weight. Both the compressive and the flexural strength of concrete are important in the design of concrete structures. The principal factors affecting the strength of concrete, assuming sound aggregates, are the water–cement ratio and the extent to which hydration has progressed. *Hydration* is the chemical reaction that takes place between the water and cement while the concrete is hardening. The strength of concrete at the end of 28 days is the generally accepted standard for evaluating the strength properties of concrete. Laboratory compressive strengths are usually obtained by testing cylinders 6 in. (15.24 cm) in diameter and 12 in. (30.48 cm) in height under gradually increased compressive loading until failure occurs. On the other hand, flexural strengths are generally obtained by loading a suitably configured test specimen transverse to the longitudinal axis at the third points, until failure occurs. That is, the force is applied simultaneously at one-half and two-thirds of the distance between the end supports.

Durability of concrete is the ability of the concrete to resist the forces of disintegration due to freezing and thawing and chemical attack. These two factors will be discussed in more detail in Chapter 7.

Workability

Workability of concrete may be defined as a composite characteristic indicative of the ease with which the mass of plastic material may be deposited in its final place without segregation during placement, and its ability to conform to fine forming detail. Workability is a term for which there is no perfectly satisfactory definition. The size and gradation of the aggregate, the amount of mixing water, the time of mixing, and the size and shape of the forms are all factors that affect workability.

No satisfactory measures of workability have been defined. However, the term *consistency* is generally considered a descriptive term for workability. Consistency measures the fluidity or the lack ot it. The most generally used test for the measurement of consistency is ASTM C143 (Slump of Portland Cement Concrete).

Economy

Economy takes into account effective use of materials, effective operation, and ease of handling. The cost of producing good-quality concrete is an important consideration in the overall cost of the construction project.

INFLUENCES OF INGREDIENTS ON PROPERTIES OF CONCRETE

The amount of each ingredient used in concrete must be proportioned very carefully to produce desired effects such that the concrete may be used for its intended purposes. If the proportions are not as designed, adverse effects may result. Table 6-1 shows the influence of each principal ingredient on the properties of concrete.

TABLE 6-1 Influence of Ingredients on the Properties of Concrete[a]

Ingredient	*Quality*	*Workability*	*Economy*
Aggregate	Increases	Decreases	Increases
Portland cement	Increases	Increases	Decreases
Water	Decreases	Increases	Increases

[a]From W. A. Cordon, *Properties, Evaluation, and Control of Engineering Materials*, McGraw-Hill Book Company, New York, 1979.

Aggregates

Almost any type of aggregate can be used for the making of concrete. The most commonly used aggregates are sand, gravel, crushed stone, and air-cooled blast-furnace slag. These aggregates produce normal-weight concrete ranging from 135 to 160 lb/ft^3. Shales, clay, slag, and slate can also be used in the making of concrete. These aggregates are used to make lightweight concrete weighing between 85 and 115 lb/ft^3.

Aggregates exhibit a variety of physical and chemical characteristics; hence, their influence on concrete mixtures is varied. Physical characteristics include size and shape of the aggregate, surface texture, gradation, and top size of aggregate. Chemical characteristics of aggregates are those which may result in aggregate reactivity with the hardened concrete. In any concrete structure the maximum amount of aggregate should be used.

Water

In many concrete mix design procedures, water is considered an influential part of the total mix, as the water allows the concrete to be handled easily.

In other words, it allows the concrete mixture to be workable. Water is also important from the standpoint of hydration. Too much water added to the mixture results in poor-quality concrete. Water reacts with the cement particles, resulting in a chemical change that binds the paste to the fine and coarse aggregate particles. The combination of the water and the cement to form a paste may be considered to act as a glue binding the fine and coarse aggregate particles. If too much water is added, the glue becomes diluted, and this leads to a weak bond between the paste (matrix) and the aggregate. In addition to this, excess water produces segregation of the aggregate particles from the paste, and this results in a nonuniform mix.

Too little water produces a dry mix which easily crumbles under its design load, resulting in failure of the structure.

A normal bag of cement requires 2½ to 3 gallons of water to produce a mix that hydrates at the appropriate rate, resulting in a concrete mix of good quality.

Portland Cement

When water is added to portland cement a chemical reaction (hydration) takes place and a calcium silicate hydrate is produced. The amount of water needed to complete this reaction is approximately 30 percent of the cement by weight. However, if this is the only amount of water added to the mix, the mix will be stiff and unworkable. Therefore, additional water is added to the mix to make it more plastic and workable. The ratio of water to cement (w/c) determines the quality of the paste and controls the strength of the concrete. The w/c ratio is the most significant item affecting the strength of the concrete mix. A discussion of this fact was published by Duff A. Abrams in 1918 and is frequently called the *Abrams' water–cement ratio law*. Extensive research has proved this law to be valid, and graphs of such experimental work

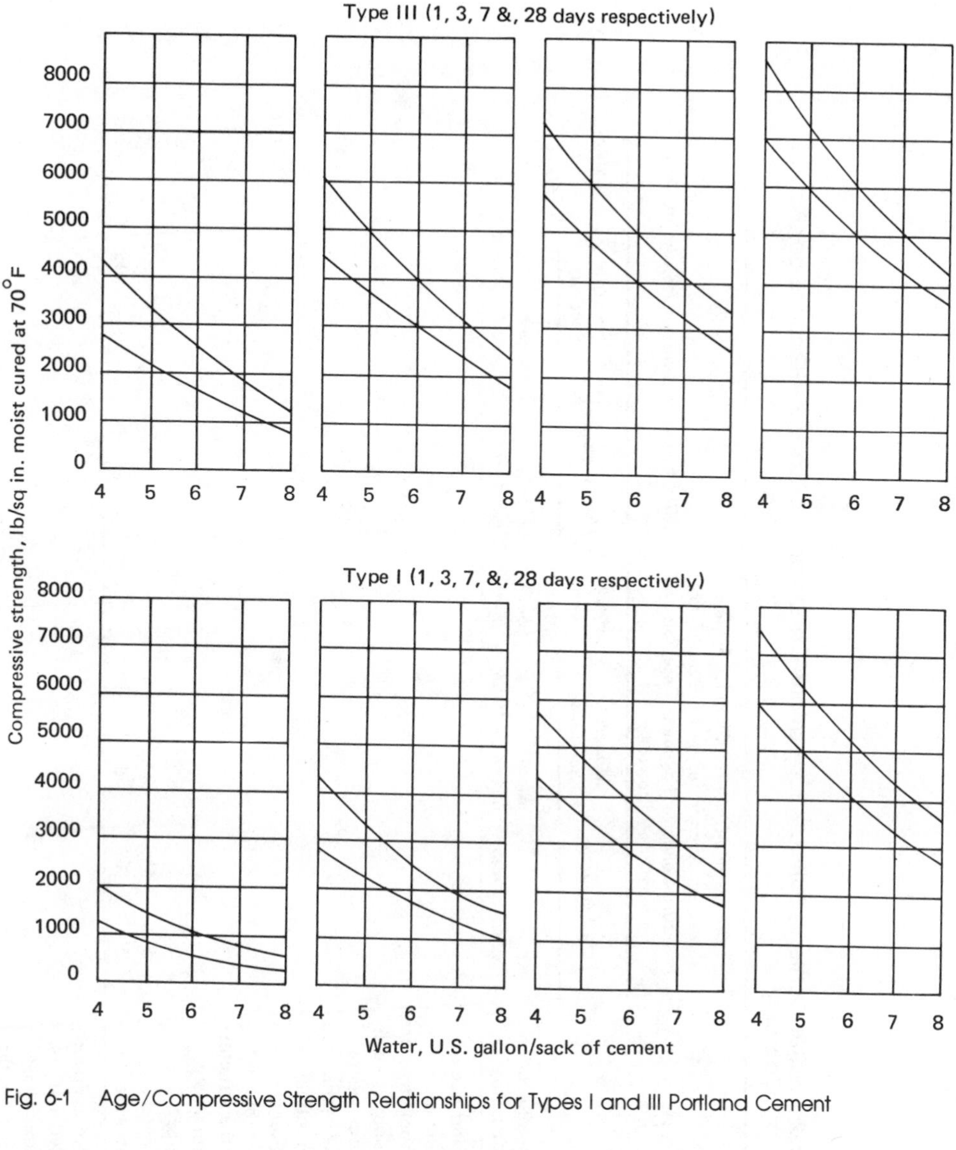

Fig. 6-1 Age/Compressive Strength Relationships for Types I and III Portland Cement

(Figure 6-1) are available for various types and combinations of materials. This type of graph is used as a beginning point in mix proportioning when an individual job design is not justifiable. Table 6-2 gives the recommended water–cement ratios for various types of structures and degrees of exposure.

Table 6-2 Water Contents Suitable for Various Conditions of Exposure (Gallons per Bag of Cement)

	Severe or Moderate Climate: Wide Range of Temperatures, Rain and Long Freezing Spells or Frequent Freezing and Thawing					*Mild Climate: Rain, or Rarely Snow or Frost*				
	Thin Sections		*Moderate Sections*		*Heavy and Mass Sections*	*Thin Sections*		*Moderate Sections*		*Heavy and Mass Sections*
Types or Location of Structure	*Reinforced*	*Plain*	*Reinforced*	*Plain*		*Reinforced*	*Plain*	*Reinforced*	*Plain*	
At the water line in hydraulic or waterfront structures or portions of such structures where complete saturation or intermittent saturation is possible, but not where the structure is continuously submerged										
In seawater	5	5½		6		5	5½		6	
In fresh water	5½	6	6	½		5½	6		6½	
Portions of hydraulic or waterfront structures some distance from the										

water line, but subject to frequent wetting								
By seawater	5½	6		6	5½	6½		7
By fresh water	6	6½		6½	6	7		7½
Ordinary exposed structures, buildings, and portions of bridges not coming under groups above	6	6½		7	6	7		7½
Complete continuous submergence								
In seawater	6	6½		7	6	6¼		7
In fresh water	6¼	7		7½	6¼	7		7¼
Concrete deposited through water	[a]	[a]	5½	5½	a	a	5½	5½
Pavement slabs directly on ground								
Wearing slabs	5½	6	[a]	[a]	6	6½	a	a
Base slabs	6½	7	[a]	[a]	7	7½	a	a

Special cases:

(a) For concrete exposed to strong sulfate groundwaters or other corrosive liquids or salts, the maximum water content should not exceed 5 gal per bag.

(b) For concrete not exposed to the weather, such as the interior of buildings and portions of structures entirely below ground, no exposure hazard is involved and the water content should be selected on the basis of the strength and workability requirements.

[a]These sections not practicable for the purpose indicated.

PROPORTIONING CONCRETE MIXES

In proportioning concrete mixes we will briefly discuss four methods: trial-batch method, job-curve method, mortar-voids method, and the method of Goldbeck and Gray.

Trial-Batch Method

The purpose of the *trial-batch method* is to produce a given water–cement ratio such that the factors quality, workability, and economy are balanced for the most desirable combination of aggregates. The size and gradation of the aggregate, surface texture of the fine and coarse aggregate, and the proportions of fine and coarse aggregate are the most important factors in determining the combination that will give the best quality and most desirable workability at the lowest cost.

A rough procedure for a typical trial-batch method of proportioning follows. Select the desired water–cement ratio. Weigh out a definite amount of cement and place it in the mixing apparatus, with the proper amount of water, to obtain the required water–cement ratio. Weigh out definite quantities of fine and coarse aggregate and place them in a container. Add fine and coarse aggregate from the weighed quantities to the cement–water paste until the desired consistency is obtained for the plastic state.

The remaining fine and coarse aggregate not used to make the plastic mixture is weighed and subtracted from the original quantities. The proportions of cement to sand to coarse aggregate (by weight) may be changed to volumetric proportions by dividing the weight of each material by its unit weight.

To measure the consistency of the mix, the slump-cone test is utilized. Table 6-3 shows the recommended consistency for various classes of concrete structures.

After the desired proportion has been established and the desired consistency is satisfactory, further experimenting may be carried out by finding a desirable ratio of fine to coarse aggregate. Concrete mixtures with a low cement–sand mortar (one that does not fill the spaces between pebbles) are hard to work with and result in a honeycombed surface. A concrete mixture with a high cement–sand mortar produces a very porous concrete.

The trial-batch method may be used for making a batch of any size. The accuracy obtained is dependent on the care exercised in producing the batch.

TABLE 6-3 Recommended Mix Consistencies for Cement

Type of Structure	Slump (in.)	
	Minimum	Maximum
Massive sections, pavement and floors laid on ground	1	4
Heavy slabs, beams, or walls	3	6
Thin walls and columns, ordinary slabs or beams	4	8

Job-Curve Method

The *job-curve method* utilizes Abrams' water–cement-ratio law but allows variations due to the differences caused by cements and aggregates. Trial batches are prepared utilizing different water–cement ratios, using the cement and fine and coarse aggregate to be used on the job. Various test specimens are molded, cured, and tested. A job curve is plotted showing compressive strength versus gallons of water per sack of cement. If a strength of 3500 psi were required for a specific job after 28 days of curing, one would simply find 3500 psi on the job curve, read across until the job curve was intersected, and read down to the required number of gallons per sack of concrete. A further trial batch would be run with the new data to refine the properties of the concrete and the necessary quantities of materials for the design mixture. Figure 6-2 shows a typical job curve in which the desired strength would be 3500 psi and the necessary water in gallons per sack of cement would be 6.

Mortar-Voids Method

The *mortar-voids method* was developed by A. N. Talbot in 1923. His method involves the proportioning of concrete by the determination of voids in the mortar. It was his belief that the strength of concrete depends upon the composition of the cement paste (matrix). The matrix occupies all that space not occupied by the aggregate. Therefore, concrete is a composite of a matrix and of an aggregate. The method is not used that much but deserves mentioning.

Method of Goldbeck and Gray

The *method of Goldbeck and Gray* has also been called the b/b_0 *method* and is sometimes referred to as the *ACI method*. This method is

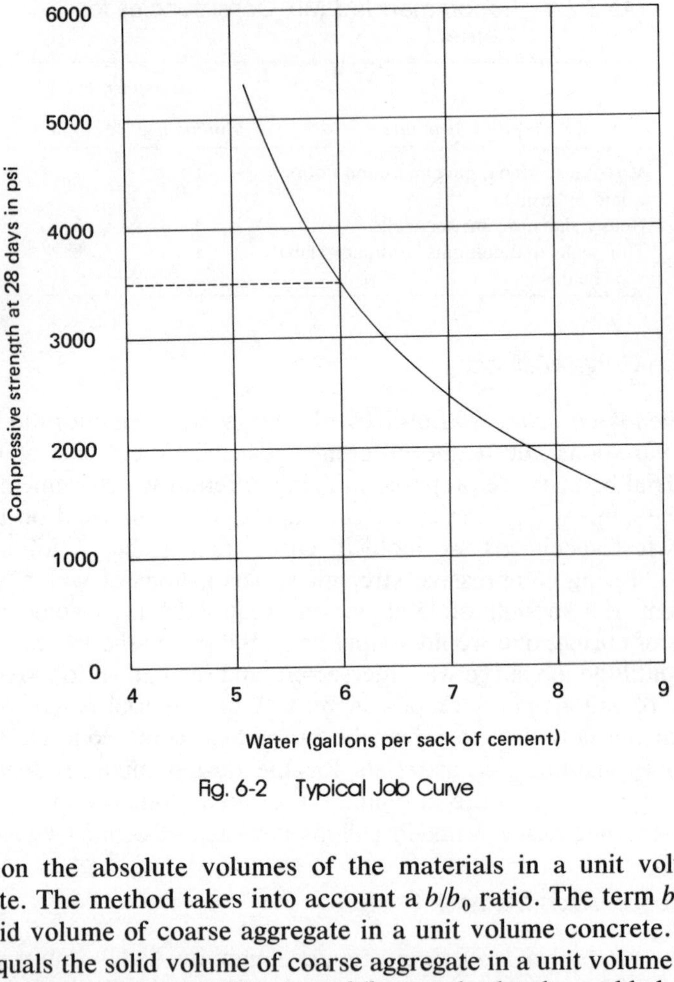

Fig. 6-2 Typical Job Curve

based on the absolute volumes of the materials in a unit volume of concrete. The method takes into account a b/b_0 ratio. The term b equals the solid volume of coarse aggregate in a unit volume concrete. The b_0 term equals the solid volume of coarse aggregate in a unit volume of dry-rodded coarse aggregate. The term b/b_0 equals the dry-rodded volume (bulk volume) of coarse aggregate in a unit of concrete. The procedure for designing concrete mixtures involves the use of data from Tables 6-4 to 6-6 which have been prepared by Goldbeck and Gray. These tables are utilized for typical materials. Table 6-7 is for pavement concrete and should be used for such.

The steps in performing a concrete mix design will be listed and then an example will be given. The steps are as follows:

1. Perform the following ASTM specifications on the fine and coarse aggregates.

TABLE 6-4 Dry-rodded Volume b/b_0 of Coarse Aggregate (any type) per Unit Volume of Concrete[a]

Size of Coarse Aggregate (Square-opening Laboratory Sieves)	*Fine Sand*		*Medium Sand*			*Coarse Sand*		
	Fineness Modulus of Sand							
	2.4	*2.5*	*2.6*	*2.7*	*2.8*	*2.9*	*3.0*	*3.1*
	Values[b] for b/b_0							
No. 4 to ½ in.	0.59	0.58	0.57	0.56	0.55	0.54	0.53	0.52
No. 4 to ¾ in.	0.66	0.65	0.64	0.63	0.62	0.61	0.60	0.59
No. 4 to 1 in.	0.71	0.70	0.69	0.68	0.67	0.66	0.65	0.64
No. 4 to 1½ in.	0.75	0.74	0.73	0.72	0.71	0.70	0.69	0.68
No. 4 to 2 in.	0.78	0.77	0.76	0.75	0.74	0.73	0.72	0.71
No. 4 to 2½ in.	0.80	0.79	0.78	0.77	0.76	0.75	0.74	0.73

[a]From A. T. Goldbeck and J. E. Gray, "A Method of Proportioning Concrete for Strength, Workability, and Durability," *Bull. 11*, National Crushed Stone Association, Nov. 1953.

[b]For concrete that is to be assisted in place by internal vibration under very rigid inspection, increase tabulated values of b/b_0 approximately 10%.

a. Determine the bulk specific gravities of the fine and coarse aggregate.
b. Determine the dry-rodded unit weight of the coarse aggregate.
c. Determine the gradation and fineness modulus of the fine aggregate and the gradation of the coarse aggregate.

2. Compute the solid weights per cubic foot of the cement and the fine and coarse aggregates. This is computed by multiplying the bulk specific gravity by 62.4 lb.
3. Pick the size of coarse aggregate desired along with the determined fineness modulus of sand and enter Table 6.3 to determine b/b_0.
4. Calculate the solid volume of coarse aggregate per cubic foot of dry-rodded material.

$$b_0 = \frac{\text{dry-rodded weight per cubic foot}}{\text{solid weight per cubic foot}}$$

5. Calculate b, which is equal to $b/b_0 \times b_0$.
6. Knowing the kind and size of the coarse aggregate desired as well as the 28-day compressive strength desired and the chosen slump, determine from Table 6-5 (if non-air-entrained concrete is to be used) or Table 6-6 (if air-entrained concrete is to be used) the cement factor and the water content. Also select from the appro-

Table 6-5 Non-air-entraining Structural Concrete Cement Factors Required for 28-Day Compressive Strengths Listed[a,b]

Size of coarse Aggregate Square-opening Laboratory Sieves):	*No. 4 to ½ in.*		*No. 4 to ¾ in.*		*No. 4 to 1 in.*		*No. 4 to 1½ in.*		*No. 4 to 2 in.*		*No. 4 to 2½ in.*	
Slump (in.):	*3*	*6*	*3*	*6*	*3*	*6*	*3*	*6*	*3*	*6*	*3*	*6*
Water[c] (gal yd^3 of concrete)												
Angular coarse aggregate	*42*	*44*	*40*	*42*	*38*	*40*	*36*	*38*	*35*	*37*	*34*	*36*
Rounded coarse aggregate	*38*	*40*	*36*	*38*	*34*	*36*	*32*	*34*	*31*	*33*	*30*	*32*

28-Day Compressive Strength (psi)	*Cement (sacks/yd³ of concrete)*											
2000	4.6	4.8	4.4	4.6	4.2	4.4	4.0	4.2	3.9	4.0	3.8	3.9
2500	5.0	5.2	4.8	5.0	4.5	4.8	4.2	4.5	4.1	4.3	4.0	4.2
3000	5.4	5.7	5.2	5.4	4.9	5.2	4.6	4.9	4.4	4.7	4.3	4.6
3500	5.9	6.3	5.6	5.9	5.3	5.6	5.0	5.3	4.9	5.2	4.8	5.0
4000	6.5	6.9	6.2	6.5	5.8	6.2	5.5	5.8	5.4	5.7	5.2	5.5
4500	7.2	7.5	6.8	7.1	6.4	6.8	6.1	6.4	5.9	6.3	5.7	6.1
5000	8.1	8.5	7.7	8.1	7.3	7.7	6.9	7.3	6.7	7.1	6.5	6.9
Entrapped air (approx. %)	2.5		2		1.5		1		1		1	

[a]From A. T. Goldbeck and J. E. Gray, "A Method of Proportioning Concrete for Strength, Workability, and Durability," *Bull. 11*, National Crushed Stone Association, Nov. 1953.

[b]For concrete to be assisted in place by internal vibration, use 3-in. slump and decrease tabulated water contents by approximately 4 gal. No reduction in cement factor is suggested.

[c]This is water actually effective as mixing water.

Table 6-6 Air-entraining Structural Concrete Cement Factors Required for 28-Day Compressive Strengths Listed[a,b]

Size of Coarse Aggregate Square-opening Laboratory Sieves):	*No. 4 to ½ in.*		*No. 4 to ¾ in.*		*No. 4 to 1 in.*		*No. 4 to 1½ in.*		*No. 4 to 2 in.*		*No. 4 to 2½ in.*	
Slump (in.):	*3*	*6*	*3*	*6*	*3*	*6*	*3*	*6*	*3*	*6*	*3*	*6*
Water[c] (gal yd³ of concrete)												
Angular coarse aggregate	*38*	*40*	*36*	*38*	*34*	*36*	*32*	*34*	*31*	*33*	*30*	*32*
Rounded coarse aggregate	*35*	*37*	*33*	*35*	*31*	*33*	*29*	*31*	*28*	*30*	*27*	*29*

28-Day Compressive Strength (psi)	Cement (sacks/yd³ of concrete)											
2000	4.4	4.7	4.2	4.4	3.9	4.2	3.7	3.9	3.6	3.8	3.5	3.7
2500	4.9	5.2	4.6	4.9	4.4	4.7	4.2	4.4	4.0	4.3	3.9	4.2
3000	5.6	5.9	5.3	5.6	5.0	5.3	4.7	5.0	4.5	4.8	4.3	4.7
3500	6.3	6.7	6.0	6.3	5.6	6.0	5.3	5.6	5.1	5.4	4.9	5.3
4000	7.2	7.5	6.8	7.2	6.4	6.8	6.0	6.4	5.8	6.2	5.6	6.0
4500	8.1	8.5	7.6	8.1	7.2	7.6	6.8	7.2	6.6	7.0	6.4	6.8
5000	9.2	9.7	8.7	9.2	8.2	8.7	7.7	8.2	7.4	8.0	7.2	7.7
Optimum entrained-air content (%)	6.0		6.0		5.5		5.0		5.0		4.5	

[a]From A. T. Goldbeck and J. E. Gray, "A Method of Proportioning Concrete for Strength, Workability, and Durability," *Bull. 11,* National Crushed Stone Association, Nov. 1953.
[b]This table should always be used to proportion concrete subject to freezing.
[c]This is water actually effective as mixing water.

TABLE 6-7 Pavement Concrete Use: Dry-rodded Volume of Coarse Aggregate and Mixing Water Required[a]

Size of Coarse Aggregate (Square-opening Laboratory Sieve)	Fine Sand		Medium Sand			Coarse Sand			Water[b] (gal/yd³ of concrete for 2-in. slump)			
	Fineness Modulus of Sand								Non-air-entrained[c]		Air-entrained	
	2.4	2.5	2.6	2.7	2.8	2.9	3.0	3.1	Coarse Aggregate			
	Values for b/b_o								Angular	Rounded	Angular	Rounded
No. 4 to 1 in.	0.75	0.74	0.73	0.72	0.71	0.70	0.69	0.68	35	31	31	28
No. 4 to 1½ in.	0.79	0.78	0.77	0.76	0.75	0.74	0.73	0.72	33	29	29	26
No. 4 to 2 in.	0.82	0.81	0.80	0.79	0.78	0.77	0.76	0.75	32	28	28	25
No. 4 to 2½ in.	0.84	0.83	0.82	0.81	0.80	0.79	0.78	0.77	31	27	27	24

[a]From A. T. Goldbeck and J. E. Gray, "A Method of Proportioning Concrete for Strength, Workability, and Durability," *Bull. 11*, National Crushed Stone Association, Nov. 1953.

[b]This is to water actually effective as mixing water.

[c]Since the maximum size of coarse aggregate used in highways is 1½ in. or larger, constant value of 1% may be taken for entrapped air, or 27 ft³/yd³ of concrete.

priate tables the percentage of entrapped air or optimum entrapped air and calculate its solid volume per cubic yard of concrete.

7. Determine and sum up the solid volumes of cement, coarse aggregate, water, and air.
8. Determine the solid volume of sand in a cubic yard of concrete by subtracting from 27 ft³ from item 7.
9. Convert solid volumes to pounds per cubic yard of concrete.
10. Calculate the additional water requirement.

The following example should illustrate the procedure that could be used to design a concrete mix and to calculate the batching weight.

Example 6.1: Determine the batch quantities for a concrete mix design to have a 28-day compressive strength of 3000 psi using angular aggregate from a No. 4 sieve opening to ½ in. in size and a medium fine aggregate (sand) with a fineness modulus of 2.70. A slump of 3 in. is desired.

The material descriptions are as follows:

Coarse Aggregate (CA)

Size, No. 4 to ½ in., angular aggregate
Absorption (%) = 0.1
SG (bulk dry) = 2.7
SG (bulk saturated surface dry) = 2.73
Solid weight (lb/ft³) = SG (bulk dry) × unit weight of water
= 2.7 × 62.4 = 168.48
Dry-rodded unit weight (lb/ft³) = 101.5

b_0 = dry-rodded unit weight ÷ solid weight = 101.5 ÷ 168.48
= 0.60

Fine Aggregate (FA)

Fineness modulus (FM) = 2.7
Absorption (%) = 0.8
SG (bulk dry) = 2.5
SG (bulk saturated surface dry) = 2.53
Solid weight (lb/ft³) = SG (bulk dry) × unit weight of water
= 2.5 × 62.4 = 156.00

Cement

SG = 3.14
Solid weight (lb/ft³) = SG × unit weight of water
= 3.14 × 62.4 = 195.9
Weight per sack (lb) = 94
Solid volume per sack (ft³) = weight per sack ÷ solid weight
= 94 ÷ 195.9 = 0.48

Calculations of Proportions

b/b_0 (from Table 6-4) = 0.56

$b = b/b_0 \times b_0 = 0.56 \times 0.60 = 0.34$

Designing for a 3-in. slump and 3000-psi compressive strength for non-air-entrained concrete (2.5%) using Table 6-5.

Cements (sacks/yd³) = 5.4

Water (gal/yd³) = 42

	Solid Volume (ft^3/yd^3 of concrete)	Quantities (lb/yd^3 of concrete)
Cement: $5.4 \times 0.48 =$	$2.59 \times 195.9 =$	507.38
CA: $0.34 \times 27 =$	$9.18 \times 168.5 =$	1546.83
Water: $42 \div 7.5 =$	$5.60 \times 62.4 =$	349.44
Air: $0.025 \times 27 =$	0.68×0	0
	Σ 18.05	
FA: $27 - 18.05 =$	$8.96 \times 156.00 =$	1397.76

Notice that the weights for fine and coarse aggregate are dry, and thus the water content or amount of water must be increased. This increase in water is equivalent to the amount absorbed by the aggregates. In this example the increase would be as follows:

$$\text{water absorbed by CA} = 0.001 \times 1546.83 = 1.547 \text{ lb}$$

$$\text{water absorbed by FA} = 0.008 \times 1397.76 = 11.180 \text{ lb}$$

$$\text{total} = 12.73 \text{ lb}$$

Thus, the actual amount of mixing water would be increased by 12.73 lb, for a total of 362.17 lb. In most cases the amount of water added is insignificant, but in large jobs it may be significant. In situations in which the aggregate is wet, the mixing water is decreased by the amount of free water.

Keep in mind that pavement concrete is a less workable mix and therefore requires different b/b_0 values and slumps than those given here. Use Table 6-6.

PROPORTIONING MATERIALS

Cement

Cement is usually purchased in bags or sacks. A sack contains 94 lb and is assumed to be 1 ft^3 in volume. Cement may also be purchased in the form of a barrel, which is equal to four sacks weighing 376 lb. Cement may also be purchased by bulk weight. The reason for measuring cement by weight rather than volume is that if cement were purchased by the cubic foot, the weight might vary due to different amounts of compaction.

Water

The accurate weighing of water is important due to the fact that the cement and water form a paste which binds the aggregate and thus influences its strength. In stationary mixing plants, the water is usually weighed out. Volumetric measurements are usually made with modern concrete mixers.

Aggregates

In modern proportioning plants aggregates are weighed, thus eliminating the errors of volumetric measurement. Volumetric measurements are common but are likely to be inaccurate, owing to the differences in compaction.

MIXING AND DEPOSITING

Mixing

When the ingredients have been measured in the proper proportion, the next step in the manufacturing of concrete is to mix the ingredients until the aggregate particles are coated with cement and a homogeneous mixture is obtained. Mixing may be done by hand or by a power-driven machine.

Most concrete, regardless of the size of the job, is machine-mixed in batch mixers of 3 to 4 yd^3 capacity. Most batch mixers operate with a revolving drain fitted with blades projecting inward. The drum revolves and the ingredients are carried part way around and turned over as they drop to the bottom, thus producing a thorough mix. The time of mixing varies depending on the consistency of the mix.

Transporting Concrete

After being mixed, concrete is transported to the place of deposit in such a way as to prevent segregation of the ingredients. The means of transportation is a question of economy. The only important requirement is that the concrete arrive at the place of deposit and be of good quality and uniformity. On small projects, wheelbarrels or two-wheeled carts are used for transportation. On large projects, industrial cars, cable cars, conveyor belts, or towers may be used. Sometimes, concrete is pumped through pipelines from the mixer to the forms. Concrete may also be transported and placed by trucks.

Depositing Concrete

The proper placement of concrete in the forms is an important factor in obtaining durable concrete structures. Before the placement of the concrete, the forms should be free of debris and accumulated water. To obtain uniform concrete it should be placed in the forms evenly and without segregation of materials. If the concrete is of a dry consistency, it should be placed in layers 6 to 8 in. thick. If the concrete is of a wet consistency, it may be placed in layers 12 to 15 in. thick. After placement the concrete is spaded to remove any entrapped air, thus producing a smooth surface. The same results may be obtained by vibration, but care should be taken not to segregate the ingredients.

Depositing Concrete under Water

When one places concrete under water, care should be taken to prevent the cement from being washed away or cements from segregating. The common method of depositing concrete under water is by use of a *tremie*, a long pipe about 1 ft (30.5 cm) in diameter with a hopper at the top and a slightly flarring bottom. It is plugged at the bottom, filled, then lowered to position and kept filled at all times. The only problem is that one never knows if the concrete is being placed uniformly.

Vibrated Concrete

Vibration is a mechanical method (high-frequency electric or pneumatic) of puddling concrete which is used extensively for large masses. Vibration is usually recommended for mixes that are stiff in consistency and may need help to effectively consolidate. Vibration may be internal, or may be achieved by vibration of forms, or by vibrating floats on top of the concrete.

TABLE 6-8 Comparison of Compressive, Flexural, and Tensile Strength

Strength of Plain Concrete (psi)			*Ratio (%)*		
Compressive	*Modulus of Rupture*	*Tensile*	*Modulus of Rupture to Compressive Strength*	*Tensile Strength to Compressive Strength*	*Tensile Strength to Modulus of Rupture*
1000	230	110	23	11	48
2000	375	200	19	10	53
3000	485	275	16	9	57
4000	580	340	14	8	59
5000	675	400	14	8	60
6000	765	460	13	8	61
7000	855	520	12	7	61
8000	930	580	12	7	62
9000	1010	630	11	7	63

Bonding New Concrete to Old Concrete

In most construction projects it is best to try and avoid the bonding of new concrete to old concrete, and thus a continuous pour is recommended. However, it is sometimes impossible to avoid joints, and thus precautions should be taken to make the bond between the new concrete and the old concrete as strong as possible.

In massive work with horizontal joints it will probably be sufficient to wet the old concrete and continue with the new concrete. When the walls are thin, the old concrete is roughened and cleaned of foreign matter and laitance and slushed with grout of neat cement or mortar.

Finishing Concrete

After the concrete has been deposited, it may be finished off by several methods: steel trowel, spading, canvas belt, wooden float, burlap, or a grinding tool.

Curing

Concrete sets and hardens as a result of a chemical reaction that takes place between the cement and the water. This process continues as long as water is present. Thus, a strong concrete results if water is present to carry on the reaction. One other factor that is important to curing is temperature. If the temperature is high, undue cracking will occur. If the temperature is too low, the curing process will not continue. Chapter 5 discusses some of the curing techniques.

COMPRESSIVE, TENSILE, AND FLEXURAL STRENGTH

The major factors that affect concrete strength have been discussed in Chapter 5. Only a comparison table is shown here (Table 6-8, p. 165).

PROBLEMS

6.1. Define the term "concrete."

6.2. Impurities in mixing water may cause undesirable effects in the final concrete. List five undesirable effects.

6.3. List four undesirable impurities in mixing water and discuss each in detail.

6.4. In the design of concrete mixes, three principal requirements for concrete are of importance. Name them and discuss each in detail.

6.5. For the three requirements requested in Problem 6.4, list the ways in which the ingredients of a concrete mix affect each requirement. Explain.

6.6. List four methods of proportioning concrete mixes and discuss each in detail.

6.7. What is meant by the term b/b^0?

6.8. List the steps in performing a concrete mix design according to the method of Goldbeck and Gray.

6.9. In the method of Goldbeck and Gray, how is the b/b^0 ratio affected if the concrete is to be assisted by internal vibration?

6.10. Determine the batch quantities for a concrete mix design that is to have a 28-day compressive strength of 2500 psi using rounded aggregate from a No. 4 sieve opening to 1 in. in size and a medium-fine aggregate with a fineness modulus of 2.65. A slump of 3 in. is desirable. The following descriptions apply to the coarse and fine aggregates.

Coarse Aggregate

Size, No. 4 to 1 in., rounded
Absorption (%) = 0.1
SG (bulk dry) = 2.7
SG (bulk saturated surface dry) = 2.74
Dry-rodded unit weight (lb/ft^3)

Fine Aggregate

Fineness modulus = 2.65
Absorption (%) = 0.8
SG (bulk dry) = 2.4
SG (bulk saturated surface dry) = 2.45

REFERENCES

Abrams, D. A., Design of Concrete Mixtures, *Lewis Inst. Struct. Mater. Res. Lab. Bull. 1*, Chicago, 1918.

American Society for Testing and Materials, "Significance of Tests and Properties of Concrete and Concrete-making Material," *ASTM Spec. Tech. Publ. 169B*, 1979.

BANER, E. E., *Plain Concrete,* 3rd ed., McGraw-Hill Book Company, New York, 1949.

CORDON, W. A., *Properties, Evaluation, and Control of Engineering Materials,* McGraw-Hill Book Company, New York, 1979.

CORDON, W. A., and MERRILL, D., "Requirements for Freezing and Thawing Durability for Concrete," *Proc. ASTM, 63,* (1963), pp. 1026–1035.

FULLER, W. B., and THOMPSON, S. E., "The Laws of Proportioning Concrete," *Trans. Am. Soc. Civil Eng.* (1907), p. 67.

GOLDBECK, A. T., and GRAY, J. E., "A Method of Proportioning Concrete for Strength, Workability and Durability," *Bull. 11,* National Crushed Stone Association, Nov. 1953.

KREBS, R. D., and WALKER, R. D., *Highway Materials,* McGraw-Hill Book Company, New York, 1971.

MILLS, A. P., HAYWARD, H. W., and RADER, L. F., *Materials of Construction,* 6th ed., John Wiley & Sons, Inc., New York, 1955.

POWERS, T. C., "The Air Requirements in Frost-resistant Concrete," *Proc. Highway Res. Board* (1949), p. 184.

POWERS, T. C., "The Mechanism of Frost Action in Concrete," Stanton Walker Lecture Series on the Material Sciences, Lecture No. 3, National Ready-Mixed Concrete Association, Silver Spring, Md., 1965.

TALBOT, A. N., "Proposed Method of Estimating Density and Strength of Concrete and Proportioning Materials by Experimental and Analytical Considerations of Voids in Mortar and Concrete," *Proc. ASTM* (1921).

WALKER, S., BLOEM, D. L., and GAYNOR, R. D., "Relationships of Concrete Strength to Maximum Size of Aggregate," *Proc. Highway Res. Board* (1959).

PERFORMANCE CHARACTERISTICS OF CONCRETE 7

As indicated in Chapter 5, concrete has high compressive strength but low shear and tensile strength. It is porous, extensible, and fire-resistant, but can be damaged by intense heat. Under ordinary loading conditions, concrete is elastic, although it will creep under sustained, heavy loads. The shear and tensile strength of concrete can be increased by the addition of reinforcing bars or by prestressing with steel wires that are under tension. This fact allows much design flexibility, yet deterioration of concrete can pose significant problems. In the paragraphs that follow, the failure mechanism of concrete, the factors that effect deterioration of concrete, and those admixtures that may aid the concrete mixture will be discussed.

In discussions of concrete failure, cracking is usually the principal criterion upon which failure is based. However, not all cracking is an indication of concrete failure, as the concrete may still be able to perform a useful function. In determining whether or not the concrete structure has failed, considerable judgment must be exercised.

FAILURE MECHANISM OF CONCRETE

The failure mechanism of concrete has been under discussion for many years. Many investigators have implied that the failure mechanism is associated with internal microcracking. The formation and propagation of such microcracks have been studied indirectly by sonic velocity, acoustic methods, and by the observation of macrocracks on the surface of the models. Robinson and Hsu, Slate, Sturman, and Winter have directly observed the formation and propagation of microcracks by x-ray analysis. (Because of the limitations of the techniques employed, the detection of microcracks has been somewhat uncertain.) Derucher also directly observed the formation and propagation of microcracks by utilizing the scanning electron microscope.

According to the research of Robinson and Hsu, Slate, Sturman, Winter, and Derucher, three types of microcracks were found to exist: bond, matrix, and aggregate microcracks. Further, bond microcracks (microcracks between the cement mortar matrix and aggregate particles) exist in the form of shrinkage microcracks (Figure 7-1) prior to the application of compressive stress fields.

Shrinkage Microcracks

The results obtained by these workers indicate the existence of shrinkage microcracks in concrete containing both rounded and angular aggregate. These initial shrinkage microcracks, which were in reality bond microcracks (microcracks between the cement mortar matrix and aggregate particles), exist for a variety of reasons. Sturman suggests that shrinkage microcracks may be formed by a variety of processes, including carbonation shrinkage, hydration shrinkage, segregation due to settlement, and drying shrinkage. All concrete, regardless of the measures or procedures used to eliminate shrinkage microcracks, will contain them.

Carbonation shrinkage occurs when any cement compound is stored in air and decomposed by carbon dioxide. The portland cement used in most construction projects is of sound character, so that carbonation shrinkage is unlikely to occur.

Hydration shrinkage occurs when the primary cement–paste volume decreases during hydration, resulting in the formation of microcracks. In most cases hydration shrinkage is responsible for a small amount of shrinkage microcracking in concrete. This type of microcracking is believed to be controlled by using an expansive cement.

The influence of segregation due to settlement on the formation of

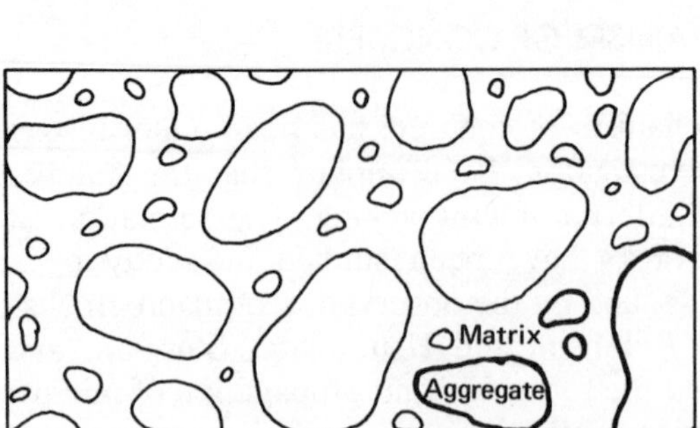

Fig. 7-1 Shrinkage Microcracks

microcracks may be analyzed by applying Stokes' law to the viscous material first formed when sand, cement, and water are mixed to form mortar. Stokes found that for very small solid particles suspended in a viscous fluid, the steady-state or terminal velocity acquired by the larger and denser particles is greater than that of the smaller, less dense particles. Thus, in a sand–cement–water mixture, the large sand particles will settle first, the fines next, and the extremely fine, flocculated particles last. This leads to a condition in which there is a thin film of fluid adjacent to the aggregate which has an extremely high water-to-solids ratio. Eventually, this water is absorbed by the adjoining cement paste, which hydrates continuously, leaving a thin space on the aggregate. When this sedimentation occurs at the exposed horizontal surface of freshly poured concrete, it is referred to as *blading*. This phenomenon is likely to cause most of the shrinkage microcracks in concrete and is probably the leading reason for eventual concrete failure.

Drying shrinkage occurs in plastic concrete if the rate of evaporation exceeds 0.1 lb/ft^2 of surface area per hour. In other words, hydrostatic tension is present, resulting in shrinkage microcracks. This is a minor cause of shrinkage microcracks, as most concrete is cured in a water-saturated atmosphere.

Bond Microcracks

Bond microcracks (Figure 7-2) are extensions of shrinkage microcracks. As the compressive stress field increases, the shrinkage microcracks widen but do not propagate into the matrix. According to Derucher, this widening of the shrinkage microcracks occurs at 15 to 20 percent of the ultimate strength of concrete (f'_c). It should be kept in mind that the weakest point in concrete is the bond between the aggregate and the matrix.

Fig. 7-2 Bond Microcracks

Matrix Microcracks

Matrix microcracks (Figure 7-3) are microcracks that occur in the matrix which are extensions of the bond microcracks into the matrix. Again according to Derucher, the bond microcracks start to propagate at/or around 20 percent of f'_c. They propagate to the point where they begin to bridge one another at about 30 to 45 percent of the ultimate strength of concrete. At 75 percent of f'_c, matrix microcracks start to bridge one another.

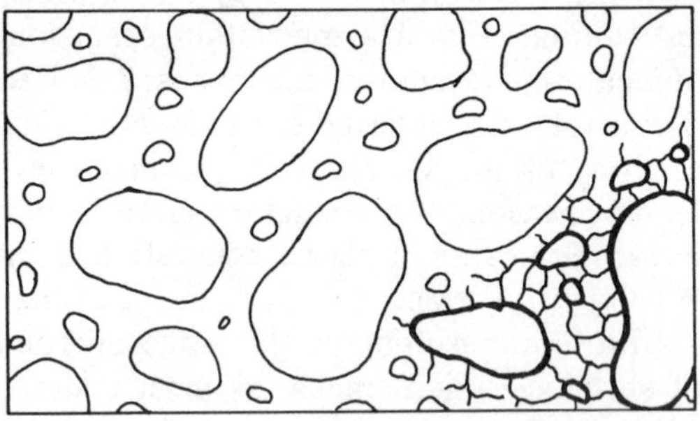

Fig. 7-3 Matrix Microcracks

Aggregate Microcracks

Aggregate microcracks occur just before failure, at about 90 percent of f'_c; failure will definitely occur.

FACTORS THAT CAUSE DETERIORATION

The factors that cause deterioration may be grouped under the following areas: poor design details, construction deficiencies and operations, temperature variations, chemical attack, reactive aggregates (alkali–silica, alkali–carbonate, Kansas–Nebraska sand–gravel, and silicification of carbonate aggregate) and high-alkali cement, moisture absorption, wear and abrasion, shrinkage and flexure forces, collision damage, scouring, shock waves, overstress, fire damage, foundation movement, corrosion of reinforcing bars and prestressing wires, and aggregates that cause deterioration. The visual systems of deterioration consist of cracking, scaling, spalling, rust stains, surface disintegration, efflorescence and exudation, and vehicular damage. Each will be examined with the form of deterioration most responsible for that form of disintegration.

Poor Design Details

One of the poor design details that can cause concrete to crack is insufficient drainage. In bridge construction, scuppers may not be provided (or improperly provided) with downspouts to keep the water discharge away from concrete surfaces, including caps and decks. The scuppers may be too small and, as a result, easily clogged. They should also be provided at low spots. Weep holes, if provided, may be too small, too few, or discharging over another concrete surface. Spalling may result when not enough space is provided at an expansion joint (Figure 7-4). Insufficient cover over rebars may cause corrosion of the rebars, or, in the case of prestressed concrete, of the prestressing wires. Of all these considerations, the need for sufficient expansion space cannot be overstated.

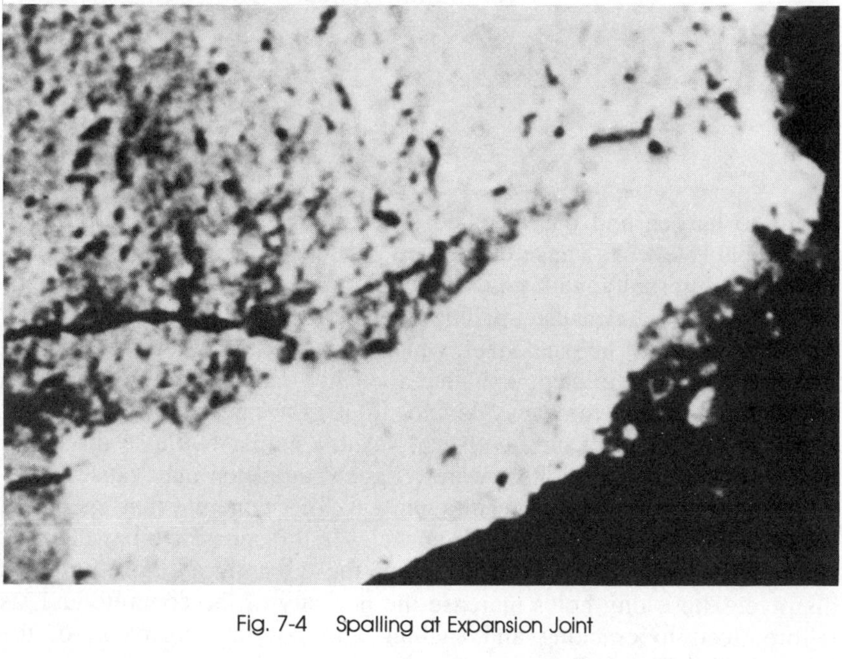

Fig. 7-4 Spalling at Expansion Joint

Construction Deficiencies

Construction deficiencies may result in concrete deterioration regardless of the care taken in the design procedure. Soft spots in the subgrade of a foundation will cause settlement, resulting in cracking (Figure 7-5). If formwork is removed between the time the concrete

Fig. 7-5 Cracking Due to Foundation Settlement

begins to harden and the specified time for formwork removal, cracks will probably occur. These cracks are especially dangerous, since they may occur internally, with no external manifestations. Water can collect in these cracks and cause spalling due to freezing and thawing or can cause corrosion of internal steel, which may eventually lead to cracking and spalling. If sufficient spacing does not exist between rebars in reinforced sections, voids may develop if the mix is not properly vibrated. These voids collect water, with end results similar to those discussed earlier for cracks that collect water. Excess vibration may cause segregation of the concrete mix, resulting in a weaker concrete than specified. The inclusion of clay or soft shale particles in the concrete mix will cause small holes to appear in the surface of the concrete as these particles dissolve. These tiny holes increase the porosity of the concrete and, as before, lead to cracking and spalling and possible corrosion of the internal steel (Figure 7-6).

Temperature Variations (Freezing and Thawing)

Freezing and thawing (temperature variation) is a common cause of concrete deterioration. Most ordinary concrete contains 1 or 2 percent air in the form of voids. This entrappped air makes the concrete very

Fig. 7-6 Surface Pitting Due to Included Clay

porous. Porous concrete absorbs water, and if allowed to freeze, the water will expand approximately 9 percent. In the process of expansion, unfrozen water is forced through small capillaries in the cement paste. Resistance to this movement of water through these small capillaries produces hydrostatic pressure. This internal expansive pressure often produces cracking (Figure 7-7), spalling (Figure 7-8), or scaling.

Three factors influence the deteriorating effect of freezing and thawing:

1. Concrete permeability.
2. Ratio of freezable water to the air voids.
3. Rate of temperature decline upon freezing.

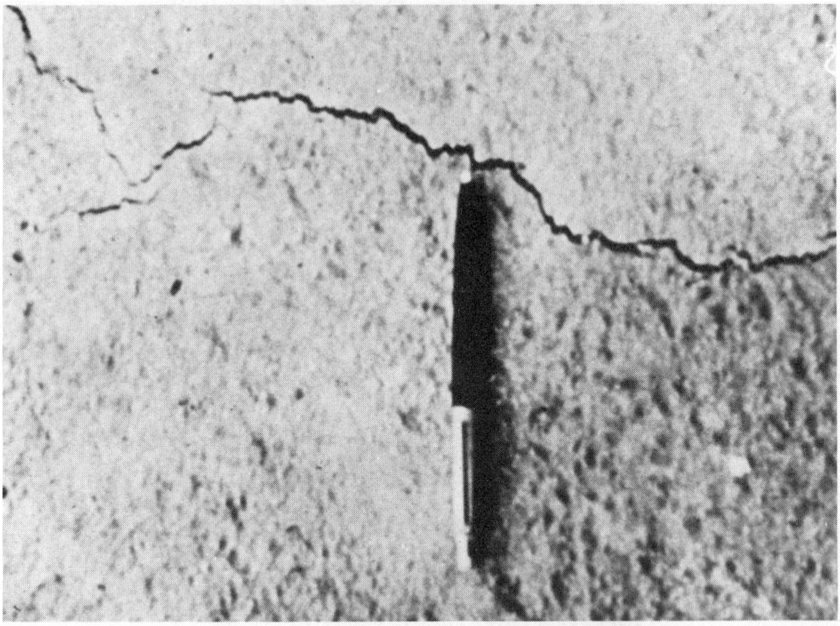

Fig. 7-7 Freeze–Thaw Cracking

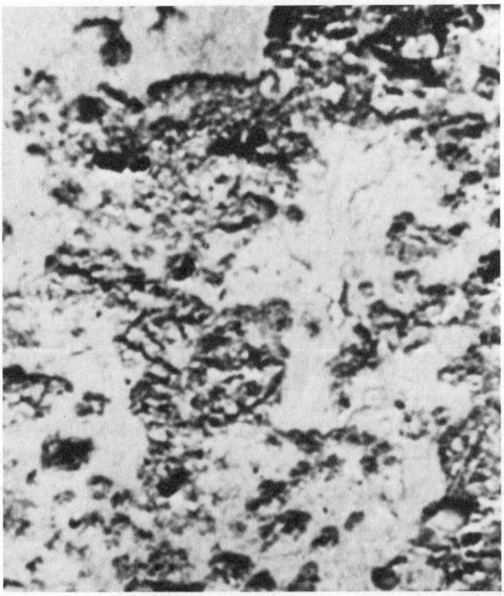

Fig. 7-8 Surface Deterioration

Factors 1 and 2 are influenced by the character and type of aggregate, the quality of the cement, mix procedures, placement, curing, and the number of air bubbles in the concrete. Factor 3 has little effect in practical field operations but plays a substantial role in laboratory testing. That is, the environment does not allow a temperature decline as fast as that which can be produced in the laboratory.

In most situations the effects of freezing and thawing can be eliminated by providing sound drainage under the concrete slab. Another method is the utilization of an air-entrained concrete, which allows for thousands of additional air voids. In this case, as water begins to freeze and expand, the unfrozen water is forced out of that space and into one that possibly has no water in it, thus minimizing the effect of hydrostatic pressure.

Temperature fluctuations may further affect concrete integrity if the coefficient of thermal expansion differs significantly from that of the mortar. Aggregates with lower coefficients may cause high tensile stresses, resulting in cracking and spalling. Further problems arise if the concrete section is restrained from expansion or contraction. The internal forces set up under these circumstances are sufficient to result in cracking and spalling and eventual failure of the member.

Chemical Attack

Chemical attack of concrete may come from two sources. The use of salts or chemical deicing agents contributes to weathering through recrystalization, and salt may increase the water retention. The results are similar to the effects of freezing and thawing (cracking and spalling). Sodium and calcium chloride are the principal salts applied directly to the concrete pavement to keep ice and snow from the pavement. This result is most serious in concrete bridge decks, as they generally freeze before the adjacent roadway; hence, more salt is applied to the bridge deck. As a consequence, bridge decks deteriorate more quickly than the adjacent roadway. Several coats of linseed oil applied to the bridge deck help to prolong deck life.

Deicing agents generally result in scaling of the concrete surface. These agents lower the freezing point of water, which in effect increases the number of freeze–thaw cycles during a prolonged cold spell.

Further chemical attack may come from chemicals in the soil or water surrounding the concrete members. Ammonium and magnesium ions react with the calcium in the cement paste. Sodium, magnesium, and calcium sulfate react with the tricalcium aluminate in the cement paste, producing additional quantities of gypsum. Gypsum, you will recall, is added to the cement clinker to react with tricalcium aluminate in the portland cement to prevent a flash set (a situation in which the concrete gives the impression of being set but is not). In other words, gypsum acts as a retarder in preventing the concrete from setting too quickly, giving workers the time to mix, place, and work with the concrete. However, if too much gypsum is interground with the cement clinker or allowed to form during the hydrolysis process, the concrete may not set for an indefinite period of time. Thus, if additional sulfate is introduced to the system during the hydration process, additional gypsum is produced and the process of sulfate attack occurs. Magnesium sulfate is one of the more important attacking agents. It reacts with lime [$Ca(OH)_2$] as follows:

$$Ca(OH)_2 + MgSO_4 \cdot 7H_2O \rightarrow CaSO_4 \cdot 2H_2O + Mg(OH)_2 + 5H_2O \quad (7.1)$$

The calcium sulfate and the water combine to form gypsum, which then reacts with the tricalcium aluminate. Sulfate-resistant cement will restrict this type of action.

Reactive Aggregates and High-Alkali Cement

Reactive aggregates and high-alkali cements cause swelling, map cracking (Figure 7-9), and popouts (Figure 7-10) in portland cement concrete. The literature points out four types of reactions that may result due to reactive aggregates:

1. Alkali–silica reaction.
2. Alkali–carbonate reaction.
3. Kansas–Nebraska sand–gravel reaction.
4. Silicification of carbonate aggregate.

The alkali–silica reaction was first reported by Thomas E. Stanton in 1940. Stanton noticed that certain forms of silica and siliceous material may produce deterioration of concrete structures by expansion and cracking of concrete through reaction with alkalies in cements in the form of sodium and potassium oxides. The expansion is caused by the formation of a gel-like material around the reactive aggregate. This material dries to a white amorphous substance around the aggregate (reactive rim). The reaction between the cement and aggregate may be

Fig. 7-9 Map Cracking

Fig. 7-10 Popouts Due to Reactive Aggregate

explained as follows. During the initial mixing of the concrete and for several hours thereafter, the water in the concrete acquires alkalies (sodium oxide and potassium oxide) from the cement by preferential solubility, a solution being formed. As the hydration process continues, the calcium aluminates and the silicates of the cement paste extract water. Thus, less water is left in solution, resulting in an increase in the alkalies in the remaining water. If this results and if the cement is high in alkali, a rather caustic solution is formed. During curing this caustic solution reacts chemically with susceptible aggregates to form a silica-gel reaction rim around the aggregate particle. This rim has a great affinity for water, thus attracting water from the cement paste and reducing its viscosity. Osmotic pressures are set up by the drawing of the water from the cement paste, resulting in fracture of the cement paste close to the reaction rim. As the silica gel continues to grow along the fracture, cracks develop and extend until the entire concrete surface is a series of cracks. The cracks themselves may be detrimental to the concrete structure, but an even more serious problem results if the cracks allow water to enter and then go through various cycles of freezing and thawing. The most reactive aggregates have been found to contain opal, chalcedony, certain forms of chert, glass in felsites, rhyolite, basalts, cristobalite, and tridymite.

Three factors are necessary before the alkaki–silica reaction can occur:

1. A reactive aggregate must be used.
2. Alkalies in the cement must be high.
3. The concrete must be partially or totally wet.

Obviously, to avoid the consequences of the alkali–silica reaction in concrete, three principles must be recognized:

1. Do not use reactive aggregates.
2. Do not use cement with a high alkali content.
3. If it becomes necessary to bring a reactive aggregate together with a high-alkali cement, use an admixture (such as a pozzolan) that will mitigate the deleterious consequences of the reaction.

To avoid the use of a reactive aggregate, it first becomes necessary to know which aggregates are reactive. If it is known that an aggregate has a satisfactory performance characteristic, continue to use it. If there is doubt or if the performance characteristics of a specific aggregate are not known, certain ASTM tests should be made on the aggregate. These tests entail a petrographic examination, a chemical test, and a proof test. ASTM C295 (Recommended Practice for Petrographic Examination of Aggregate for Concrete) provides specific information with reference to deleterious materials. Some reactive aggregates can be recognized readily, whereas others require more testing such as ASTM C289 [Test for Potential Reactivity of Aggregates (Chemical Method)] and/or ASTM C227 [Test for Potential Alkali Reactivity of Cement–Aggregate Combinations (Mortar-Bar Method)].

The second way of avoiding an alkali–silica reaction is to limit the alkali content in the cement to a value of 0.60 percent, as suggested by ASTM C150 (Specifications for Portland Cement). This value was arrived at empirically by correlation with field indications of distress.

The third method of avoiding an alkali–silica reaction is to use a pozzolanic or other admixture to mitigate the effects of alkali–aggregate reactions. This method of avoidance is covered by ASTM C441 (Test for the Effectiveness of Mineral Admixtures in Preventing Excess Expansion of Concrete Due to Alkali–Aggregate Reaction).

The alkali–carbonate reaction was first recognized by E. G. Swanson in 1957. Unlike the alkali–silica reaction, no visible gel, reaction product, or rim is apparent. In this reaction certain argillaceous (dolomitic limestones) aggregates react with the alkalies in the cement, which results in a destructive expansion. This expansion is referred to as *dedolomitization,* in which the dolomite is replaced by the formation of calcite and brucite. According to D. W. Hedley, the reaction is as follows:

$$CaMg(CO_3)_2 + 2MOH = Mg(OH)_2 + CaCO_3 + M_2CO_3 \tag{7.2}$$

in which M represents potassium, sodium, or lithuim. In concrete, the alkali carbonate produced by this reaction would react with hydration products of portland cement to regenerate alkali:

$$M_2CO_3 + Ca(OH)_2 \rightarrow 2MOH + CaCO_3 \tag{7.3}$$

This process continues until all the dolomite has reacted or all the alkali has been used up.

Another important factor in the alkali–carbonate reaction is the texture of the aggregate. If the dolomite has a coarse grain, the reaction will not take place to any significant degree. For the reaction to occur, the dolomite must consist of fine-grained particles.

If it is known that an aggregate has a satisfactory performance characteristic, continue to use it. If there is doubt or if the performance characteristics of a specific aggregate are not known, certain ASTM tests should be made on the aggregate. These tests include ASTM C289 [Test for Potential Reactivity of Aggregates (Chemical Method)], ASTM C227 [Test for Potential Reactivity of Cement–Aggregate Combinations (Mortar-Bar Method)], and the draft form of another possible ASTM test (Test for Length Change of Concrete Due to Alkali–Carbonate Rock Reaction).

Concrete made with the sand–gravel aggregate (natural aggregate) found in the Kansas–Nebraska area as well as surrounding areas has been known to result in expansion, cracking, and eventually early destruction of the concrete. This reaction is known as the Kansas–Nebraska sand–gravel reaction. The problem is not a serious problem nationally, but where it does occur it is a major one. The cause of the reaction has not yet been determined. Many suspect that it could be an alkali–silica reaction or an as-yet-undefined chemical or physical phenomenon.

The final type of aggregate reaction is the silicification of carbonate aggregate. It has been observed in Iowa by J. Lemish, who proved that certain aggregates develop siliceous rims around the aggregate when used in portland cement concrete. This type of reaction is still under investigation by researchers.

Moisture Absorption

Moisture absorption will cause concrete to swell. Concrete cylinders 13 ft in diameter have grown as much as 6 in. in a marine environment. If restrained, the restraining material will break apart or the concrete will crack and spall. In addition, saltwater will tend to leach out the lime in the cement, leaving a powdery residue.

Wear and Abrasion

Wear and abrasion cause the surface of concrete to disintegrate over extended periods. In general, roadway pavements may become a problem if the aggregate is susceptible to polishing. In waterway structures made of concrete, wind and water-driven particles, particularly sand, play a significant role in abrasion near the mud line and above the high-water line. Piles may be damaged due to the rubbing action of vessels, resulting in scaling, swelling, and cracking at joints, and scarring.

Shrinking and Flexure Forces

Shrinking and flexure forces may set up tensile stresses that exceed the capacity of the concrete section. The end result is cracking. Setting shrinkage generally causes shallow surface cracking (Figure 7-11). Drying shrinkage may take place over extended periods of time, producing tensile forces and eventually cracking. The time frame may amount to several years.

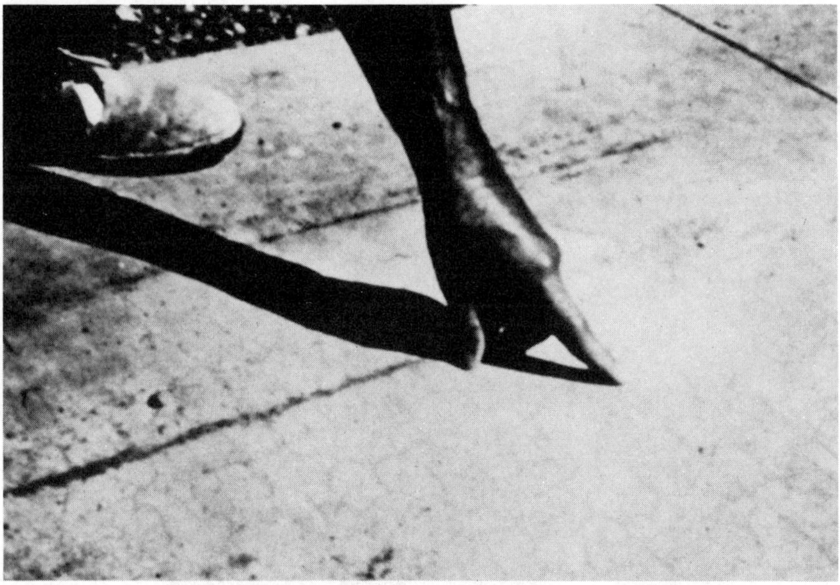

Fig. 7-11 Shallow Surface Cracking

Collision Damage

Collision damage is generally considered to be an accidental occurrence. Any concrete structure that can be reached by a moving vehicle has suffered collision damage at some time. In most instances the vehicle is more severely damaged than the structure.

Scouring

Concrete surfaces in surf zones scarred by sand and silt (Figure 7-12) are referred to as scour. Ice flows in rivers and bays are also responsible for considerable damage to concrete piling and piers (Figure 7-13). Most damage of this type occurs between the low-and high-water marks.

Shock Waves

A shock wave can damage concrete due to the varying transmission rates through the aggregate, the paste, and the reinforcing steel. The shock waves then become partially additive, setting up conditions leading to cracking and spalling of the concrete mass. Concrete piles are also vulnerable to damage from shock waves while they are being driven.

Fig. 7-12 Sand and Silt Scarring

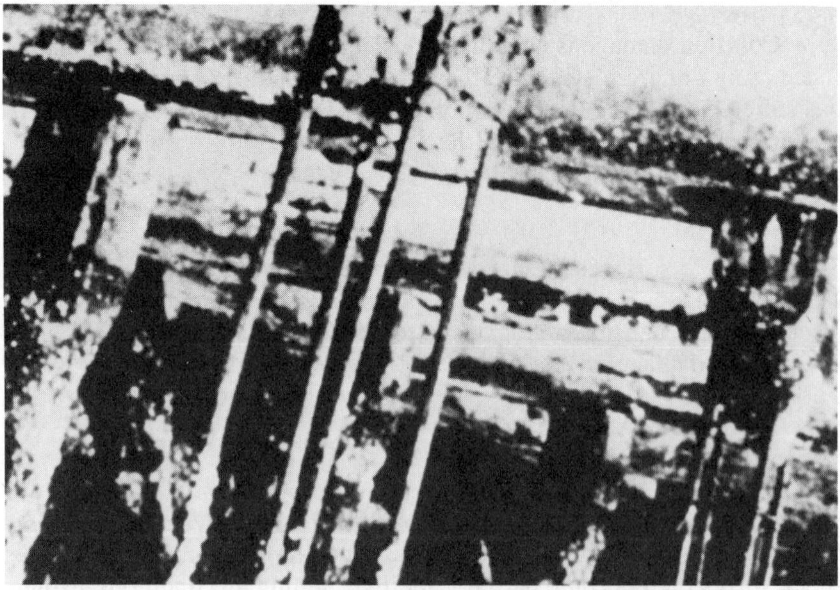

Fig. 7-13 Total Removal of Concrete Matrix from Reinforcing Rods Due to Ice Flow

Overstressing

Overstressed concrete in a deck may exhibit longitudinal and lateral cracking. Over a bearing point there may be diagonal cracking at the end of a simple beam (Figure 7-14), vertical cracking running from the bottom at the center of a simple beam to the neutral axis, and vertical cracking from the top of the beam extending downward to the neutral axis for a beam that is continuous over the bearing area (Figure 7-15).

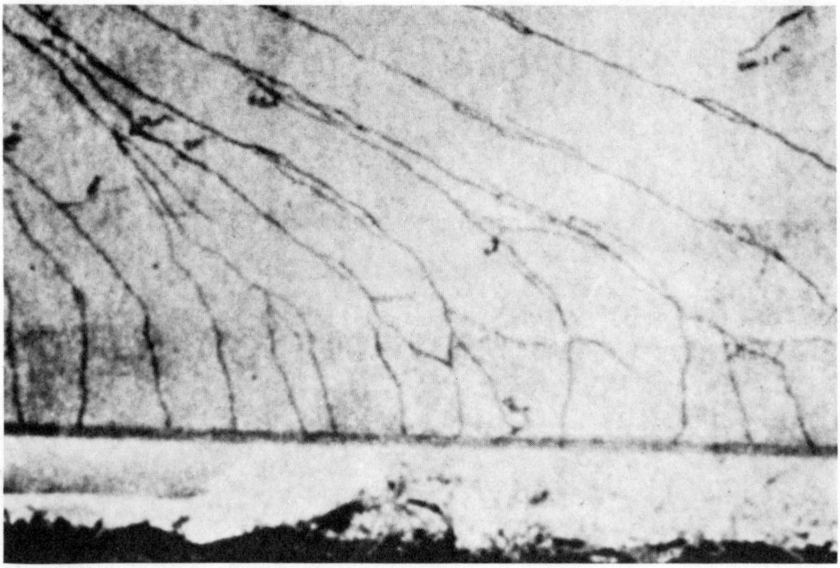

Fig. 7-14 Diagonal Cracking

Fire Damage

Fire damage results from the extreme temperatures of a large fire. Temperatures above 570°F (300°C) will cause weakening in the cement paste and lead to cracking and spalling (Figure 7-16).

Foundation Movement

Foundation movement will cause serious cracking in concrete structures if it generates a sizable tensile stress in the concrete piers or abutments (vertical cracking predominates).

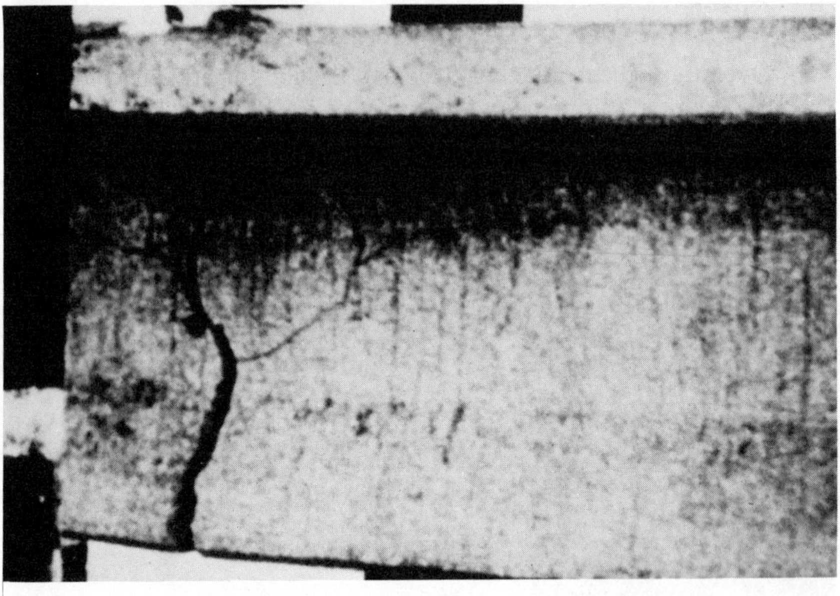

Fig. 7-15 Concrete Member in Models Laboratory

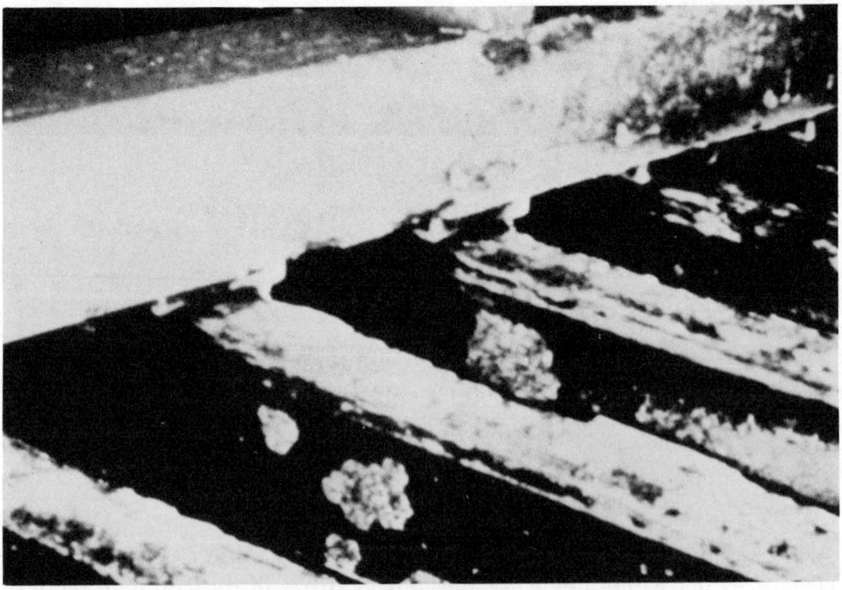

Fig. 7-16 Cracking and Spalling Due to Fire

Corrosion

Corrosion of the internal reinforcing steel causes tremendous expansion pressures within the concrete. Fully corroded steel occupies seven times the volume of uncorroded steel, causing cracking and spalling as a result of the volume increase (Figure 7-17). The steel may also corrode if water finds its way into the concrete member.

Corrosion of the prestressing wire combined with the high tensile stress required for prestressing may result in failure of a member due to stress corrosion. Failure of a significant number of prestressing wires in this manner will result in loss of tensile strength in the member and could lead to failure under heavy loading conditions. Further deterioration of the tensile strength of a member may result from creep of the prestressing steel; shrinkage of the concrete, causing a relaxation in the prestressing wires; and creep of the concrete, shortening the overall length of the member and thereby relaxing the prestressing wires. Other causes for a loss of prestress include elastic deformation of the concrete, drawing in of the anchorages, and friction loss in the post-tensioning operations. The combined loss of prestress due to such causes can amount to as much as 25 to 35 percent of the member's design strength. The result of loss of prestress is cracking, particularly near the anchorages and in the

Fig. 7-17 Piles Cracking

compression face. Unlike cracks in high-tension areas of reinforced-concrete members, the appearance of cracks in a prestressed member may have a serious effect on its structural integrity. A prestressed member is usually under high compressive stress; consequently, cracking should not be expected.

Aggregates that Cause Deterioration

The importance of identifying aggregates that cause deterioration of concrete cannot be overstated. ASTM C88 (Soundness of Aggregates by Use of Sodium Sulfate or Magnesium Sulfate) covers the testing of aggregates to determine their resistance to disintegration by saturated solutions of sodium sulfate or magnesium sulfate. It further furnishes information helpful in judging the soundness of aggregates subjected to weathering action, particularly when adequate information is not available from service records of the material exposed to actual weathering conditions.

There are basically three classifications of distress in concrete due to unsound aggregate:

1. Pitting and popouts.
2. D-line deterioration.
3. Map cracking.

Pitting is the disintegration of weak, friable particles of aggregate as a result of frost action. It is a gradual deterioration near the surface of the deleterious particles. *Popouts* are caused by the rapid disruption of harder but saturated aggregate particles. Chert is the most common cause of popouts. The usual result of pitting and popouts is the poor appearance of the structure. However, the structure generally remains sound because of the insufficient quantity of the particles causing pitting or popouts. A considerable amount of unsound aggregate would be required to cause structural deficiencies. However, if water were to enter into crevices provided by pitting or popouts, severe damage could result if various cycles of freezing and thawing were encountered. If the unsound aggregate is sufficient in quantity, strength and appearance will suffer. In pavement, the concrete slab becomes susceptible to structural failure through heavy truckloads and blowups. Blowups are caused by the expansion that occurs within a pavement immediately after precipitation on a hot day.

D-line cracking is a result of a change in volume in coarse-aggregate particles, resulting in the failure of the aggregate–matrix bond. Further deterioration occurs as water enters the cracks. D-line cracks appear on first sight as very fine cracks along transverse joints in concrete pavements. As they progress, they appear along longitudinal joints and free edges.

Map cracking is the disintegration of concrete in which the cracking is random and appears over an entire surface. It is caused mainly by the property of unsound aggregates to expand while the cement mortar is shrinking. Freezing and thawing, alkali–silica, alkali–carbonate, and excess heat may cause map cracking.

ADMIXTURES

Admixtures are generally added to the basic ingredients that make up concrete in order to modify various properties of the concrete to make it more suitable for use. Admixtures are used only when the desired properties cannot be achieved by altering the design mix. ASTM C494 (Chemical Admixtures for Concrete) covers materials for use as chemical admixtures to be added to portland cement concrete mixtures in the field for the purpose or purposes indicated for the five types as follows:

Type A. Water-reducing admixtures.
Type B. Retarding admixtures.
Type C. Accelerating admixtures.
Type D. Water-reducing and -retarding admixtures.
Type E. Water-reducing and -accelerating admixtures.

This specification stipulates tests for an admixture with suitable concreting materials or with cement, pozzolan, aggregates, and an air-entraining admixture proposed for specific work.

Water-reducing and -retarding Agents

There are five classes of water-reducing and -retarding agents:

1. Lignosulfonic acids and their salts.
2. Modifications and derivations of lignosulfonic acids and their salts.
3. Hydroxylated carboxylic acids and their salts.
4. Modifications and derivations of hydroxylated carboxylic acids and their salts.
5. Carbohydrates and modifications and derivatives thereof.

Water-reducing admixtures reduce the quantity of mixing water required to produce concrete of a given consistency. The most common water-reducing admixtures are of organic material. The water-reducing admixtures yield good strength advantages for a given cement content by reducing the water content of the concrete mix. The compressive strength increases 10 to 20 percent while the water content is reduced 5 to 15 percent. Water-reducing admixtures are effective with all types of portland cement, portland blast-furnace-slag cement, and portland–pozzolan cements.

Retarding admixtures increases the setting time for concrete, allowing a permissible period between batching and the final placement.

Accelerating Admixtures

An accelerating admixture accelerates the setting and early-strength development of concrete. Accomplishing this results in better scheduling of the workload, earlier removal of construction forms, and compensation effects of low temperature on early-strength development. The most common accelerator is calcium chloride ($CaCl_2$).

In addition to increasing strength, calcium chloride has the following secondary effects:

1. Increases workability by a small amount.
2. Increases air content and air-bubble size in air-entrained concrete.
3. Reduces bleeding.
4. Increases drying shrinkage slightly.
5. Lowers resistance to sulfate attack.

Water-reducing and -retarding Admixtures

Water-reducing and -retarding admixtures reduce the quantity of mixing water required to produce concrete of a given consistency and retard the setting of concrete.

Water-reducing and -accelerating Admixtures

Water-reducing and -accelerating admixtures reduce the quantity of mixing water required to produce concrete of a given consistency and accelerate the setting and early-strength development of concrete.

Air-entraining Agents and Admixtures

Admixtures may also be used to entrain air in the concrete. Air-entrained concrete is more plastic and workable than non-air-entrained concrete. It can be placed and compacted with less bleeding and less aggregate segregation. It improves both workability and durability, but reduces strength.

Air-entraining agents can also be interground with cement clinker at the mill to form air-entraining cements. The only advantage of air-entraining cements is the convenience of knowing that you do not have to include an admixture.

Air-entraining agents and admixtures are covered by ASTM standards. ASTM C233 (Air-Entraining Agents for Concrete) covers the testing of materials proposed for use as air-entraining admixtures to be added to concrete mixtures in the field. These tests are based on arbitrary stipulations permitting highly standardized testing in the laboratory and are not intended to simulate actual job conditions. ASTM C260 (Air-Entraining Admixtures for Concrete) covers materials proposed for use as air-entraining admixtures to be added to concrete mixtures in the field.

Other Admixtures

Of the large number of admixtures available for concrete, an important group falls into the category of finely divided mineral admixtures, which are divided into three classifications:

1. Those which are chemically inert.
2. Those which are pozzolanic.
3. Those which are cementitious.

The inert finely divided material includes quartz, limestone, bentonite, and hydrated lime. These finely divided materials are often substituted for portland cement in concrete to save on portland cement. To have the necessary workability and plasticity, most concrete must have large amounts of cement. Substituting finely divided material for portland cement saves a considerable amount of portland cement.

Pozzolanic or cementitious finely divided material contributes to the strength development of concrete and reduces the amount of portland cement required for a given strength. A pozzolan, as described by ASTM C618, is a siliceous or siliceous and aluminous material which itself has little or no cementitious value but which will, in finely divided form and in the presence of moisture, chemically react with calcium hydroxide at ordinary temperatures to form compounds that do have cementitious properties. Pozzolans include diatomaceous earth, opaline cherts, shales, volcanic ashes, fly ash, and clays. The most common pozzolan is fly ash, a finely divided residue that results from the combustion of ground or powdered coal. ASTM C618 (Fly Ash and Raw or Calcined Natural Pozzolan for Use as a Mineral Admixture in Portland Cement Concrete) covers fly ash and raw or calcined natural pozzolan for use as a mineral admixture in concrete where cementitious or pozzolanic action or both is desired; where other properties normally attributed to finely divided mineral admixtures may be desired; or where both objectives are to be achieved. Four classifications exist under ASTM C618: Classes N, F, C, and S. Class N comprises raw or calcined natural pozzolans. They include diatomaceous earths; opaline cherts and shales; tuffs and volcanic ashes or pumices, any of which may or may not be processed by calcination; and various materials requiring calcination to induce satisfactory properties, such as some clays and shales.

Class F is fly ash normally produced from burning anthracite or bituminous coal which meets the applicable requirements for this class. This class of fly ash has pozzolanic properties.

Class C is fly ash normally produced from lignite or subbituminous coal that meets the applicable requirements for this class. This class of fly ash, in addition to having pozzolanic properties, has some cementitious properties. Some Class C fly ashes may have a lime content exceeding 10 percent.

Class S is any pozzolan that meets the applicable requirements for this class, examples of which are certain processed pumices and certain calcined and ground shales, clays, and diatomites.

PROBLEMS

7.1. Explain the failure mechanism of concrete according to Derucher.

7.2. Why does concrete shrink?

7.3. List the factors that cause deterioration of concrete.

7.4. What factors influence the deteriorating effect of freezing and thawing?

7.5. Explain the alkali–silica reaction.

7.6. Explain the alkali–carbonate reaction.

7.7. List and discuss three classifications of distress in concrete that result from the use of unsound aggregate.

7.8. List the five types of chemical admixtures outlined by the ASTM.

7.9. Calcium chloride is an accelerator in concrete. What is the primary advantage of using it? The secondary advantages?

7.10. List and discuss pozzolanic admixtures.

REFERENCES

BRANDTZAEG, A., "Study of the Failure of Concrete under Combined Compressive Stresses," *UIEES Bull. 185*, 1928.

BURG, O., "The Factors Controlling the Strength of Concrete," *Constr. Rev., 33,* No. 11 (1950), p. 19.

DERUCHER, K. N., "Applications of the Scanning Electron Microscope to Fracture Studies of Concrete," *Build. Environ., 13,* No. 2 (1978), pp. 135–141.

DERUCHER, K. N., and HEINS, C. P., *Bridge and Pier Protective Systems and Devices* Civil Engineering Series, Marcel Dekker Inc., New York, 1979.

HANSEN, T. C., "Microcracking of Concrete," *J. Am. Concr. Inst., 64,* No. 2 (1968), pp. 9–12.

HERMITE, J., "Present Day Ideas on Concrete Technology," *University of Illinois Bull. 18,* 1954, pp. 27–39.

HSU, T., SLATE, F., STURMAN, G., and WINTER, G., "Microcracking of Plain Concrete and the Shape of the Stress Strain Curve," *J. Am. Concr. Inst., 60,* No. 2 (1963), pp. 209–224.

JONES, R., "A Method of Studying the Formation of Cracks in a Material Subjected to Stress," *Br. J. Appl. Phys., 3* (1952), p. 329.

ROBINSON, J., "X-ray Analysis of Concrete Fracture," *J. Am. Concr. Inst., 27,* No. 4 (1955).

RUSCH, H., "Physical Problems in the Testing of Concrete," *Cement-Chalk, 12,* No. 1 (1959), pp. 1–9.

STANTON, T. E., "California Experience with the Expansion of Concrete through Reaction Between Cement and Aggregate," *J. Am. Concr. Inst., 39* (Jan. 1942).

STURMAN, G., "Shrinkage Microcracks in Concrete," Ph.D. dissertation, Cornell University, 1969.

SWENSON, E. F., "A Reactive Aggregate Undetected by ASTM Tests," *ASTM Bull. 226,* Dec. 1957, pp. 48–50.

TIMBER 8

Timber has been one of the basic materials of construction since the earliest days of humankind. Today it has been largely superseded by concrete and steel. However, the use of timber remains quite extensive.

In the United States 600 different tree species exist, of which 15 are utilized in the building trades (ash, birch, cedar, cypress, fir, gum, hemlock, hickory, maple, oak, pine, poplar, redwood, spruce, and walnut). A number of these include several varieties, each showing different characteristics. One variety of walnut, for example, might make a better building product than another.

Therefore, the engineer should have some knowledge of the classification of trees, as well as of their growth and structural ability, to understand the physical and mechanical properties of each.

CLASSES OF TREES

Exogens and Endogens

Botanists classify trees into two main categories: exogens and endogens. These terms refer to the pattern of growth of a particular species. *Exogenous trees* are those which grow diametrically, by adding new cells in a layer between the existing wood and the bark. Almost all wood that is of commercial use falls in this category. The exogenous growth pattern forms a characteristic transverse section with concentric annual rings which correspond to each year's growth. At the center of a cross section of a tree is pith, a dark area of small diameter that does not increase in size after the first year of growth. Encircling the outermost annual ring is a layer of microscopic thickness called the cambium ring, and it is here that new growth is formed. The outmost layer is known as the bark.

Endogenous trees add new living fiber to the old by allowing new fiber to intermingle with the old, thus producing growth both diametrically and longitudinally. Most endogens are fairly small plants, such as corn, sugar cane, palm, and bamboo. The latter two plants are used extensively throughout the Orient as a structural material.

Hardwoods and Softwoods

Exogenous trees are further classified as hardwoods or softwoods. These are somewhat misleading terms, since they refer not to a quality of the wood, but to the types of tree the wood comes from. Hardwoods comprise the broadleafed trees, mostly deciduous, and include the many varieties of oak, maple, ash, hickory, cherry, poplar, gum, walnut, elm, beech, birch, basswood, whitewood or tulip, and ironwood. These trees are in general slow-growing, forming heavy and hard wood that is used more extensively for furniture, interior finishing, and cabinetwork than for structural purposes. The softwoods are the conifers, such as pine, spruce, fir, hemlock, cedar, cypress, and redwood.

WOOD STRUCTURE

Many of the qualities that are unique to wood for use as a building material arise from the structure of wood. Wood is formed mostly of filamentous cells, fairly tubular in shape, varying in length from 1 to 3 mm and having a diameter that is about $^1/_{100}$ of their length. The shape of the cell in transverse section is approximately rectangular with rounded corners. Perforated pits in the cells make possible the flow of sap between cells.

Wood cells are composed primarily of the polymers cellulose and hemicellulose. These are similar to the sugar molecule, but form long chains that run in bundles helically. They are cemented together by a third polymer, lignin. Wood is typically 50 to 60 percent cellulose material and 20 to 35 percent lignin, with small amounts of other carboxydrates, such as resin and gum, present. It is the cellulose that gives wood its axial strength and elastic properties, and the lignin that is responsible for its compressive strength. These carboxydrates, although indigestible by man, are a favorite food for some other forms of life, such as termites, marine borers, and fungi, and measures must often be taken to protect wooden structures from these species.

Most fibers run parallel to the axis of the tree, and these cells, which are more or less empty in wood that is ready for use, act as a collection of empty tubes under stress. Since the tree develops from a sapling by adding new longitudinal cells in layers around the existing wood, differences in periods of growth give rings of cells with different characteristics. Growth is rapid in the spring and slows down in the summer, stopping altogether in the fall and winter. During the spring growth period, cells that are formed have relatively thin walls and an open texture, producing what is known as spring wood. The cross-sectional area of spring wood may form as little as 10 percent of the cell wall. The summer growth is slower and the wood formed is denser and stronger, with thicker cell walls. This summer wood may have cell walls constituting up to 90 percent of the cross-sectional area. The pattern of a layer of spring wood followed by increasingly dense wood that is formed over each year gives the wood its annual rings. These rings vary in size with the type of tree involved and the growth conditions during a particular year. A typical annual thickness is approximately $^1/_{10}$ in., with ½ in. being about the maximum. In tropical trees, the growth rings correspond to the rainy and dry seasons.

Over a period of several years the wood ceases to function as part of the living tree and has only structural value for the tree. The living part is called *sapwood*; the older wood is *heartwood*. Heartwood is easily distinguished from sapwood because it is darker in color. It is more highly valued since it is more resistant to decay than sapwood, although the strength of each is about the same.

Also parallel to the axis of the tree and running the length of the tree are vessels formed by the fusion of chains of cells. These vessels are characteristic of hardwoods, and are sometimes visible in cross section without a microscope. Other groups of cells lie in radial arrangements, and these form what are known as radial, pith, or medullary rays. Medullary rays are present in all trees, but are especially prominent in woods such as oak. In the living tree they carry food from the inner bark to the cambrium layer and add strength in a radial direction.

The terminology of the tree is best illustrated by a cross section such as the one shown in Figure 8-1.

PHYSICAL CHARACTERISTICS OF WOOD

Season of Cutting; Slash and Rift Cutting

The first step in the manufacturing process of lumber is to select and fell the trees. This is normally done in the fall and winter when the flow of sap is minimal and destructive fungi and insects are least active. The logs are cut into standard-size boards at sawmills, as illustrated in Figure 8-2. The lumber must then be dried or seasoned.

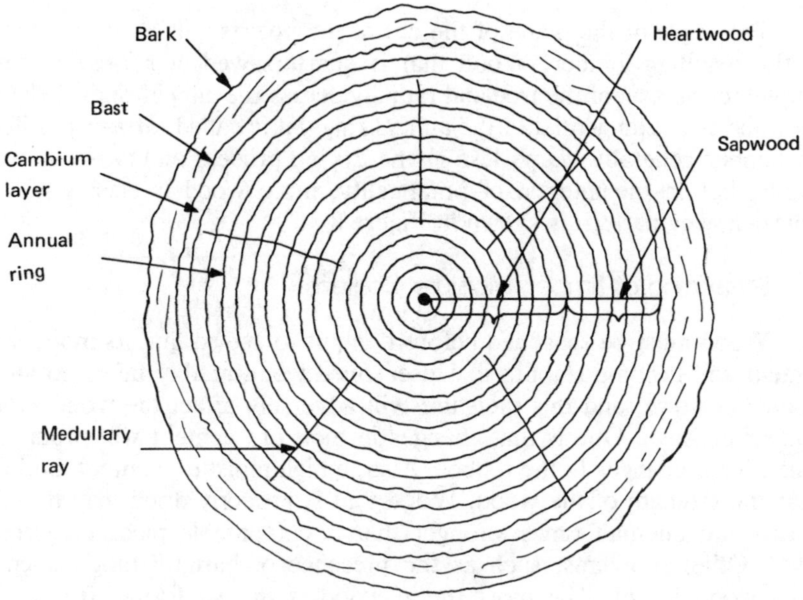

Fig. 8-1 Tree Terminology

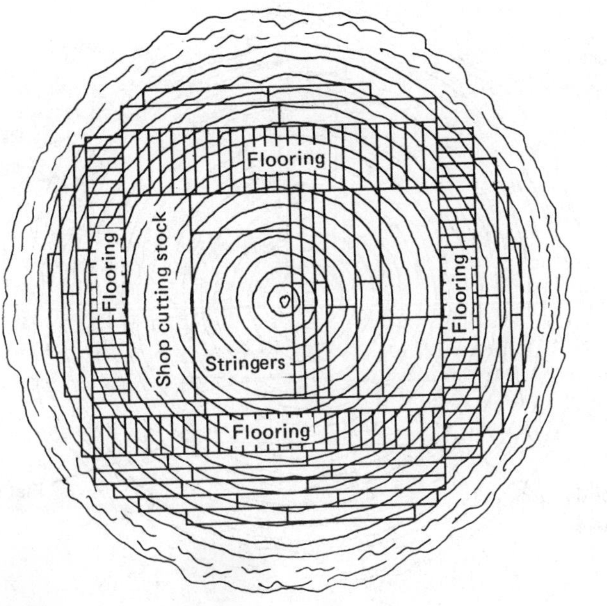

Fig. 8-2. Log Cutting into Standard Size

The angle of the plane of the cut of the boards will have an effect on the resulting lumber. Wood that is quartersawed or *rift-cut* is cut parallel to the axis of the tree and radially across the annual rings. When the wood is cut tangential to the annual rings, it is called *slash-cut*. Rift-cut lumber generally warps less in the drying process and wears more evenly, but in the interest of practicality, more wood is slash-cut. A typical sawing pattern is shown in Figure 8-3.

Seasoning of Timber, Moisture Content

Wood must be seasoned before it is put to use so that its moisture content will become stabilized. Unseasoned wood readily takes up and retains moisture, and this moisture will adversely affect the wood in a number of ways. One is that changes in moisture content will produce dimensional changes in the timber. Also, a high moisture content diminishes the strength of the wood. Wood that is properly dried may be up to two and one-half times stronger than a comparable piece of green wood. Other problems, such as the presence of harmful fungi, occur when wood is cut. The moisture in wood is in two forms: free and combined. *Free moisture* is moisture outside the cell, and *combined moisture* is water that has been incorporated into the cell walls. In green

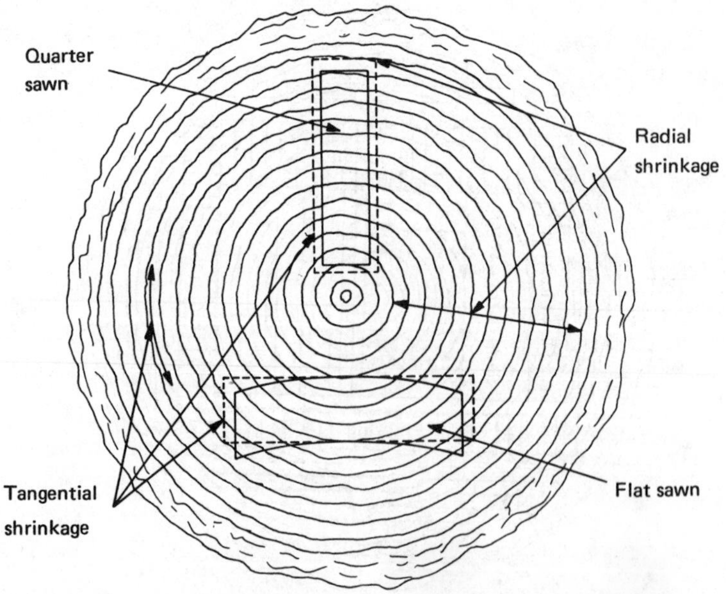

Fig. 8-3 Sawing Pattern

wood, moisture will constitute from 25 to 200 percent of the oven-dry weight of the lumber, and most of this must be evaporated from the wood in the seasoning process.

The timber can be either air-dried or kiln-dried. As the wood dries, the free moisture is evaporated from the surface and the moisture from the interior of the porous lumber equalizes with the drier surface. When the moisture content stabilizes, the drying process is complete.

$$\text{moisture content (\%)} = \frac{\text{original amount of moist wood—oven-dry weight}}{\text{oven-dry weight}}$$

Air drying will reduce the moisture content to about 15 percent, a level that is acceptable for most construction purposes. Lumber that is to be used for purposes such as furniture, flooring, and cabinetwork is kiln-dried in a few days using temperatures of 155 to 180°F (68 to 82°C). A 1-in. pine board, for example, can be dried in 4 days. Hardwoods warp more easily, so the drying process is more gradual. They are normally air-dried for several months before they are kiln-dried. Temperatures used need to be lower, 100 to 120°F (38 to 49°C) for those woods that warp most easily. At the hotter kiln temperatures, a 1-in. hardwood board may take 6 to 10 days to dry.

All wood has the ability to reabsorb moisture from the atmosphere, depending on the humidity of its surroundings. This can cause a certain amount of dimensional instability, which may lead to problems in certain applications. Wood that has been subjected to temperatures of 212°F (100°C) for a length of time during boiling, steaming, or soaking will, with time, become less able to absorb humidity from the atmosphere.

Shrinkage, Warping, and Checking

Shrinkage will occur in the drying of wood when the combined moisture begins to be evaporated from the cell walls. The contraction will be across the cell walls rather than longitudinal, cells shrinking proportional to the thickness of their cell walls. The heavy-celled summerwood thus shrinks far more than the thin-walled cells of springwood. In a tangential direction, the bands of heavy summerwood force the lighter springwood to contract with them, and for this reason there is a variation in the amount of shrinkage that a piece of wood will exhibit in different directions. The usual limits in shrinkage between a green condition and an over-dry condition are:

Volumetric shrinkage 7–21%

Longitudinal shrinkage	0.1–0.3%
Radial shrinkage	2–8%
Tangential shrinkage	4–14%

When shrinkage occurs unequally for any reason, warping results. If drying proceeds at unequal rates over different portions of a piece of lumber, warping will occur. When wood is cut tangentially, the dried piece will be convex because of unequal tangential shrinkage. Irregularities in the grain of the wood will produce more pronounced irregularities in the dried boards.

Checks are longitudinal cracks across the growth rings that can occur during drying of timber. Unequal shrinkage produces strains within the wood, and checks can result. They can be temporary, occurring when the outer portion of a piece of wood dries too rapidly and contracts over the inner portion. These checks that appear in the outer layer of the wood may close up and become imperceptible as the wood of the inner part of the lumber dries and contracts. Other checks of a more serious nature can occur when the difference in tangential shrinkage is too great relative to the radial shrinkage to be accommodated. Large radial checks such as those sometimes seen in posts can then occur.

Defects in Lumber

Anything that adversely affects the strength, durability, or utility of a piece of wood is a defect, and these may be within the wood itself or be produced by warping.

Shakes are longitudinal cracks in the wood that follow the growth rings and develop prior to the lumber's being cut. They are sometimes a result of heavy winds.

Checks, as described above, are longitudinal splits across the growth rings resulting from uneven drying.

Knots are formed at the base of branches where they extend into the wood of the tree. Only wood that comes from the base of the tree where there are no branches is free of knots. If the branch was dead, a *loose knot* is formed. A *spike knot* occurs when the cut is longitudinal to the branch. The effect of knots depends on the use of the wood and the location and size of the knots. They have little effect on wood in shear or in compression members. For beams in bending, however, a knot in the part of the beam subject to tension can significantly reduce its maximum load.

Pitch pockets are accumulations of resins in openings between the annual rings.

Bark pockets are formed when bark is wholly or partially encased in wood.

Waynes are areas where the lumber has been cut too close to the edge of the log and there is bark on the boards.

Compression wood is formed on the lower side of branches or leaning tree trucks. It is darker than normal wood, has a high lignin content, higher specific gravity, greater longitudinal shrinkage, and is not as tough as normal wood. The strength of compression wood is not predictable, and many failures in wood members have been found to be cause by compression wood. Tension wood, like compression wood, has higher specific gravity and greater longitudinal shrinkage. It is sometimes stronger and sometimes weaker than normal wood.

The effects of fungus or insect attack can be considered defects and are addressed later in the chapter.

Unequal shrinkage during the drying process produces warped lumber. The different types of warping are bow, crook, twist, and cup, as illustrated in Figure 8-4.

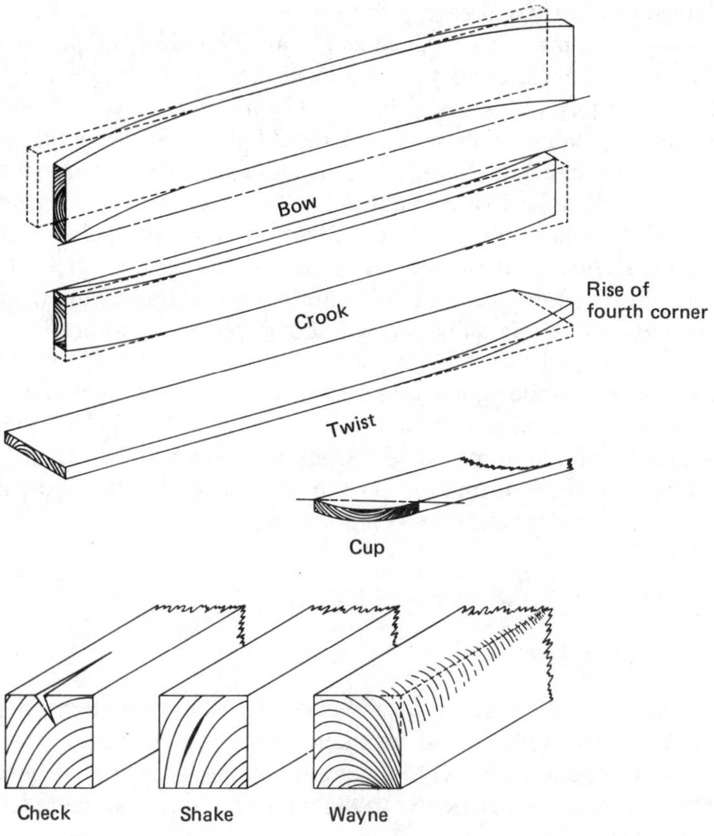

Fig. 8-4 Unequal Shrinkage and Various Defects

Grading of Lumber

The specific gravity of the material in wood fiber is about 1.5 for all woods, but because of the many voids in wood cells, the density of wood itself is usually much less. Woods range from balsa, with a density of 0.12, to lignum vitae, with a density of 1.3. The strength of wood is proportional to its density, all other things being equal. Chief factors influencing density are the number of annual rings and proportion of summer wood to spring wood, and for hardwood, the size and number of pores. As hardwood trees increase in age, the size of the pores increases, and this must be considered in some species.

Wood that is commercially available is first graded by criteria established by the lumber associations in order to provide a reasonably uniform product. Grading standards are given for both hardwoods and softwoods. For softwoods, the density and presence or absence of defects determine the classification; hardwood gradings are based on the percentage of clear wood each piece shows.

Softwood lumber is classified as (1) stress-grade structural lumber, (2) yard lumber, or (3) factory/shop lumber. *Stress-grade lumber* is used for planks and joists (2 × 4's or larger), beams and stringers (5 × 8's or larger), and posts and timbers (5 × 5's and larger). Stress-grade lumber is graded by its strength. Yard lumber is classified as *select* or *common.* Select grades A and B are suitable for finishes such as varnish and shellac, while select grades C and D are suitable for paint finishing. *Common yard lumber* is for general use and is graded from No. 1 to No. 5 by the proportion that can be utilized in building without waste. *Factory/shop lumber* is suitable for building purposes where it will be cut up.

Hardwood lumber grades, as specified by the National Hardwood Association, are as follows: First, Seconds, Selects, No. 1 Common, No. 2 Common, Sound Wormy, No. 3A Common, and No. 3B Common.

More detailed information on the grading of lumber is available from lumber manufacturers' associations.

MECHANICAL PROPERTIES OF TIMBER

Strength of Wood

Because of the structure of wood, the fibers act like so many cylindrical tubes firmly bound together in the way that they withstand stress. Loads applied parallel to the grain are carried by the strongest of the fibers; in loads perpendicular to the grain, the weakest fibers have to take their share of the load.

Tensile Strength

Tensile strength in wood parallel to the grain is much higher (three times as much) than compressive strength. Thus, the limiting factor in tension members is usually compression or shear at the points of concentration. Thus far, no means have been developed to connect a tension member without it somehow compressing and eventually resulting in failure.

Tensile loads perpendicular to the grain of wood fibers cause the fibers to split apart. The tensile loads that can be carried are only one-tenth or less of the tensile load that can be withstood by the wood in a parallel direction with the grain. An example of the differences is shown in Figure 8-5, which shows Douglas fir and its appropriate values for various loading conditions.

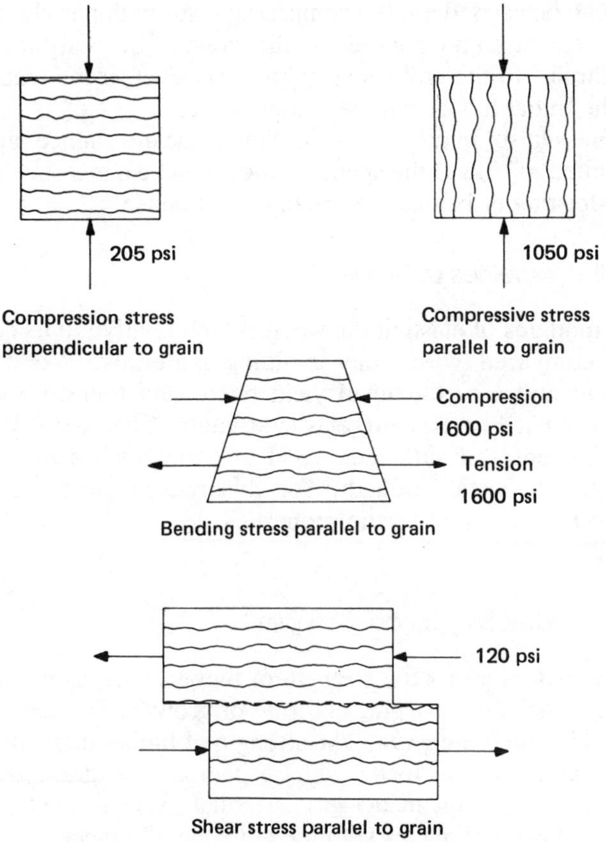

Fig. 8-5 Allowable Stresses for Douglas Fir

Compression Strength

Compression loads parallel to the grain can be carried by the strongest fibers, whereas compression loads perpendicular to the grain are carried by both weak and strong fibers. Wood in compression parallel to the grain can carry three to four times the load that wood in compression perpendicular to the grain can carry.

Compression failure of wood perpendicular to the grain involves the complete crushing of the wood fiber. Compression failure of wood parallel to the grain involves the bending or buckling of the wood fibers.

Flexural Strength and Stiffness

For timber beams in flexure, the critical factors in evaluating the load that can be carried are the compressive strength parallel to the grain and the shear strength parallel to the grain. The resistance of shear strength parallel to the grain is very low. However, if the wood is free of defects, the initial failure will be compressive.

Stiffness also plays a role in timber beams, since deflection is usually limited to $^1/_{360}$ of the span. Stiffness can be a measure of strength, in that stiffer timber beams are usually more dense.

Elastic Properties of Timber

The modulus of elasticity in wood is high relative to its compressive strength, compared with other building materials. Wood does not, however, exhibit a well-defined yield point, and therefore the proportional limit is used to measure elastic strength. Therefore, for a specific load requirement and sufficient wood to carry such a load, there is a high degree of elastic strength. For this reason, wood members are considered to have good elastic properties.

Factors Affecting Timber Strength

Many factors affect the strength of timber, such as direction of the wood fibers, moisture, weight, and rate of growth. The direction of the wood fibers is called the *grain*. The strength of timber depends so heavily on the direction of the loads with respect to the grain that it is an important consideration in design. Diagonal grain is measured by the tangent of the angle it makes with the cut edge of a piece of lumber. The

TABLE 8-1 Strength Ratios Corresponding to Various Slopes of Grain[a]

Slope of Grain	Maximum Strength Ratio (%): For Stress in Extreme Fiber in Bending (Beams and Stringers of Joists and Planks)	For Stress in Compression Parallel to Grain (Posts and Timbers)
1 in 6	—	56
1 in 8	53	66
1 in 10	61	74
1 in 12	69	82
1 in 14	74	87
1 in 15	76	100
1 in 16	80	—
1 in 18	85	—
1 in 20	100	—

[a]ASTM D245-49T.

diagonal grain or direction of the wood fibers will affect the strength ratios of lumber as shown in Table 8-1.

As indicated earlier, wood that is properly dried is far stronger than wood that has a high moisture content. When all the free water is evaporated and all the moisture is removed from the cell walls, the fiber saturation point is said to have been reached. The presence or absence of free moisture does not, in general, affect the strength properties. However, if drying were to occur beyond the fiber saturation point, there will be an increase in strength. The increase in strength beyond the fiber saturation point may be two or three times the value of wet wood.

All other things being equal, the denser the wood, the stronger it will be. This is so true that a formula is based on it:

$$\text{modulus of rupture} = 26{,}200 \times (\text{specific gravity})^{1.25} \qquad \text{(air-dry timber)}$$

$$\text{modulus of rupture} = 18{,}500 \times (\text{specific gravity})^{1.25} \qquad \text{(green timber)}$$

These formulas were developed by the Forest Products Laboratory and are very accurate, regardless of species.

Finally, rate of growth greatly affects the strength of timber. In general, most timber used for construction will exhibit an optimal number of annual rings per inch of wood (Table 8-2). Usually, the greater number of rings per inch will be the stronger one.

TABLE 8-2 Growth Rings for Various Species

Species	*Rings per Inch*
Douglas fir	24
Shortleaf pine	12
Loblolly pine	6
Western hemlock	18
Redwood	30

Tabulation of the Mechanical Properties for Wood

Tables 8-3 and 8-4 show the mechanical properties for wood. The tables are based upon clear wood (wood that is free of defects) and a 100 percent strength ratio. The tables also have a factor of safety built into them for design purposes. Further, various grades of lumber have a strength ratio stamped on them, and by multiplying this number times that in the tables, basic stresses for that particular grade of lumber can be obtained.

DECAY, DURABILITY, AND PRESERVATION OF TIMBER

Mechanism of Decay

Various factors that cause damage to a wooden structure, the durability of the wooden structure, and its preservation will be discussed in detail. Molds, stains, and decay in timber are caused by fungi, microscopic plants that require organic material to live on. Reproduction occurs through thousands of small windblown particles called *spores*. Spores send out small arms that destroy timber through the action of enzymes. Fungi require a temperature of between 50 and 90°F (10 and 32°C), food, moisture, and air. Timber that is soaked or dried normally will not decay. Molds cause little direct staining because the color caused is largely superficial. The cottony or powdery surface growths range in color from white to black and are easily brushed or planed off the wood. Stains, on the other hand, cannot be removed by surface techniques. They appear as specks, spots, streaks, or patches of varying shades and colors, depending upon the organism that is infecting the timber, Although stains and molds should not be considered as stages of decay since the fungi do not attack the wood substance to any great degree, timber infected with mold and stain fungi has a greater capacity to absorb water and is therefore more susceptible to decay fungi. Decay fungi may attack any part of the timber, causing a fluffy surface condition indicative of decay or rot. Early stages of decay may show signs of discoloration or mushroom-type growths (Figure 8-6).

Fig. 8-6 Early Stages of Decay at Base of Timber Member

In some types the color differs only slightly from the normal color, giving the appearance of being water-soaked. Later stages of decay are easily recognized because of the definite change in color and physical properties of the timber (Figure 8-7). The surface becomes spongy, stringy, or crumbly, weak, and highly absorbent. It is further characterized by a lack of resonance when struck with a hammer. Brown, crumbly rot in a dry condition is known as *dry rot*. Serious decay problems are indicative of faulty design or construction or a lack of reasonable care in the handling of the timber. Principles that assure long service life and avoid decay hazards in construction include building with dry timber,

TABLE 8-3 Mechanical Properties of a Few Important Woods Grown in the United States[a]

Common and Botanical Name	Weight (lb/ft^3)		Shrinkage from Green to Oven-dried (% of green volume)	Modulus of Rupture (psi)		Modulus of Elasticity (1000 psi)	
	Green	Air-dried[b]		Green	Air-dried	Green	Air-dried
Hardwoods							
Ash, white (*Fraxinius* sp.)	48	41	12.8	9,500	14,600	1,410	1,680
Elm, American (*Ulmus americana*)	54	35	14.6	7,200	11,800	1,110	1,340
Hickory, true (*Carya* sp.)	63	51	17.9	11,300	19,700	1,570	2,190
Maple, red (*Acer rubrum*)	50	38	13.1	7,700	13,400	1,390	1,640
Maple, sugar (*Acer saccharum*)	56	44	14.9	9,400	15,800	1,550	1,830
Oak, red (*Quercus* sp.)	64	44	14.8	8,500	14,400	1,360	1,810
Oak, white (*Quercus* sp.)	63	47	16.0	8,100	13,900	1,200	1,620
Walnut, black (*Juglans nigra*)	58	38	11.3	9,500	14,600	1,420	1,680
Conifers							
Cedar, western red (*Thuja plicata*)	27	23	7.7	5,100	7,700	920	1,120
Cypress, bald (*Taxodium distichum*)	51	32	10.5	6,600	10,600	1,180	1,440
Douglas fir, coast type (*Pseudotsagu taxifolia*)	38	34	11.8	7,600	12,700	1,570	1,950
Fir, white (*Abies* sp.)	46	27	9.8	5,900	9,800	1,150	1,490
Hemlock, eastern (*Tsuga canadensis*)	50	28	9.7	6,400	8,900	1,070	1,200
Pine, longleaf (*Pinus palustris*)	55	41	12.2	8,700	14,700	1,600	1,990
Pine, shortleaf (*Pinus echinata*)	52	36	12.3	7,300	12,800	1,390	1,760
Pine, western white (*Pinus monticola*)	35	27	11.8	5,200	9,500	1,170	1,510
Redwood (old growth) (*Sequoia sempervirens*)	50	28	6.8	7,500	10,000	1,180	1,340
Spruce, Sitka (*Picea sitchensis*)	33	28	11.5	5,700	10,200	1,230	1,570
Spruce, eastern (*Picea* sp.)	34	28	12.6	5,600	10,100	1,120	1,450
Tamarack (*Larix laricina*)	47	37	13.6	7,200	11,600	1,240	1,640

TABLE 8-3 (Continued)

Common and Botanical Name	Maximum Crushing Strength Parallel to Grain (psi)		Compression Perpendicular to Grain, Proportional Limit (psi)		Maximum Shearing Strength, Parallel to Grain (psi)	
	Green	Air-dried	Green	Air-dried	Green	Air-dried
Hardwoods						
Ash, white (*Fraxinius* sp.)	4,060	7,280	860	1,510	1,350	1,920
Elm, American (*Ulmus americana*)	2,910	5,520	440	850	1,000	1,510
Hickory, true (*Carya* sp.)	4,570	8,970	1,080	2,310	1,360	2,130
Maple, red (*Acer rubrum*)	3,280	6,540	500	1,240	1,150	1,850
Maple, sugar (*Acer saccharum*)	4,020	7,830	800	1,810	1,460	2,330
Oak, red (*Quercus* sp.)	3,520	6,920	800	1,260	1,220	1,830
Oak, white (*Quercus* sp.)	3,520	7,040	850	1,410	1,270	1,890
Walnut, black (*Juglans nigra*)	4,300	7,580	600	1,250	1,220	1,370
Conifers						
Cedar, western red (*Thuja plicata*)	2,750	5,020	340	610	710	860
Cypress, bald (*Taxodium distichum*)	3,580	6,360	500	900	810	1,000
Douglas fir, coast type (*Pseudotsagu taxifolia*)	3,860	7,430	440	870	930	1,160
Fir, white (*Abies* sp.)	2,830	5,480	360	620	770	990
Hemlock, eastern (*Tsuga canadensis*)	3,080	5,410	440	800	850	1,060
Pine, longleaf (*Pinus palustris*)	4,300	8,440	590	1,190	1,040	1,500
Pine, shortleaf (*Pinus echinata*)	3,430	7,070	440	1,000	850	1,310
Pine, western white (*Pinus monticola*)	2,650	5,620	290	540	640	850
Redwood (old growth) (*Sequoia sempervirens*)	4,200	6,150	520	860	800	940
Spruce, Sitka (*Picea sitchensis*)	2,670	5,610	340	710	760	1,150
Spruce, eastern (*Picea* sp.)	2,600	5,620	290	590	710	1,070
Tamarack (*Larix laricina*)	3,480	7,100	480	990	860	1,280

[a]Compiled from R1903-10, June, 1952, Forest Products Laboratory, U.S. Dept. of Agriculture.

[b]Air-dried lumber contained 12% moisture.

TABLE 8-4 Basic Stresses for Clear Material[a,b]

Species	*Extreme Fiber in Bending*	*Modulus of Elasticity*	*Compression Parallel to Grain*[c] *(**L/d** = 11 or less)*	*Compression Perpendicular to Grain*	*Maximum Horizontal Shear*
Hardwoods					
Ash, commercial white	2,050	1,500,000	1,450	500	185
Elm, white	1,600	1,200,000	1,050	250	150
Hickory, true and pecan	2,800	1,800,000	2,000	600	205
Maple, sugar and black	2,200	1,600,000	1,600	500	185
Oak, commercial red and white	2,050	1,500,000	1,350	500	185
Conifers					
Cedar, western red	1,300	1,000,000	950	200	120
Cypress, southern	1,900	1,200,000	1,450	300	150
Douglas fir, coast region	2,200	1,600,000	1,450	320	130
Fir, commercial white	1,600	1,100,000	950	300	100
Hemlock, eastern	1,600	1,100,000	950	300	100
Pine, western white, eastern white, ponderosa, and sugar	1,300	1,000,000	1,000	250	120
Pine, Norway	1,600	1,200,000	1,050	220	120
Pine, southern yellow (longleaf or shortleaf)	2,200	1,600,000	1,450	320	160
Redwood	1,750	1,200,000	1,350	250	100
Spruce, red, white, and Sitka	1,600	1,200,000	1,050	250	120
Tamarack	1,750	1,300,000	1,350	300	140

[a]From Forest Products Laboratory, U.S. Dept. of Agriculture. See also ASTM D245-49T.

[b]All values in pounds per square inch and for material under long-time service conditions at maximum design load.

[c]*L*, unsupported length; *d*, least dimension of cross section.

Fig. 8-7 Advanced Stage of Decay

using designs that will keep the wood dry and accelerate rain runoff, and using preservative-treated wood where the wood must be in contact with water.

Insect damage may result from infestation of the timber members with any of a number of insects, among them powder-post beetles, termites, and marine borers. Powder-post beetles are reddish brown to black, hard-shelled insects from ⅛ to ½ in. long. The life cycle of the beetle includes four distinct stages: egg, larva, transformation, and adult. The adult bores into the timber, producing a cylindrical tunnel just under the surface in which the eggs are laid. The larvae burrow through the wood, leaving tunnels packed with a fine powder $1/16$ to ⅛ in. in diameter. Powder-post damage is indicated by this fine powder, either fallen from the timber or packed into tunnels within the wood and by the tunnels and holes in the timber. The beetles attack both sound and decayed wood but are not active in decayed wood that is water-soaked. They may also cause damage by transmitting destructive fungi from one site to another, thus spreading decay.

Termites resemble ants in size and general appearance and live in similarly organized colonies. Destruction is done by the workers only, not by the soldiers or winged sexual adults. Subterranean termites are responsible for most of the damage done in the United States. They are more prevalent in the southern states but are found in varying numbers throughout the states. Termites build dark, damp tunnels well below the surface of the ground, with some of these tunnels leading to the wood, which is the termites' source of food. Termites also require a constant water supply in order to survive.

Subterranean termites do not infest structures by being carried to the construction site. They must establish a colony in the soil before they are able to attack the timber. Tell-tale signs are the tunnels in the earth leading to unprotected timber and swarms of male and female winged adults in the early spring and fall (Figure 8-8). When termites

Fig. 8-8 Swarming Termites

successfully enter the wood, they make tunnels that follow the wood grain, leaving a shell of sound wood to conceal the tunnels (Figure 8-9). Methods of controlling termites include breaking the path from the timber to the ground, although the best method is to treat the timber with a preservative.

Wood-inhabiting termites are found in a narrow strip around the southern boundary of the United States. They are most common in southern California and southern Florida. They are fewer in number than the subterranean termites, do not multiply as fast, and do not cause as much damage. But because they can live without contact with the ground and in either damp or dry wood, they are a definite problem and do considerable damage in the coastal states. They are carried to a

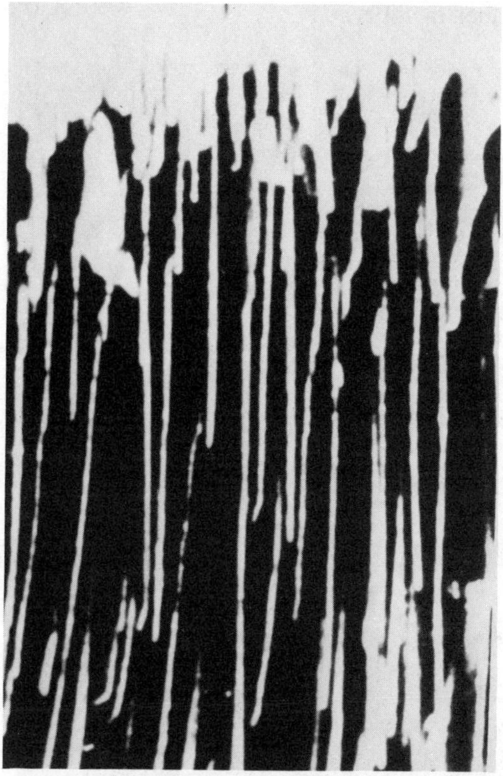

Fig. 8-9 Characteristic Termite Tunnel

building site in timber that has been infested prior to delivery, thereby making inspection prior to delivery a necessity. Full-length treatment with a good wood preservative is recommended.

Carpenter ants are usually found in stumps, trees, or logs, but are sometimes found in structural timbers. They range in color from brown to black and in size from large to small. Although carpenter ants are often confused with termites, they can be distinguished by comparison of the wings and thorax (waist) sizes of the insects. The carpenter ant has short wings and a narrow waist, whereas the termite has long wings and a thicker waist. Carpenter ants use wood as shelter, not food. They prefer naturally soft or decayed wood and construct tunnels that are very smooth and free of dust. Carpenter ants tunnel across the wood grain and cut small exterior openings for access to the food supplies (Figure 8-10). A large colony takes from 3 to 6 years to develop. Prevention of ant infestation can be accomplished by the use of preservatives along the length of the timber member.

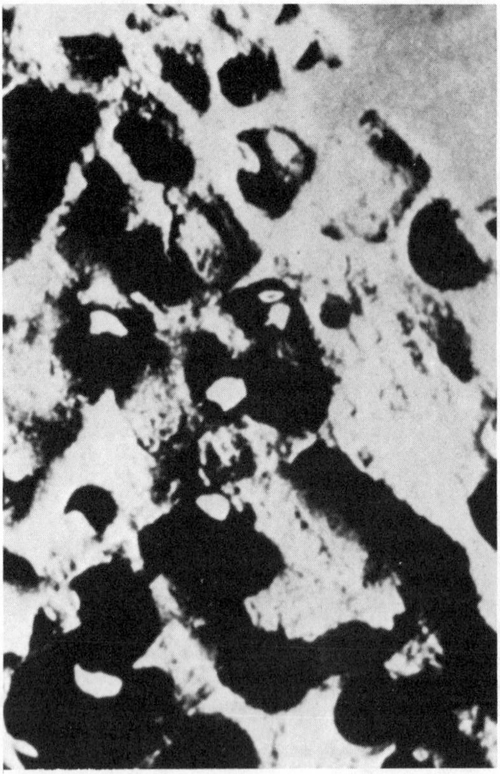

Fig. 8-10 Carpenter Ant Tunnels

By far the most damaging insect pest is the marine borer (Figure 8-11). Borers have been known to ruin piles and framing within a few months. No ocean is completely free of borers and, although some waters may be relatively free, the status of an area may change drastically within a relatively short period of time. The main point of attack is generally between the high tide level and the mud line. Submergence

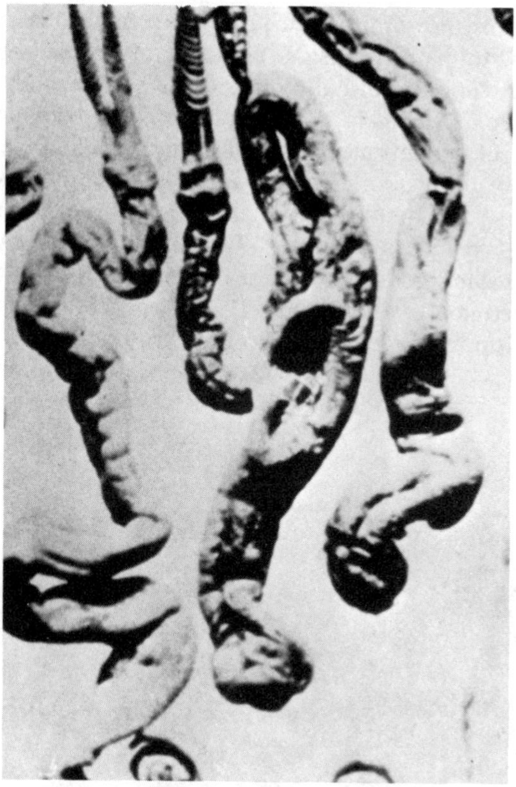

Fig. 8-11 Marine Borers

often hides tell-tale signs of infestation, so that the first sign of attack may be the failure of the structure. The borers that do the most damage are the mollusks, related to clams and oysters, and the crustaceans, related to lobsters and crabs.

The mollusk borers consist of the "shipworm" teredo, the "shipworm" bankia, and the Pholadidae. There are other types of shipworms throughout the world, but they all live and survive in much the same way, although size and environmental requirements may vary. The teredo has a wormlike, slimy gray body with two shells attached to the head which are used for boring. Two tubes that resemble a forked tail and normally remain outside the burrow are the only external indications that a shipworm is present. The shipworm can seal its entrance and thereby protect the burrow from intruders or foreign substances. The size of the teredo varies from 3/8 to 1 in. in diameter and from 6 in. to 6 ft in length. The size of the bankia is generally larger than the teredo, but its other characteristics are the same. Both types bore tiny holes when they are young and grow to maturity inside the wood. Once the young shipworm has entered the wood, it normally turns down and expands its burrow to its full diameter. Extremely careful observation with a hand lens is required to detect the entrance. The only way to detect the extent of the damage is to cut the wood (Figure 8-12) or take borings by some other method. Because the first sign of marine borer infestation may be the failure of timber members (Figure 8-13), the shipworm should be considered extremely dangerous. Pholadidae resembles a clam with its body entirely enclosed in a two-part shell. It is of particular danger because it can bore holes up to 1½ in. deep into the hardest timber.

The most common crustacean borer is the limnoria or "wood louse." Its body is slipper-shaped, from ⅛ to ¼ in. long and from $^1/_{16}$ to

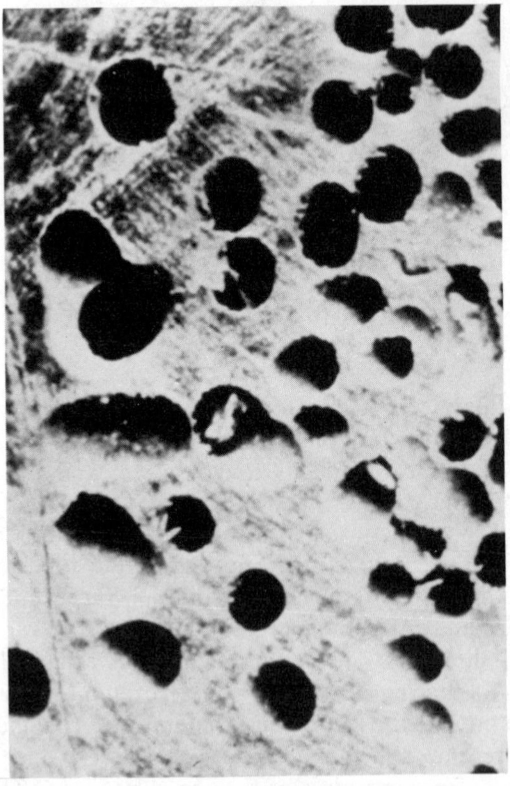

Fig. 8-12 Cross-Cut Timber Member Showing Extensive Damage

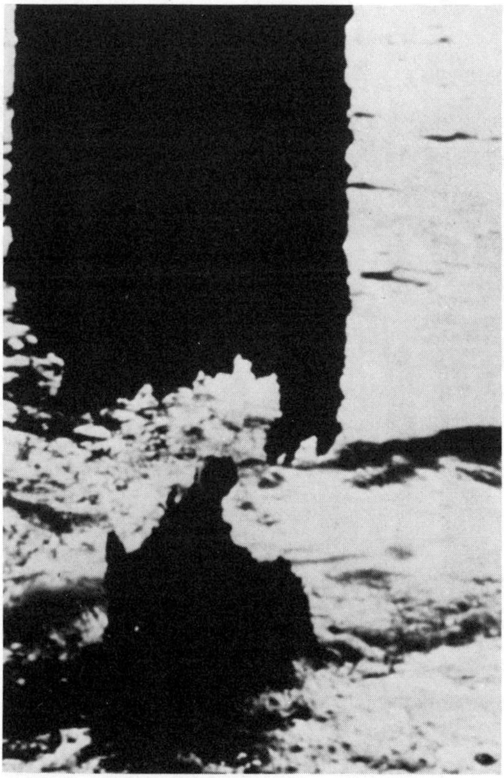

Fig. 8-13 Insect Attack Resulting in Complete Failure

⅛ in. wide. It has a hard boring mouth, two sets of antennae, and seven sets of legs with sharp claws. The limnoria is able to roll itself into a ball, to swim, and to crawl. It will gnaw interlacing branching holes in the surface of the wood with as many as 200 to 300 holes per square inch. These holes follow the softer wooden rings and are 0.05 to 0.025 in. in diameter and seldom over ¾ in. long. As a result, the wood becomes a mass of thin walls between burrows which break away, exposing new areas to attack. In this manner the pile is slowly reduced in diameter (Figure 8-14). In soft woods such as pine and spruce, the diameter may be reduced as much as 2 in. per year. The chief area of attack is between the low-water level and the mud line, with occasional activity up to the high-water level. The limnoria does not seem to be affected by small environmental changes and may be found in salt or brackish water of any temperature, polluted or clean.

Complete protection from borer attack is essential. Metal armoring

Fig. 8-14 Reduction of Water Line Due to Insect Attack

is falling into disuse, as is concrete casing. The very best method may prove to be jacketing creosoted piles. The most practical method involves heavy treatment with high-quality coal-tar creosote by the full-cell method to the point of near saturation. Although shipworms will generally not attack a creosoted member, they may attack any area that has been damaged, and any untreated area. Limnoria attack the creosoted timber directly but at a decelerated rate. Other types of wood preservatives in use include a plastic outer wrap, which is successful as long as the plastic is not damaged, and cyanide treatment, which is messy to apply, leaches out with time, and may cause damage to the environment.

Fire damage to wood is a real problem on bridge and wharf structures. Treatment of wood with preservatives may protect the wood from fungi and insect attack, but it generally results in making the wood more susceptible to fire. Damage due to fire is readily apparent due to the charred appearance and burnt odor of fire-damaged wood.

Chemical damage to wood is very difficult to determine because it often resembles damage done by other factors. Fungi atttack employs a chemical reaction between the enzymes particular to each fungus and the wood fiber. Fire damage involves the oxidation of the wood fiber, which is again a chemical process. It is therefore important to report any apparent damage to the timber as soon as possible.

Impact damage may occur in the event of a high-energy collision. Because wood has good impact characteristics, it may show only limited signs of external damage and must therefore be inspected carefully after an accident or suspected accident. The timbers may have a shattered appearance as opposed to the sagging appearance caused by overload. Compression failure will resemble wrinkled skin; tension failure looks as if the fibers have been pulled apart.

Another method of protecting timber is by injecting preservatives into the timber. There are two methods utilized: the pressure process and the nonpressure process. There are two types of *pressure processes*: the full cell and the empty cell. In the *full-cell process* it is intended that the cells remain filled with the preservative. The initial treatment removes most of the air and water before impregnation. The timber is then covered with preservatives under pressure. In the *empty-cell process,* the preservative merely forms a film over the cell walls. In the *nonpressure process* (hot and cold), the timber is immersed in the preservative in an open tank and heated to 200°F (93°C). The heat of the preserving fluid expands the air within the cells. The cooling of the bath on the immersing of the timber in a cold medium causes a contraction of the air still remaining in the cells, which tends to produce infiltration by the preservative.

PROBLEMS

8.1. List the two principal categories of trees and describe each.

8.2. How do hardwood trees differ from softwood trees?

8.3. Discuss the role of the following polymers: cellulose, hemicellulose, and lignin.

8.4. Explain the differences between slash cut and rift cut.

8.5. What role does moisture content play in timber used for structural purposes?

8.6. What are some defects in lumber? Describe each.

8.7. How is lumber graded?

8.8. How does compressive strength parallel to the grain differ from compressive strength perpendicular to the grain?

8.9. Explain the mechanism of decay in lumber.

8.10. How do marine borers damage timber?

REFERENCES

American Society for Testing Materials, *Book of Standards,* Part 22, 1979.

BROWN, H. P., PANSHIN, A. J., and FORSAITH, C. C., *Textbook of Wood Technology,* McGraw-Hill Book Company, New York, 1949.

BRUST, A. W., and BERKLEY, E. E., "The Distributions and Variations of Certain Strength and Elastic Properties of Clear Southern Yellow Pine Wood," *Proc. ASTM, 35,* Part II (1935), pp. 643–693.

CLAPP, W. F., "Recent Increases in Marine Borer Activity," *Civil Eng., 7,* No. 12 (Dec. 1937), pp. 836–838; and *17,* No. 6 (June 1947), pp. 324–327.

"Defects in Timber Caused by Insects," *U.S. Dept. Agr. Farmers Bull. 1490,* 1927.

DESCH, H.E., *Timber, Its Structure and Properties,* 2nd ed., Macmillan Publishing Co., Inc., New York, 1947.

DIETZ, A. G. H., *Materials of Construction, Wood, Plastics, Fabrics,* D. Van Nostrand Company, New York, 1949, pp. 1–189.

DIETZ, A. G. H., and GRINSFELDER, A., "Fatigue Tests on Compressed and Laminated Wood," *ASTM Bull. 129,* Aug. 1944.

ELMENDORF, A., "The Uses and Propertiesof Water-resistant Plywood," *Proc. ASTM, 20,* Part II (1920), p. 324.

Forest Products Laboratory, *Wood Handbook,* rev. ed., Superintendent of Documents, Government Printing Office, Washington, D.C., 1940.

FREAS, A. D., "Studies of the Strength of Glued Laminated Wood Construction," *ASTM Bull. 70,* Dec. 1950; and *Bull. 73,* Apr. 1953.

GAY, C. M., and PARKER, H., *Materials and Methods of Architectural Construction,* 2nd ed., John Wiley & Sons, Inc., New York, 1943, pp. 39–50, pp. 304–371.

HANSEN, H. J., *Timber Engineers Handbook,* John Wiley & Sons, Inc., New York, 1948.

HOLTMAN, D. F., *Wood Constructon,* McGraw-Hill Book Company, New York, 1929.

HUNT, G. M., and GARRATT, G. A., *Wood Preservation,* McGraw-Hill Book Company, New York, 1938.

KOFOID, C. A., *Termites and Termite Control,* University of California Press, Berkeley, Calif., 1934.

Kuerrzi, E. Q., "Testing of Sandwich Construction at the Forest Products Laboratory," *ASTM Bull. 164,* Feb. 1950.

Lewis, W. C., "Fatigue of Wood and Glued Wood Construction," *Proc. ASTM, 46* (1946), p. 814.

Markwardt, L. J., and Wilson, T. R. C., "Strength and Related Properties of Woods Grown in the United States," *U.S. Dept. Agr., Tech. Bull. 479,* 1935.

Mills, A. P., Hayword, H. W., and Rader, L. F., *Materials of Construction,* 6th ed., John Wiley & Sons, Inc., New York, 1955, pp. 508–543.

"Preventing Damage by Termites and White Ants," *U.S. Dept. Agr. Farmers Bull. 1972,* 1927.

Truax, T. R., "The Gluing of Wood," *U.S. Dept. Agr. Bull. 1500,* 1929.

United States Coast Guard, *The State of the Art: Bridge Protective Systems and Devices,* A Report by the U.S.C.G. Bridge Modification Branch, R. T. Mancill, Editor, 1979.

ASPHALT CEMENTS 9

Asphalt materials have been utilized since 3500 B.C. in building and road construction. Their main uses have been as adhesives, waterproofing agents, and as mortars for brick walls. These early asphalt materials were native asphalts. *Native asphalts* occur when the petroleum rises to the earth's crust and the volatile oils are evaporated. These native asphalts were found in pools and asphalt lakes. One of the more well-known deposits of native asphalts is the "Trinidad Lake" deposit on the island of Trinidad off the north coast of South America. Prior to the development of the processes for producing asphalt cement from crude petroleum products, native asphalts were the only sources of supply for early pavement projects. The first asphaltic pavement was built in 1869 in London, England. A year later construction of road and street pavements began in the United States in Newark, New Jersey.

With the invention of the automobile, which required smooth all-weather pavements, the demand for asphaltic products for pavements grew. Thus, the processes for producing asphalt cement from crude petroleum products grew, which led to the development of modern asphalt cement.

ASPHALT CEMENTS

Asphalt cement, according to ASTM D8 (Materials for Roads and Pavements), is fluxed or unfluxed asphalt specially prepared as to quality and consistency for direct use in the manufacture of bituminous pavements, and having a penetration at 77°F (25°C) of between 5 and 300, under a load of 0.2 lb (100 grains) applied for 5 sec. Asphalt cements fall within a broad category known as bitumens. *Bitumen,* according to ASTM D8, is a class of black or dark-colored (solid, semisolid, or viscous) cementitious substances, natural or manufactured, composed principally of high-molecular-weight hydrocarbons, of which asphalts, tars, pitches, and asphaltites are typical.

Bitumen by definition is soluble in carbon disulfide. The hydrocarbons that make up bitumen can be in general made up of the following:

1. Asphaltenes.
2. Resins.
3. Oils.

Asphaltenes are large, high-molecular-weight hydrocarbon fractions precipitated from asphalt by a designated paraffinic naphtha solvent at a specified solvent–asphalt ratio. Asphaltenes have a carbon-to-hydrogen ratio of 0.8. Asphaltenes constitute the body of the asphalt. *Resins* are hydrocarbon molecules with a carbon-to-hydrogen ratio of more than 0.6 but less than 0.8. Resins affect the adhesiveness and ductility properties of asphalt. *Oils* are hydrocarbon molecules with a carbon-to-hydrogen ratio of less than 0.6. Oils influence the viscosity and flow of the asphalt.

Ductility and adhesiveness are two properties that make asphalt cement attractive as a highway material. Oxidation of an asphalt cement causes a loss of ductility and adhesiveness, resulting in an asphalt cement that is harder and less ductile and adhesive. Oxidation results in the creation of more asphaltenes at the expense of resins. Thus, oxidation is a serious problem.

Production and Distillation

Asphalt cement is a valuable by-product obtained when petroleums are processed to obtain gasoline, kerosene, fuel oil, motor oil (diesel and lubricating), and other asphalt products. There are three types of petroleum found in the earth's crust:

1. Asphaltic-base crude oils.
2. Paraffin-base crude oils.
3. Mixed-base crude oils.

Asphalt cement is easily obtained from asphaltic-base crude oils by a staight-sun distillation process. Bituminous products may be obtained from paraffin-base crude oils by a destructive distillation process involving chemical changes. These bituminous materials should not be classified as an asphalt. Asphalt cement may also be obtained from mixed-base crude oils but the process is complicated.

All distillation of asphalt-base petroleum is fractional. During the distillation of petroleum several fractions are separated from the petroleum as given in Table 9-1.

TABLE 9-1 Fractions of Petroleum

Fraction	Product Type	Boiling-Point Range (°F)
Light distillate	Gasoline	100–400
Medium distillate	Kerosene	350–575
Heavy distillate	Diesel oil	425–700
Very heavy	Lubricating oil	Over 650
Residue	Asphalt	

Figure 9-1 shows a flowchart of the refining process necessary for the production of asphalt cement. When the process is controlled to prevent overheating and eventual chemical changes, the asphalt cement that remains as a residue is a straight-run asphalt. In the first operation the petroleum is pumped through the tube heater, in which the petroleum is placed under pressure and heated to a temperature of about 550°F (288°C). This crude product, released through the bottom of the tube heater, enters an atmospheric fractionating column (tower distillation). Upon being exposed to atmospheric pressure within the column the more volatile fractions rise to the top of the column, where the vapors pass through condensers and coolers. Traps are arranged in the column at different levels where each trap is just below the boiling point of the liquid at which it is to collect. Once the vapor passes through the appropriate traps, the vapors are condensed and cooled. These condensed vapors are referred to as distillates (light, medium, and heavy). After further processing, gasoline, kerosene, and diesel oil result. The product that settles down in the atmospheric fractionating column is *residuum,* sometimes called hot topped crude.

The crude then enters a second fractionation column. This column allows steam to be introduced at the bottom of the fractionation column in such a way as to become mixed with the hot topped crude. The boiling point of the various fractions becomes a combination of the boiling points of the oils being vaporized plus water. A second procedure (which is an aid for the use of lower temperatures) is the application of a partial vacuum in the fractionation column. The greater the vacuum applied, the lower the boiling point of the fractions to be separated from the crude. By controlling the steam, the temperature, and the partial vacuum, the quality of asphalt cement is maintained. The residual that results from the second fractionation column is asphalt cement. The volatiles are light vacuum distillate, nonvolatile oils, and heavy vacuum distillate.

Air is sometimes used to improve an asphalt material. The air is blown through an asphalt stock at temperatures of 400 to 550°F (205 to 288°C). A reaction results in which the oxygen from the air combines

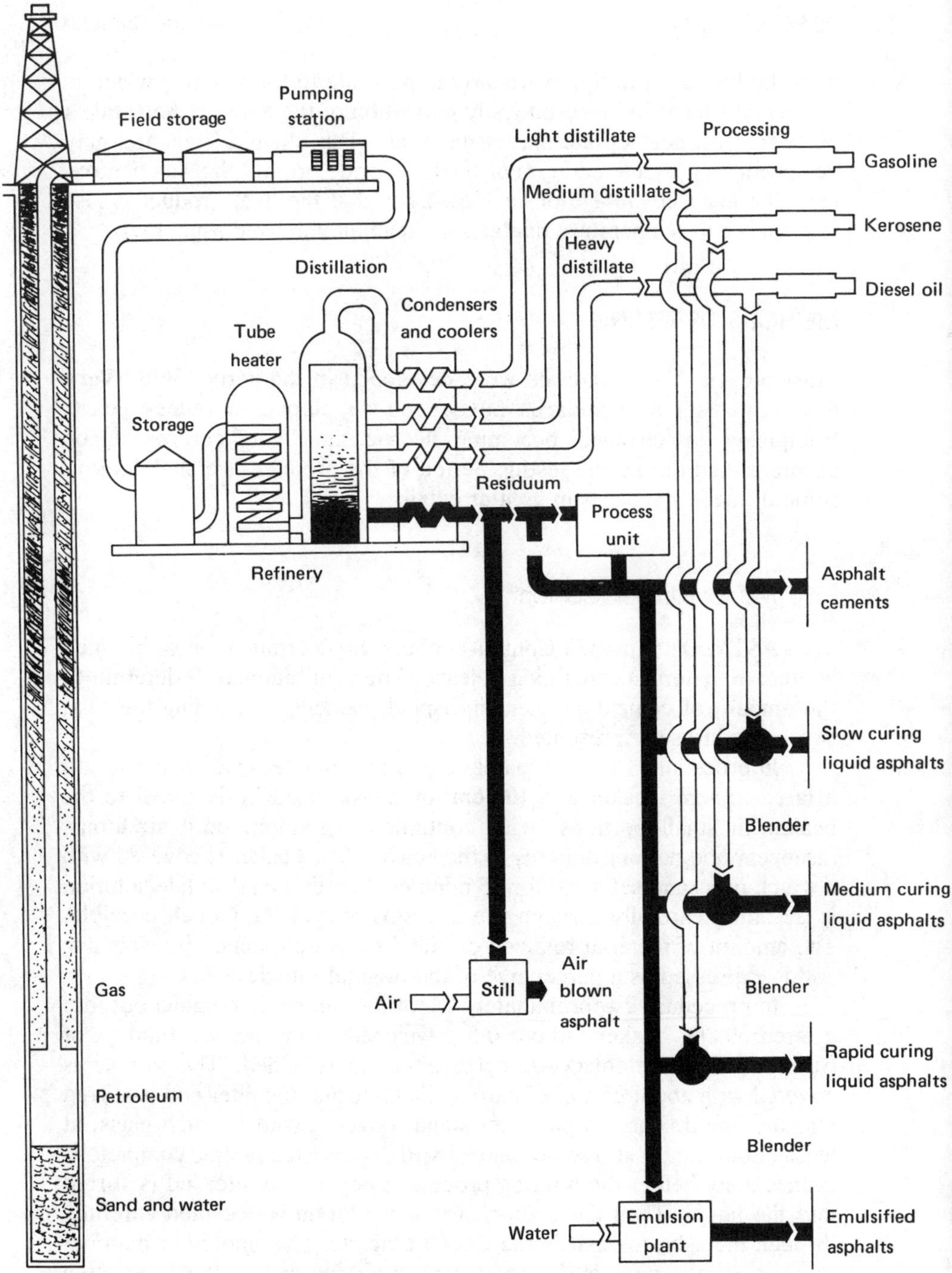

Fig. 9-1 Flowchart of the Refining Process for Asphalt Cement (Courtesy of Asphalt Inst.)

with the hydrogen in the hydrocarbon molecule to form water, which is emitted as steam. As a result, polymerization of the hydrocarbons takes place to form heavier and harder materials. This chemical reaction may be speeded up by the addition of ferric chloride or phosphorous pentoxide. The main advantage of air blowing is that the final product is less susceptible to temperature fluctuations than an unheated asphalt.

METHODS OF TESTING

Most present clay standards were developed in the early 1900s. Very few new tests or standards have been developed in recent years. Equipment has changed, becoming electric, and better control can be accomplished during the testing period of the product. We shall look at some of these standards in greater detail.

Test for Bitumen Content

ASTM D4 (Bitumen Content) covers the determination of bitumen content in materials containing at least 25 percent bitumen. It determines the amount of cementing agent in asphalt cement. In running the test, two procedures are presented.

In procedure 1 approximately 2 g of the sample is weighed out into a tared 150-cm^3 beaker and 100 cm^3 of carbon disulfide is added to the beaker in small portions, with continuous agitation, until all lumps disappear and nothing adheres to the beaker. The beaker is covered with a watch glass and set aside for 15 minutes. The carbon disulfide solution is decanted carefully through the asbestos mat in the Gooch crucible. The amount of material retained on the filter is determined by weighing and is expressed as a percentage of the original sample.

In procedure 2 approximately 2 g of the sample is weighed out into a tared 30-cm^3 beaker. About 0.5 g weighed to the nearest 0.001 g of freshly ignited diatomaceous earth filter aid is added. The sample is covered with about 25 cm^3 of carbon disulfide and the filter end is stirred into the liquid. The sample is let stand, covered with a watch glass, at least 1 hour, and is stirred occasionally to dissolve the sample completely. Immediately before the filtering process is begun, the filter aid is stirred into the liquid. Then the carbon disulfide solution is decanted carefully through the asbestos mat in the Gooch crucible. The amount of material retained on the filter is determined by weighing and is expressed as a percentage of the original sample.

The bitumen content, mineral matter, and differences are calculated as follows:

$$\text{bitumen content (\%) } X = 100\left[\frac{A - (B + D)}{A}\right]$$

$$\text{mineral matter (\%) } Y = 100\left(\frac{C + D}{A}\right)$$

$$\text{difference} = 100 - (X + Y)$$

where A = mass of water-free sample

B = net mass of insoluble residue

C = net mass of ignited mineral matter

D = total mass of correction

Bitumen content may usually be determined expeditiously and accurately using procedure 1. However, some bituminous materials that contain finely divided mineral matter may clog the filter or the mineral residue may not be easily retained—therefore the need for procedure 2.

For routine testing, benzene or carbon tetrachloride may be substituted for the carbon disulfide as carbon disulfide is flammable. If the specimen fails using either benzene or carbon tetrachloride, the test should be repeated using carbon disulfide. A material soluble in carbon disulfide and insoluble in carbon tetrachloride is called a *carbene*. A carbene content of 0.5 percent or more indicates that the asphalt has been overheated.

Penetration of Bituminous Materials

ASTM D5 (Penetration of Bituminous Materials) covers determination of the penetration of semisolid and solid bituminous materials. In other words, it measures the hardness or softness of the material. It is a consistency test for bituminous material expressed as the distance in tenths of a millimeter that a standard needle vertically penetrates a sample of the material under known conditions of loading, time, and temperature.

The procedure for the standard test, which is implied unless other conditions are stated, is for pressure to be applied to a load of 100 grams for a period of 5 sec at a temperature of 77°F (25°C).

In running the test at least three determinations should be made on the surface of the sample at points not less than 10 mm from the side of the container and not less than 10 mm apart.

The loss on weight with today's manufactured products is quite low, and thus the test has lost much of its meaning. The test back in the early 1900s was extremely valuable in characterizing early steam-refined asphalts, which had low flash points and high losses on heating.

(Asphalt Cement for Use in Pavement Constructions). The most common grades are 60 to 70 and 85 to 100. Lower penetration grades are generally needed in warmer climates to avoid softening in the summer. Obviously, higher penetration grades will be used in colder climates so that excessive brittleness does not occur during cold winter weather.

The penetration test has in the past provided long experience and service records associated with asphalts classified in this matter. However, the test is empirical and many engineers would like to replace it with ASTM D2171 (Test for Absolute Viscosity of Asphalts).

Loss on Heating

ASTM D6 (Loss on Heating of Oil and Asphaltic Compounds) covers the determination of the loss in mass (exclusive of water) of oil and asphaltic compounds when heated as prescribed.

In this test 50 g of the sample of the water-free material to be tested is placed in a tared container (55 mm in diameter by 35 mm high) and heated at 325°F (163°C) for a period of 5 hours. In the oven, the sample is placed on a rotating shelf (5 to 6 rpm). The loss of weight is reported after heating.

The loss on weight with today's manufactured products is quite low, and thus the test has lost much of its meaning. Back in the early 1900s the test was extremely valuable in characterizing early steam-refined asphalts, which had low flash points and high losses on heating.

Softening Point of Bitumen

ASTM D36 [Softening Point of Bitumen (Ring and Ball Apparatus)] covers the determination of the softening point of bitumen having a softening point in the range 85 to 395°F (30 to 200°C) by means of the ring and ball apparatus.

Bituminous materials do not have a definite melting point, but the change from a solid to a liquid is a very slow process over a variety of temperatures. The test is a consistency test and is used for control in refining operations, especially in the production of airblown asphalts. A high softening point means that the material will not flow while in service.

Two disks of bitumen, cast in shouldered rings, are heated from 41°F (5°C) at a rate of 9°F/min (5°C/min) in a water or glycerin bath. The softening point is defined as the mean of the temperatures at which the bitumen disks soften and sag downward a distance of 25 mm under the weight of a steel ball.

This method is useful in determining the consistency of bitumen as one element in establishing the uniformity of shipments or sources of supply.

Specific Gravity

ASTM D70 (Specific Gravity of Semi-solid Bituminous Materials) covers the determination of the specific gravity of semisolid bituminous materials, asphalt cements, and soft tars pitches by use of a pycnometer.

In running this test, the sample is heated until it can be poured. The material is placed in a pycnometer. The asphalt volume is determined by taking the difference between the total volume of the bottle and the volume of water required to complete the filling. From this information the specific gravity can be expressed as the ratio of the weight of a given volume of the material at 77°F (25°C) or at 60°F (15.6°C) to that of an equal volume of water at the same temperature. The following formula may be used to calculate the specific gravity:

$$\text{specific gravity} = \frac{C - A}{(B - A) - (D - C)}$$

where A = weight of pycnometer (plus stopper)

B = weight of pycnometer filled with water

C = weight of pycnometer partially filled with asphalt

D = weight of pycnometer plus asphalt plus water

The specific gravity is reported to the nearest third decimal place at 77/77°F (25/25°C) or 60/60°F (15.6°/15.6°C).

Specific gravity is usually not a requirement of asphalt specifications but is of value in the design of bituminous mixtures as well as for determining costs.

Density of Asphalt

ASTM D71 [Density of Solid Pitch and Asphalt (Displacement Method)] covers the determination of density by water displacement of hard pitches and asphalts with softening points above 158°F (70°C).

In this procedure the sample is suspended from a thin wire and weighed, first in air, then submerged in water at 77°F (25°C). The density is calculated from these masses as follows:

$$\text{density at } 77°\text{F } (25°\text{C}) = \frac{A}{A - B} \times 997$$

where A = mass of the specimen in air, mg

B = weight of the specimen in water at 77°F (25°C), mg

997 = density of water, in kg/m^3 at 77°F (25°C)

Flash and Fire Points

ASTM D92 (Flash and Fire Points By Cleveland Open Cup) covers the determination of the flash and fire points of all petroleum products except fuel oils and those having an open cup flash below 175°F (79°C).

The test cup is filled to a specified level with the sample. The temperature of the sample is increased rapidly at first and then at a slow constant rate as the flash point is approached. At specified intervals a small test flame is passed across the cup. The lowest temperature at which application of the test flame causes the vapors above the surface of the liquid to ignite is taken as the flash point. To determine the fire point, the test is continued until the application of the test flame causes the oil to ignite and burn for at least 5 seconds.

The purpose of this test is to determine the temperature at which asphalt can be heated safely. Sometimes it is used to determine or detect small quantities of contaminants in the asphalt, such as gasoline or kerosene.

Another method of determining the flash point is given by ASTM D93 (Flash Point By Pensky–Martens Closed Tester) which covers the determination of the flash point of fuel oils, lube oils, suspensions of solids, liquids that tend to form a surface film under test conditions, and other liquids by the Pensky–Martens closed-cup tester.

In this test the sample is heated at a slow, constant rate with continual stirring. A small flame is directed into the cup at regular intervals with simultaneous interruption of stirring. The flash point is the lowest temperature at which application of the test flame causes the vapor above the sample to ignite.

The Pensky–Martens closed-cup tester gives lower flash point values for a given material than does the Cleveland cup procedure. The Pensky–Martens tester method can detect smaller amounts of the lighter hydrocarbon vapors than can be obtained using the Cleveland procedure.

Ductility

ASTM D113 (Ductility of Bituminous Materials) covers the procedure for determining the ductility. The ductility of a bituminous material is measured by the distance to which it will elongate before breaking when two ends of a briquette specimen of the material are pulled apart at a specified speed and temperature.

The results of the ductility test are controversial. The test is believed to measure the adhesiveness and elasticity of the asphalt. The property of adhesiveness is most important, since asphalt cement is used to bind stone, sand, and filler to make bituminous concrete. Also of concern is the temperature at which the product is run, 77°F (25°C). It is believed that this temperature is too high and that a much lower temperature should be used, because lack of ductility, or brittleness, is more serious in cold weather.

Float Test

ASTM D139 (Float Test for Bituminous Materials) is a consistency test used for materials that are too soft to undergo the standard penetration test and too hard for use with the Saybolt–Furol viscosity test.

The test is performed by filling a brass collar with the asphalt or asphalt product to be tested, cooling the holder with its contents to 41°F (5°C), screwing the collar to a float, and floating the apparatus in water at a specified temperature. The float-test results are reported as the time in seconds from placing the float in water until the water displaces the material in the collar.

Sampling Bituminous Materials

ASTM D140 (Sampling Bituminous Materials) covers the method used to sample bituminous materials at points of manufacture, storage, or delivery. The purpose of this procedure is to determine the true nature and condition of the material. Samples are taken as a representation of the bulk of the material and to ascertain the maximum variation in characteristics that the material possesses. This procedure covers sampling of semisolid or uncrushed solid material and of crushed or powdered material at place of manufacture, from containers, tankcars, vehicle tanks, distributor trucks, recirculating storage tanks, tankers, barges, pipelines, drums, and barrels.

Viscosity

Viscosity of asphalt materials can be determined by one of three methods: ASTM D2170 [Kinematic Viscosity of Asphalt (Bitumens)], ASTM D88 (Saybolt Viscosity); and ASTM D2171 (Viscosity of Asphalts by Vacuum Capillary Viscometer).

ASTM D2170 [Kinematic Viscosity of Asphalt (Bitumens)] covers determination of the kinematic viscosity of liquid asphalts (bitumens), road oils, and distillation residues of liquid asphalts (bitumens), all at 140°F (60°C), and of asphalt cements at 275°F (135°C) in the range 30 to 100,000 centistokes. Kinematic viscosity is the ratio of the viscosity to the density of a liquid. It is a measure of the resistance to flow of a liquid under gravity. The SI unit of kinematic viscosity is m^2/s; for practical use, a submultiple (mm^2/s) is more convenient. The centistoke (cSt) is 1 mm^2/s and is customarily used. *Viscosity* as referred to herein is the ratio between the applied shear stress and the rate of shear, called the *coefficient of dynamic viscosity.* This coefficient is a measure of the resistance to flow of a liquid. The SI unit of viscosity is the pascal-second; for practical use, a submultiple (mPa·s) is more convenient. The centipoise, 1 mPa·s, is customarily used. Density as referred to herein is the mass per unit volume of liquid. The cgs unit of density is 1 g/cm^3 and the SI unit of density is 1 kg/m^3.

The method includes measuring the time for a fixed volume of the liquid to flow through the capillary of a calibrated glass capillary viscometer under an accurately reproducible head and at a closely controlled temperature. The kinematic viscosity is then calculated by multiplying the efflux time in seconds by the viscometer calibration factor.

The results of this test can be used to calculate viscosity when the density of the test material at the test temperature is known or can be determined.

ASTM D88 (Saybolt Viscosity) covers procedures for the empirical measurement of the Saybolt viscosity of petroleum products at specified temperatures between 70 and 210°F (21 and 99°C).

In the Saybolt viscosity test the efflux time in seconds of 60 ml of sample, flowing through a calibrated orifice, is measured under carefully controlled conditions. This time is corrected by an orifice factor and is reported as the viscosity of the sample at that temperature.

The final method, ASTM D2171 (Viscosity of Asphalts by Vacuum Capillary Viscometer), covers procedures for the determination of viscosity of asphalt (bitumen) by vacuum capillary viscometers at 140°F

(60°C). It is applicable to materials having viscosities in the range from 0.036 to over 200,000 poises (P).

In this procedure the time is measured for a fixed volume of the liquid to be drawn up through a capillary tube by means of vacuum, under closely controlled conditions of vacuum and temperature. The viscosity, in poise, is calculated by multiplying the flow time in seconds by the viscometer calibration factor.

As previously indicated, there has been a strong movement to change the system of grading asphalt cements from penetration units at 77°F (25°C) to absolute viscosity units at 140°F (160°C), in accordance with ASTM D2171. Many researchers feel that this temperature represents a critical service temperature in asphalt pavements and that its use tends to minimize viscosity differences at higher and lower temperatures among asphalts of the same penetration grade having different temperature susceptibilities. However, opponents indicate that the 140°F (60°C) temperature is too high—that the asphalt begins to behave as a Newtonian liquid. In any event, some manufacturers are now utilizing both classification techniques. It is doubtful that the penetration method of classification will be dropped entirely for several years to come, as this would result in the loss of many years of accumulated field experience.

MODIFICATIONS OF ASPHALT CEMENTS

Asphalt cements may be modified into several different products to make it easier for distribution and use. Asphalt cement is generally hard and relatively solid. It must therefore be heated or treated before it can be mixed with aggregates to produce an asphaltic concrete (pavement). Asphalt cement can be made into a liquid at lower temperatures by mixing it with a volatile oil, or it can be emulsified with water to produce liquids at normal temperatures. When the volatile oils evaporate or the emulsion breaks, a hard stable asphalt cement remains which with the aggregate particles make up the pavement.

Liquid Asphalts (Cutbacks)

When volatile oils are mixed with asphalt cement to make a liquid product, the product is referred to as a *cutback*. The purpose of the cutback is to allow relatively easy placement of the asphalt product without the use of high temperatures. After the material has been placed, the product reverts back to its natural penetration value through the evaporation of the volatile oils. In general, there are three types of liquid asphalt in the cutback category:

1. Rapid-curing (RC).
2. Medium-curing (MC).
3. Slow-curing (SC; road oils).

Rapid-curing (RC) cutbacks are made by diluting gasoline or naphtha with asphalt cement having a penetration between 70 and 110 at 77°F (25°C). The grades of RC cutbacks range from RC 0 to RC 5, depending on the amount of gasoline or naphtha used. Careful control must be maintained when utilizing gasoline or naphtha to keep the flash point of the material above 80°F (27°C). The percentage of the total volatile oils driven off at 500°F (260°C) varies from 75 percent for RC 0 to 25 percent for RC 5. Obviously, the viscosity increases from RC 0 to RC 5.

Medium-curing (MC) cutbacks are similar to rapid-curing cutbacks, with the exception that kerosene is used rather than gasoline or naphtha to liquefy the asphalt cement. The asphalt cement used to make medium-curing cutbacks should have a penetration of 70 to 250 at 77°F (25°C).

Comparing rapid-curing cutbacks with medium-curing cutbacks, it can be shown that rapid-curing cutbacks have a harder base asphalt and that the gasoline or naphtha will evaporate at lower temperatures, resulting in a material that is believed to cure rapidly. The medium-curing cutbacks have a softer base asphalt and a less volatile solvent (kerosene), and as a result will cure much more slowly than will a rapid-curing cutback.

Therefore, if one is constructing a pavement in a northern region, one might well choose the medium-curing cutback, as it has a softer base and after curing the material would be less brittle in the winter and subject to less cracking. In a southern region, one might well choose the rapid-curing cutback, to obtain a harder asphalt base. This harder base may well provide a much more stable pavement under the hot sun.

As previously mentioned, there are six grades of cutbacks (Table 9-2). These grades are based upon the ASTM D88 (Saybolt Viscosity) test.

A cutback of grade 0 will pour and flow freely at room temperature; a cutback of grade 5 will have to be heated in order to flow. Grades 0 and 1 are used as tack coats or prime coats. A tack coat is an application of bituminous material to an existing relatively nonabsorptive surface to provide a thorough bond between old and new surfacing. A prime coat is an application of a low-viscosity bituminous material to an absorptive

TABLE 9-2 Cutback Grades

Saybolt Viscosity	*Grade*
15–30	0
40–80	1
100–200	2
250–500	3
600–1200	4
1500–3000	5

surface, designed to penetrate, bond, and stabilize this existing surface and to promote adhesion between it and the construction course that follows. Grades 2, 3, 4, and 5 have progressively higher grades and are used in surface treatments and penetration applications. This type of grading system has been replaced by a viscosity system.

Many state specifications for rapid-curing and medium-curing cutbacks are based upon viscosity tests run at 140°F (60°C). In this system RC and MC asphalt cutbacks have the following grades; 70, 250, 800, and 3000 for RC and 30, 70, 250, 800, and 3000 for MC. These numbers refer to the minimum allowable range of viscosity as determined by the kinematic viscosity test, in units of centistokes.

Cutback liquid asphalts are governed by ASTM D2027 [Cutback Asphalt (Medium-Curing Type)] and D2028 [Cutback Asphalt (Rapid-Curing Type)].

Road Oils

Slow-curing liquid asphalts *(road oils)* were originally manufactured by a straight-run distillation process and were really liquid asphalt cements. In other words, they were manufactured like asphalt cements, but the distillation process was cut off earlier and many of the volatile oils remained as part of the asphalt. Today, slow-curing (SC) road oils are cutback asphalts. However, they are not cutback with gasoline, naphtha, or kerosene but are fluxed with nonvolatile oils.

The grading system for SCs is the same as that for RC and MC materials. Their main applications are for dust control (dust binding). *Dust binding* is a light application of bituminous material for the express purpose of laying and bonding loose dust. Slow-curing road oils are governed by ASTM D2026 [Cutback Asphalt (Slow-Curing Type)].

Asphalt Emulsions

An *asphalt emulsion* is a suspension of minute globules of water or of an aqueous solution in a liquid bituminous material. It is a mixture of asphalt cement and water in which water is the continuous phase and asphalt cement is the dispersed phase. Asphalt emulsions are another way of liquefying asphalt cements. The two most commonly used types of emulsified asphalts are anionic emulsions and cationic emulsions. Anionic emulsions are a type of emulsion such that a particular emulsifying agent establishes a predominance of negative charges on the discontinuous phase. *Cationic emulsions* are a type of emulsion such that a particular emulsifying agent establishes a predominance of positive charges on the discontinuous phase. Originally, all asphalt emulsions were the anionic type. The type of asphalt emulsion is determined by the kind of emulsifying agent or soap used. An anionic emulsion works well with aggregates that have a positive charge (thus, unlikes attract). Cationic emulsions work well with all types of aggregate. Cationic emulsions can extend the paving season, as they can better handle cold weather and are not damaged by sudden rain.

The setting time of each of the three general grades of asphalt emulsions is dependent upon the breaking time of the emulsion and the evaporation of the water. Each type of emulsion has two grades, 1 and 2. There also exists an MS-2h grade, which is suitable for cold and hot plant mix. The other types are used cold, although in special cases they may be used hot.

The advantages of asphalt emulsions are as follows:

1. Can be used with cold or hot aggregate.
2. Can be used with aggregate that is dry, damp, or wet.
3. Eliminates the fire and toxicity hazards of cutback liquid asphalt.

Testing of Asphalt Cements Emulsions

Most of the ASTM test procedures previously discussed for asphalt cements may be used for emulsions. However, ASTM D244 (Emulsified Asphalt) covers specific testing for emulsified asphalts. The method covers the compositions, consistency, stability, and examination of residue of asphalt emulsions composed principally of a semisolid or liquid asphaltic base, water, and an emulsifying agent. The composition test includes water content, residue by distillation, identification of oil distillation by microdistillation, residue by evaporation, and particle charge of emulsified asphalts. The consistency test is the Saybolt viscosity test. The stability test includes demulsibility, settlement, cement mixing, sieve test, coating, miscibility with water, modified miscibility with water, freezing, coating ability and water resistance, and storage stability of asphalt emulsions. The final test is examination of residue.

ROAD TARS

Road tars were the most common paving materials until the development of the petroleum industry. With the widespread use of gasoline for automobile consumption, the asphalt by-product of the refining process has become so readily available that it has put tars out of use. However, in England, road tars are still in use because of the large bituminous coal industry.

Manufacture

Tars are brown or black bituminous material, liquid or semisolid in consistency, in which the predominating constituents are bitumens obtained as condensates in the destructive distillation of coal, petroleum, oil shale, wood, or other organic materials, and which yields substantial quantities of pitch when distilled. *Road tars* are the product of straight-run distillation of crude tars. The two most common methods are the coke-oven process and the water-gas method.

In the *coke-oven method,* bituminous coal is processed to produce coke in which crude coke-oven tar is the by-product. Approximately 10 tones of bituminous coal is heated to 2500°F (1370°C) in a brick-lined oven. The crude tar is part of the volatile product that is removed in the heating process, the residue being coke. With this method the properties of the tar vary with the makeup of the coal, the kind of oven, the temperature, the length of time the temperature is applied, and the pressure of the system. When the vapors are removed from the oven, condensed and cooled, crude tar is formed. The crude tar is then placed through a straight-run distillation process to form tar.

Although most of the road tars are produced by destructive distillation of bituminous coal in the coke-oven tar procedure, a very small amount is produced by the *water-gas operation* of cracking petroleum. In this method, the crude tar is a by-product of the second stage of producing heating gas. Initially, steam is passed over a bed of incandescent coke which decomposes the steam; carbon monoxide and hydrogen gases are formed, which are known as water gas. This water-gas tar is too low in Btu for heating purposes, and hydrocarbon gases must be added for enrichment. These enriching hydrocarbon gases result from the passing of petroleum oils over hot firebrick in a carburetor. The high temperature in this process causes cracking (destructive distillation) of the petroleum oils, and in addition to the hydrocarbon gases, which enrich the water gas, a heavy residue of hydrocarbons is condensed to form crude tar. The tar is referred to as water-gas tar if a light petroleum fraction is used, and as residuum tar if a heavier residue is used.

In the manufacture of tar, the crude tar is stored until the water settles out. At this point the crude tar is transported to the refining plant, where the small amount of water that remains is heated out. In the tar refining process, many other products are produced besides road tars. The crude tar is treated by fractional distillation in much the same manner as petroleum is treated to produce asphalt. The viscosity of tars is produced by straight-run distillation. The distillation is carried to that extent required for the grade desired. Road tars are graded from RT 1 up to RT 12. RT 1 is a light fraction of tar and RT 12 is as hard in consistency as the No. 200 penetration asphalt. Cutback tars also exist in two grades, RTCB 5 and RTCB 6. These two grades are made by using light or middle oils with RT 10, 11, or 12 tar.

RT 1 is used for dust control. RT 2 and 3 are light prime coats. RT 4 is for prime coats and sometimes surface treatments. RT 5, 6, and 7 are used for surface treatments and road mixes. RT 8 and 9 are for use as surface treatments, seal coats, and road mixes. RT 10 and 11 are used in seal coats, tar concrete, and hot repairs. RT 12 is used in penetration macadam work, tar concrete, and hot repairs.

The RT 1-6 and RTCB 5-6 can be used at temperatures up to 150°F (66°C). RT 7 and above may be used at higher temperatures.

Methods of Test

Road tars are tested in much the same manner as asphalt cements, but the tests differ somewhat, primarily because the tests were developed at different times. In general, the tars are many years older.

PROPORTIONING ASPHALTIC MIXES

The most suitable asphaltic concrete is one that produces a stable, durable, flexible, and skid-resistant pavement at a minimum cost with adequate aggregate. It is not possible to optimize all four properties; thus, compromises result. We will now look at the compromises. One was made in the design of bituminous-concrete mix design.

Properties

The term *stability* is related to strength and refers to the ability of a pavement to resist deformation under application of loads. The stability depends on the distribution of loads by point-to-point contact of the aggregate particles. The forces of this distribution are affected by aggregate interlocking developed among aggregate particles and the cohesiveness supplied by the asphalt cement.

In maximizing stability, the aggregate would have to be of crushed angular particle shape with a rough surface texture. It should also be hard and dense, its grade approaching the Fuller curve. Further, there should be just enough asphalt to coat the aggregate particles such that all would benefit from the adhesiveness.

Rounded particles would result in low stability and load distribution. Rounded particles tend to slide over one another, whereas angular particles interlock with one another. If soft aggregate particles were used, they would break and wear under the impact of vehicular loads and reduce stability greatly.

Durability refers to the resistance of the pavement to disintegrate under traffic loads. In other words, the pavement will remain smooth and serviceable during summer heat and will not crack or ravel during the winter cold. It will further resist the forces of freezing and thawing. In general, the greater protection of the aggregate provided by the asphalt cement, the more durable the pavement. Thus, in maximizing durability, one would not only wish to coat all the aggregate particles with asphalt but also to fill the voids within it. In this matter, the aggregate would be immune to the forces of freezing and thawing, for no water could enter the mix. Further stripping would not occur, because the water could not get between the aggregate and the asphalt. In addition, oxygen would not penetrate the pavement; thus, oxidation cannot take place and durability and adhesiveness would last longer.

In maximizing durability, stability and skid resistance are compromised. In a thick coating of asphalt over the aggregate particles, the aggregates tend to float in the asphalt. The result is that no interlocking of the aggregate particles takes place, and stability is lost.

As traffic tends to compact a pavement in which there is more asphalt than that needed to cover the aggregate, skid resistance is compromised. As the pavement compacts, bleeding of the asphalt occurs (asphalt comes to the surface), making a very slippery pavement.

Flexibility refers to the ability of a pavement to withstand deflections and bending without cracking. To maximize flexibility, one would use an open-graded aggregate mixture. However, in this case stability is compromised. The stability might be improved by increasing the pavement thickness or by increasing such flexibility decreases.

Skid resistance is a form of stability, and there are two categories that cause slippery pavements:

1. Pavement bleeding.
2. Aggregate polishing.

In *pavement bleeding* too few voids are left in the mix and compaction occurs under traffic loads on a hot day, forcing asphalt to the surface of the roadway. This covers the exposed rough aggregate and results in a slippery pavement. With voids in pavements as low as ½ or 1 percent, slippery pavements will not occur; however, most specifications require 2 or 3 percent air voids to prevent surface bleeding.

Aggregate polishing results after a pavement receives continuous wear from traffic and the surface aggregates polish, forming a slippery pavement. One method to prevent aggregate polishing is to use relatively hard aggregates. Another is to use a mixture of varying aggregates of different hardness. In this way, as one type of aggregate polishes, the other, harder aggregates do not, so that the road does not totally polish.

Mix Design

A good asphalt pavement that compromises among stability, durability, flexibility, and skid resistance for a given gradation of aggregate with just enough asphalt to cover the aggregate particles for good adhesive properties. In proportioning asphaltic concrete, seven steps should be followed:

1. Evaluate and select the aggregates to be used.
2. Select the aggregate gradation.
3. Make trial mixes with various amounts of asphalt cement.
4. Measure the relative stability of each mix.
5. Compute the void content of each compacted mix.
6. Compute the voids in the mineral aggregate and voids in the aggregate filled with asphalt.
7. Select the optimum design.

In step 1, the contractor is allowed to select the most convenient and economical aggregate source that meets the general specifications. These specifications usually demand an aggregate of the highest quality that is economically feasible.

In step 2, aggregate gradation, many highway and government agencies have specifications that define aggregate gradings for various types of mixes and sizes of aggregates. If in doubt, the Asphalt Institute publishes various manuals that give minimum recommendations on grading and sizes.

Step 3, the making of the trial mixes with different amounts of asphalt cements (2 to 6 percent), is probably the most important step. As previously discussed, the optimum mix should be one in which a compromise is made among stability, durability, flexibility, and skid resistance with sufficient asphalt to coat the aggregate particles of a given size and gradation.

Step 4, which allows for the measurement of stability, may utilize any one of four laboratory tests that measure the deformation under load of compacted laboratory specimens. These tests are all governed by ASTM specifications and are as follows:

1. Hubbard, ASTM D1138.
2. Marshall, ASTM D1559.
3. Unconfined Compression, ASTM D1974.
4. Hucem Stabilometer, ASTM D1560.

The most common test of these four is the Marshall test, ASTM D1138.

In step 5, the void content of the compacted specimen is computed. The percentage of voids in a compacted asphaltic concrete influences the stability and can be used as a specification to prevent pavement bleeding and thus slippery roadways.

The void content is computed by comparing the volume of the compacted mix with the volume occupied by each of the ingredients. The volume of any amount of material can be computed by knowing its specific gravity or density.

$$\text{specific gravity} = \frac{\text{density of the material}}{\text{density of water}} \tag{9.1}$$

$$\begin{matrix}\text{solid volume} \\ \text{of each ingredient}\end{matrix} = \frac{\text{weight of the ingredient in the specimen}}{\text{specific gravity of the ingredient}} \tag{9.2}$$

$$\text{percent voids} = 100\frac{V_S - V_I}{V_S} \tag{9.3}$$

where V_S = volume of specimen

V_I = combined solid volumes of all ingredients in the specimen

In step 6, we compute the voids in the mineral aggregate. When one discusses the total voids within the compacted aggregate particles, it becomes clear that asphalt occupies most of the void space. As indicated, if the voids are completely filled with asphalt, the pavement loses its stability. Therefore, it is useful to compute the voids in the mineral aggregate (VMA) and the percentage of voids filled with asphalt. The VMA and the voids filled with asphalt are usually controlled by specifications. The VMA may be computed as follows:

$$\text{VMA} - \text{volume of specimen} - \text{volume occupied by the aggregate} \quad (9.4a)$$

Therefore,

$$\text{VMA} = V_S - V_{\text{agg}} \quad (9.4b)$$

$$\text{percent VMA filled with asphalt} = 100 \frac{V_a}{\text{VMA}} \quad (9.5)$$

where V_a is the volume of asphalt in the aggregate voids.

The final step, 7, is to select the optimum asphalt content according to the design specifications.

Example 9.1: Calculate the percent voids and VMA in a compacted mixture with a unit weight of 144 lb/ft³ with the following criteria:

Ingredients	*Percent of Total Aggregate*	*Effective Specific Gravity*
Coarse aggregate	70	2.70
Fine aggregate	25	2.65
Filler aggregate	5	2.60

Percent asphalt = 5 with an effective specific gravity = 1.00

percent asphalt = 5 with an effective specific gravity = 1.00

Solution:

	Weight		*Volume*	
	0	Air	$0.06 = 1 - 0.94$	
	$0.05 \times 144 = 7.2$	Asphalt	$0.12 = \dfrac{7.2}{1 \times 62.4}$	
144 lb	$0.047 \times 144 = 6.7$	Filler	$0.04 = \dfrac{6.8}{2.60 \times 62.4}$	1 ft³

$$0.24 \times 144 = 34.1 \quad \text{Fine aggregate} \quad 0.21 = \frac{34.1}{2.65 \times 62.4}$$

$$0.67 \times 144 = 95.9 \quad \text{Coarse aggregate } 0.57 = \frac{95.9}{2.70 \times 62.4}$$

1. Assume a 1-ft³ volume and determine the total weight of the mix.
2. Compute the percent of aggregate based upon 95 percent of total ingredients.
3. Determine the weights of each ingredient.
4. Determine the volume of each ingredient by eq. (9.2).
5. Use eq. (9.3) to determine the volume of the air.
6. Use eq. (9.4b) to determine the VMA.

Answer: 6 percent air and 18 percent VMA.

In Example 9.1 the term "effective specific gravity" was used, and needs an explanation. The specific gravity of aggregates is an important consideration in the determination of void contents in compacted bituminous materials. There are three types of specific gravity values used in the field: apparent, bulk, and effective.

The *apparent specific gravity* is the ratio of the weight of an aggregate particle to the weight of a volume of water equal to the volume of solid aggregate and pores impermeable to water. The diagram below illustrates the term of apparent specific gravity:

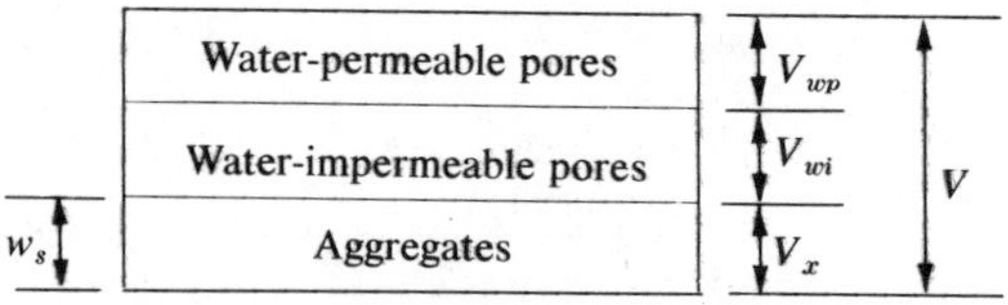

The apparent specific gravity is as follows:

$$\text{apparent specific gravity} = \frac{W_s}{(V_s + V_{wi})\gamma_w} \tag{9.6}$$

where W_s = weight of the aggregate particles (dry)

V_s = volume of solid aggregate

V_{wi} = volume of water-impermeable pores

γ_w = unit weight of water

V_{wp} = volume of water-permeable pores

ASTM C127 and C128 give the procedure for the determination of the specific gravity (apparent) for coarse aggregate and fine aggregate, respectively.

The *bulk specific gravity* is the ratio of the weight of aggregate particles to the weight of a volume of water equal to volume of solid aggregate, pores impermeable to water, and pores permeable to water. In viewing the previous diagram, the bulk specific gravity is as follows:

$$\text{bulk specific gravity} = \frac{W_s}{(V_{wp} + V_{wi} + V_s)\gamma_w} \quad \text{or} \quad \frac{W_s}{V_w} \tag{9.7}$$

where V is the total volume of aggregate. The same ASTM procedures, ASTM C127 and C128, may be used to determine the bulk specific gravity.

In the determination of the air-void content in bituminous concrete mixtures, the use of the apparent or bulk specific is incorrect. If the apparent specific gravity is used, this is incorrect because this assumes that the permeable voids are filled with bitumen to the same extent as with water. On the other hand, if the bulk specific gravity is used, the asphalt is assumed not to penetrate into the permeable voids. Therefore, effective specific gravity is used.

The *effective specific gravity* is the ratio of the weight of aggregate particle to the weight of a volume of water equal to the volume of solid aggregate and pores impermeable to asphalt. Again using a diagram, the effective specific gravity is illustrated.

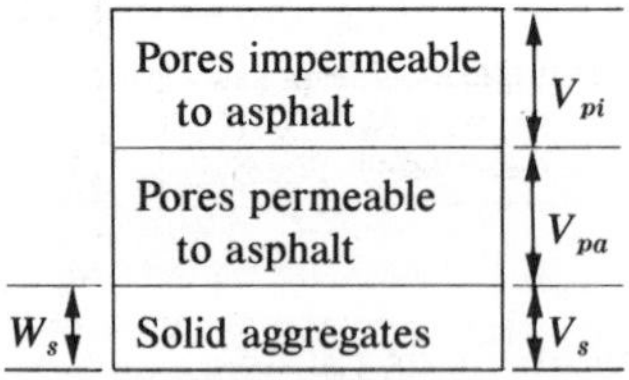

$$\text{effective specific gravity} = \frac{W_s}{(V_s + V_{pi})\gamma_w} \tag{9.8}$$

where V_{pi} = pores impermeable to asphalt

V_{pa} = pores permeable to asphalt

Generally speaking, an average effective specific gravity can be obtained as follows:

$$\text{effective specific gravity} = \frac{\text{apparent SG} + \text{bulk SG}}{2} \tag{9.9}$$

Marshall Mix Design

The Marshall mix design procedure is probably the most widely used method of bituminous mix design. It is governed by ASTM C1559 (Resistance to Plastic Flow of Bituminous Mixtures Using Marshall Apparatus). This method covers the measurement of the resistance to plastic flow of cylindrical specimens of bituminous paving mixtures loaded on the lateral surface by means of the Marshall apparatus. This method is for use with mixtures containing asphalt cement, asphalt cutback or tar, and aggregate up to 1 in. (2.54 cm) maximum size.

In this procedure, a 4-in. (10.16-cm) diameter by 2.5-in. (6.35-cm)-high specimen is prepared by compacting in a mold with a compaction hammer that weighs 10 lb (4.54 kg) and has a free fall of 18 in. (45.72 cm). Depending upon the expected design traffic, either 35, 50, or 75 blows of the hammer are applied to each side of the specimen. After 24 hours of curing, the density and voids are determined and the specimen is heated to 140°F (60°C) for the Marshall stability and flow tests. The specimen is situated in a cylindrical-shaped half-split breaking head and the specimen is loaded at a rate of 2 in./min (5.08 cm/min). The maximum load registered, in pounds (kilograms) is referred to as the Marshall stability of the specimen. The amount of movement between no load and maximum load is referred to as *strain* and is measured in units of 0.01 in.; it is generally referred to as the *flow.*

Figure 9-2 illustrates test property curves for hot mix-design data by the Marshall method. As shown, five curves are presented: unit weight, precent air voids, Marshall stability, percent VMA, and flow. The stability value obtained for each test specimen is modified if the thickness is greater or less than a height of 2 in. (5.08 cm).

Table 9-3 illustrates the Marshall design criteria. This information is based upon the Asphalt Institute Marshall design criteria. These criteria are applied to the curves and a suitable asphalt percentage is determined. The most common procedure for doing this is to take the most desirable asphalt percentages for stability, unit weight, and percent voids and to average them. This average value should satisfy the required criteria.

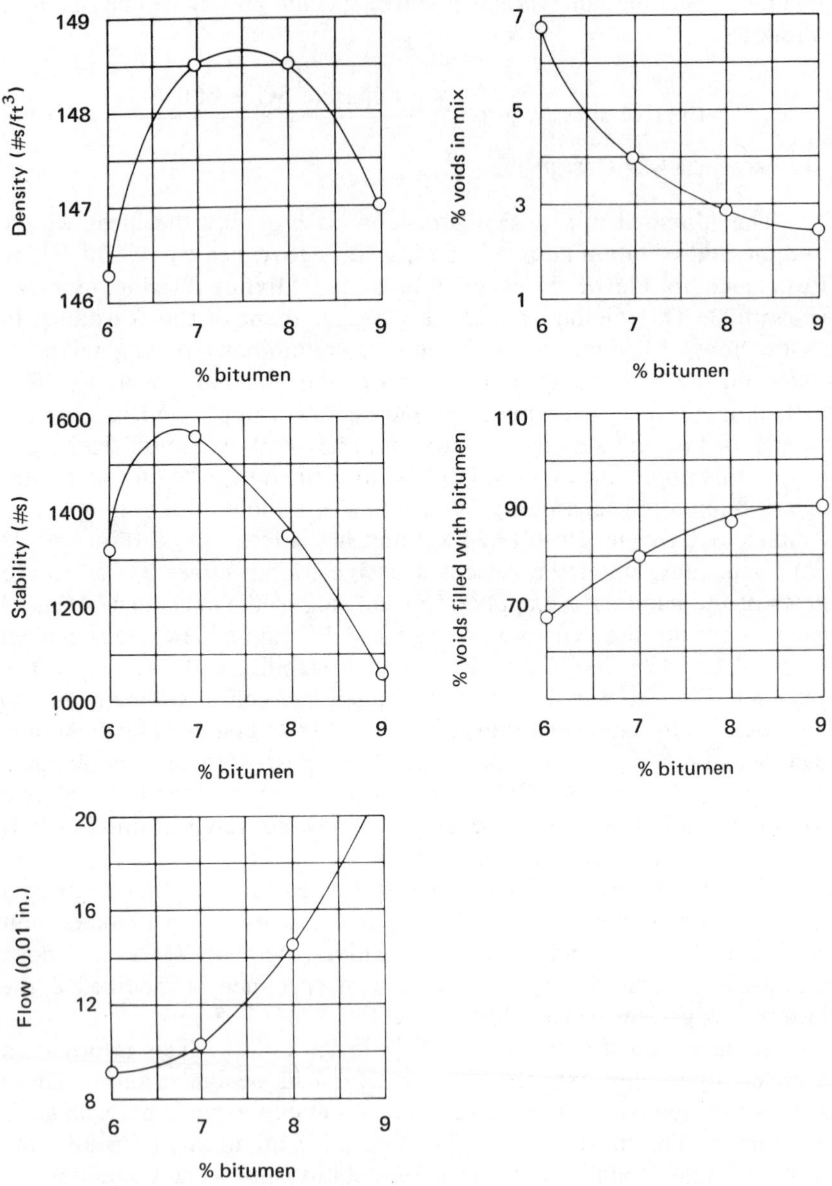

Fig. 9-2 Typical Test Data for Marshall Mix

TABLE 9-3 Marshall Design Criteria

Traffic Category:	*Heavy and Very Heavy*		*Medium*		*Light*	
Number of Compaction Flows Each End of Specimen:	*75*		*50*		*35*	
Test Property	*Min.*	*Max.*	*Min.*	*Max.*	*Min.*	*Max.*
Stability, all mixtures	750	—	500	—	500	—
Flow, all mixtures	8	16	8	18	8	20
Percent air voids						
Surfacing or leveling	3	5	3	5	3	5
Sand or stone sheet	3	5	3	5	3	5
Sand asphalt	5	8	5	8	5	8
Binder or base	3	8	3	8	3	8
Percent voids in mineral aggregate	Varies with particle size					

PROBLEMS

9.1. Define asphalt cement.

9.2. List and explain the purpose of the hydrocarbons that make up bitumen.

9.3. What is the best type of petroleum found in the earth's crust? Why?

9.4. Explain the distillation process of crude oil.

9.5. Discuss the specific ASTM standards for asphalt cements. Give the purpose and procedure for each test.

9.6. Explain liquid asphalts and discuss in detail.

9.7. Describe the manufacturing process for road tars.

9.8. List and describe the four essential properties of a good bituminous mix design.

9.9. What are the seven basic steps in proportioning asphaltic concrete? Explain.

9.10. Calculate the percent voids and VMA in a compacted mixture with a unit weight of 135 lb/ft^3 with the following criteria:

Ingredient	*Percent of Total Aggregate*	*Effective Specific Gravity*
Coarse aggregate	76	2.67
Fine aggregate	20	2.65
Filler aggregate	4	2.55

Percent asphalt = 6 with an effective specific gravity = 1.00

REFERENCES

Asphalt as a Material, *Asphalt Inst. Inform. Ser. 93,* 1965.

Asphalt Institute, *Asphalt Paving Manual,* 1962.

Asphalt Institute, "Brief Introduction to Asphalt," *Manual MS-5,* 1974.

BROOME, D. C., "Native Bitumens," in Arnold J. Holberg, *Bituminous Materials: Asphalts, Tars and Pitches,* vol. 2, part 1, Interscience Publishers, John Wiley & Sons, Inc., New York, 1965.

GOETZ, W. H., and WOOD, L. E., "Bituminous Materials and Mixtures," in K. B. Woods, *Highway Engineering Handbook,* sec. 18, McGraw-Hill Book Company, New York, 1960.

KREBS, R. D., and WALKER, R. D., *Highway Materials,* McGraw-Hill Book Company, New York, 1971.

WHITEHURST, E. A., and GOODWIN, W. A., *Bituminous Materials and Bitumen Aggregate Mixes,* Pitman Publishing Corp., New York, 1958.

PLASTICS 10

A textbook on materials for civil and highway engineers would not be complete without a chapter on plastics, as plastics comprise an important group of materials of construction. Many plastic substances have combinations of properties that cannot be duplicated by other materials. In general, plastics exhibit a number of outstanding characteristics:

1. Lightness in weight (generally half as light as aluminum).
2. High dielectric strength (electrical insulation).
3. Low heat conductivity (heat insulation).
4. Special properties toward lights (colorability).
5. Extremely resistant toward chemicals and organic liquids.
6. Metal inserts may be molded into the plastic (since plastics are inert toward such materials).
7. Many high-quality products can be developed by using a lathe, sawing, punching, and drilling.

CLASSIFICATION

A *plastic* is a polymeric material (usually organic) of high molecular weight which can be shaped by flow. The term usually refers to the final product, with fillers, plasticizers, pigments, and stabilizers included (as opposed to the resin, the homogeneous polymeric starting material). Most plastics are synthetic organic compounds which derive their coherence and strength from large chain-linked macromolecules formed from one or more single molecules (monomers) to a macromolecule (polymer) by a chemical reaction called *polymerization*. Examples include polyvinyl chloride, polyethylene, and urea–formaldehyde. Most organic plastics contain a *binder,* which imparts the plastic properties to the composition. In addition, many plastics contain a *filler,* an inert extender that adds hardness, strength, and other desirable properties. In general, organic plastics can be divided into three general classifications:

1. Thermoplastics.
2. Thermosetting plastics.
3. Chemically setting plastics.

Table 10-1 shows the classification of organic plastics.

Thermoplastics

Thermoplastics are organic plastics, either natural or synthetic, which remain permanently soft at elevated temperatures. Upon cooling, they again become hard. These materials can be shaped and reshaped any number of times by heating and cooling and repeating. Natural thermoplastics include asphalts, bitumen, pitches, and resin, to name some of the most familiar.

Thermosetting Plastics

Thermosetting plastics are organic plastics that were originally soft or soften at once upon heating, but upon further heating, they harden permanently. Thermosetting plastics are hardened by chemical changes due to heat, a catalyst, or to both. Thermosetting plastics remain hardened without cooling and do not soften appreciably when reheated. The most common thermosetting plastic is polyester.

Chemically Setting Plastics

Chemically setting plastics are those that harden by the addition of a suitable chemical to the composition just before molding or by subsequent chemical treatment following fabrication.

TYPES OF PLASTICS

Polymerization and Condensation

Polymerization involves unsaturated molecules that contain double or triple bonds between carbon atoms which are weaker than single bonds. Unsaturated molecules are unstable and they react in such a way as to break the multiple bond. The mechanism by which polymerization takes place is grouped into two categories:

1. Addition polymerides.
2. Condensation polymerides.

TABLE 10-1 Classification of Organic Plastics

Plastic Binder

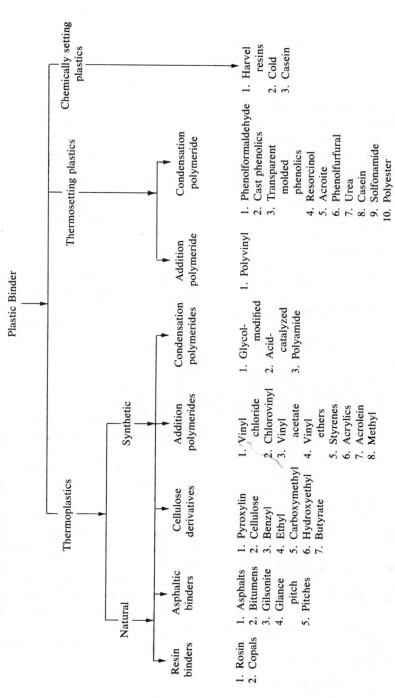

Thermoplastics					Thermosetting plastics		Chemically setting plastics
Natural			Synthetic				
Resin binders	Asphaltic binders	Cellulose derivatives	Addition polymerides	Condensation polymerides	Addition polymeride	Condensation polymeride	
1. Rosin 2. Copals	1. Asphalts 2. Bitumens 3. Gilsonite 4. Glance pitch 5. Pitches	1. Pyroxylin 2. Cellulose 3. Benzyl 4. Ethyl 5. Carboxymethyl 6. Hydroxyethyl 7. Butyrate	1. Vinyl chloride 2. Chlorovinyl 3. Vinyl acetate 4. Vinyl ethers 5. Styrenes 6. Acrylics 7. Acrolein 8. Methyl	1. Glycol-modified 2. Acid-catalyzed 3. Polyamide	1. Polyvinyl	1. Phenolformaldehyde 2. Cast phenolics 3. Transparent molded phenolics 4. Resorcinol 5. Acroite 6. Phenolfurfural 7. Urea 8. Casein 9. Solfonamide 10. Polyester	1. Harvel resins 2. Cold 3. Casein

Addition polymerides are mixtures of polymers that have been formed by addition of like molecules. The molecules are added to increase the average molecular size and weight. This reaction occurs by the breaking of double bonds between atoms in monomers and the forming of two single bonds in their place.

Condensation polymerides are formed by chemical reactions in which two or more different molecules combine with the separation of water or other simple substances in the formation of resins. Condensation polymerides are produced by a by-product as well as the growing polymer molecules. The nonpolymerizable molecule is usually water or some simple molecule.

Copolymers are produced by mixing monomers and then polymerizing. In other words, copolymers are produced by the simultaneous polymerization of two or more chemically different monomers.

Thermoplastics

Properties and uses of common plastics of the thermoplastic type are shown in Table 10-2. Thermoplastic types of binders are divided into two groups: natural and synthetic. Of the *natural* type two categories exist: resin and asphaltic binders. The resin binders at this point in materials of construction are of little importance. The asphaltic binders were thoroughly discussed in Chapter 9.

The *synthetic* group consists of three categories: cellulose derivatives, addition polymerides, and condensation polymerides. *Cellulose* plastic is the term employed for plastics made from derivatives of cellulose, such as pyroxylin, cellulose acetate, benzyl cellulose, ethyl cellulose, carboxymethyl cellulose, hydroxyethyl cellulose, and cellulose acetate butyrate. Cellulose plastics are true thermoplastics and exhibit the greatest toughness and resilience of any of the plastics. They are used for objects having thin-walled sections where other plastics may be too brittle.

Addition and condensation polymerides are grouped together and referred to as *noncellulose* type. This group consist of vinyl chloride, chlorovinyl chloride, vinyl acetate, vinyl ethers, styrene, acrylics, acrolein, methyl methacrylate, cumarone–indene, glycol-modified alkyds, acid-catalyzed phenol–formaldehyde, and polyamide (nylon). The noncellulose plastics are all oderless, tasteless, nontoxic, and transparent. These plastics are strong, tough, and chemically inert with little water absorption. The noncellulose plastics are suited for electrical insulation or cable coatings, safety glass, nylon, and Saran (covering to prevent corrosion and/or chemical attack on pipes and tubing).

Thermosetting Plastics

Thermosetting plastics can be divided into two groups: addition polymerides and condensation polymerides. Addition polymerides consist of any polyvinyl types of resins; condensation polymerides, the largest group, consist of phenolformaldehyde, cast phenolics, transparent molded phenolics, resorcinol–formaldehyde, acroite, phenolfurfural, urea, casein–formaldehyde, sulfonamide resins, and polyesters.

Plastics of the thermosetting group have excellent mechanical and electrical properties and are highly resistant to heat (Table 10-3). These plastics are also highly resistant to water, oil, alkalies, and acids. They also exhibit very little shrinkage. Thus, they are excellent where high first degree of precision is required. In this group the phenol–formaldehyde resins are probably the most important resins. The phenol–formaldehyde group can be subdivided into four general classes:

1. Cellulose-filled compositions.
2. Mineral-filled compositions.
3. Molding sheet.
4. Impact-resistant materials.

The *cellulose-filled composition* utilizes wood flour as the filler and the final product has high dielectric strength, mechanical strength, and it is very light. Uses have been found in the aerospace and automotive industries.

The *mineral-filled composition* utilizes asbestos as the filler. Thus, articles made of these plastics resist chemicals and heat. Uses are for insulation of high-voltage transmissions.

Molding sheets are made by impregnating paper with phenolic resins. These sheets can be softened by heat and become brittle in the cold weather. These sheets are used for molding applications where the material must be able to give.

In the final group, *impact-resistant materials,* the plastic is made by impregnating paper and fabric fillers, built up in layers. This product is called laminated plastic.

Chemically Setting Plastics

In the chemically setting group three categories of groups exist: harvel resins, cold-molded plastics, and casein. These plastics are generally used where resistance to heat and arcing are of primary importance. Thus, they are utilized for electrical insulating parts.

(*text continues on page 262*)

TABLE 10-2 Properties and Uses of Common Plastics (Thermoplastics)

Property	*Shellac*	*Polyethylene*	*Polymono-chloro-trifluro-ethylene*	*Vinylidene Chloride Molding*	*Polystyrene*	*Methyl methacryl-ate, Cast*	*Polyamide (Nylon) Molding*	*Cellulose Acetate Molding*	*Cellulose Nitrate (Pyroxylin)*
Injection molding pressure (1000 psi)	1.00–1.20	8–15	20–60	10–30	10–30	—	10–25	8–32	
Specific gravity	1.1–2.7	0.92	2.10	1.65–1.72	1.05–1.07	1.18–1.20	1.14	1.27–1.37	1.35–1.40
Tensile strength (1000 psi)	0.9–2.0	1.5–1.8	5.7	3–5	5–9	6–7	7–9	1.9–8.5	7–8
Elongation (% in 2 in.)	—	50–400	28–36	20–250	1–3.6	2–7	40–100	6–50	40–45
Modulus of elasticity in tension (100,000 psi)	5–6	0.19	1.9	0.5–0.8	4–6	3.5–5	2.6–4	0.86–4.0	1.9–2.2
Compressive strength, (1000 psi)	10–17	—	32–80	7.5–8.5	11.5–16	11–19	7.2–13	13–36	22–35
Impacts strength, Izod test on ½ × ½-in. notched bar (ft-lb./in. width of notch)	2.6–2.9	Less than 16	3.6	0.3–1.0	0.26–0.50	0.4–0.5	1.0	0.4–5.2	5–7
Hardness, Rockwell[a]	—	R11	R110–115	M50–65	M65–90	M90–100	M111–118	R50–125	R95–115

Highest usable temperature continuous (°F)	150–190	212	390	160–200	150–205	140–200	270–300	140–220	140
Thermal conductivity 10^{-4} cal/ (sec) (cm^3) (°C)	—	8	1.4	3	2.4–3.3	4–6	5.2–5.5	4–8	3.1–5.5
Thermal expansion (10^{-5} in./in. °C)	—	16–18	4.5–7.0	19	6–8	9	10–15	8–16	8–12
Dielectric strength, short time, ⅛ in. thickness (V/mil)	200–600	400	2500	350	500–700	450–500	385–470	250–365	300–600
Water absorption 24 hr, ⅛ in. thick (%)	0–0.1	Less than 0.01	0.00	0–0.1	0.03–0.05	0.3–0.4	0.4–1.5	1.9–6.5	1.0–2.0
Effect of strong acids	Deteriorated	Attacked by oxidizing acids	None	Highly resistant	Attacked by oxidizing acids	Attacked by oxidizing acids	Attacked	Decomposed	Decomposed
Color possibilities	Limited	Unlimited	Unlimited	Extensive	Unlimited	Unlimited	Unlimited	Unlimited	Unlimited
Common uses	Phonograph records, electrical insulation	Bottle stoppers, flexible bottles, wire insulation, textiles, tablewear	Filter disks, insulators, gaskets	Screening, chemical tubing, auto seat covers	Electrical insulators, battery boxes, lenses, toys, boxes	Windows, furniture, dentures, picture frames	Bearings, cups, fabrics, bristles	Fountain pens, tools, toys, spectacles, packaging	Packaging, foils, glazing, materials, photographic film

[a]Rockwell scales: M, ¼-in.-diameter ball, 100-kg major load; R, ½-in.-diameter ball: 60-kg major load.

TABLE 10-3 Properties and Uses of Common Plastics (Thermosetting)

Property	*Phenol–Formaldehyde Resin*							
	Macerated Cotton Fabric or Cord Filler, Molded	*Mechanical Grade, No Filler, Cast*	*Urea–Formaldehyde, α-Cellulose, Molded*	*Melamine–Formaldehyde, Asbestos, Paper or Fabric Laminate*	*Polyester, Glass Fiber, Mat, Laminate*	*Silicone Glass Fabric Laminate*	*Cold-molded[a] Cement Binder Asbestos-filled*	*Hard Rubber,[b] No Filler*
Compression molding pressure (1000 psi)	2.00–8.00	0	2.00–8.00	1.00–1.80	0.01–0.15	1.00–2.00	1.00–10.00	1.20–1.80
Specific gravity	1.34–1.47	1.25–1.30	1.45–1.55	1.75–1.85	1.5–1.8	1.6–1.8	1.6–2.2	1.4
Tensile strength (1000 psi)	2–9	4–7	6–13	6.5–12	10–20	10–25	1.6–2.2	8–10
Elongation (% in 2 in.)	0.4–0.6	Very small	0.5–1.0	Very small	Very small	Very small	Very small	5–7.5
Modulus of elasticity in tension (100,000 psi)	9–13	5–7	12–15	16–39	10–19	20	—	3.0
Compressive strength (1000 psi)	15–30	15–20	25–35	27–50	30–50	35–46	16	8–12
Impact strength, Izod test on ½ × ½ in. notched bar (ft-lb/in. width of notch)	1–8	0.3–0.4	0.24–0.36	0.7–5.0	11–25	5–22	0.4	0.5
Hardness, Rockwell[c]	M110–120	M70–110	M115–120	M110–115	M90–100	M100	M75–95	HR95
Highest usable temperature, continuous (°F)	250	250	170	225–245	300–400	400–480	900–1300	—

Thermal conductivity [10^{-4} cal/(sec) (cm^3) (°C)]	4–7	3–5	7–10	10–17	8–12	3.5	—	2.9
Thermal expansion (10^{-5} in./in. °C)	1–4	8–11	2.5–4.5	2.0–4.8	1.0–3.0	0.5	—	7.7
Dielectric strength, short time ⅛ in. thickness (V/mil)	200–400	—	300–400	40–150	250–400	200–480	45	470
Water absorption, 24 hr, ⅛ in. thick (%)	0.04–1.8	0.2–0.4	0.4–0.8	1–5	0.3–1.0	0.2–0.7	0.5–15	0.02
Effect of strong acids	Decomposed by oxidizing acids	Decomposed by oxidizing acids	Decomposed	Decomposed	Some attack	Very slight	Decomposed	Attacked by oxidizing acids
Color possibilities	Limited	Limited	Unlimited	Limited	Unlimited	Limited	Gray and black	Limited
Common uses	Serving trays, radio cabinets, electrical parts	Punches and dies	Tablewear, electrical controls, housings	Aircraft structural parts, high-strength electrical parts, stove switches	Aviation and automotive structures, decorative applications	High-temperature-resisting electrical insulation	Arc-shield terminal insulators, electric heater elements	Beakers, funnels, etc., for chemicals, combs

[a]The cement binder is not strictly a thermosetting material but is set by chemical combination with water from steam (hydration).
[b]Hard rubber is not usually classified as thermosetting but as vulcanizing.
[c]Rockwell scales: M, ¼-in.-diameter ball, 100-kg major load; HR (hard rubber), ¼-in.-diameter ball, 60-kg major load.

TABLE 10-4 Mechanical Properties of Molded Compositions

Type of Material	*Tensil Strength (psi)*	*Impact Strength Notched Bar, (ft-lb./in.)*	*Modulus of Rupture (psi)*	*Modulus of Elasticity (psi)*	*Specific Gravity*	*Heat Distortion (°C)*	*Hardness*	*Coefficient of Linear Expansion per °C*
Phenolic laminated, paper base	6,000–13,000	0.8–2.4C[a]	13,000–20,000	$1.0–2.0 \times 10^{6}$	1.34–1.55	100–140	85–125R[b]	$20–50 \times 10^{-6}$
Phenolic laminated, canvas base	8,000–12,000	1.6–10.4C	12,000–19,000	$1.0–2.0 \times 10^{6}$	1.34–1.55	100–140	95–115R	$30–70 \times 10^{-6}$
Phenolic molded, woodflour filled	6,000–11,000	0.26–0.50C	8,000–15,000	$0.8–1.5 \times 10^{6}$	1.25–1.52	120–140	95–120R	$35–80 \times 10^{-6}$
Phenolic molded, cellulose filled	6,000–11,000	0.40–0.80C	8,000–15,000	$0.8–1.5 \times 10^{6}$	1.32–1.48	120–140	95–115R	$35–80 \times 10^{-6}$
Phenolid molded, fabric filled	6,000–8,000	0.80–6.0C	8,000–13,000	$0.8–1.5 \times 10^{6}$	1.35–1.40	120–140	90–115R	$35–80 \times 10^{-6}$
Phenolic molded, mineral filled	5,000–9,000	0.26–1.0C	8,000–18,000	$1.0–5.0 \times 10^{6}$	1.70–2.05	120–150	100–120R	$25–50 \times 10^{-6}$
Cast phenolics	3,000–7,000	0.30–0.50C	3,000–14,000	$0.25–0.75 \times 10^{6}$	1.26–1.70	40–80	70–110R	$70–160 \times 10^{-6}$
Phenolic molded, transparent	8,000	—	1,600	—	1.27	107	—	—
Furfuryl-phenol molded, woodflour filled	5,000–12,000	1.0–6.5I[c]	10,000–16,000	—	1.3–1.4	131	35–40B[d]	30×10^{-6}
Furfuryl-phenol molded, asbestos filled	4,000–12,000	1.0–6.0I	8,000–14,000	—	1.6–2.0	136	44–46B	20×10^{-6}

Furfuryl-phenol molded, fabric filled	5,000–10,000	20–39I	10,000–16,000	—	1.3–1.4	—	30–35B	—
Furfuryl-phenol laminated, paper base	10,000–20,000	5.0–20I	20,000–30,000	—	1.3–1.4	—	—	—
Furfuryl-phenol laminated, cloth base	9,000–12,000	10.0–50I	—	—	1.3–1.4	—	—	—
Urea molded	9,000–12,000	0.28–0.36C	10,000–14,000	$1.2–1.9 \times 10^6$	1.45–1.55	95–130	110–125R	—
Polystyrene molded	5,000–7,000	0.40–0.60C	6,000–8,000	$0.40–0.60 \times 10^6$	1.05–1.07	75–80	82–92R	$65–75 \times 10^{-6}$
Acrylate molded	4,000–8,000	0.3–4.0C	9,000–16,000	$0.4–0.6 \times 10^6$	1.18–1.19	51–60	18–20B	80×10^{-6}
Methyl methacrylate molded	9,000–12,000	0.2–3.0C	12,000–14,000	—	1.18–1.20	60–135	17–20B	$70–90 \times 10^{-6}$
Cellulose nitrate molded								
Cellulose acetate molded	4,900–8,500	10–11.5I	5,000–8,000	$0.2–0.4 \times 10^6$	1.35–1.60	71–91	7–12B	$120–160 \times 10^{-6}$
	4,000–5,000	1.7–3.0C	5,000–7,000	$0.20–0.40 \times 10^6$	1.27–1.63	50–80	85–120R	$140–160 \times 10^{-6}$
Ethyl cellulose	5,000–9,000	3.2–9.6I	—	0.3×10^6	1.14	100–130	—	—
Shellac	900–2,000	—	—	—	1.1–2.7	66–90	—	—
Cold-molded composition	700–1,700	1.5–4.5	3,500–7,800	—	1.9–2.12	182–260	—	—
Hard rubber	1,500–10,000	0.5I	—	0.33×10^6	1.12–1.80	—	31B	80×10^{-6}

[a]C, Charpy.
[b]R, Rockwell.
[c]I, Izod.
[d]B, Brinell.

TABLE 10-5 Electrical Properties of Molded Compositions

	Power Factor			Dielectric Constant		
Type of Material	*60 cycles/sec*	*1000 cycles/sec*	*10^6 cycles/sec*	*60 cycles/sec*	*1000 cycles/sec*	*10^6 cycles/sec*
Phenolic laminated, paper base	0.02–0.15	0.02–0.10	0.02–0.06	4.5–6.5	4.5–6.0	4.0–5.5
Phenolic laminated, canvas base	0.05–0.30	0.05–0.20	0.04–0.10	5.0–9.0	4.5–8.0	4.0–6.0
Phenolic molded, woodflour filled	0.02–0.30	0.02–0.15	0.035–0.08	4.5–10.0	4.5–10.0	4.0–8.0
Phenolic molded, cellulose filled	0.05–0.30	0.04–0.15	0.04–0.10	4.5–10.0	4.5–10.0	4.0–8.0
Phenolic molded, fabric filled	0.05–0.30	0.04–0.15	0.04–0.10	4.5–15	4.5–15	4.0–10
Phenolic molded, mineral filled	0.020–0.50	0.01–0.20	0.005–0.10	4.5–20	4.5–15	4.0–10
Cast phenolics	0.070–0.50	0.030–0.30	0.04–0.13	7.0–30	6.0–25	5.5–15
Phenolic molded, transparent	0.06	0.04	0.019	5.5	5.0	4.5
Furfuryl-phenol molded, woodflour filled	—	0.04–0.15	0.01–0.06	—	4.0–8.0	6–7.5
Furfuryl-phenol molded, asbestos filled	—	0.1–0.15	0.06–0.15	—	4.5–20	5–18
Furfuryl-phenol molded, fabric filled	—	0.08–0.2	0.05–0.08	—	4.5–6	6
Furfuryl-phenol laminated, paper base	—	—	0.15–0.48	—	4.5	—
Furfury-phenol laminated, cloth base	—	—	0.41–0.64	—	4.5	—
Urea molded	0.050–0.13	0.035–0.07	0.035–0.040	8.0–10	8.0–9.0	6.9–7.5
Polystyrene molded	0.0001–0.0002	0.0001–0.0002	Under 0.0002	2.6	2.6	2.6
Acrylate molded	0.05–0.06	—	—	3.4–3.6	—	—
Methyl methacrylate, molded	0.06–0.08	—	0.02	4.0–4.4	—	2.8
Cellulose nitrate, molded	0.062–0.149	—	0.07–0.09	6.7–7.3	—	6.2
Cellulose acetate, molded	0.03–0.05	0.035–0.07	0.035–0.07	4.5–7.0	4.5–6.5	4.0–4.5
Ethyl cellulose, molded	0.03–0.06	0.025	—	2.6–2.9	3.9	—
Shellac, molded filled	0.004–0.018P[a]	—	—	3–4	—	—
Cold-molded composition	0.2	—	0.07	15.0	—	6.0
Hard rubber	—	—	0.003–0.008	2.8	2.9–3.0	3.0

[a]P, Paper filled.

Type of Material	Loss Factor			Resistivity	Dielectric Strength Step Test
	60 cycles/sec	*1000 cycles/sec*	*10^6 cycles/sec*	*(megohm-cm)*	*(V/mil)*
Phenolic laminated, paper base	0.1–1.0	0.1–0.6	0.10–0.30	2×10^6–1×10^8	400–600
Phenolic laminated, canvas base	0.25–2.0	0.15–1.0	0.15–1.0	3×10^5–4×10^7	250–400
Phenolic molded, woodflour filled	0.10–3.0	0.20–1.5	0.15–0.80	10^4–10^6	200–300
Phenolic molded, cellulose filled	0.25–3.0	0.20–1.5	0.15–1.0	10^4–10^5	200–300
Phenolic molded, fabric filled	0.25–4.5	0.20–2.0	0.15–1.0	10^3–10^5	200–300
Phenolic molded, mineral filled	0.10–4.0	0.045–3.0	0.020–1.0	10^3–10^8	200–375
Cast phenolics	0.70–16.0	0.20–4.0	0.20–2.0	10^4–10^7	120–300
Phenolic molded, transparent	—	—	—	7.5×10^5	325
Furfuryl-phenol molded, woodflour filled	—	—	—	10^{10}–10^{12}	400–600
Furfuryl-phenol molded, asbestos filled	—	—	—	10^9–10^{11}	200–500
Furfuryl-phenol molded, fabric filled	—	—	—	10^{10}	200–500
Furfuryl-phenol laminated, paper base	—	—	—	—	900–1800
Furfury-phenol laminated, cloth base	—	—	—	—	300–700
Urea molded	0.40–1.2	0.28–0.65	0.24–0.30	1–10×10^6	275–325
Polystyrene molded	0.00026–0.00053	0.00026–0.00053	Under 0.0005	Over 10^{10}	500–525
Acrylate molded	—	—	—	1.0×10^{15}	500
Methyl methacrylate, molded	—	—	—	1.0×10^{15}	480
Cellulose nitrate, molded	—	—	—	2–30×10^{10}	660–780
Cellulose acetate, molded	0.13–0.30	1.15–0.40	0.15–0.40	10^6–10^8	300–350
Ethyl cellulose, molded	—	—	—	—	1500
Shellac, molded filled	0.016–0.16P	—	—	10^8–10^{16}	200–600
Cold-molded composition	—	—	—	1.3×10^{12}	85–100
Hard rubber	—	—	—	10^{12}–10^{15}	250–1000

MANUFACTURE OF ORGANIC PLASTICS

In general, four main steps are required in the manufacturing of articles made from organic plastics:

1. The production of intermediate materials (chemical) from the raw materials of coal, petroleum, and cotton.
2. The manufacture of synthetic resins from the above.
3. The preparation of molding powders, fillers, rods, and sheets from step 2.
4. Molding the articles from the powders, fillers, rods, and sheets.

Methods of Forming and Fabricating Plastics

Most of the methods utilized in the forming and fabrication of plastic have their counterpart in the processing of other materials of engineering. Examples include the following:

1. Casting.
2. Compression molding.
3. Injection molding.
4. Transfer molding.
5. Extruding.
6. Blowing.
7. Laminating.

Casting is the simplest molding method available. Casting utilizes low-cost lead–antimony dies (molds) where the melted resins are poured into open molds and cured. The cast resins are baked in an oven at about 167°F (75°C) for several days, until a permanent set is obtained. Both thermoplastics and thermosetting plastics may be cast.

Compression molding is the most widespread molding operation for thermosetting plastics. Compression molding requires the use of pressure on the molding compound placed in the mold of heated platens of the press. Thermoplastics may be compression-molded, but the mold must be cooled after each molding before the article is removed from the press. This is time-consuming and uneconomical, so other methods are utilized for thermoplastic plastics.

Injection molding is one of the most widely used and most rapid methods of producing articles of intricate shape. Injection molding consists of forcing softened plastic materials into a closed mold maintained at a temperature below the softening point of the compression. This method is widely used for thermoplastic materials. The thermoplastic material is fed into a pressure chamber and a plunger compresses the material and forces it into a heating chamber, where it is progressively heated to a uniform temperature. The thermoplastic material emerges through a nozzle in a thoroughly softened condition and is then forced into a cold mold, where pressure causes it to take the shape of the cavity. The plastic is cooled, the mold is opened, and the part removed.

Transfer molding is used for thermosetting materials when the article is to include delicate metal inserts for any reason. In this method the molding compound is heated under pressure until it is soft enough to flow. The compound is then placed into a cavity in which the part is formed and allowed to cool.

Extruding is basically the opposite of injection molding. Extruding of thermoplastic plastics is a widely used practice for producing rods, tubes, and other cylindrical shapes.

Blowing of thermoplastics is a process by which hollow objects are made. It is accomplished in much the same manner as the blowing of glass.

Laminating is used to produce hard boards or sheets of resin-impregnated papers, wood veneers, and/or fabrics. Laminates are made into stock sizes of flat sheets under medium to high pressure in a press. The process involves building up individual resin-coated fabric or veneer over a form, inserting the assembly into a large rubber bag (called bag-

TABLE 10-6 Strength–Weight Relationships for Plastic and Some Common Structural Materials[a]

Material	*Specific Gravity*	*Tensile Strength (psi)*	*Tensile Strength / Specific Gravity*	*Tensile Strength / (Specific Gravity)²*
Chrome–vanadium steel	7.85	164,000	20,800	2,700
Structural steel	7.83	65,000	8,300	1,060
Cast iron, gray	7.0	35,000	5,000	720
Titanium alloy	4.7	145,000	31,000	6,600
Aluminum alloy	2.8	56,000	20,000	7,200
Magnesium alloy	1.81	44,000	24,300	13,400
Glass fabric laminate	1.9	45,000	23,600	12,500
Asbestos cloth laminate	1.7	9,000	5,300	3,100
Paper laminate	1.33	20,000	15,000	11,300
Cellulose acetate	1.3	5,000	3,900	3,000
Methyl methacrylate	1.18	8,500	7;200	6,100
Polystrene	1.06	5,500	5,200	4,900
Polyethylene	0.92	1,300	1,400	1,500
Sitka spruce	0.40	17,000	42,500	106,000

[a]The values given here are average values. Different compositions, heat treatments, etc., of a given material may differ widely from the values shown here.

molding process), withdrawing the air from the bag, and subjecting the bag and contents to pressure in an autoclave.

PROPERTIES OF ORGANIC PLASTICS

Table 10-4 shows the mechanical properties of molded compositions. Table 10-5 shows the electrical properties of molded compositions which most plastics are used for. Table 10-6 illustrates the strength–weight relationships for plastics and some common structural materials.

PROBLEMS

10.1. List seven outstanding characteristics of plastics.
10.2. Define organic plastic.
10.3. List the three general classifications of organic plastics.
10.4. Describe a thermoplastic plastic and list several.
10.5. Describe a thermosetting plastic and list several.
10.6. Describe a chemically setting plastic and list several.
10.7. Explain the difference between polymerization and condensation.
10.8. Describe the procedure for manufacturing organic plastics.
10.9. List and describe the methods of forming and fabricating plastics.
10.10. List the mechanical properties of organic plastics and discuss how they compare to other materials of construction.

REFERENCES

American Society for Testing Materials, *Symposium on Plastics,* Philadelphia, 1938.

GILMORE, G., and SPENCER, R., "Injection Molding Process," *Modern Plastics,* No. 27 (Apr. 1950), p. 143.

MACTAGGORT, E. F., and CHAMBERS, H. H., *Plastics and Building,* Sir Isaac Pitman & Sons Ltd., London, 1951.

MILLS, A. P., HAYWARD, H. W., and RADER, L. F., *Materials of Construction,* John Wiley & Sons, Inc., New York, 1955.

MOORE, H. F., and MOORE, M. B., *Materials of Engineering,* McGraw-Hill Book Company, New York, 1953.

APPENDICES

American National Standard Z168.13-1971
American National Standards Institute

Designation: E 8 – 69ε

Standard Methods of

TENSION TESTING OF METALLIC MATERIALS[1]

This Standard is issued under the fixed designation E 8; the number immediately following the designation indicates the year of original adoption or, in the case of revision, the year of last revision. A number in parentheses indicates the year of last reapproval.

These methods have been approved by the Department of Defense to replace method 211.1 of Federal Test Method Standard No. 151b and for listing in DoD Index of Specifications and Standards. Future proposed revisions should be coordinated with the Federal Government through the Army Materials and Mechanics Research Center, Watertown, Mass. 02172.

ε NOTE 1—Because detailed drawings for jig construction are no longer available from ASTM, Footnote 3 was editorially deleted in January 1973.

NOTE 2—Editorial changes were made in Notes 1 and 9 in March 1974.

1. Scope

1.1 These methods cover the tension testing of metallic materials in any form.

NOTE 1—Exceptions to the provisions of these methods may need to be made in individual specifications or methods of test for a particular material. For example, see ASTM Methods and Definitions A 370, Mechanical Testing of Steel Products,[2] and ASTM Methods B 557, Tension Testing Wrought and Cast Aluminum and Magnesium Alloy Products.[3]

NOTE 2—The values stated in U.S. customary units are to be regarded as the standard. The metric equivalents of U.S. customary units may be approximate.

2. Definitions

2.1 The definitions of terms relating to tension testing appearing in ASTM Definitions E 6, Terms Relating to Methods of Mechanical Testing,[4] shall be considered as applying to the terms used in these methods of tension testing.

3. Apparatus

3.1 *Testing Machines:*

3.1.1 Machines used for tension testing shall conform to the requirements of ASTM Methods E 4, Verification of Testing Machines.[5] The loads used in determining tensile strength and yield strength or yield point shall be within the loading range of the testing machine as defined in Methods E 4.

3.2 *Gripping Devices:*

3.2.1 *General*—Various types of gripping devices may be used to transmit the measured load applied by the testing machine to the test specimens. To ensure axial tensile stress within the gage length, the axis of the test specimen should coincide with the center line of the heads of the testing machine. Any departure from this requirement may introduce bending stresses that are not included in the usual stress computation (load divided by cross-sectional area).

NOTE 3—The effect of this eccentric loading may be illustrated by calculating the bending moment and stress thus added. For a standard ½-in. (12.5-mm) diameter specimen, the stress increase is 1.5 percentage points for each 0.001 in. (0.025 mm) of eccentricity. This error increases to 2.24 percentage points/0.001 in. (0.025 mm) for a 0.350-in. (8.90-mm) diameter specimen and to 3.17 percentage points/0.001 in. (0.025 mm) for a 0.250-in. (6.40-mm) diameter specimen.

3.2.2 *Wedge Grips*—Testing machines usually are equipped with wedge grips. These wedge grips generally furnish a satisfactory means of gripping long bars of ductile metal and flat plate test specimens such as those shown in Fig. 6. If, however, for any reason, one grip of a pair advances farther than the other as the grips tighten, an undesirable

---

[1] These methods are under the jurisdiction of ASTM Committee E-28 on Mechanical Testing.
Current edition effective Nov. 14, 1969. Originally issued 1924. Replaces E 8 – 68.

[2] *1974 Annual Book of ASTM Standards*, Parts 1, 2, 3, 4, 5, and 10.

[3] *1974 Annual Book of ASTM Standards*, Part 7.

[4] *1974 Annual Book of ASTM Standards*, Part 10.

[5] *1974 Annual Book of ASTM Standards*, Parts 10, 14, 32, 35, and 41.

bending stress may be introduced. When liners are used behind the wedges, they must be of the same thickness and their faces must be flat and parallel. For best results, the wedges should be supported over their entire length by the heads of the testing machine. This requires that liners of several thicknesses be available to cover the range of specimen thickness. For proper gripping, it is desirable that the entire length of the serrated face of each wedge be in contact with the specimen. Proper alignment of wedge grips and liners is illustrated in Fig. 1. For short specimens and for specimens of many materials it is generally necessary to use machined test specimens and to use a special means of gripping to ensure that the specimens, when under load, shall be as nearly as possible in uniformly distributed pure axial tension (see 3.2.3, 3.2.4, and 3.2.5).

3.2.3 *Grips for Threaded and Shouldered Specimens and Brittle Materials*—A schematic diagram of a gripping device for threaded-end specimens is shown in Fig. 2, while Fig. 3 shows a device for gripping specimens with shouldered ends. Both of these gripping devices should be attached to the heads of the testing machine through properly lubricated spherical-seated bearings. The distance between spherical bearings should be as great as feasible.

3.2.4 *Grips for Sheet Materials*—The self-adjusting grips shown in Fig. 4 have proved satisfactory for testing sheet materials that cannot be tested satisfactorily in the usual type of wedge grips.

3.2.5 *Grips for Wire*—Grips of either the wedge or snubbing types as shown in Figs. 4 and 5 or flat wedge grips may be used.

3.3 *Dimension-Measuring Devices*—Micrometers and other devices used for measuring linear dimensions shall be accurate to at least one half the smallest unit to which the individual dimension is required to be measured.

4. Test Specimens

4.1 *General:*

4.1.1 Test specimens shall be either substantially full size or machined, as prescribed in the product specifications for the material being tested.

NOTE 4—Dimensions of specimens are given in U.S. customary units and metric units. The metric units are not always exact arithmetic equivalents but have been adjusted to provide practical equivalents for critical dimensions while retaining geometric proportionality.

4.1.2 Improperly prepared test specimens often cause unsatisfactory test results. It is important, therefore, that care be exercised in the preparation of specimens, particularly in the machining, to assure good workmanship.

4.1.3 It is desirable to have the cross-sectional area of the specimen smallest at the center of the gage length to ensure fracture within the gage length. This is provided for by the taper in the gage length permitted for each of the specimens described in the following sections.

4.1.4 For brittle materials it is desirable to have fillets of large radius at the ends of the gage length.

4.2 *Plate-Type Specimens:*

4.2.1 The standard plate-type test specimen is shown in Fig. 6. This specimen is used for testing metallic materials in the form of plate, shapes, and flat material having a nominal thickness of $^3/_{16}$ in. (4.8 mm) or over. When product specifications so permit, other types of specimens may be used, as provided in 4.3, 4.4, and 4.5.

4.3 *Sheet-Type Specimens:*

4.3.1 The standard sheet-type test specimen is shown in Fig. 6. This specimen is used for testing metallic materials in the form of sheet, plate, flat wire, strip, band, hoop, rectangles, and shapes ranging in nominal thickness from 0.005 to $^5/_8$ in. (0.13 to 16 mm). When product specifications so permit, other types of specimens may be used, as provided in 4.2, 4.4, and 4.5.

NOTE 5—ASTM Methods E 345, Tension Testing of Metallic Foil,[4] may be used for tension testing of materials in thicknesses up to 0.0059 in. (0.150 mm).

4.3.2 Pin ends as shown in Fig. 7 may be used. In order to avoid buckling in tests of thin and high-strength materials, it may be necessary to use stiffening plates at the grip ends.

4.4 *Round Specimens:*

4.4.1 The standard 0.500-in. (12.5-mm) diameter round test specimen shown in Fig. 8 is used quite generally for testing metallic materials, both cast and wrought.

4.4.2 Figure 8 also shows small-size specimens proportional to the standard specimen. These may be used when it is necessary to test material from which the standard specimen or specimens shown in Fig. 6 cannot be prepared. Other sizes of small round specimens may be used. In any such small-size specimen it is important that the gage length for measurement of elongation be four times the diameter of the specimen.

4.4.3 The shape of the ends of the specimen outside of the gage length shall be suitable to the material and of a shape to fit the holders or grips of the testing machine so that the loads are applied axially. Figure 9 shows specimens with various types of ends that have given satisfactory results.

4.5 *Specimens for Sheet, Strip, Flat Wire, and Plate:*

4.5.1 In testing sheet, strip, flat wire, and plate one of the following types of specimens shall be used:

4.5.1.1 For material ranging in nominal thickness from 0.005 to $^5/_8$ in. (0.13 to 16 mm), the sheet-type specimen described in 4.3.

NOTE 6—Attention is called to the fact that either of the flat specimens described in 4.2 and 4.3 may be used for material from $^3/_{16}$ to $^5/_8$ in. (4.8 to 16 mm) in thickness, and one of the round specimens described in 4.4 may also be used for material $^1/_2$ in. (12.5 mm) or more in thickness.

4.5.1.2 For material having a nominal thickness of $^3/_{16}$ in. (4.8 mm) or over (Note 6), the plate-type specimen described in 4.2.

4.5.1.3 For material having a nominal thickness of $^1/_2$ in. (12.5 mm) or over (Note 6), the largest practical size of specimen described in 4.4.

4.6 *Specimens for Wire, Rod, and Bar:*

4.6.1 For round wire and rod, test specimens having the full cross-sectional area of the wire shall be used wherever practicable. The gage length for the measurement of elongation of wire less than $^1/_4$ in. (6 mm) in diameter shall be as prescribed in product specifications. In testing wire of $^1/_4$-in. (6-mm) or larger diameter, unless otherwise specified, a gage length equal to four times the diameter shall be used. The total length of the specimens shall be at least equal to the gage length plus the length of wire required for the full use of the grips employed.

4.6.2 For wire of octagonal, hexagonal, or square cross section, for rod or bar of round cross section where the specimen called for in 4.6.1 is not practicable, and for rod or bar of octagonal, hexagonal, or square cross section, one of the following types of specimens shall be used:

4.6.2.1 *Full Cross Section* (Note 7)—It is permissible to reduce the section slightly throughout the test section with abrasive cloth or paper, or machine it sufficiently to ensure fracture within the gage marks. For material not exceeding 0.188 in. (4.8 mm) in diameter or distance between flats, the cross-sectional area may be reduced to not less than 90 percent of the original area without changing the shape of the cross section. For material over 0.188 in. (4.8 mm) in diameter or distance between flats, the diameter or distance between flats may be reduced by not more than 0.010 in. (0.25 mm) without changing the shape of the cross section. Square, hexagonal, or octagonal wire or rod not exceeding 0.188 in. (4.8 mm) between flats may be turned to a round having a cross-sectional area no smaller than 90 percent of the area of the maximum inscribed circle. Fillets, preferably with a radius of $^3/_8$ in. (10 mm), but not less than $^1/_8$ in. (3 mm), shlll be used at the ends of the reduced sections. Square, hexagonal, or octagonal rod over 0.188 in. (4.8 mm) between flats may be turned to a round having a diameter no smaller than 0.010 in. (0.25 mm) less than the original distance between flats.

NOTE 7—The ends of copper or copper alloy specimens may be flattened 10 to 50 percent from the original dimension in a jig similar to that shown in Fig. 10, to facilitate fracture within the gage marks. In flattening the opposite ends of the test specimen care shall be taken to ensure that the four flattened surfaces are parallel and that the two parallel surfaces on the same side of the axis of the test specimen shall lie in the same plane.

4.6.2.2 For rod and bar, the largest practical size of round specimen as described in 4.4 may be used in place of a test specimen of full cross section. Unless otherwise specified in the product specification, specimens shall be parallel to the direction of rolling or extrusion.

4.7 *Specimens for Rectangular Bar:*

4.7.1 In testing rectangular bars one of the following types of specimens shall be used:

4.7.1.1 *Full Cross Section*—It is permissible to reduce the width of the specimen throughout the test section with abrasive cloth or paper, or by machining sufficiently to facilitate fracture within the gage marks, but in no case shall the reduced width be less than 90 percent of the original. The edges of the midlength of the reduced section not less than $^{3}/_{4}$ in. (20 mm) in length shall be parallel to each other and to the longitudinal axis of the specimen within 0.002 in. (0.05 mm). Fillets, preferably with a radius of $^{3}/_{8}$ in. (10 mm) but not less than $^{1}/_{8}$ in. (3 mm) shall be used at the ends of the reduced sections.

4.7.1.2 Rectangular bars of thickness small enough to fit the grips of the testing machine but of too great width may be reduced in width by cutting to fit the grips, after which the cut surfaces shall be machined or cut and smoothed to ensure failure within the desired section. The reduced width shall be not less than the original bar thickness. Also, one of the types of specimens described in 4.2, 4.3, and 4.4 may be used.

4.8 *Shapes, Structural and Other:*

4.8.1 In testing shapes other than those covered by the preceding sections, one of the types of specimens described in 4.2, 4.3, and 4.4 shall be used.

4.9 *Specimens for Pipe and Tube* (Note 8):

4.9.1 For all small tube (Note 8), particularly sizes 1 in. (25 mm) and under in nominal outside diameter, and frequently for larger sizes, except as limited by the testing equipment, it is standard practice to use tension test specimens of full-size tubular sections. Snug-fitting metal plugs shall be inserted far enough into the ends of such tubular specimens to permit the testing machine jaws to grip the specimens properly. The plugs shall not extend into that part of the specimen on which the elongation is measured. Figure 11 shows a suitable form of plug, the location of the plugs in the specimen, and the location of the specimen in the grips of the testing machine.

NOTE 8—The term "tube" is used to indicate tubular products in general, and includes pipe, tube, and tubing.

4.9.2 For large-diameter tube that cannot be tested in full section, longitudinal tension test specimens shall be cut as indicated in Fig. 12. Specimens from welded tube shall be located approximately 90 deg from the weld. If the tube-wall thickness is under $^{3}/_{4}$ in. (20 mm), either a specimen of the form and dimensions shown in Fig. 13 or one of the small-size specimens proportional to the standard $^{1}/_{2}$-in. specimen, as mentioned in 4.4.2 and shown in Fig. 8, shall be used. Specimens of the type shown in Fig. 13 may be tested with grips having a surface contour corresponding to the curvature of the tube. When grips with curved faces are not available, the ends of the specimens may be flattened without heating. If the tube-wall thickness is $^{3}/_{4}$ in. (20 mm) or over, the standard specimen shown in Fig. 8 shall be used.

4.9.3 Transverse tension test specimens for tube may be taken from rings cut from the ends of the tube as shown in Fig. 14. Flattening of the specimen may be either after separating as in *A*, or before separating as in ***B***. Transverse tension test specimens for large tube under $^{3}/_{4}$ in. (20 mm) in wall thickness shall be either of the small-size specimens shown in Fig. 8 or of the form and dimensions shown in Fig. 15. When using the specimen shown in Fig. 15, either or both surfaces of the specimen may be machined to secure a uniform thickness, provided not more than 15 percent of the normal wall thickness is removed from each surface. For large tube $^{3}/_{4}$ in. (20 mm) and over in wall thickness, the standard specimen shown in Fig. 8 shall be used for transverse tension tests. Specimens for transverse tension tests on large welded tube to determine the strength of welds shall be located perpendicular to the welded seams, with the welds at about the middle of their lengths.

4.10 *Specimens for Forgings:*

4.10.1 In testing forgings the largest practical size of specimen described in 4.4 shall be used.

4.11 *Specimens for Castings:*

4.11.1 In testing castings either the standard specimen shown in Fig. 8 or the specimen shown in Fig. 16 shall be used unless otherwise provided in the product specifications.

NOTE 9—Test coupons for castings shall be made as shown in Fig. 3 of Methods and Definitions A 370.

4.12 *Specimen for Malleable Iron:*

4.12.1 For testing malleable iron the test specimen shown in Fig. 17 shall be used, unless otherwise provided in the product specifications.

4.13 *Specimen for Die Castings:*

4.13.1 For testing die castings the test specimen shown in Fig. 18 shall be used unless otherwise provided in the product specifications.

4.14 *Specimens for Powdered Metals:*

4.14.1 For testing powdered metals the test specimens shown in Figs. 19 and 20 shall be used, unless otherwise provided in the product specifications.

4.15 *Gage Length of Test Specimens:*

4.15.1 The gage length for the determination of elongation shall be in accordance with the product specifications for the material being tested. Gage marks shall be stamped lightly with a punch, scribed lightly with dividers or drawn with ink as preferred. For material that is sensitive to the effect of slight notches and for small specimens, gage marks located with the aid of layout dope will minimize the possibility of fracture in the gage marks.

4.16 *Location of Test Specimens:*

4.16.1 The test specimens shall be selected from the material as prescribed in the various product specifications. Unless otherwise specified, the axis of the test specimen shall be located as follows:

4.16.1.1 At the center for products 1 $^1/_2$ in. (38 mm) or less in thickness, diameter, or distance between flats.

4.16.1.2 Midway from the center to the surface for products over 1 $^1/_2$ in. (38 mm) in thickness, diameter, or distance between flats.

5. Definitions and Procedures

5.1 *Measurement of Dimensions of Test Specimens:*

5.1.1 To determine the cross-sectional area of a tension test specimen, measure the dimensions of the cross section at the center of the gage length except that for referee testing of specimens under $^3/_{16}$ in. (4.8 mm) in their least dimension, measure the dimensions where the least cross-sectional area is found. Measure and record the cross-sectional dimensions of tension test specimens 0.200 in. (5.0 mm) and over to the nearest 0.001 in. (0.025 mm); the cross-sectional dimensions less than 0.200 in. (5.0 mm) and not less than 0.100 in. (2.5 mm) to the nearest 0.0005 in. (0.01 mm); the cross-sectional dimensions less than 0.100 in. (2.5 mm) and not less than 0.020 in. (0.50 mm), to the nearest 0.0001 in. (0.025 mm); and when practical, the cross-sectional dimensions less than 0.020 in. (0.50 mm), to at least the nearest 1 percent but in all cases to at least the nearest 0.0001 in. (0.0025 mm).

NOTE 10—Measurements of dimensions presume smooth surface finishes for the specimens. Rough surfaces due to the manufacturing process such as hot rolling, metallic coating, etc., may lead to inaccuracy of the computed areas greater than the measured dimensions would indicate. Therefore, cross-sectional dimensions of tension test specimens with rough surfaces due to processing may be measured and recorded to the nearest 0.001 in. (0.025 mm).

5.1.2 Determine cross-sectional areas of full-size tension test specimens of nonsymmetrical cross sections by weighing a length not less than 20 times the largest cross-sectional dimension and using the value of density of the material. Determine the weight to the nearest 0.5 percent or less.

5.2 *Speed of Testing:*

5.2.1 Speed of testing may be defined (*a*) in terms of free-running crosshead speed (rate of movement of the crosshead of the testing machine when not under load), (*b*) in terms of rate of separation of the two heads of the testing machine during a test, (*c*) in terms of the elapsed time for completing part or all of the test, (*d*) in terms of rate of stressing the specimen, or (*e*) in terms of rate of straining the specimen. For some materials the first of these, which is the least accurate, may be adequate, while for other materials one of the others, listed in increasing order of precision, may be necessary in order to obtain test values within acceptable limits. Suitable limits for speed of testing should be specified for materials for which the differences resulting from the use of different speeds are of such magnitude that the test results are unsatisfactory for determining the acceptability of the material. In such instances, depending upon the material and the use for which it is intended, one or more of the methods described in the following paragraphs is recommended for specifying speed of testing.

5.2.2 *Free-Running Crosshead Speed*—The allowable limits for the rate of movement of the crosshead of the testing machine, when not under load, shall be specified in inches per inch (or millimeters per millimeter) of gage length per minute. The limits for the crosshead speed may be further qualified by specifying different limits for various types and sizes of specimens. The average crosshead speed can be experimentally determined by using a suitable measuring device and a stop watch.

5.2.3 *Rate of Separation of Heads During Tests*—The allowable limits for rate of separation of the heads of the testing machine during a test shall be specified in inches per inch (or millimeters per millimeter) of gage length per minute. The limits for the rate of separation may be further qualified by specifying different limits for various types and sizes of specimen. Many testing machines are equipped with pacing devices for the measurement and control of the rate of separation of the heads of the machine during a test, but in the absence of such a device the average rate of separation of the heads can be experimentally determined by using a suitable measuring device and a stop watch.

5.2.4 *Elapsed Time*—The allowable limits for the elapsed time from the beginning of loading (or from some specified stress) to the instant of fracture, to the maximum load, or to some other stated stress, shall be specified in minutes or seconds. The elapsed time can be determined with a stop watch.

5.2.5 *Rate of Stressing*—The allowable limits for rate of stressing shall be specified in pounds per square inch (or kilogram-force per square millimeter) per minute. Many testing machines are equipped with pacing devices for the measurement and control of the rate of stressing, but in the absence of such a device the average rate of stressing can be determined with a stop watch by observing the time required to apply a known increment of stress.

5.2.6 *Rate of Straining*—The allowable limits for rate of straining shall be specified in inches per inch (or millimeters per millimeter) per minute. Some testing machines are equipped with pacing devices for the measurement and control of rate of straining, but in the absence of such a device the average rate of straining can be determined with a stop watch by observing the time required to effect a known increment of strain.

5.2.7 Unless otherwise specified, any convenient speed of testing may be used up to one half the specified yield strength or yield point, or up to one quarter the specified tensile strength, whichever is smaller. The speed above this point shall be within the limits specified. If different speed limitations are required for use in determining yield strength, yield point, tensile strength, elongation, and reduction of area, they should be stated in the product specifications. In the absence of any more specified limitations on speed of testing the following general rules shall apply:

5.2.7.1 The speed of testing shall be such that the loads and strains used in obtaining the test results are accurately indicated.

5.2.7.2 During the conduct of the test to determine yield strength or yield point the rate of stress application shall not exceed 100,000 psi (690 MPa)/min. The speed may be increased after removal of the extensometer, but it shall not exceed 0.5 in./in. (0.5 mm/mm) of gage length (or distance between grips for specimens not having reduced sections) per min.

NOTE 11—In writing new standards, or in revising old standards, the ASTM product specifications committee will have the responsibility of deciding for any given material what method of measuring speed of testing is to be used, and of specifying suitable numerical limits for free-running crosshead speed, rate of separation of heads during a test, elapsed time, rate of stressing, or rate of straining.

5.3 *Determination of Yield Strength:*

5.3.1 Determine yield strength by either of the methods described in 5.3.1.1 or 5.3.1.2.

5.3.1.1 *Offset Method*—To determine the yield strength by the "offset method," it is necessary to secure data (autographic or numerical) from which a stress-strain diagram may be drawn. Then on the stress-strain diagram (Fig. 21) lay off *Om* equal to the specified value of the set, draw *mn* parallel to *OA*, and thus locate *r*, the intersection of *mn* with the stress-strain diagram (Note 12). In reporting values of yield strength obtained by this method, the specified value of "offset" used should be stated in parentheses after the term yield strength. Thus:

Yield strength (offset = 0.2 percent)
= 52,000 psi (360 MPa)

In using this method a Class B1 extensometer[6] would be sufficiently sensitive for most materials.

[6] See ASTM Method E 83, Verification and Classification of Extensometers, which appears in the *1974 Annual Book of ASTM Standards*, Parts 10 and 41.

NOTE 12 Automatic devices are available that determine offset yield strength without plotting a stress-strain curve. Such devices may be used if their accuracy has been demonstrated.

NOTE 13 If the load drops before the specified offset is reached, technically the material does not have a yield strength (for that offset), but the stress at the maximum load before the specified offset is reached may be reported instead of the yield strength.

NOTE 14 The proportional limit may be regarded as a special value of the yield strength. It is the highest stress for which the offset is not measurable with the instrument used.

5.3.1.2 *Extension-Under-Load Method*—For tests to determine the acceptance or rejection of material whose stress-strain characteristics are well known from previous tests of similar material in which stress-strain diagrams were plotted, the total strain corresponding to the stress at which the specified offset occurs will be known within satisfactory limits; therefore, in such tests a specified total strain may be used, and the stress on the specimen, when this total strain is reached, is the value of the yield strength. The total strain can be obtained satisfactorily by use of a Class B1 extensometer.[6] It is recommended that this approximate method be used only after agreement between the manufacturer and the purchaser, with the understanding that check tests be made for obtaining stress-strain diagrams for use with the offset method to settle any misunderstandings.

5.4 *Determination of Yield Point:*

5.4.1 For material having a "sharp-kneed" stress-strain diagram, determine the yield point by one of the methods described in 5.4.1.1 or 5.4.1.2.

5.4.1.1 *Drop-of-the-Beam or Halt-of-the-Pointer Method*—Apply an increasing load to the specimen at a uniform rate. When a lever and poise machine is used, keep the beam in balance by running out the poise at approximately a steady rate. When the yield point of the material is reached, the increase of the load stops. Then run the poise a trifle beyond the balance position until the beam of the machine drops for a brief but appreciable interval of time. When a machine equipped with a load-indicating dial is used there is a halt or hesitation of the load-indicating pointer corresponding to the drop of the beam. Note the load at the "drop of the beam" or the "halt of the pointer" and record the corresponding stress as the yield point.

5.4.1.2 *Autographic Diagram Method*—Obtain a stress-strain diagram by an autographic recording device, and record the stress corresponding to the top of the knee or the point at which the curve drops as the yield point (see Fig. 22).

5.4.2 When test specimens do not exhibit a well-defined disproportionate deformation that characterizes a yield point as measured by the drop-of-the-beam, halt-of-the-pointer, or autographic diagram methods described above, a value equivalent to the yield point in its practical significance may be determined by the following methods and may be substituted for the yield point:

5.4.2.1 *Divider Method*—With a pair of dividers or similar device, note any detectable elongation between two gage marks on the specimen. When detectable elongation occurs, note the load at that instant, and record the stress corresponding to the load as the yield point. The gage marks selected for this determination shall be not more than 2 in. (50 mm) apart, even on specimens of longer gage length.

5.4.2.2 *Strain Rate Method*—Attach a Class B1 extensometer to the specimen at the gage marks. When the specimen is in place and the extensometer attached, increase the load at a reasonably uniform rate. Watch the elongation of the specimen as shown by the extensometer and note for this determination the load at which the rate of elongation shows a sudden increase.

5.4.2.3 *Total-Extension-Under-Load Method*—Attach a Class C extensometer to the specimen. When the load producing a specified extension is reached, record the stress corresponding to this load as the yield point, and remove the extensometer. This same value may be obtained from an autographic stress-strain diagram (Fig. 23).

NOTE 15 The appropriate value of the total extension should be specified. For steel with yield point specified not over 80,000 psi (550 MPa), an appropriate value is 0.005 in./in. (0.005 mm/mm) of gage length. For higher strength steels, yield strength should be specified in preference to yield point.

NOTE 16 A suitable device that automatically determines the load at the specified extension without plotting a stress-strain curve may be used if its accuracy has been demonstrated.

5.5 *Tensile Strength:*

5.5.1 Calculate the tensile strength by dividing the maximum load carried by the specimen during a tension test by the original cross-sectional area of the specimen.

5.6 *Elongation:*

5.6.1 Fit ends of the fractured specimen together carefully and measure the distance between the gage marks to the nearest 0.01 in. (0.25 mm) for gage lengths of 2 in. (50 mm) and under, and to the nearest 0.5 percent of the gage length for gage lengths over 2 in. (50 mm). A percentage scale reading to 0.5 percent of the gage length may be used. The elongation is the increase in length of the gage length, expressed as a percentage of the original gage length. In reporting elongation values, give both the percentage increase and the original gage length.

5.6.2 If any part of the fracture takes place outside of the middle half of the gage length or in a punched or scribed mark within the reduced section, the elongation value obtained may not be representative of the material. If the elongation so measured meets the minimum requirements specified, no further testing is indicated, but if the elongation is less than the minimum requirements, discard the test and retest.

5.6.3 In determining extension at fracture (elastic plus plastic extension) autographic or extensometer methods may be employed.

5.7 *Reduction of Area:*

5.7.1 Fit the ends of the fractured specimen together and measure the mean diameter or the width and thickness at the smallest cross section to the same accuracy as the original dimensions. The difference between the area thus found and the area of the original cross section expressed as a percentage of the original area, is the reduction of area.

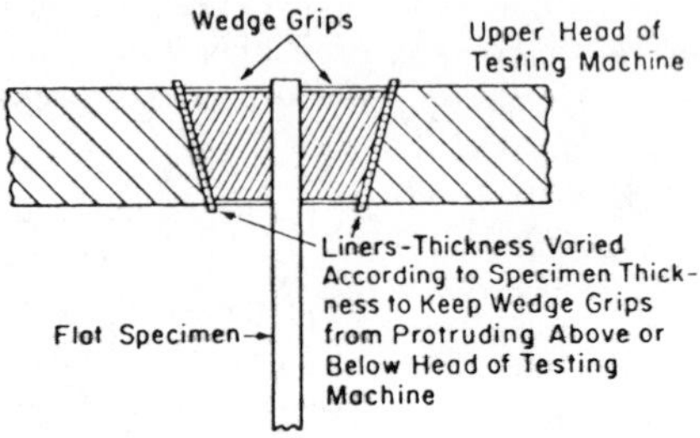

FIG. 1 Wedge Grips with Liners for Flat Specimens.

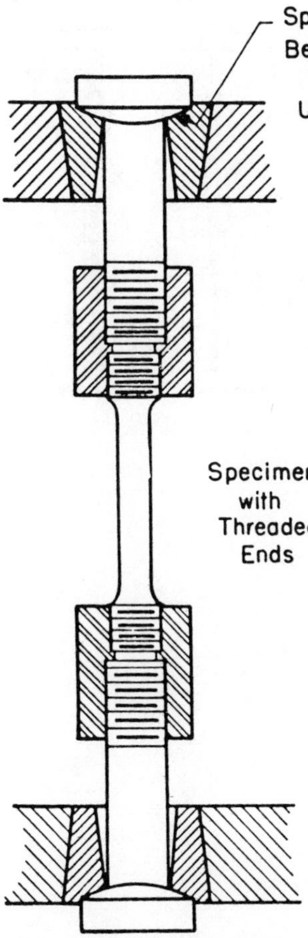

FIG. 2 Gripping Device for Threaded-End Specimens.

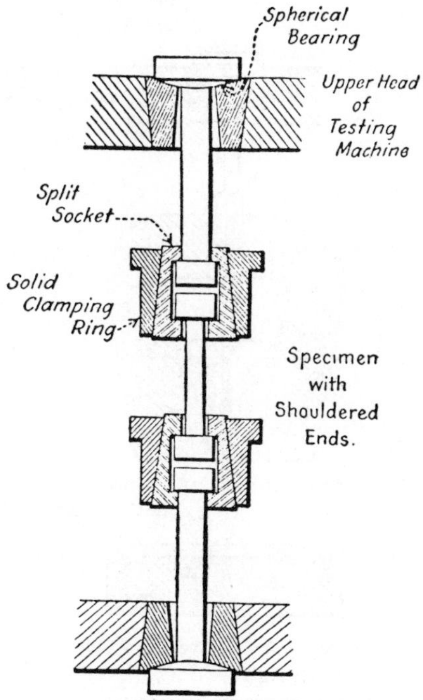

FIG. 3 Gripping Device for Shouldered-End Specimens.

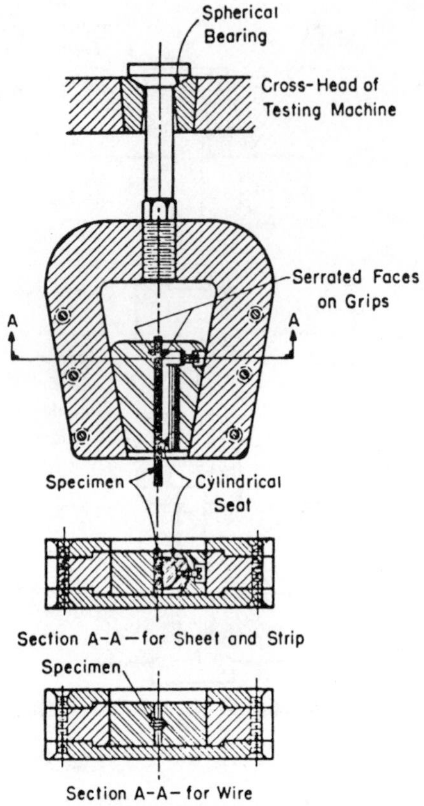

FIG. 4 Gripping Devices for Sheet and Wire Specimens.

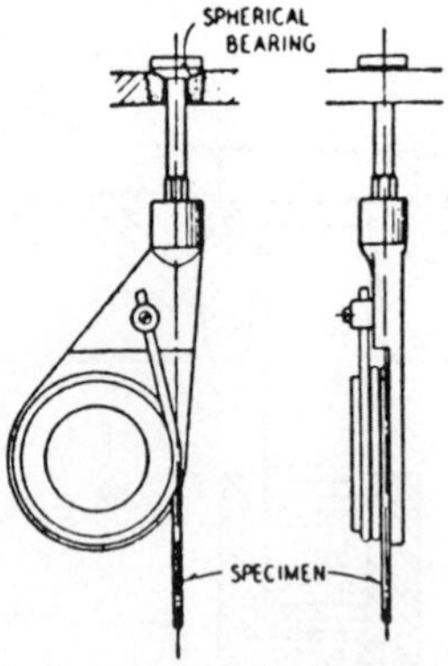

FIG. 5 Snubbing Device for Testing Wire.

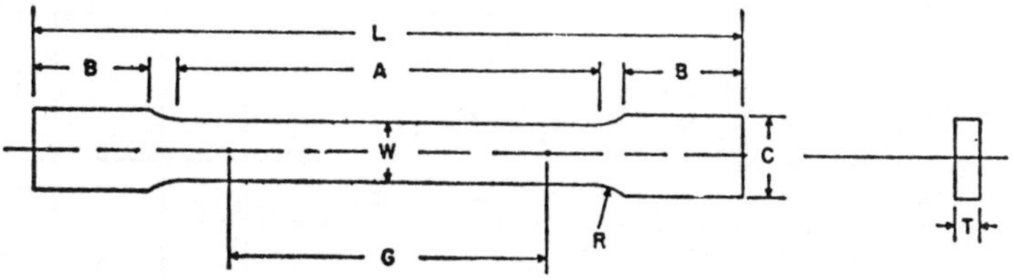

Dimensions

	Standard Specimens				Subsize Specimen	
	Plate-Type, $1\frac{1}{2}$-in. Wide		Sheet-Type, $\frac{1}{2}$-in. Wide		$\frac{1}{4}$-in. Wide	
	in.	mm	in.	mm	in.	mm
G—Gage length (Notes 1 and 2)	8.00 ± 0.01	2.00 ± 0.25	2.000 ± 0.005	50.0 ± 0.10	1.000 ± 0.003	25.0 ± 0.08
W—Width (Notes 3 and 4)	$1\frac{1}{2}$ + $\frac{1}{8}$ − $\frac{1}{4}$	40 ± 2	0.500 ± 0.010	12.5 ± 0.25	0.250 ± 0.002	6.25 ± 0.05
T—Thickness (Note 5)			thickness of material			
R—Radius of fillet, min (Note 6)	1	25	$\frac{1}{2}$	13	$\frac{1}{4}$	6
L—Over-all length, min (Notes 2 and 7)	18	450	8	200	4	100
A—Length of reduced section, min	9	225	$2\frac{1}{4}$	60	$1\frac{1}{4}$	32
B—Length of grip section, min (Note 8)	3	75	2	50	$1\frac{1}{4}$	32
C—Width of grip section, approximate (Notes 4 and 9)	2	50	$\frac{3}{4}$	20	$\frac{3}{8}$	10

NOTE 1—For the $1\frac{1}{2}$-in. (40-mm) wide specimen, punch marks for measuring elongation after fracture shall be made on the flat or on the edge of the specimen and within the reduced section. Either a set of nine or more punch marks 1 in. (25 mm) apart, or one or more pairs of punch marks 8 in. (200 mm) apart may be used.

NOTE 2—When elongation measurements of $1\frac{1}{2}$-in. (40-mm) wide specimens are not required, a gage length (*G*) of 2.000 in. ± 0.005 in. (50.0 mm ± 0.10 mm) with all other dimensions similar to the plate-type specimen may be used.

NOTE 3—For the three sizes of specimens, the ends of the reduced section shall not differ in width by more than 0.004, 0.002 or 0.001 in. (0.10, 0.05 or 0.025 mm), respectively. Also, there may be a gradual decrease in width from the ends to the center, but the width at either end shall not be more than 0.015 in., 0.005 in., or 0.003 in. (0.40, 0.10 or 0.08 mm), respectively, larger than the width at the center.

NOTE 4—For each of the three sizes of specimens, narrower widths (*W* and *C*) may be used when necessary. In such cases the width of the reduced section should be as large as the width of the material being tested permits; however, unless stated specifically, the requirements for elongation in a product specification shall not apply when these narrower specimens are used. If the width of the material is less than *W*, the sides may be parallel throughout the length of the specimen.

NOTE 5—The dimension *T* is the thickness of the test specimen as provided for in the applicable material specifications. Minimum thickness of $1\frac{1}{2}$-in. (40-mm) wide specimens shall be $\frac{3}{16}$ in. (5 mm). Maximum thickness of $\frac{1}{2}$-in. (12.5-mm) and $\frac{1}{4}$-in. (6.25-mm) wide specimens shall be $\frac{5}{8}$ in. (16 mm) and $\frac{1}{4}$ in. (6.25 mm), respectively.

NOTE 6—For the $1\frac{1}{2}$-in. (40-mm) wide specimen, a $\frac{1}{2}$-in. (13-mm) minimum radius at the ends of the reduced section is permitted for steel specimens under 100,000 psi (690 MPa) in tensile strength when a profile cutter is used to machine the reduced section.

NOTE 7—To aid in obtaining axial loading during testing of $\frac{1}{4}$-in. (6-mm) wide specimens, the over-all length should be as large as the material will permit.

NOTE 8—It is desirable, if possible, to make the length of the grip section large enough to allow the specimen to extend into the grips a distance equal to two thirds or more of the length of the grips. If the thickness of $\frac{1}{2}$-in. (13-mm) wide specimens is over $\frac{3}{8}$ in. (10 mm), longer grips and correspondingly longer grip sections of the specimen may be necessary to prevent failure in the grip section.

NOTE 9—For the three sizes of specimens, the ends of the specimen shall be symmetrical with the center line of the reduced section within 0.10, 0.01 and 0.005 in. (2.5, 0.25 and 0.13 mm), respectively. However, for steel if the ends of the $\frac{1}{2}$-in. (12.5-mm) wide specimen are symmetrical within 0.05 in. (1.0 mm), a specimen may be considered satisfactory for all but referee testing.

FIG. 6 Rectangular Tension Test Specimens.

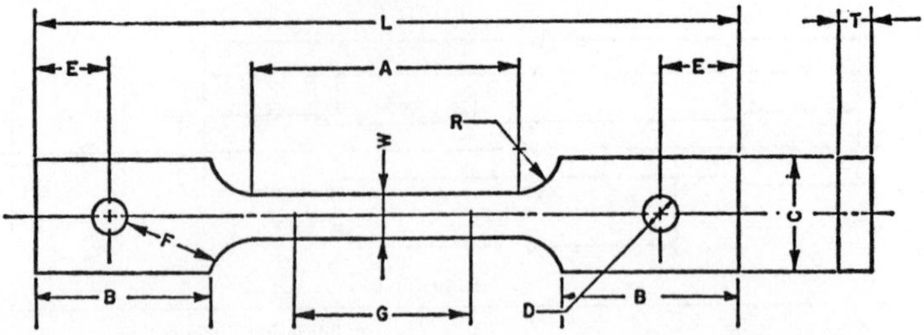

Dimensions

	in.	mm
G—Gage length	2.000 ± 0.005	50.0 ± 0.10
W—Width (Note 1)	0.500 ± 0.010	12.5 ± 0.25
T—Thickness, max (Note 2)	$^5/_8$	16
R—Radius of fillet, min (Note 3)	$^1/_2$	13
L—Over-all length, min	8	200
A—Length of reduced section, min	$2^1/_4$	60
B—Length of grip section, min	2	50
C—Width of grip section, approximate	2	50
D—Diameter of hole for pin, min (Note 4)	$^1/_2$	13
E—Edge distance from pin, approximate	$1^1/_2$	38
F—Distance from hole to fillet, min	$^1/_2$	13

NOTE 1—The ends of the reduced section shall differ in width by not more than 0.002 in. (0.05 mm). There may be a gradual taper in width from the ends to the center, but the width at either end shall be not more than 0.005 in. (0.10 mm) greater than the width at the center.

NOTE 2—The dimension *T* is the thickness of the test specimen as stated in the applicable product specifications.

NOTE 3—For some materials, a fillet radius *R* larger than $^1/_2$ in. (13 mm) may be needed.

NOTE 4—Holes must be on center line of reduced section, within ± 0.002 in. (± 0.05 mm).

NOTE 5—Variations of dimensions *C*, *D*, *E*, *F*, and *L* may be used that will permit failure within the gage length.

FIG. 7 Pin-Loaded Tension Test Specimen with 2-in. (50-mm) Gage Length.

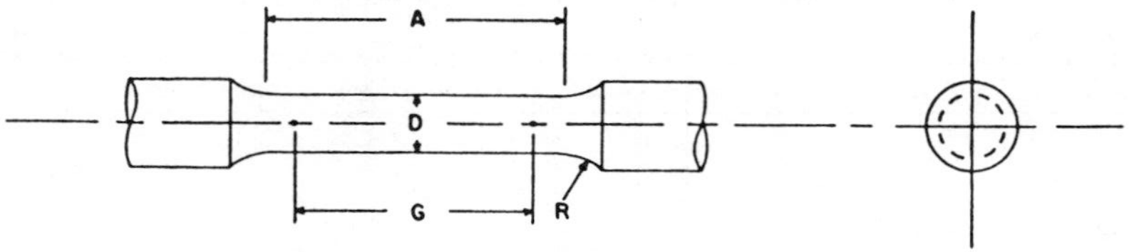

Dimensions

	Standard Specimen		Small-Size Specimens Proportional to Standard							
	in.	mm	in.	mm	in.	mm	in.	mm	in.	mm
Nominal Diameter	0.500	12.5	0.350	8.75	0.250	6.25	0.160	4.00	0.113	2.50
G—Gage length	2.000 ± 0.005	50.0 ± 0.10	1.400 ± 0.005	35.0 ± 0.10	1.000 ± 0.005	25.0 ± 0.10	0.640 ± 0.005	16.0 ± 0.10	0.450 ± 0.005	10.0 ± 0.10
D—Diameter (Note 1)	0.500 ± 0.010	12.5 ± 0.25	0.350 ± 0.007	8.75 ± 0.18	0.250 ± 0.005	6.25 ± 0.12	0.160 ± 0.003	4.00 ± 0.08	0.113 ± 0.002	2.50 ± 0.05
R—Radius of fillet, min	3/8	10	1/4	6	3/16	5	5/32	4	3/32	2
A—Length of reduced section, min (Note 2)	2 1/4	60	1 3/4	45	1 1/4	32	3/4	20	5/8	16

NOTE 1—The reduced section may have a gradual taper from the ends toward the center, with the ends not more than 1 percent larger in diameter than the center (controlling dimension).

NOTE 2—If desired, the length of the reduced section may be increased to accommodate an extensometer of any convenient gage length. Reference marks for the measurement of elongation should, nevertheless, be spaced at the indicated gage length.

NOTE 3—The gage length and fillets shall be as shown, but the ends may be of any form to fit the holders of the testing machine in such a way that the load shall be axial (see Fig. 9). If the ends are to held in wedge grips it is desirable, if possible, to make the length of the grip section great enough to allow the specimen to extend into the grips a distance equal to two thirds or more of the length of the grips.

NOTE 4—On the round specimens in Figs. 8 and 9, the gage lengths are equal to four times the nominal diameter. In some product specifications other specimens may be provided for, but unless the 4-to-1 ratio is maintained within dimensional tolerances, the elongation values may not be comparable with those obtained from the standard test specimen.

NOTE 5—The use of specimens smaller than 0.250-in. (6.25-mm) diameter shall be restricted to cases when the material to be tested is of insufficient size to obtain larger specimens or when all parties agree to their use for acceptance testing. Smaller specimens require suitable equipment and greater skill in both machining and testing.

NOTE 6—Five sizes of specimens often used have diameters of approximately 0.505, 0.357, 0.252, 0.160, and 0.113 in., the reason being to permit easy calculations of stress from loads, since the corresponding cross-sectional areas are equal or close to 0.200, 0.100, 0.0500, 0.0200, and 0.0100 $in.^2$, respectively. Thus, when the actual diameters agree with these values, the stresses (or strengths) may be computed using the simple multiplying factors 5, 10, 20, 50, and 100, respectively. (The metric equivalents of these five diameters do not result in correspondingly convenient cross-sectional areas and multiplying factors.)

FIG. 8 Standard 0.500-in. (12.5-mm) Round Tension Test Specimen with 2-in. (50-mm) Gage Length and Examples of Small-Size Specimens Proportional to the Standard Specimen.

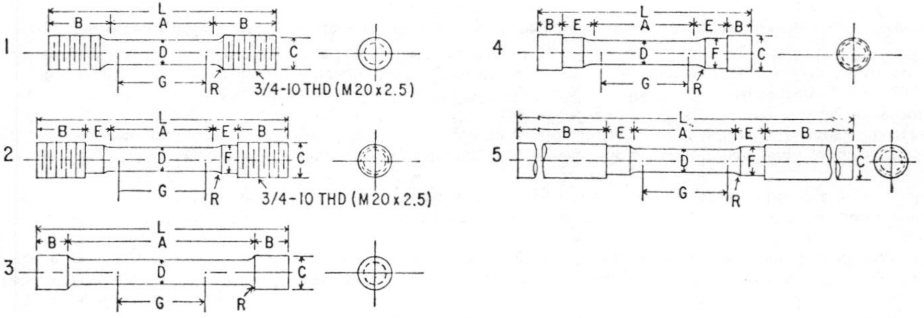

Dimensions

	Specimen 1		Specimen 2		Specimen 3		Specimen 4		Specimen 5	
	in.	mm	in.	mm	in.	mm	in.	mm	in.	mm
G—Gage length	2.000 ± 0.005	50.0 ± 0.10	2.000 ± 0.005	50.0 ± 0.10	2.000 ± 0.005	50.0 ± 0.10	2.000 ± 0.005	50.0 ± 0.10	2.000 ± 0.005	50.0 ± 0.10
D—Diameter (Note 1)	0.500 ± 0.010	12.5 ± 0.25	0.500 ± 0.010	12.5 ± 0.25	0.500 ± 0.010	12.5 ± 0.25	0.500 ± 0.010	12.5 ± 0.25	0.500 ± 0.010	12.5 ± 0.25
R—Radius of fillet, min	3/8	10	3/8	10	1/16	2	3/8	10	3/8	10
A—Length of reduced section	2 1/4, min	60, min	2 1/4, min	60, min	4, approximately	100, approximately	2 1/4, min	60, min	2 1/4, min	60, min
L—Over-all length, approximate	5	125	5 1/2	140	5 1/2	140	4 3/4	120	9 1/2	240
B—Length of end section (Note 2)	1 3/8, approximately	35, approximately	1, approximately	25, approximately	3/4, approximately	20, approximately	1/2, approximately	13, approximately	3, min	75, min
C—Diameter of end section	3/4	20	3/4	20	23/32	18	7/8	22	3/4	20
E—Length of shoulder and fillet section, approximate	...	...	5/8	16	...	...	3/4	20	5/8	16
F—Diameter of shoulder	...	...	5/8	16	...	...	5/8	16	19/32	15

NOTE 1—The reduced section may have a gradual taper from the ends toward the center with the ends not more than 0.005 in. (0.10 mm) larger in diameter than the center.

NOTE 2—On Specimen 5 it is desirable, if possible, to make the length of the grip section great enough to allow the specimen to extend into the grips a distance equal to two thirds or more of the length of the grips.

FIG. 9 Various Types of Ends for Standards Round Tension Test Specimen.

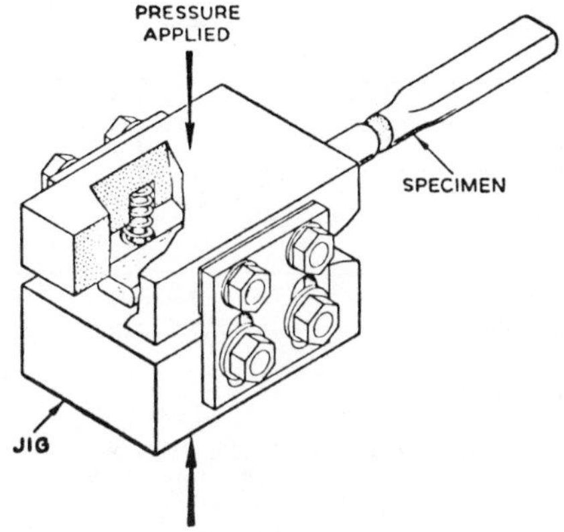

FIG. 10 **Squeezing Jig for Flattening Ends of Full-Size Tension Test Specimens.**

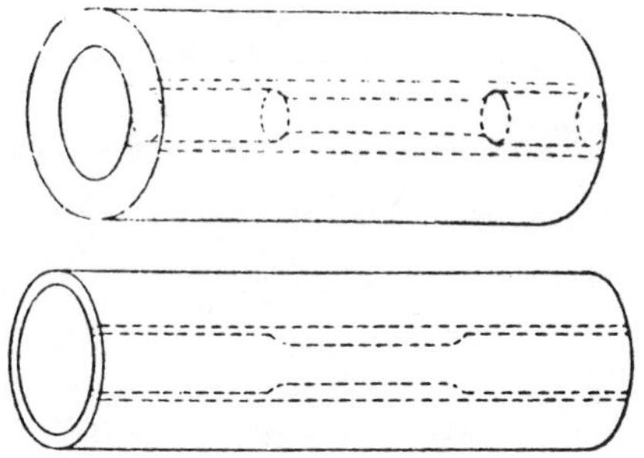

NOTE—The edges of the strip specimen shall be cut parallel to each other.

FIG. 12 Location from Which Longitudinal Tension Test Specimens Are to be Cut from Large-Diameter Tube.

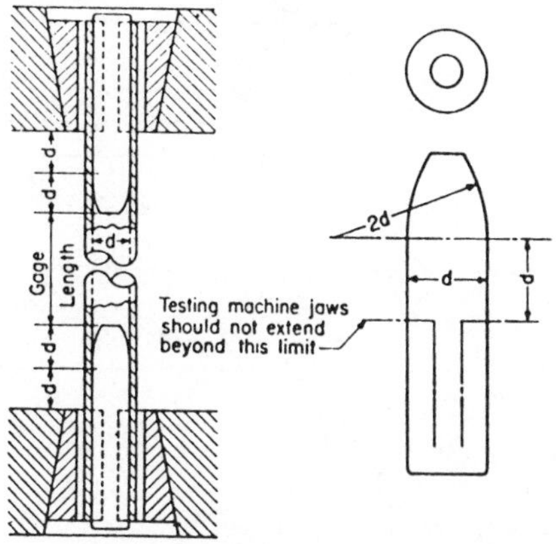

NOTE—The diameter of the plug shall have a slight taper from the line limiting the testing machine jaws to the curved section.

FIG. 11 Metal Plugs for Testing Tubular Specimens, Proper Location of Plugs in Specimen and of Specimen in Heads of Testing Machine.

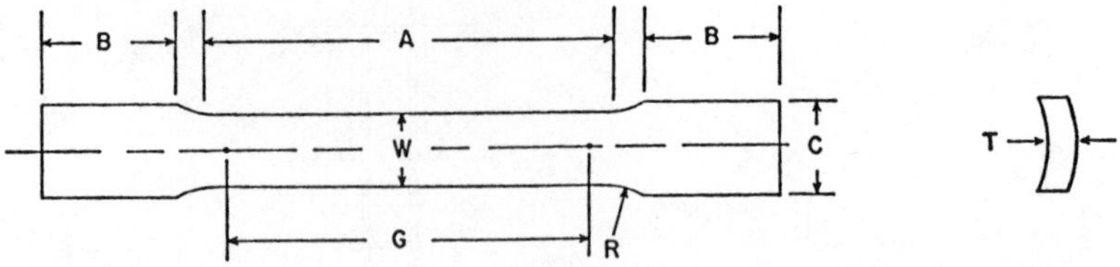

Dimensions

	Specimen 1		Specimen 2	
	in.	mm	in.	mm
W—Width (Note 1)	0.500 ± 0.010	12.5 ± 0.25	$1\frac{1}{2} + \frac{1}{8}$ $- \frac{1}{4}$	40.0 ± 2.0
G—Gage length (*a*)	2.000 ± 0.005	50.0 ± 0.10	2.000 ± 0.005	50.0 ± 0.10
(*b*)	...	...	8.00 ± 0.01	200.0 ± 0.25
T—Thickness (Note 2)		thickness of tubular section		
R—Radius of fillet, min	$\frac{1}{2}$	13	1	25
A—Length of reduced section, min (*a*)	$2\frac{1}{4}$	60	$2\frac{1}{4}$	60
(*b*)	...	...	9	230
B—Length of grip section, min (Note 3)	3	75	3	75
C—Width of grip section, approximate (Note 4)	$\frac{11}{16}$	18	2	50

NOTE 1—The ends of the reduced section shall differ in width by not more than 0.004 in. (0.10 mm). There may be a gradual taper in width from the ends to the center, but the width at either end shall be not more than 0.005 in. (0.10 mm) greater than the width at the center for 2-in. (50-mm) gage length specimens and not more than 0.015 in. (0.40 mm) greater for 8-in. (200-mm) gage length specimens.

NOTE 2—The dimension *T* is the thickness of the tubular section as provided for in the applicable material specifications.

NOTE 3—It is desirable, if possible, to make the length of the grip section great enough to allow the specimen to extend into the grips a distance equal to two thirds or more of the length of the grips.

NOTE 4—The ends of the specimen shall be symmetrical with the center line of the reduced section within 0.05 in. (1.0 mm) for Specimen No. 1 and 0.10 in. (2.5 mm) for Specimen No. 2.

NOTE 5—Cross-sectional area may be calculated by multiplying *W* and *T*. However, if the ratio of the dimension *W* to the diameter of the tubular section is larger than about $\frac{1}{4}$, the error in this method of calculating the cross sectional area may be appreciable and it may be desirable to use a more exact method of determining the area.

FIG. 13 Longitudinal Tension Test Specimens for Large-Diameter Tubular Products.

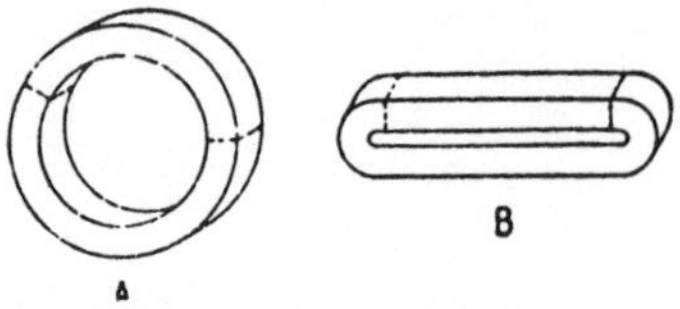

FIG. 14 Location of Transverse Tension Test Specimen in Ring Cut from Tubular Products.

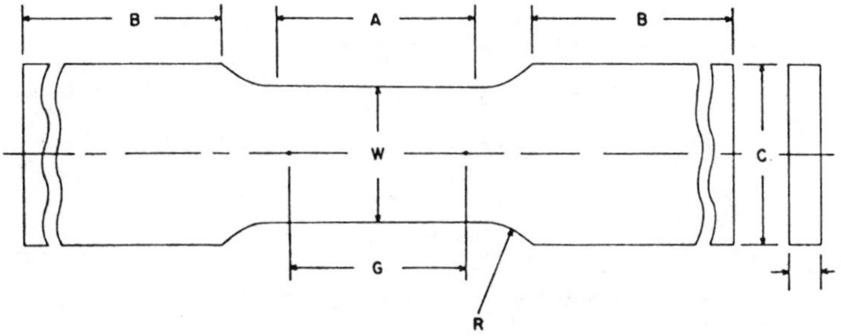

Dimensions

	in.	mm
G—Gage length	2.000 ± 0.005	50.0 ± 0.10
W—Width (Note 1)	1 1/2 + 1/8, − 1/4	40 ± 2
T—Thickness (Note 2)	thickness of tubular section	
R—Radius of fillet, min	1	25
A—Length of reduced section, min	2 1/4	60
B—Length of grip section, min (Note 3)	3	75
C—Width of grip section, approximate (Note 4)	2	50

NOTE 1—The ends of the reduced section shall differ in width by not more than 0.002 in. (0.05 mm). There may be a gradual taper from the ends toward the center, but the width of either end shall be not more than 0.005 in. (0.10 mm) greater than the width at the center.

NOTE 2—The dimension *T* is the thickness of the test specimen as provided for in the applicable material specifications.

NOTE 3—It is desirable, if possible, to make the length of the grip section great enough to allow the specimen to extend into the grips a distance equal to two thirds or more of the length of the grips.

NOTE 4—The ends of the specimen shall be symmetrical with the center line of the reduced section within 0.05 in. (1.0 mm).

FIG. 15 Transverse Tension Test Specimen Machined from Ring Cut from Tubular Products.

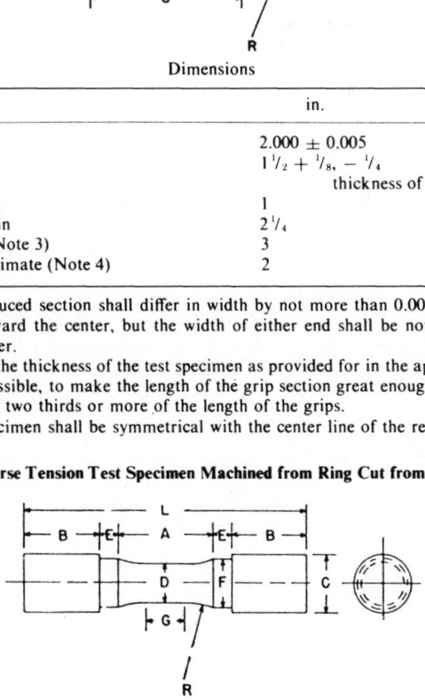

Dimensions

	Specimen 1		Specimen 2		Specimen 3	
	in.	mm	in.	mm	in.	mm
G—Length of parallel	Shall be equal to or greater than diameter *D*					
D—Diameter	0.500 ± 0.010	12.5 ± 0.25	0.750 ± 0.015	20.0 ± 0.40	1.25 ± 0.025	30.0 ± 0.60
R—Radius of fillet, min	1	25	1	25	2	50
A—Length of reduced section, min	1 1/4	32	1 1/2	38	2 1/4	60
L—Over-all length, min	3 3/4	95	4	100	6 3/8	160
B—Length of end section, approximate	1	25	1	25	1 3/4	45
C—Diameter of end section, approximate	3/4	20	1 1/8	30	1 7/8	48
E—Length of shoulder, min	1/4	6	1/4	6	5/16	8
F—Diameter of shoulder	5/8 ± 1/64	16.0 ± 0.40	15/16 ± 1/64	24.0 ± 0.40	1 7/16 ± 1/64	36.5 ± 0.40

NOTE—The reduced section and shoulders (dimensions *A*, *D*, *E*, *F*, *G*, and *R*) shall be as shown, but the ends may be of any form to fit the holders of the testing machine in such a way that the load shall be axial. Commonly the ends are threaded and have the dimensions *B* and *C* given above.

FIG. 16 Standard Tension Test Specimen for Cast Iron.

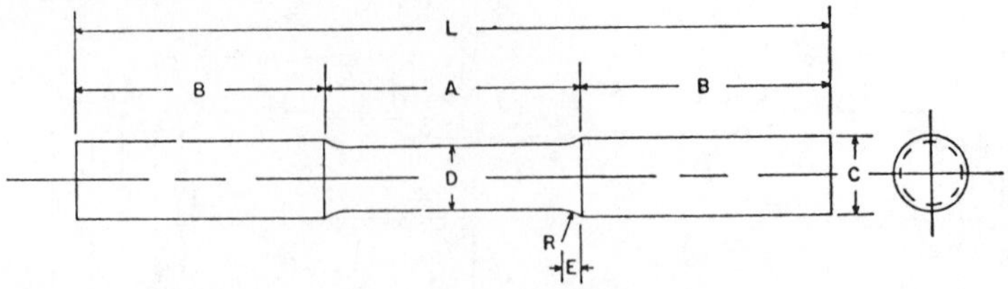

Dimensions

	in.	mm
D—Diameter	5/8	16
R—Radius of fillet	5/16	8
A—Length of reduced section	2 1/2	64
L—Over-all length	7 1/2	190
B—Length of end section	2 1/2	64
C—Diameter of end section	3/4	20
E—Length of fillet	3/16	5

FIG. 17 Standard Tension Test Specimen for Malleable Iron.

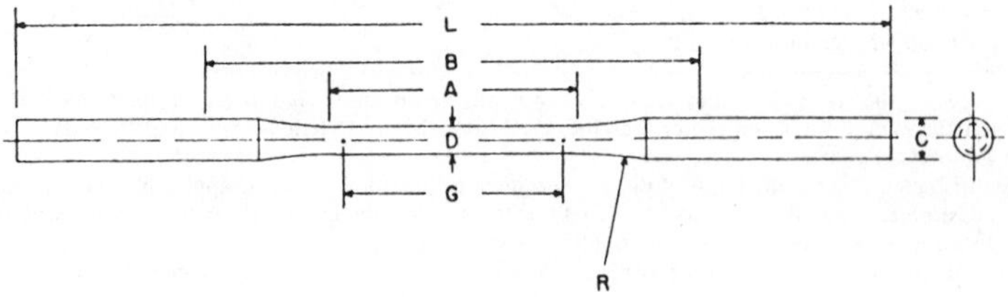

Dimensions

	in.	mm
G—Gage length	2.000 ± 0.005	50.0 ± 0.10
D—Diameter (see Note)	0.250 ± 0.005	6.4 ± 0.10
R—Radius of fillet, min	3	75
A—Length of reduced section, min	2 1/4	60
L—Over-all length, min	9	230
B—Distance between grips, min	4 1/2	115
C—Diameter of end section, approximate	3/8	10

NOTE—The reduced section may have a gradual taper from the ends toward the center, with the ends not more than 0.005 in. (0.10 mm) larger in diameter than the center.

FIG 18 Standard Tension Test Specimen for Die Castings.

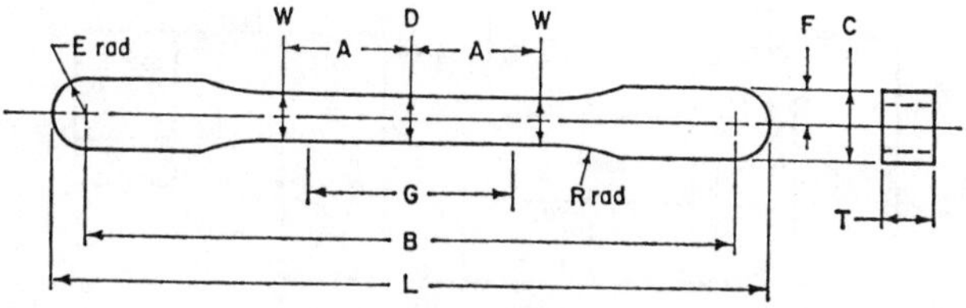

Pressing Area = 1.00 in² (645mm²)
Dimensions Specified are Those of the Die.

Dimensions

	in.	mm
G—Gage length	1	25
D—Width at center	0.225 ± 0.001	5.72 ± 0.025
W—Width at end of reduced section	0.235 ± 0.001	5.97 ± 0.025
T—Compact to this thickness	0.200 to 0.250	5.08 to 6.40
R—Radius of fillet	1	25
A—Half-length of reduced section	5/8	16
B—Grip length	3.187 ± 0.001	80.95 ± 0.025
L—Over-all length	3.529 ± 0.001	89.64 ± 0.025
C—Width of grip section	0.343 ± 0.001	8.71 ± 0.025
F—Half width of grip section	0.1715 ± 0.0010	4.356 ± 0.025
E—End radius	0.171 ± 0.001	4.34 ± 0.025

FIG. 19 Standard Tension Test Specimen for Powdered Metal Products—Flat Unmachined Tension Test Bar.

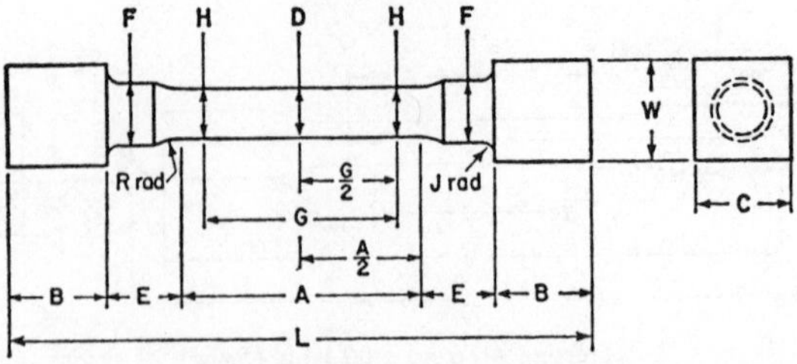

Pressing Area of Unmachined · Compact = 1.5 in.² (968 mm²)

Machining Recommendations

1. Rough Machine to 5/16 in. dia
2. Finish Turn 0.250 in. dia with Radii and Taper
3. Polish with 00 Emery Cloth
4. Lap with Crocus Cloth

Dimensions

	in.	mm
G—Gage length	1	25
D—Diameter at center of reduced section	0.250 ± 0.001	6.25 ± 0.025
H—Diameter at ends of gage length	0.250 + (0.001/0.002)	6.25 + (0.025/0.050)
R—Radius of fillet	1/4	6
A—Length of reduced section	1 1/4	32
L—Over-all length (die cavity length)	3	75
B—Length of end section	1/2	13
C—Compact to this end thickness	0.500 ± 0.050	12.5 ± 1.0
W—Die cavity width	1/2	13
E—Length of shoulder and fillet	3/8	10
F—Diameter of shoulder	5/16	8
J—End fillet radius, max	1/16	1.6

NOTE—The gage length and fillets of the specimen shall be as shown. The ends as shown are designed to provide a total pressing area of 1.00 in.[2] Other end designs are acceptable, and in some cases are required for high-strength sintered materials. Some suggested alternative end designs include:

1. Longer ends, of the same general shape and configuration as the standard, provide more surface area for gripping.
2. Shallow transverse grooves, or ridges, may be pressed in the ends to be gripped by jaws machined to fit the bar contour.

FIG. 20 Standard Tension Test Specimen for Powdered Metal Products—Round Machined Tension Test Bar.

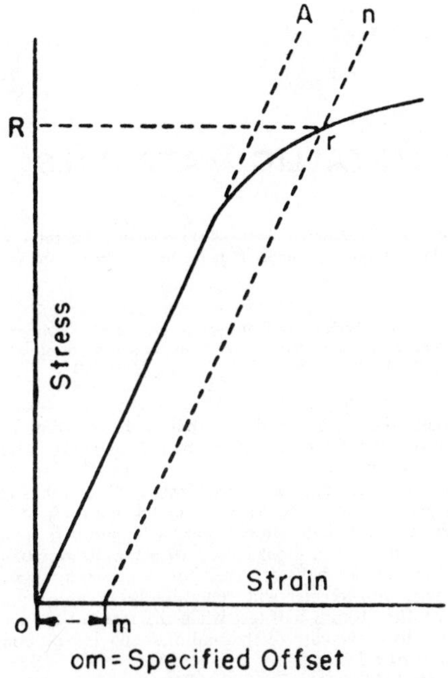

FIG. 21 Stress - Strain Diagram for Determination of Yield Strength by the Offset Method.

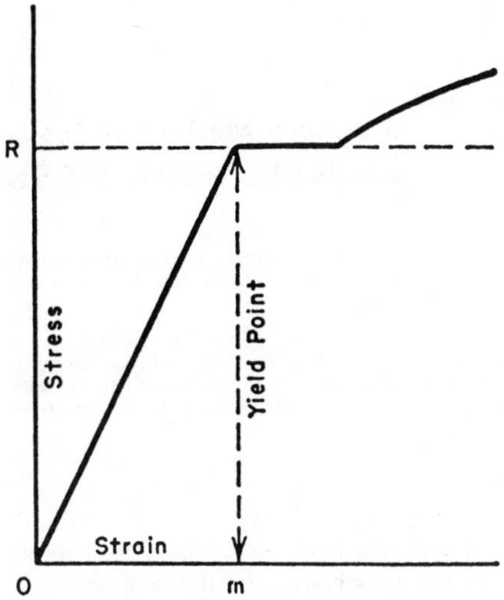

FIG. 22 Stress - Strain Diagram Showing Yield Point Corresponding with Top of Knee.

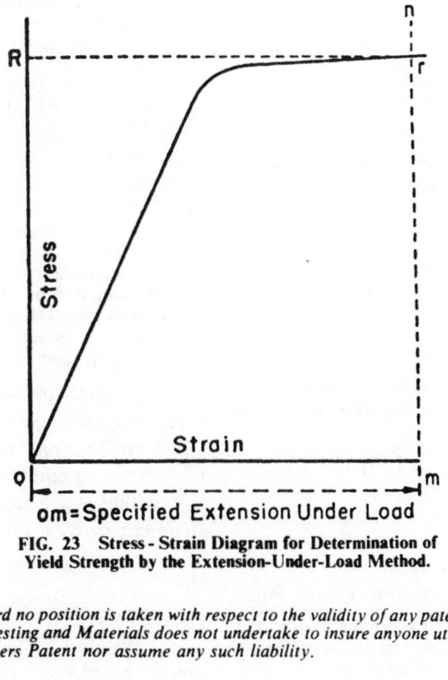

FIG. 23 Stress - Strain Diagram for Determination of Yield Strength by the Extension-Under-Load Method.

By publication of this standard no position is taken with respect to the validity of any patent rights in connection therewith, and the American Society for Testing and Materials does not undertake to insure anyone utilizing the standard against liability for infringement of any Letters Patent nor assume any such liability.

Designation: E 10 - 73

American National Standard Z115.5
American National Standards Institute

Standard Method of Test for
BRINELL HARDNESS OF METALLIC MATERIALS[1]

This Standard is issued under the fixed designation E 10, the number immediately following the designation indicates the year of original adoption or, in the case of revision, the year of last revision. A number in parentheses indicates the year of last reapproval.

This method has been approved by the Department of Defense to replace method 242.1 of Federal Test Method Standard No. 151b and for listing in DoD Index of Specifications and Standards. Future proposed revisions should be coordinated with the Federal Government through the Army Materials and Mechanics Research Center, Watertown, Mass. 02172.

1. Scope

1.1 This method covers the determination of the Brinell hardness of metallic materials, including methods for the verification of Brinell hardness testing machines (Part B) and the calibration of standardized hardness test blocks (Part C). Two general classes of standard tests are recognized:

1.1.1 *Verification, Laboratory, or Referee Tests*, where a high degree of accuracy is required, and

1.1.2 *Routine Tests*, where a somewhat lower degree of accuracy is permissible.

NOTE 1—The values stated in U.S. customary units are to be regarded as the standard.

2. Definitions

2.1 *Brinell hardness test*—an indentation hardness test using calibrated machines to force a hard ball, under specified conditions, into the surface of the material under test and to measure the diameter of the resulting impression after removal of the load.

2.2 *Brinell hardness number, HB*—a number related to the applied load and to the surface area of the permanent impression made by a ball indenter computed from the equation:

$$HB = 2P/\pi D(D - \sqrt{D^2 - d^2})$$

where:

P = applied load, kgf,
D = diameter of the ball, mm, and
d = mean diameter of the impression, mm.

NOTE 2—It is a recognized experimental fact that the Brinell hardness number of nearly all materials is influenced by the magnitude of the indenting load, the diameter of the ball indenter, and the elastic characteristics of the ball. In general, a ball 10 mm in diameter of suitable composition should be used with applied loads 3000, 1500, or 500 kgf, depending upon the hardness of the material to be tested. Although Brinell hardness numbers may vary with the test load used with the 10-mm ball, when smaller balls are used for tests on thin specimens, test results will generally be in agreement with the 10-mm ball test when the ratio of the test load to the square of the ball diameter is held constant (see 3.2 and 6.1).

In Table 1 are given the Brinell hardness numbers corresponding to various diameters of impression for 3000, 1500, and 500-kgf loads, making it unnecessary to calculate for each test the value of the Brinell hardness number by the above equation when these loads are used with a ball 10 mm in diameter.

NOTE 3—The Brinell hardness number followed by the symbol HB without any suffix numbers denotes the following test conditions:

Ball diameter	10 mm
Load	3000 kgf
Duration of loading	10 to 15 s

For other conditions, the hardness number and symbol HB is supplemented by numbers indicating the test conditions in the following order: diameter of ball, load, and duration of loading.

Example—63 HB 10/500/30 indicates a Brinell hardness of 63 measured with a ball of 10-mm diameter and a load of 500 kgf applied for 30 s.

2.3 *verification*—checking or testing to assure conformance with the specification.

2.4 *calibration*—determination of the values of the significant parameters by comparison with values indicated by a reference instrument or by a set of reference standards.

[1] This method is under the jurisdiction of ASTM Committee E-28 on Mechanical Testing.

Current edition approved Aug. 27, 1973. Published October 1973. Originally published as E 10 – 24T. Last previous edition E 10 – 66 (1972).

A. GENERAL DESCRIPTION AND TEST PROCEDURE FOR BRINELL HARDNESS TESTS

3. Apparatus

3.1 *Testing Machine*—Equipment for Brinell hardness testing usually consists of a testing machine which supports the test specimen and applies a predetermined indenting load to a ball in contact with the specimen. The magnitude of the indenting load is limited to certain discrete values. The design of the testing machines shall be such that no rocking or lateral movement of the indenter or specimen occurs while the load is being applied. In the use of machines employing a dead-weight system, precautions shall be taken to prevent a momentary overload caused by the inertia of the dead-weight system. The loading system shall be operated with the utmost care as the maximum value of the indenting load is approached to prevent a large acceleration of the dead-weight system.

3.2 *Brinell Balls:*

3.2.1 The standard ball for Brinell hardness testing shall be 10.000 mm in diameter with a deviation from this value of not more than 0.005 mm in any diameter. Smaller balls having the diameters and tolerances indicated in Table 2 may also be used provided the precautions set forth in 6.1 are observed.

3.2.2 A steel ball having a Vickers hardness (HV) of at least 850 (see 3.2.2.1) using a 10-kgf load may be used on material having a HB not over 450, or a carbide ball on material not over 630. The Brinell test is not recommended for material having a HB over 630. The ball shall be polished and free of defects. For laboratory or referee tests the type of ball used shall be specifically reported where Brinell hardness values exceed 200.

3.2.2.1 The maximum value of the mean of the two diagonals of the impression on the ball made by the Vickers indenter shall not exceed the values listed in Table 3.

3.2.3 The ball shall not show a permanent change in diameter greater than 0.005 mm when pressed with the test load against the test specimens. If a ball is used in a test of a specimen which shows a Brinell hardness number greater than the limit for the ball as detailed in 3.2.2, the ball shall be remeasured after the test. Should the ball show a permanent change in diameter greater than that specified above, the results of the test shall be considered unreliable and the ball shall be unsuitable for further use.

3.3 *Measuring Microscope*—The divisions of the micrometer scale of the microscope or other measuring devices used for the measurement of the diameter of the impression shall be such as to permit the direct measuring of the diameter to 0.1 mm and the estimation of the diameter to 0.02 mm.

NOTE 4—This requirement applies to the construction of the microscope only and is not a requirement for measurement of the impression.

4. Test Specimens

4.1 Specimens used in Brinell hardness testing vary greatly in form since it is frequently desirable to make the impression upon a part to be used in the finished product rather than upon a sample test specimen.

4.1.1 *Thickness*—The thickness of the piece tested shall be such that no bulge or other marking showing the effect of the load appears on the side of the piece opposite the impression. In any event the thickness of the specimen shall be at least ten times the depth of the indentation (Table 4). The minimum width shall conform with the requirements of 6.3.

4.1.2 *Finish*—When necessary, the surface on which the impression is to be made shall be filed, ground, machined, or polished with abrasive material so that the edge of the impression shall be clearly enough defined to permit the measurement of the diameter to the specified accuracy (see 7.1). Care should be taken to avoid overheating or cold working the surface.

5. Verification of Apparatus

5.1 *Verification Methods*—The hardness testing machine shall be verified as specified in Part B.

5.1.1 Two acceptable methods of verifying Brinell hardness testing machines are given in Part B.

5.2 *Loading Range*—A Brinell hardness testing machine for laboratory or referee tests is acceptable for use over a loading range within which the machine error does not exceed ±1 percent. A Brinell hardness testing machine used for routine testing is acceptable for use over a loading range within which the machine error does not exceed ±2 percent.

6. Procedure

6.1 *Magnitude of Test Load* The load in the standard Brinell test shall be 3000, 1500, or 500 kgf. It is desirable that the test load be of such magnitude that the diameter of the impression be in the range 2.50 to 6.00 mm (25.0 to 60.0 percent of ball diameter). A lower limit in impression diameter is necessary because of the reduction in sensitivity of the test with reduction in impression size. The upper limit is influenced by limitations of travel of the indenter in certain types of testers. The thickness and spacing requirements of 4.1.1 and 6.3 may determine the maximum permissible diameter of impression for a specific test. Table 5 gives standard test loads and approximate Brinell hardness numbers for the above range of impression diameters. It is not mandatory that the Brinell test conform to these HB ranges, but it should be realized that different Brinell hardness numbers may be obtained for a given material by using different loads on a 10-mm ball. For the purpose of obtaining a continuous scale of values it may be desirable, however, to use a single load to cover the complete range of hardness for a given class of materials. For softer metals, loads of 250, 125, or 100 kgf are sometimes used. The load used shall be specifically stated in the test report (Note 4). For testing thin or small specimens a ball less than 10 mm in diameter is sometimes used. Such tests (which are not to be regarded as standard tests) will approximate the standard tests more closely if the relation between the applied load, P, measured in kilograms-force, and the diameter of the ball, D, measured in millimetres, is the same as in the standard tests, where:

P/D^2 = 30 for 3000-kgf load and 10-mm ball
P/D^2 = 15 for 1500-kgf load and 10-mm ball
P/D^2 = 5 for 500-kgf load and 10-mm ball

Example A 125-kgf test load on a 5-mm diameter ball would approximate a standard 500-kgf test load on a 10-mm diameter ball.

6.1.1 Special tests for soft metals which do not simulate the standard tests are made with the following load-diameter ratios:

$$P/D^2 = 2.5$$
$$P/D^2 = 1.25$$
$$P/D^2 = 1.0$$

When balls smaller than 10 mm in diameter are used, both the test load and ball size shall be specifically stated in the test report (see 2.2, Note 3, and 9.1). Balls differing in size from the standard 10-mm ball shall conform to the requirements for the material and the permissible variations in diameter specified for the standard ball.

NOTE 5 It is the responsibility of the ASTM committees having jurisdiction over product specifications to specify the load to be used. If no load is specified, it shall be assumed that a load of 3000 kgf is intended.

6.2 *Radius of Curvature* When indentations are made on a curved surface, the minimum radius of curvature of the surface shall be not less than 1 in. (25 mm) for a 10-mm ball. The diameter of the indentation shall be taken as the mean of the two principal diameters.

6.3 *Spacing of Indentations* The distance of the center of the indentation from the edge of the specimen or edge of another indentation shall be at least two and one half times the diameter of the indentation.

6.4 *Application of Test Load* Apply the load to the specimen steadily, without a jerk. In machines employing a dead-weight system take precautions to prevent a momentary overload caused by the inertia of the weights. Also operate the loading system with the utmost care as the maximum value of the indenting load is approached to prevent a large acceleration of the dead-weight system. Apply the full test load for 10 to 15 s except for certain soft metals.

NOTE 6 If for some materials a duration of loading other than 10 to 15 s is necessary, the product specification should specify a definite time of load application. The results of the test should be reported using the nomenclature outlined in Note 3 of this method.[2]

6.5 *Alignment* The angle between the load line and the normal to the specimen shall not exceed 2 deg.

[2] See discussion of the Report of Committee E-1, *Proceedings*, ASTEA, Am. Soc. Testing Mats., Vol. 24, Part I, 1924, pp. 729 and 730.

7. Measurement of Impression

7.1 *Diameter*—In the Brinell hardness test, two diameters of the impression at right angles to each other shall be measured and their mean value used as a basis for calculation of the Brinell hardness number. For verification, laboratory, or referee tests the diameter of the impression shall be read to at least 0.02 mm. For routine tests the diameter of the impression may be read to 0.05 mm.

NOTE 7—When the measuring microscope is used to measure the diameter of the impression, the sharpness of definition of the edge of the impression can be increased by the use of a movable lamp for illuminating the specimens, placing the lamp so that the contrast of light and shade will bring first one edge of the impression, then the other into sharp definition. In testing very hard material, the sharpness of definition of the impression can be somewhat increased by use of a pigment, such as Prussian blue, on the ball. In testing material in which there is considerable recovery of shape, the material may first be coated with a dull black pigment, such as drawing ink or a mixture of graphite and alcohol. The edge of the impression is rendered clear on a surface so coated.

8. Conversion to Other Hardness Scales or Tensile Strength Values

8.1 There is no general method for converting accurately Brinell hardness numbers to other hardness scales or tensile strength values. Such conversions are, at best, approximations and therefore should be avoided, except for special cases where a reliable basis for the approximate conversion has been obtained by comparison tests.

NOTE 8—The ASTM Standard Hardness Conversion Tables E 140, for Metals[3] give approximate conversion values for specific materials such as steel, austenitic stainless steel, nickel and high-nickel alloys, and cartridge brass.

9. Report

9.1 The report shall include the following information:

9.1.1 The Brinell hardness number,

9.1.2 The test conditions when the Brinell hardness number is determined from loads other than 3000 kgf, ball diameters other than 10 mm, and loading times other than 10 to 15 s (see 2.2, Note 3), and

9.1.3 The type ball used if the Brinell hardness number exceeds 200 (see 3.2.2).

[3] *1974 Annual Book of ASTM Standards*, Part 10.

B. VERIFICATION OF BRINELL HARDNESS TESTING MACHINES

10. Scope

10.1 Part B covers two procedures for the verification of Brinell hardness testing machines. These are:

10.1.1 Separate verification of load application, indenter, and the measuring microscope for measuring the diameter of the impression.

10.1.2 Verification by standardized test block method.

10.2 New or rebuilt machines, and machines used for laboratory or referee tests shall be checked by the separate verification method (see 10.1.1).

10.3 Machines used for routine testing may be checked by either verification method.

11. General Requirements

11.1 Before a Brinell hardness testing machine is verified, the machine shall be examined to ensure that:

11.1.1 The machine is properly set up.

11.1.2 The ball holder, with a new steel ball, whose nominal diameter has been checked (see 12.1.2), is firmly mounted in the plunger.

11.1.3 The load can be applied and removed without shock or vibration and in such a manner that the readings are not influenced.

11.2 If the measuring device is integral with the machine, the machine shall be examined to ensure that:

11.2.1 The change from loading to measuring does not influence the readings.

11.2.2 The method of illumination does not affect the readings.

11.2.3 The center of the impression is in the center of the field of view.

12. Verification

12.1 *Separate Verification of Load Application, Indenter, and Measuring Microscope:*

12.1.1 *Load Application*—Brinell hardness testing machines shall be verified at applied loads of 3000, 1500, and 500 kgf. If tests are to be made using other loads, the machine should be verified also at each of these loads.

The applied load shall be checked periodically with a proving ring, by the use of dead weights and proving levers, or by an elastic calibration device or springs in the manner described in ASTM Methods E 4, Verification of Testing Machines.[4, 5] A Brinell hardness testing machine for laboratory or referee tests is acceptable for use over a loading range within which the machine error does not exceed ±1 percent. A Brinell hardness testing machine used for routine testing is acceptable for use over a loading range within which the machine error does not exceed ±2 percent.

12.1.2 *Indenter*—The indenter to be verified shall be a new steel ball selected at random from a lot meeting the hardness requirements specified in Table 3. The diameter of each ball shall be measured to an accuracy of ±0.0005 mm at not less than three positions and the mean of these readings shall not differ from the nominal diameter by more than the tolerance specified in Table 2.

12.1.3 *Measuring Microscope*—The measuring microscope or other device for measuring the diameter of the impression shall be verified at five intervals over the working range by the use of an accurate scale such as a stage micrometer. The adjustment of the micrometer microscope shall be such that, throughout the range covered, the difference between the scale divisions of the microscope and of the calibrating scale does not exceed 0.01 mm.

12.2 *Verification by Standardized Test Block Method:*

12.2.1 A Brinell hardness testing machine used only for routine testing may be checked by making a series of impressions on standardized hardness test blocks (Part C).

12.2.2 If the machine is to be used at conditions other than 10/3000/15, the machine shall also be verified at those other conditions.

12.2.3 The Brinell hardness testing machine shall be considered verified if the mean diameter of any hardness impression differs by no more than 3 percent from the mean diameter corresponding to the hardness value of the standardized hardness test block.

[4] *1974 Annual Book of ASTM Standards*, Parts 10, 14, 32, 35, and 41.

[5] See also Jones, J. L., and Marshall, C. H., "A New Method for Calibrating Brinell Hardness Testing Machines," *Proceedings*, Am. Soc. Testing Mats., Vol. 20, Part II, 1920, p. 392.

C. CALIBRATION OF STANDARDIZED HARDNESS TEST BLOCKS FOR BRINELL HARDNESS TESTING MACHINES

13. Scope

13.1 Part C covers the calibration of standardized hardness test blocks for the verification of Brinell hardness testing machines as described in Part B.

14. Manufacture

14.1 Each metal block to be calibrated shall be not less than $^5/_8$ in. (16 mm) in thickness for 10-mm balls, $^1/_2$ in. (12 mm) for 5-mm balls, and $^1/_4$ in. (6 mm) for smaller balls.

14.2 Each block shall be specially prepared and heat treated to give the necessary homogeneity and stability of structure.

14.3 Each block, if of steel, shall be demagnetized by the manufacturer and maintained demagnetized by the user.

14.4 The supporting surface of the test block shall have a ground finish.

14.5 The test surface shall be free from scratches which would interfere with measurements of the diameters of the impression.

14.5.1 The mean surface roughness height rating shall not exceed 12 μin. (0.0003 mm) center line average for the standard 10-mm ball. For smaller balls a mean surface roughness height rating of 6 μin. (0.00015 mm) is recommended.

14.6 To ensure that no material is subsequently removed from the test surface of the standardized test block, an official mark or the thickness at the time of calibration shall be stamped or engraved on the test surface to an accuracy of ±0.005 in. (±0.1mm).

15. Standardizing Procedure

15.1 The standardized blocks shall be calibrated on a Brinell hardness testing machine verified in accordance with the requirements of 12.1.

15.2 The mechanism that controls the application of the load shall ensure that the speed of approach immediately before the ball touches the specimen and the speed of penetration does not exceed 1 mm/s.

15.3 The full load shall be applied for 15 s.

16. Indenter

16.1 A new steel ball conforming to the requirements of 12.1.2 shall be used for calibrating standardized hardness test blocks.

17. Number of Impressions

17.1 At least five randomly distributed impressions shall be made on each test surface of the block. If the area of the test block is greater than 100 cm^2 the number of randomly distributed impressions on the test surface should be increased to eight.

18. Measurement of the Diameters of the Impression

18.1 The illuminating system of the measuring microscope shall be adjusted to give uniform intensity over the field of view and maximum contrast between the impressions and the undisturbed surface of the block.

18.2 The measuring microscope shall be graduated to read 0.002 mm for impressions made with balls of 5-mm diameter or larger and 0.001 mm for impressions made with balls of smaller diameter.

18.3 The measuring microscope shall be checked by a stage micrometer, or by other suitable means to ensure that the difference between readings corresponding to any two divisions of the instrument is correct within ±0.001 mm for balls of less than 5-mm diameter and within ±0.002 mm for balls of larger diameter.

18.4 It is recommended that each impression be measured at least by two observers.

19. Repeatability

19.1 If d_1, $d_2 \ldots d_n$ are the mean values of the measured diameters as determined by one observer and arranged in increasing order of magnitude the repeatability of the hardness readings, on the block under the particular conditions of standardization, is defined as $d_n - d_1$ where $n = 5$ or 8.

20. Uniformity of Hardness

20.1 Unless the repeatability of the mean diameters of each of the five or eight impressions is within 2 percent of the mean value for the five or eight mean readings, the block cannot be regarded as sufficiently uniform for standardization purposes.

21. Marking

21.1 Each block shall be marked with the following:

21.1.1 The arithmetic mean of the hardness values found in the standardizing test (see also 2.2, Note 3),

21.1.2 The name or mark of the supplier,

21.1.3 The serial number of the block,

21.1.4 The thickness of the block or an official mark on the top surface (see 14.6).

NOTE 10—All of the markings except the official mark or thickness shall be placed on the side of the block, the markings being upright when the test surface is the upper face.

TABLE 1 Brinell Hardness Numbers[a]

(Ball 10 mm in Diameter; Applied Loads of 500, 1500, and 3000 kgf)

NOTE—The values given in this table for Brinell hardness numbers are merely solutions of the equation given in the definition in 2.2, and include values for impression diameters outside the ranges recommended in 6.1. These values are indicated by italics.

Diameter of Indentation, mm	Brinell Hardness Number			Diameter of Indentation, mm	Brinell Hardness Number			Diameter of Indentation, mm	Brinell Hardness Number			Diameter of Indentation, mm	Brinell Hardness Number		
	500-kgf Load	1500-kgf Load	3000-kgf Load		500-kgf Load	1500-kgf Load	3000-kgf Load		500-kgf Load	1500-kgf Load	3000-kgf Load		500-kgf Load	1500-kgf Load	3000-kgf Load
2.00	*158*	*473*	*945*	2.30	*119*	*356*	*712*	2.60	92.6	278	555	2.90	74.1	222	444
2.01	*156*	*468*	*936*	2.31	*118*	*353*	*706*	2.61	91.8	276	551	2.91	73.6	221	441
2.02	*154*	*463*	*926*	2.32	*117*	*350*	*700*	2.62	91.1	273	547	2.92	73.0	219	438
2.03	*153*	*459*	*917*	2.33	*116*	*347*	*694*	2.63	90.4	271	543	2.93	72.5	218	435
2.04	*151*	*454*	*908*	2.34	*115*	*344*	*688*	2.64	89.7	269	538	2.94	72.0	216	432
2.05	*150*	*450*	*899*	2.35	*114*	*341*	*682*	2.65	89.0	267	534	2.95	71.5	215	429
2.06	*148*	*445*	*890*	2.36	*113*	*338*	*676*	2.66	88.4	265	530	2.96	71.0	213	426
2.07	*147*	*441*	*882*	2.37	*112*	*335*	*670*	2.67	87.7	263	526	2.97	70.5	212	423
2.08	*146*	*437*	*873*	2.38	*111*	*332*	*665*	2.68	87.0	261	522	2.98	70.1	210	420
2.09	*144*	*432*	*865*	2.39	*110*	*330*	*659*	2.69	86.4	259	518	2.99	69.6	209	417
2.10	*143*	*428*	*856*	2.40	*109*	*327*	*653*	2.70	85.7	257	514	3.00	69.1	207	415
2.11	*141*	*424*	*848*	2.41	*108*	*324*	*648*	2.71	85.1	255	510	3.01	68.6	206	412
2.12	*140*	*420*	*840*	2.42	*107*	*322*	*643*	2.72	84.4	253	507	3.02	68.2	205	409
2.13	*139*	*416*	*832*	2.43	*106*	*319*	*637*	2.73	83.8	251	503	3.03	67.7	203	406
2.14	*137*	*412*	*824*	2.44	*105*	*316*	*632*	2.74	83.2	250	499	3.04	67.3	202	404
2.15	*136*	*408*	*817*	2.45	*104*	*313*	*627*	2.75	82.6	248	495	3.05	66.8	200	401
2.16	*135*	*404*	*809*	2.46	*104*	*311*	*621*	2.76	81.9	246	492	3.06	66.4	199	398
2.17	*134*	*401*	*802*	2.47	*103*	*308*	*616*	2.77	81.3	244	488	3.07	65.9	198	395
2.18	*132*	*397*	*794*	2.48	*102*	*306*	*611*	2.78	80.8	242	485	3.08	65.5	196	393
2.19	*131*	*393*	*787*	2.49	*101*	*303*	*606*	2.79	80.2	240	481	3.09	65.0	195	390
2.20	*130*	*390*	*780*	2.50	100	301	601	2.80	79.6	239	477	3.10	64.6	194	388
2.21	*129*	*386*	*772*	2.51	99.4	298	597	2.81	79.0	237	474	3.11	64.2	193	385
2.22	*128*	*383*	*765*	2.52	98.6	296	592	2.82	78.4	235	471	3.12	63.8	191	383
2.23	*126*	*379*	*758*	2.53	97.8	294	587	2.83	77.9	234	467	3.13	63.3	190	380
2.24	*125*	*376*	*752*	2.54	97.1	291	582	2.84	77.3	232	464	3.14	62.9	189	378
2.24	*124*	*372*	*745*	2.55	96.3	289	578	2.85	76.8	230	461	3.15	62.5	188	375
2.26	*123*	*369*	*738*	2.56	95.5	287	373	2.86	76.2	229	457	3.16	62.1	186	373
2.27	*122*	*366*	*732*	2.57	94.8	284	569	2.87	75.7	227	454	3.17	61.7	185	370
2.28	*121*	*363*	*725*	2.58	94.0	282	564	2.88	75.1	225	451	3.18	61.3	184	368
2.29	*120*	*359*	*719*	2.59	93.3	280	560	2.89	74.6	224	448	3.19	60.9	183	366

TABLE 1—*Continued*

Diameter of Indentation, mm	Brinell Hardness Number 500-kgf Load	1500-kgf Load	3000-kgf Load	Diameter of Indentation, mm	Brinell Hardness Number 500-kgf Load	1500-kgf Load	3000-kgf Load	Diameter of Indentation, mm	Brinell Hardness Number 500-kgf Load	1500-kgf Load	3000-kgf Load	Diameter of Indentation, mm	Brinell Hardness Number 500-kgf Load	1500-kgf Load	3000-kgf Load
3.20	60.5	182	363	3.50	50.3	151	302	3.80	42.4	127	255	4.10	36.2	109	217
3.21	60.1	180	361	3.51	50.0	150	300	3.81	42.2	127	253	4.11	36.0	108	216
3,22	59.8	179	359	3.52	49.7	149	298	3.82	42.0	126	252	4.12	35.8	108	215
3.23	59.4	178	356	3.53	49.4	148	297	3.83	41.7	125	250	4.13	35.7	107	214
3.24	59.0	177	354	3.54	49.2	147	295	3.84	41.5	125	249	4.14	35.5	106	213
3.25	58.6	176	352	3.55	48.9	147	293	3.85	41.3	124	248	4.15	35.3	106	212
3.26	58.3	175	350	3.56	48.6	146	292	3.86	41.1	123	246	4.16	35.1	105	211
3.27	57.9	174	347	3.57	48.3	145	290	3.87	40.9	123	245	4.17	34.9	105	210
3.28	57.5	173	345	3.58	48.0	144	288	3.88	40.6	122	244	4.18	34.8	104	209
3.29	57.2	172	343	3.59	47.7	143	286	3.89	40.4	121	242	4.19	34.6	104	208
3.30	56.8	170	341	3.60	47.5	142	285	3.90	40.2	121	241	4.20	34.4	103	207
3.31	56.5	169	339	3.61	47.2	142	283	3.91	40.0	120	240	4.21	34.2	103	205
3.32	56.1	168	337	3.62	46.9	141	282	3.92	39.8	119	239	4.22	34.1	102	204
3.33	55.8	167	335	3.63	46.7	140	280	3.93	39.6	119	237	4.23	33.9	102	203
3.34	55.4	166	333	3.64	46.4	139	278	3.94	39.4	118	236	4.24	33.7	101	202
3.35	55.1	165	331	3.65	46.1	138	277	3.95	39.1	117	235	4.25	33.6	101	201
3.36	54.8	164	329	3.66	45.9	138	275	3.96	38.9	117	234	4.26	33.4	100	200
3.36	54.4	163	326	3.67	45.6	137	274	3.97	38.7	116	232	4.27	33.2	99.7	199
3.38	54.1	162	325	3.68	45.4	136	272	3.98	38.5	116	231	4.28	33.1	99.2	198
3.39	53.8	161	323	3.69	45.1	135	271	3.99	38.3	115	320	4.29	32.9	98.8	198
3.40	53.4	160	321	3.70	44.9	135	269	4.00	38.1	114	229	4.30	32.8	98.3	197
3.41	53.1	159	319	3.71	44.6	134	268	4.01	37.9	114	228	4.31	32.6	97.8	196
3.42	52.8	158	317	3.72	44.4	133	266	4.02	37.7	113	226	4.32	32.4	97.3	195
3.43	52.5	157	315	3.73	44.1	132	265	4.03	37.5	113	225	4.33	32.3	96.8	194
3.44	52.2	156	313	3.74	43.9	132	263	4.04	37.3	112	224	4.34	32.1	96.4	193
3.45	51.8	156	311	3.75	43.6	131	262	4.05	37.1	111	223	4.35	32.0	95.9	192
3.46	51.5	155	309	3.76	43.4	130	260	4.06	37.0	111	222	4.36	31.8	95.5	191
3.47	51.2	154	307	3.77	43.1	129	259	4.07	36.8	110	221	4.37	31.7	95.0	190
3.48	50.9	153	306	3.78	42.9	129	257	4.08	36.6	110	219	4.38	31.5	94.5	189
3.49	50.6	152	304	3.79	42.7	128	256	4.09	36.4	109	218	4.39	31.4	94.1	188

[a] Prepared by the Engineering Mechanics Section, National Bureau of Standards.

TABLE 1 *Continued*

Diameter of Indentation, mm	Brinell Hardness Number			Diameter of Indentation, mm	Brinell Hardness Number			Diameter of Indentation, mm	Brinell Hardness Number			Diameter of Indentation, mm	Brinell Hardness Number		
	500-kgf Load	1500-kgf Load	3000-kgf Load		500-kgf Load	1500-kgf Load	3000-kgf Load		500-kgf Load	1500-kgf Load	3000-kgf Load		500-kgf Load	1500-kgf Load	3000-kgf Load
4.40	31.2	93.6	187	4.70	27.1	81.4	163	5.00	23.8	71.3	143	5.30	20.9	62.8	126
4.41	31.1	93.2	186	4.71	27.0	81.0	162	5.01	23.7	71.0	142	5.31	20.9	62.6	125
4.42	30.9	92.7	185	4.72	26.9	80.7	161	5.02	23.6	70.7	141	5.32	20.8	62.3	125
4.43	30.8	92.3	185	4.73	26.8	80.3	161	5.03	23.5	70.4	141	5.33	20.7	62.1	124
4.44	30.6	91.8	184	4.74	26.6	79.9	160	5.04	23.4	70.1	140	5.34	20.6	61.8	124
4.45	30.5	91.4	183	4.75	26.5	79.6	159	5.05	23.3	69.8	140	5.35	20.5	61.5	123
4.46	30.3	91.0	182	4.76	26.4	79.2	158	5.06	23.2	69.5	139	5.36	20.4	61.3	123
4.47	30.2	90.5	181	4.77	26.3	78.9	158	5.07	23.1	69.2	138	5.37	20.3	61.0	122
4.48	30.0	90.1	180	4.78	26.2	78.5	157	5.08	23.0	68.9	138	5.38	20.3	60.8	122
4.49	29.9	89.7	179	4.79	26.1	78.2	156	5.09	22.9	68.6	137	5.39	20.2	60.6	121
4.50	29.8	89.3	179	4.80	25.9	77.8	156	5.10	22.8	68.3	137	5.40	20.1	60.3	121
4.51	29.6	88.8	178	4.81	25.8	77.5	155	5.11	22.7	68.0	136	5.41	20.0	60.1	120
4.52	29.5	88.4	177	4.82	25.7	77.1	154	5.12	22.6	67.7	135	5.42	19.9	59.8	120
4.53	29.3	88.0	176	4.83	25.6	76.8	154	5.13	22.5	67.4	135	5.43	19.9	59.6	119
4.54	29.2	87.6	175	4.84	25.5	76.4	153	5.14	22.4	67.1	134	5.44	19.8	59.3	119
4.55	29.1	87.2	174	4.85	25.4	76.1	152	5.15	22.3	66.9	134	5.45	19.7	59.1	118
4.56	28.9	86.8	174	4.86	25.3	75.8	152	5.16	22.2	66.6	133	5.46	19.6	58.9	118
4.57	28.8	86.4	173	4.87	25.1	75.4	151	5.17	22.1	66.3	133	5.47	19.5	58.6	117
4.58	28.7	86.0	172	4.88	25.0	75.1	150	5.18	22.0	66.0	132	5.48	19.5	58.4	117
4.59	28.5	85.6	171	4.89	24.9	74.8	150	5.19	21.9	65.8	132	5.49	19.4	58.2	116
4.60	28.4	85.4	170	4.90	24.8	74.4	149	5.20	21.8	65.5	131	5.50	19.3	57.9	116
4.61	28.3	84.8	170	4.91	24.7	74.1	148	5.21	21.7	65.2	130	5.51	19.2	57.7	115
4.62	28.1	84.4	169	4.92	24.6	73.8	148	5.22	21.6	64.9	130	5.52	19.2	57.5	115
4.63	28.0	84.0	168	4.93	24.5	73.5	147	5.23	21.6	64.7	129	5.53	19.1	57.2	114
4.64	27.9	83.6	167	4.94	24.4	73.2	146	5.24	21.5	64.4	129	5.54	19.0	57.0	114
4.65	27.8	83.3	167	4.95	24.3	72.8	146	5.25	21.4	64.1	128	5.55	18.9	56.8	114
4.66	27.6	82.9	166	4.96	24.2	72.5	145	5.26	21.3	63.9	128	5.56	18.9	56.6	113
4.67	27.5	82.5	165	4.97	24.1	72.2	144	5.27	21.2	63.6	127	5.57	18.8	56.3	113
4.68	27.4	82.1	164	4.98	24.0	71.9	144	5.28	21.1	63.3	127	5.58	18.7	56.1	112
4.69	27.3	81.8	164	4.99	23.9	71.6	143	5.29	21.0	63.1	126	5.59	18.6	55.9	112

TABLE 1—*Continued*

Diameter of Indentation, mm	Brinell Hardness Number 500-kgf Load	Brinell Hardness Number 1500-kgf Load	Brinell Hardness Number 3000-kgf Load	Diameter of Indentation, mm	Brinell Hardness Number 500-kgf Load	Brinell Hardness Number 1500-kgf Load	Brinell Hardness Number 3000-kgf Load	Diameter of Indentation, mm	Brinell Hardness Number 500-kgf Load	Brinell Hardness Number 1500-kgf Load	Brinell Hardness Number 3000-kgf Load	Diameter of Indentation, mm	Brinell Hardness Number 500-kgf Load	Brinell Hardness Number 1500-kgf Load	Brinell Hardness Number 3000-kgf Load
5.60	18.6	55.7	111	5.95	16.2	48.7	97.3	6.30	14.2	42.7	85.5	6.65	12.6	37.7	75.4
5.61	18.5	55.5	111	5.96	16.2	48.5	96.9	6.31	14.2	42.6	85.2	6.66	12.5	37.6	75.2
5.62	18.4	55.2	110	5.97	16.1	48.3	96.6	6.32	14.1	42.4	84.9	6.67	12.5	37.5	74.9
5.63	18.3	55.0	110	5. 86	16.0	48.1	96.2	6.33	14.1	42.3	84.6	6.68	12.4	37.3	74.7
5.64	18.3	54.8	110	5.99	16.0	47.9	95.9	6.34	14.0	42.1	84.3	6.69	12.4	37.2	74.4
5.65	18.2	54.6	109	6.00	15.9	47.7	95.5	6.35	14.0	42.0	84.0	6.70	12.4	37.1	74.1
5.66	18.1	54.4	109	6.01	15.9	47.6	95.1	6.36	13.9	41.8	83.7	6.71	12.3	36.9	73.9
5.67	18.1	54.2	108	6.02	15.8	47.4	94.8	6.37	13.9	41.7	83.4	6.72	12.3	36.8	73.6
5.68	18.0	54.0	108	6.03	15.7	47.2	94.4	6.38	13.8	41.5	83.1	6.73	12.2	36.7	73.4
5.69	17.9	53.7	107	6.04	15.7	47.0	94.1	6.39	13.8	41.4	82.8	6.74	12.2	36.6	73.1
5.70	17.8	53.5	107	6.05	15.6	46.8	93.7	6.40	13.7	41.2	82.5	6.75	12.1	36.4	72.8
5.71	17.8	53.3	107	6.06	15.6	46.7	93.4	6.41	13.7	41.1	82.2	6.76	12.1	36.3	72.6
5.72	17.7	53.1	106	6.07	15.5	46.5	93.0	6.42	13.6	40.9	81.9	6.77	12.1	36.2	72.3
5.73	17.6	52.9	106	6.08	15.4	46.3	92.7	6.43	13.6	40.8	81.6	6.78	12.0	36.0	72.1
5.74	17.6	52.7	105	6.09	15.4	46.2	92.3	6.44	13.5	40.6	81.3	6.79	12.0	35.9	71.8
5.75	17.5	52.5	105	6.10	15.3	46.0	92.0	6.45	13.5	40.5	81.0	6.80	11.9	35.8	71.6
5.76	17.4	52.3	105	6.11	15.3	45.8	91.7	6.46	13.4	40.4	80.7	6.81	11.9	35.7	71.3
5.77	17.4	52.1	104	6.12	15.2	45.7	91.3	6.47	13.4	40.2	80.4	6.82	11.8	35.5	71.1
5.78	17.3	51.9	104	6.13	15.2	45.5	91.0	6.48	13.4	40.1	80.1	6.83	11.8	35.4	70.8
5.79	17.2	51.7	103	6.14	15.1	45.3	90.6	6.49	13.3	39.9	79.8	6.84	11.8	35.3	70.6
5.80	17.2	51.5	103	6.15	15.1	45.2	90.3	6.50	13.3	39.8	79.6	6.85	11.7	35.2	70.4
5.81	17.1	51.3	103	6.16	15.0	45.0	90.0	6.51	13.2	38.6	79.3	6.86	11.7	35.1	70.1
5.82	17.0	51.1	102	6.17	14.9	44.8	89.6	6.52	13.2	39.5	79.0	6.87	11.6	34.9	69.9
5.83	17.0	50.9	102	6.18	14.9	44.7	89.3	6.53	13.1	39.4	78.7	6.88	11.6	34.8	69.6
5.84	16.9	50.7	101	6.19	14.8	44.5	89.0	6.54	13.1	39.2	78.4	6.89	11.6	34.7	69.4
5.85	16.8	50.5	101	6.20	14.7	44.3	88.7	6.55	13.0	39.1	78.2	6.90	11.5	34.6	69.2
5.86	16.8	50.3	101	6.21	14.7	44.2	88.3	6.56	13.0	38.9	78.0	6.91	11.5	34.5	68.9
5.87	16.7	50.2	100	6.22	14.7	44.0	88.0	6.57	12.9	38.8	77.6	6.92	11.4	34.3	68.7
5.88	16.7	50.0	99.9	6.23	14.6	43.8	87.7	6.58	12.9	38.7	77.3	6.93	11.4	34.2	68.4
5.89	16.6	49.8	99.5	6.24	14.6	43.7	87.4	6.59	12.8	38.5	77.1	6.94	11.4	34.1	68.2
5.90	16.5	49.6	99.2	6.25	14.5	43.5	87.1	6.60	12.8	38.4	76.8	6.95	11.3	34.0	68.0
5.91	16.5	49.4	98.8	6.26	14.5	43.4	86.7	6.61	12.8	38.3	76.5	6.96	11.3	33.9	67.7
5.92	16.4	49.2	98.4	6.27	14.4	43.2	86.4	6.62	12.7	38.1	76.2	6.97	11.3	33.8	67.5
5.93	16.3	49.0	98.0	6.28	14.4	43.1	86.1	6.63	12.7	38.0	76.0	6.98	11.2	33.6	67.3
5.94	16.3	48.8	97.7	6.29	14.3	42.9	85.8	6.64	12.6	37.9	75.7	6.99	11.2	33.5	67.0

TABLE 2 Brinell Hardness Balls Other Than Standard

Diameter of Ball, mm	Tolerance,[a] mm
from 1 to 3, incl.	±0.0035
over 3 to 6, incl.	±0.004
over 6 to 10, incl.	±0.0045

[a] Steel balls for ball bearings normally satisfy these tolerances.

TABLE 3 Maximum Mean Diagonal of Vickers Hardness Impression on Brinell Hardness Balls

Ball Diameter, mm	Maximum Mean Diagonal of Impression on the Ball Made with Vickers Indenter Under 10-kgf Load, mm
10	0.146
5	0.145
2.5	0.143
2	0.142
1	0.139

TABLE 4 Minimum Thickness Requirements for Brinell Hardness Tests

Minimum Thickness of Specimen		Minimum Hardness for Which the Brinell Test May Safely Be Made		
in.	mm	3000-kgf Load	1500-kgf Load	500-kgf Load
1/16	1.6	602	301	100
1/8	3.2	301	150	50
3/16	4.8	201	100	33
1/4	6.4	150	75	25
5/16	8.0	120	60	20
3/8	9.6	100	50	17

TABLE 5 Standard Test Loads

Ball Diameter, mm	Load, kgf	Recommended Range HB
10	3000	96 to 600
10	1500	48 to 300
10	500	16 to 100

By publication of this standard no position is taken with respect to the validity of any patent rights in connection therewith, and the American Society for Testing and Materials does not undertake to insure anyone utilizing the standard against liability for infringement of any Letters Patent nor assume any such liability.

Designation: E 18 - 74

American National Standard Z115.6
American National Standards Institute

Standard Methods of Test for
ROCKWELL HARDNESS AND ROCKWELL SUPERFICIAL HARDNESS OF METALLIC MATERIALS[1]

This Standard is issued under the fixed designation E 18; the number immediately following the designation indicates the year of original adoption or, in the case of revision, the year of last revision. A number in parentheses indicates the year of last reapproval.

These methods have been approved by the Department of Defense to replace method 243.1 of Federal Test Method Standard No. 151 b and for listing in DoD Index of Specifications and Standards. Future proposed revisions should be coordinated with the Federal Government through the Army Materials and Mechanics Research Center, Watertown, Mass. 02172.

1. Scope

1.1 These methods cover the determination of the Rockwell hardness (3 to 9) and the Rockwell superficial hardness (10 to 16) of metallic materials, including methods for the verification of machines for Rockwell hardness testing (Part C) and the calibration of standardized hardness test blocks (Part D).

1.2 The information in 2, 3, 4, 10, and 11 is intended to describe and define the type of test that is involved and to outline the limitations of acceptable testing machines. This descriptive material is not mandatory in connection with the test itself.

NOTE 1—The values stated in U.S. customary units are to be regarded as the standard.

2. Definitions

2.1 *Rockwell hardness test*—an indentation hardness test using a calibrated machine to force a diamond spheroconical penetrator (diamond penetrator), or hard steel ball under specified conditions, into the surface of the material under test in two operations, and to measure the permanent depth of the impression under the specified conditions of minor and major loads.

2.2 *Rockwell superficial hardness test*—same as the Rockwell hardness test except that smaller minor and major loads are used.

2.3 *Rockwell hardness number, HR*—a number derived from the net increase in the depth of impression as the load on a penetrator is increased from a fixed minor load to a major load and then returned to the minor load.

NOTE 2: *Penetrators*—Penetrators for the Rockwell hardness test include a diamond sphero-conical penetrator having an included angle of 120 deg with a spherical tip having a radius of 0.200 mm and steel balls of several specified diameters.

NOTE 3—Rockwell hardness numbers are always quoted with a scale symbol representing the penetrator, load, and dial used. The hardness number is followed by the symbol HR and the scale designation.

Examples: 64 HRC = Rockwell hardness number of 64 on Rockwell C scale. 81 HR30N = Rockwell superficial hardness number of 81 on Rockwell 30 N scale.

2.4 *verification*—checking or testing to assure conformance with the specification.

2.5 *calibration*—determination of the values of the significant parameters by comparison with values indicated by a reference instrument or by a set of reference standards.

---

[1] These methods are under the jurisdiction of ASTM Committee E-28 on Mechanical Testing.

Current edition approved April 29, 1974. Published July 1974. Originally published as E 18 - 32 T. Last previous edition E 18 - 67.

A. GENERAL DESCRIPTION AND TEST PROCEDURE FOR ROCKWELL HARDNESS TESTS

3. Apparatus

3.1 *General Principles* The general principles of the Rockwell hardness test are illustrated in Fig. 1 (diamond penetrator) and Fig. 2 (ball penetrator) and the accompanying Tables 1 and 2.

3.2 *Description of Machine and Method of Test* The tester for making Rockwell hardness determinations is essentially a machine that measures hardness by determining the depth of penetration of a penetrator into the specimen under certain arbitrarily fixed conditions of test. The penetrator may be either a diamond sphero-conical penetrator or a steel ball. The hardness value, as read from the dial, is an arbitrary number which is related to the depth of penetration caused by two superimposed impressions, and since the scales are reversed, the number is higher the harder the material. A minor load of 10 kgf (98 N) is first applied which causes an initial penetration which sets the penetrator on the material and holds it in position. The dial is set at zero on the black-figure scale, and the major load is applied. This major load is the total applied and the depth measurement depends solely on the increase in depth due to the increase from minor to major load. After the major load is applied and removed, according to standard procedure, the reading is taken while the minor load is still in position. The major load is usually 150 kgf (1471 N) when a diamond sphero-conical penetrator is employed and is customarily 60 kgf (589 N) or 100 kgf (981 N) when a steel ball is used as a penetrator, but other loads may be used when found necessary. The ball penetrator is $\frac{1}{16}$ in. (1.588 mm) in diameter normally, but other penetrators of larger diameter such as $\frac{1}{8}$, $\frac{1}{4}$, or $\frac{1}{2}$ in. (3.175, 6.350, or 12.70 mm) may be employed for soft metals. A variety of loads and penetrators are thus provided and experience decides the best combination for use.

3.3 *Rockwell Hardness Scales* Rockwell hardness values are usually determined and reported according to one of the standard scales specified in Table 3. There is no Rockwell hardness value designated by a figure alone because it is necessary to indicate which penetrator and load have been employed in making the test. In all cases the minor load is 10 kgf (98 N) and the dial is adjusted after applying the minor load so that the pointer reads at "SET" (C 0 or B 30). The diamond penetrator is not recommended for materials giving readings below 20. The use of ball penetrators for materials that give readings greater than 100 is not recommended primarily because of lack of sensitivity and possible flattening of the ball. For the Rockwell hardness test, one Rockwell number represents 0.00008-in. (0.002-mm) movement of the penetrator. Typical applications of the various scales are shown in Table 3. There may be cases in which more than one scale could be used. It is desirable to employ the smallest ball that can properly be used, because of the loss of sensitivity as the diameter of the ball increases. An exception to this is made when soft nonhomogeneous material is to be tested, in which case it may be preferable to use a larger ball which makes an impression of greater area, thus obtaining more of an average hardness. While the choice of scales is optional, the scale symbol for the combination of penetrator, load, and dial used should be as listed in Table 3.

3.4 *Penetrators:*

3.4.1 The standard penetrators, as have been mentioned in 3.3, are the diamond sphero-conical penetrator and steel balls, $\frac{1}{16}$, $\frac{1}{8}$, $\frac{1}{4}$, and $\frac{1}{2}$ in. (1.588, 3.175, 6.350, and 12.70 mm) in diameter.

3.4.1.1 The shape of the diamond sphero-conical penetrator should be a cone forming a 120-deg angle with a spherical apex of 0.200-mm radius (Fig. 1).

3.4.1.2 The steel balls used should be free from surface imperfections. The balls should be round and conform to the requirements prescribed in 19.1.3.

3.4.2 An occasional check of the contour of the penetrator should be made by examination with a magnifying glass. This will reveal any chipping of the diamond or flattening of a ball penetrator. If either of these conditions is discovered, the penetrator should be replaced.

3.4.3 Dust, dirt, grease, and scale should not be allowed to accumulate on the penetrator as this will affect the results.

3.5 *Anvils*—An anvil should be used that is suitable for the specimen to be tested. Cylindrical pieces should be tested with a V-notch anvil that will support the specimen with the axis directly under the penetrator, or on hard, parallel, twin cylinders properly positioned and clamped to their base. Flat pieces should be tested on a flat anvil that has a smooth flat bearing surface whose plane is perpendicular to the axis of the penetrator. For thin materials or specimens that are not perfectly flat, a flat anvil having an elevated "spot" about ¼ in. (6 mm) in diameter and about ¾ in. (19 mm) in height is used. This spot should be polished smooth and flat and should have a Rockwell hardness of at least C 60. The seating and supporting surfaces of all anvils should be free from pits, heavy scratches, dust, dirt, and grease. If the provisions of 4.3 on thickness of specimen are complied with, there will be no danger of indenting the anvil but, if the specimen is so thin that the impression will show through on the under side, it is possible that the anvil may be damaged. Damage may also occur from accidental contacting of the anvil with the penetrator. If the anvil is damaged from any cause, it should be replaced. Anvils showing the least perceptible dent will give inaccurate results on thin material. Very soft material should not be tested on the "spot" anvil because the applied load may cause the penetration of the anvil into the under side of the specimen regardless of its thickness.

3.6 *Test Blocks*—Test blocks meeting the requirements outlined in Part D should be used in periodically checking the accuracy of the hardness tester. The test blocks should be of uniform material, thick enough to be free from "anvil effect," of smooth surface on both top and bottom, and of not more than 4 in.^{2} (26 cm^2) surface area on the test surface. They should be prepared to a reasonably fine finish; the surface roughness height rating of the test surface should be 12 or less, that is, 12 μin. (0.0003 mm), center line average, and the blocks should be inspected for freedom from any major surface defects or imperfections that might affect the hardness readings.

3.7 *Verifying Machines for Rockwell Hardness Testing with Standardized Hardness Test Blocks*—It has been found practicable to keep machines for Rockwell hardness tests within the tolerances of the standardized hardness test blocks (see Part D). When machines used in everyday production inspection testing are verified and adjusted once a month, it has been found that they very seldom deviate more than this amount. Daily checking (see Section 7) of such machines assures the operator that the penetrator is in good condition and the machine is operating properly.

3.8 *General Precautions:*

3.8.1 *Protection Against Vibration*—If the bench or table on which the hardness tester is mounted is subject to vibration, such as experienced in the vicinity of other machines, the tester should be mounted on a metal plate on sponge rubber at least 1 in. (25 mm) in thickness or any other type of mounting that will effectually eliminate vibration from the machine, otherwise the penetrator will penetrate farther into the material than when such vibrations are absent.

3.8.2 *Preparation of Specimens* — Specimens employed for hardness determinations shall be prepared with care. Should sheet metal be employed, special care should be taken with material that is curved. The concave side of the curved metal should face toward the penetrator. If such specimens are reversed, an error will be introduced due to the flattening of the metal on the anvil.

3.8.3 *Overhang of Specimens*—Specimens that have sufficient overhang so that they do not balance themselves on the anvil shall be properly supported.

3.8.4 *Penetrator and Anvil*—The penetrator and anvil should not be brought together without a test specimen between them, otherwise the anvil will be indented and the ball flattened.

3.8.5 *Dial Gage Plunger*—The dial gage plunger shall move freely in any position.

4. Test Specimens

4.1 *Form*—Specimens used in Rockwell hardness testing vary greatly in form since it is frequently desirable to make the impression upon a part to be used in the finished product rather than upon a sample test specimen. It is recognized that all of the many conditions of test specimens, size, preparation, etc., cannot be covered specifically and the following paragraphs are intended only as a general guide in the selection of test specimens.

4.2 *Surface Conditions*—Surface conditions have a marked effect on the readings obtained on thin materials and, due to the fact that the thickness of such specimens may influence the results, it cannot be assumed that the indications on the standardized test blocks furnish any reliable measure of the errors to be expected when testing thin material. Standardized test blocks are always thick enough to eliminate the effect of the underlying anvil, but it must be remembered that errors of unknown magnitude may occur when tests are made on material so thin that the impression shows through on the reverse side. The standardized test blocks will indicate the errors of the machine when used to test specimens of similar size, shape, and surface condition, but there is at the present time no satisfactory way of checking the accuracy of the readings taken on thin material nor of evaluating the anvil effect when this is present.

4.3 *Thickness*—Rockwell hardness tests of the highest accuracy are made on specimens of sufficient thickness so that the Rockwell reading is not noticeably affected by the supporting anvil. Absence of a bulge or other marking on the surface of the test specimen opposite the impression is an indication that the specimen is sufficiently thick for precision testing. Commercially acceptable Rockwell hardness readings may be obtained on sheet materials that show some bulging or marking, and in some specifications the sheets to be tested are of a thickness and hardness such that anvil effect will be present. Values of limiting thicknesses at various hardness levels for selected Rockwell scales using the diamond penetrator and the $^1/_{16}$-in. (1.588-mm) diameter ball are given in Tables 4 and 5. Rockwell hardness tests on sheet metals are acceptable for hardness specification purposes when the tests are made on thicknesses in accordance with these tables and using the methods defined in these Methods E 18. In specifications in which the Rockwell hardness is used as an approximate indication of the tensile strength (Note 3), Tables 4 and 5 do not apply. In these cases the relations between the specification limits for tensile strength and hardness have been established for certain specified thickness limits, hence anvil effects due to testing thin sheets are incorporated in the relationship. For the E scale, material harder than E 60 may be tested as thin as $^1/_8$ in. (3.175 mm), but if softer than E 60, the minimum thickness should be $^3/_{16}$ in. (4.762 mm).

NOTE 4—See the ASTM specifications listed in Appendix A1. To obtain Rockwell hardness values completely independent of anvil effect, minimum thicknesses greater than those given in Tables 4 and 5 may be required, and the specimen should be free of marking on the side opposite the impression.

4.4 *Surface Preparation*—The preparation of the test material should be carefully controlled to avoid any alterations in hardness, such as may be caused by heating during grinding or by work hardening during machining and polishing operations. The test surface of the specimen should be such that the load can be applied normal to it. The surface should be clean, dry, free from scale, pits, and foreign material that might crush or flow under the test pressure and so affect the results. If etching of the test surface is required, it should be no deeper than necessary for metallographic study. The surface in contact with the anvil should be clean, dry and free from any condition which may affect results. In testing coated materials, if a hardness value for the metal is desired, the coatings should be thoroughly removed before determining the hardness, and this should be done in such a manner that the base metal is not affected.

4.5 *Spacing of Indentations*—An error may result if an indentation is spaced closer than $2^1/_2$ diameters from its center to the edge of the specimen or 3 diameters from another indentation measured center to center.

4.6 *Cylindrical Specimens*—Readings on cylindrical specimens are subject to a correction (see 7.2.9).

5. Verification of Apparatus

5.1 *Verification Methods*—Verify the hardness testing machine as specified in Part C.

5.1.1 Two acceptable methods of verifying machines for Rockwell hardness testing are given in Part C.

6. Adjustment of Apparatus

6.1 *Speed of Load Application*—Adjust the dash pot on the hardness tester so that the operating handle completes its travel in from 4 to 5 s with no specimen on the machine and with the machine set up to apply a major load of 100 kgf (981 N). (Manufacturers of machines without operating handles shall specify an equivalent time cycle and method of adjustment.)

6.2 *Index Lever Adjustment*—Make the following tests (and adjustments, if necessary): Place a piece of material on the anvil and turn the capstan elevating nut to bring the material against the penetrator. Keep turning to elevate the material until the hand feels positive resistance to further turning; this will be felt after the 10-kgf (98-N) minor load has been picked up and when the major load is encountered. When excessive power would have to be used to raise the work higher, take note of the position of the pointer on the dial, after setting the dial so that C 0 and B 30 are at the top. Then if the pointer stands between B 50 and B 70, no adjustment is necessary; if the pointer stands between B 45 and B 50, adjustment is advisable; and if the pointer stands anywhere else, adjustment is imperative. As the pointer revolves several times when the work is being elevated, the readings mentioned above apply to that revolution of the pointer which occurs either as the reference mark on the gage stem disappears into the sleeve or as the auxiliary hand on the dial passes beyond the zero setting on the dial. The object of this adjustment is to see that the elevation of the specimen to pick up the minor load shall not be carried so far as to cause even a partial application of the major load which, to make a proper test, shall be applied only through the release mechanism.

7. Procedure

7.1 *Determination of Machine Accuracy*—Before using the machine for Rockwell hardness tests, determine its accuracy as described in the following paragraphs:

7.1.1 *Checking Against Standardized Test Block*—Select a standardized test block as near as possible to the hardness of the material being tested (preferably within ±5 hardness numbers for the C scale or scales using the diamond penetrator and ±10 hardness numbers for the B scale using ball penetrators). Make five impressions on the test surface of the block and compare the means thereof with the mean of the five readings made in establishing the hardness value of the block. The difference between these two means is defined as the error of the machine. If the error is more than ±2 hardness numbers, examine and adjust the machine until it comes within this limit before it is used. If the error is less than ±2 hardness numbers, the machine error may be taken into account when comparing the results of two or more machines.

7.1.2 *Reinspection*—In the case of reinspection of material by the manufacturer and the purchaser (Note 5), they shall agree upon the standardized test blocks to be used in checking and calibrating the machines used for Rockwell hardness testing.

NOTE 5—When referee tests are to be made, the machine shall be thoroughly examined and all adjustments shall be carefully checked before calibrating it as described above. If readings of high accuracy are necessary, the machine may be calibrated before and after making the tests. If the machine has the same error both times, it can be safely assumed that the correction for this error will give the true Rockwell hardness, except under the conditions described in 4.2.

7.1.3 *Use of Standardized Test Blocks*—Standardized test blocks shall be used only on the test surface because this is the only one that has been checked on a machine of accepted accuracy. Moreover, each impression has a small ridge around it, and if the block is tested on its reverse side, these ridges tend to be flattened under the pressure of the major load and may result in a low reading. Blocks shall never be reground or otherwise resurfaced after being used, because it is the top surface that was originally standardized and this may be of a different hardness from the new surface. The impressions work-harden the block to a considerable depth and this may result in the new test surface being in a work hardened condition and not of the same hardness as the original test surface.

7.1.4 *Use of Single Standardized Test Block*—When a single standardized test block is used, the Rockwell hardness range to be considered verified is that which exceeds the maximum and minimum test block values by ±5 points for the C scale or scales using the diamond penetrator and by ±10 points for the B scale or scales using the ball penetrators.

7.1.5 *Checking Several Ranges*—If tests are made in several ranges of a scale, it is permissible to check that scale at the upper, middle, and lower level. Examples are 20 to 30,

35 to 55, and 59 to 65 when using the C scale; and 40 to 59, 60 to 79, and 80 to 100 when using the B scale.

7.2 *Use of Machine:*

7.2.1 Adjust the machine according to the methods described in 6.1 and 6.2.

7.2.2 Select a suitable scale and use the proper load and penetrator in accordance with 3.3, 3.4, and 4.3.

7.2.3 Select a suitable anvil in accordance with 3.5.

7.2.4 All Rockwell hardness tests shall be made on a single thickness of the material regardless of its thickness. Experience has shown that tests made on more than one thickness of material are unreliable.

7.2.5 The penetrator shall be normal to the surface to be tested.

7.2.6 *Minor Load Application*—Place the specimen to be tested on the anvil and apply the minor load gradually until the proper dial indication is obtained. This shall be understood to be when the pointer has made the proper number of complete revolutions and stands within ±5 divisions of the "SET" position at the top of the dial. The proper number of complete revolutions shall be indicated either by a reference mark on the stem of the gage or by an auxiliary hand on the dial. In bringing the penetrator and the test specimen into contact avoid all impact. The last movement of the elevating or lowering screw shall always be in a direction that will bring penetrator and test specimen together. If the proper setting is overrun, remove the minor load and select a new spot for the test. After the minor load has been applied set the dial pointer at zero on the black-figure scale.

7.2.7 *Major Load Application*—Apply the major load by tripping the operating lever without shock. Remove the major load by bringing the operating lever back to its latched position within 2 s after its motion has stopped, or in accordance with 7.2.7.1 or 7.2.7.2.

7.2.7.1 In the case of materials that exhibit little or no plastic flow after application of the major load, the pointer will come to rest before the motion of the operating lever stops, and in this case the operating lever shall be brought back to its latched position immediately after the pointer stops.

7.2.7.2 In the case of materials that exhibit plastic flow after application of the major load, the pointer will continue to move after the operating lever stops, and in this case the operating lever shall be brought back to its latched position at a specified elapsed time between tripping and removal of load (Note 6). In case the elapsed time after the operating lever stops is other than 2 s, the time shall be recorded, unless the time is specified in the product specification.

NOTE 6—For materials requiring the use of this method the time for application of the major load should be specified in the product specification.

7.2.8 *Reading Scale for Rockwell Hardness*—Take the Rockwell hardness as the reading of the pointer on the proper dial figures after the major load has been removed and while the minor load is still applied. These readings are sometimes estimated to one half of a division or to one tenth of a division, depending on the material being tested.

7.2.9 *Cylindrical Specimens*—Readings on cylindrical specimens are subject to a correction; see Tables 6 and 7.

7.2.10 *Reporting Rockwell Hardness Test Results*—All reports of Rockwell hardness test readings shall indicate the scale used, as described in 3.3. Unless otherwise specified, all readings are to be reported to the nearest whole number, rounding to be in accordance with ASTM Recommended Practice E 29, for Indicating Which Places of Figures Are to Be Considered Significant in Specified Limiting Values.[2]

8. Conversion to Other Hardness Scales or Tensile Strength Values

8.1 There is no general method of converting accurately the Rockwell hardness numbers on one scale to Rockwell hardness numbers on another scale, or to other types of hardness numbers, or to tensile strength values. Such conversions are, at best, approximations and therefore should be avoided, except for special cases where a reliable basis for the approximate conversion has been obtained by comparison tests.

NOTE 7—The ASTM Standard Hardness Conversion Tables E 140, for Metals,[3] give approximate conversion values for specific materials such as steel, austenitic stainless steel, nickel and high-

[2] *1974 Annual Book of ASTM Standards*, Parts 10, 11, 12, 35, 39, and 41.

[3] *1974 Annual Book of ASTM Standards*, Part 10.

nickel alloys, and cartridge brass.

9. Report

9.1 The report shall include the following information (see Note 3):

9.1.1 The Rockwell hardness number,

9.1.2 The Rockwell hardness scale, that is, C-scale, B-scale, etc., and

9.1.3 The time of application of the major load (only when the product specification requires that the time be reported).

B. GENERAL DESCRIPTION AND TEST PROCEDURE FOR ROCKWELL SUPERFICIAL HARDNESS TESTS

10. Apparatus

10.1 *General Principles*—The general principles of the Rockwell superficial hardness test are illustrated in Fig. 3 (diamond penetrator) and Fig. 4 (ball penetrator) and the accompanying Tables 8 and 9.

10.2 *Description of Machine and Method of Test*—The tester for making the Rockwell superficial hardness test is a specialized form of the regular tester. It measures hardness by the same principle as the regular test but employs a smaller minor load, smaller major loads, and a more sensitive depth-measuring, system. It is recommended for use where for one reason or another a very shallow impression, or one of small area is desired. The minor load applied in the Rockwell superficial hardness test is 3 kgf (29 N). The major load (total load) is 15, 30, or 45 kgf (147, 294, or 441 N). Minute indentation tests are of value in testing the hardness of thin strip or sheet material, nitrided or lightly carburized pieces, finished pieces on which large test marks would be undesirable, areas near edges, extremely small parts or sections, and shapes that would collapse under a large test load.

10.3 *Rockwell Superficial Hardness Scales*—Rockwell superficial hardness values are usually determined and reported according to one of the standard scales specified in Table 10. In all cases the minor load is 3 kgf (29 N) and the dial is adjusted after applying the minor load so that the pointer reads at "SET." Major loads of 15, 30, and 45 kgf (147, 294, and 441 N) are used. In recording results, the proper scale symbols as shown in Table 10, should be prefixed to the Rockwell superficial hardness numbers obtained by this method. For the Rockwell superficial hardness test, one Rockwell number represents 0.00004-in. (0.001-mm) movement of the penetrator. The "N" scales are used for materials similar to those tested on the Rockwell C, A, and D scales, but of thinner gage or case depth or where a minute indentation is required. The "T" scales are used for materials similar to those tested on the Rockwell B, F, or G scales, but of thinner gage or where a minute indentation is required. The W, X, and Y scales are used for very soft materials. It is desirable to employ the smallest ball that can properly be used, because of the loss of sensitivity as the size of the penetrator increases. An exception to this is when soft nonhomogeneous material is to be tested, in which case it may be preferable to use a larger ball which makes an impression of greater area, thus obtaining more of an average hardness value.

10.4 *Penetrators*—The commonly used penetrators are the superficial diamond spheroconical penetrator and steel ball $^{1}/_{16}$ in. (1.588 mm) in diameter. The shape of the diamond sphero-conical penetrator should be a cone forming a 120-deg angle with a spherical apex of 0.200-mm radius. Because of their method of preparation, the normal Rockwell diamond penetrator is not interchangeable with the superficial diamond penetrator. The steel balls used should be free from surface imperfections (see 3.4.1.2). Larger steel ball penetrators, $^{1}/_{8}$, $^{1}/_{4}$, or $^{1}/_{2}$ in. (3.175, 6.350, or 12.70 mm) in diameter, may be used.

10.5 *Anvils*—An anvil should be used that is suitable for the specimen to be tested. Cylindrical pieces should be tested with a V-notch anvil that will support the specimen with the axis directly under the penetrator or on hard, parallel, twin cylinders properly positioned and clamped to their base. Flat pieces should be tested on a flat anvil that has a smooth flat bearing surface whose plane is perpendicular to the axis of the penetrator. For thin materials (see 4.3) or specimens that are not perfectly flat, a flat anvil having an elevated "spot" about 1/4 in. (6 mm) in diameter and about 3/4 in. (19 mm) in height is used. This spot should be polished smooth and flat and should be free from pits and heavy scratches. This spot should have a Rockwell hardness of at least C 60. When testing thin material with the steel ball penetrator, the use of a diamond spot anvil will present a highly uniform and polished surface to the flow of the material under test load. This diamond spot anvil should never be used with the diamond sphero-conical penetrator because if the material under test should break and both diamonds come in contact with each other, it would probably result in damage to one or both of these parts. When tests are made with the diamond spot anvil this should be so stated in the specifications.

10.6 *Test Blocks*—Test blocks meeting the requirements outlined in Part D should be used in periodically checking the accuracy of the superficial hardness tester. The test blocks should be of uniform material, thick enough to be free from "anvil effect," of smooth surface on both top and bottom and of not more than 4 in.^{2} (26 cm^2) surface area on the test surface. They should be prepared to a reasonably fine finish; the surface roughness height rating of the test surface should be 12 or less, that is, 12 μin. (0.0003 mm), center line average, and the blocks should be inspected for freedom from any major surface defects or imperfections that might affect the hardness readings.

10.7 *Verifying Machines for Rockwell Superficial Hardness Testing with Standardized Hardness Test Blocks*—It has been found practicable to keep machines for Rockwell superficial hardness tests within the tolerances of the standardized hardness test blocks (see Part D). When machines used in everyday production inspection testing are verified and adjusted once a month, it has been found that they very seldom deviate more than this amount. Daily checking (see Section 14) of such machines assures the operator that the penetrator is in good condition and the machine is operating properly.

10.8 *Precautions*—The precautions included in 3.8 should be observed in the use of the superficial hardness tester.

11. Test Specimens

11.1 The test specimens should conform to the requirements specified in Section 4 except that smoother surfaces are desirable for the Rockwell superficial hardness impressions and greater care should be observed to avoid overhang of the specimens on the anvil.

11.2 Commercially acceptable Rockwell superficial hardness readings may be obtained on sheet materials that show some bulging or marking, and in some specifications the sheets to be tested are of a thickness and hardness such that anvil effect will be present. Tables of limiting thicknesses at various hardness levels for selected Rockwell superficial hardness scales using the diamond penetrator and the 1/16-in. (1.588-mm) diameter ball are given in Tables 11 and 12. In the case of tin mill products, the Rockwell superficial test is used for specimens thinner than those recommended in Table 12. Thickness limitations for these products appear in the product specifications. Rockwell superficial hardness tests on sheet metals are acceptable for hardness specification purpose when the tests are made on thicknesses in accordance with these tables and using the methods defined in these Methods E 18. In specifications in which the Rockwell superficial hardness is used as an approximate indication of the tensile strength (Note 4 and Appendix A1) Tables 11 and 12 do not apply. In these cases the relations between the specification limits for tensile strength and hardness have been established for certain specified thickness limits, hence anvil effects due to testing thin sheets are incorporated in the relationship. To obtain Rockwell superficial hardness values completely independent of anvil effect, minimum thicknesses greater than those given in Tables 11 and 12 may be required, and the specimen should be free of marking on the side opposite the impression.

11.3 Readings on cylindrical specimens are subject to correction (see 14.10).

12. Verification of Apparatus

12.1 *Verification Methods*—Verify the

hardness testing machine as specified in Part C.

12.1.1 Two acceptable methods of verifying the machine for Rockwell superficial hardness testing are given in Part C.

13. Adjustment of Apparatus

13.1 *Speed of Load Application* Adjust the dash pot on the superficial hardness tester so that the operating handle completes its travel in 5 to 7 s with no specimen on the machine and with the machine set up to apply a major load of 30 kgf (294 N). (Manufacturers of superficial hardness testing machines without operating handles shall specify an equivalent time cycle and method of adjustment.)

13.2 *Index Lever Adjustment*—Make the following tests (and adjustments, if necessary): Place a piece of material on the anvil and turn the capstan elevating nut to bring the material against the penetrator. Keep turning to elevate the material until the hand feels positive resistance to further turning; this will be felt after the 3-kgf (29-N) minor load has been picked up and when the major load is encountered. When excessive power would have to be used to raise the work higher, take note of the position of the pointer on the dial, after setting the dial so that "SET" is at the top. Then if the pointer stands between 45 and 55, adjustment is not necessary; if the pointer stands anywhere else, adjustment is imperative. As the pointer revolves several times when the work is being elevated, the readings mentioned above apply to that revolution of the pointer which occurs as the auxiliary hand on the dial passes beyond the zero setting on the dial. The object of this adjustment is to see that the elevation of the specimen to pick up the minor load shall not be carried so far as to cause even a partial application of the major load which, to make a proper test, shall be applied only through the release mechanism.

14. Procedure

14.1 Before using the machine, determine its accuracy as described in 7.1.

14.2 Adjust the machine in accordance with the methods described in 13.1 and 13.2.

14.3 Select a suitable scale and use the proper load and penetrator in accordance with 10.3, 10.4, and 4.3.

14.4 Select a suitable anvil in accordance with 10.5.

14.5 All Rockwell superficial hardness tests shall be made on a single thickness of the material regardless of thickness.

14.7 *Minor Load Application*—Place the specimen to be tested on the anvil and apply the minor load gradually until the proper dial indication is obtained. This shall be understood to be when the pointer has made the proper number of complete revolutions and stands within ±5 divisions of the "SET" positions at the top of the dial. The proper number of complete revolutions shall be indicated either by a reference mark on the stem of the gage or by an auxiliary hand on the dial. In bringing the penetrator and the specimen into contact avoid all impact. The last movement of the elevating screw shall always be in a direction which will bring penetrator and specimen together. If the proper setting is overrun, remove the minor load and select a new spot for the test. After the minor load has been applied, set the dial pointer at zero on the black-figure scale.

14.8 *Major Load Application*—Apply the major load by tripping the operating lever without shock. Remove the major load by bringing the operating lever back to its latched position within 2 s after its motion has stopped, or in accordance with 14.8.1 or 14.8.2.

14.8.1 In the case of materials that exhibit little or no plastic flow after application of the major load, the pointer will come to rest before the motion of the operating lever stops, and in this case the operating lever shall be brought back to its latched position immediately after the pointer stops.

14.8.2 In the case of materials that exhibit plastic flow after application of the major load, the pointer will continue to move after the operating lever stops, and in this case the operating lever shall be brought back to its latched position at a specified elapsed time between tripping and removal of load (Note 8). In case the elapsed time is other than 2 s, the time shall be recorded, unless the time is specified in the product specification.

NOTE 8—For materials requiring the use of this method the time for application of the major load should be specified in the product specification.

14.9 *Reading Scale for Rockwell Superficial Hardness*—Take the Rockwell superficial hardness number as the reading of the pointer on the dial after the major load has been removed and while the minor load is still applied. These readings are sometimes estimated to one half of a division or to one tenth of a division, depending on the material being tested.

14.10 Readings on cylindrical specimens are subject to correction; see Tables 13 and 14.

14.11 *Reporting Rockwell Superficial Hardness Test Results*—All reports of Rockwell superficial hardness test readings shall indicate the scale used, as described in 10.3. Unless otherwise specified, all readings are to be reported to the nearest whole number, rounding to be in accordance with ASTM Recommended Practice E 29, for Indicating Which Places of Figures Are to Be Considered in Specified Limiting Values.[2]

15. Conversion to Other Hardness Scales or Tensile Strength Values

15.1 There is no general method for converting accurately the Rockwell superficial hardness numbers on another scale, or to other types of hardness numbers, or to tensile strength values. Such conversions are, at best, approximations and therefore should be avoided, except for special cases where a reliable basis for the approximate conversion has been obtained by comparison tests.

NOTE 9 The ASTM Standard Hardness Conversion Tables E 140, for Metals,[3] give approximate conversion values for specific materials such as steel, nickel and high-nickel alloys, and cartridge brass.

16. Report

16.1 The report shall include the following information (see Note 3):

16.1.1 The Rockwell superficial hardness number,

16.1.2 The Rockwell superficial hardness scale, that is, 30N-scale, 30T-scale, etc., and

16.1.3 The time of application of the major load (only when the product specification requires that the time be reported).

C. VERIFICATION OF MACHINES FOR ROCKWELL HARDNESS AND ROCKWELL SUPERFICIAL HARDNESS TESTING

17. Scope

17.1 Part C covers two procedures for the verification of machines for Rockwell hardness and Rockwell superficial hardness testing and a procedure which is recommended for use to confirm that the machine has not become maladjusted in the intervals between the periodical routine checks made by the user. The two methods of verification are:

17.1.1 Separate verification of load application, penetrator, and the depth measuring device followed by a performance test (19.2). This method shall be used for new and rebuilt machines.

17.1.2 Verification by standardized test block method. This method shall be used in referee, laboratory, or routine testing to assure the operator that the machine for Rockwell hardness testing is operating properly (see 19.2).

18. General Requirements

18.1 Before a hardness testing machine is verified, the machine shall be examined to ensure that:

18.1.1 The machine is properly set up.

18.1.2 The dial gage plunger on the depth measuring device moves freely in any position.

18.1.3 The penetrator holder is properly seated in the plunger.

18.1.4 When the penetrator is a steel ball, the holder shall be fitted with a new ball whose diameter has been checked (see 19.1.3).

18.1.5 The diamond penetrator is free from cracks or flaws which would lead to incorrect readings.

18.1.6 The load can be applied and removed without shock or vibration in such a manner that the readings are not influenced.

19. Verification

19.1 *Separate Verification of Load Application, Penetrator, and Depth Measuring Device:*

19.1.1 *Load Application*—Machines for Rockwell hardness testing shall be verified at loads of 10, 60, 100, and 150 kgf (98, 589, 981, and 1472 N). Machines for Rockwell superficial hardness testing shall be verified at loads of 3, 15, 30, and 45 kgf (29, 147, 294, and 441 N). The applied load shall be checked by the use of standardized dead weights (masses), by the use of standardized dead weights and proving levers, or by the use of an elastic proving device or springs in the manner described in ASTM Methods E 4, Verification of Testing Machines.[2] Each testing machine for the determination of Rockwell hardness shall be verified at the minor load (10 ± 0.2 kgf) (98 ± 2.0 N) before application and after removal of the additional load. Each testing machine for the determination of Rockwell superficial hardness shall be verified at the minor load (3 ± 0.060 kgf) (29 ± 0.59 N) before application and after removal of the additional load. The load application shall be considered verified if the mean for three readings, at each of three positions of the power lever, for each load falls within the tolerance listed in Table 15.

19.1.2 *Diamond Sphero-Conical Penetrator (Diamond Penetrator)*—The verification of the form of the diamond penetrator shall be made by direct measurement of its shape or by measurement of its projection on a screen. Verification shall be made at not less than four sections. The diamond penetrator shall have an included angle of 120 deg ±30 min and shall have its axis in line with the axis of the penetrator within a tolerance of ±30 min. The tip of the cone shall have a nominal radius of 0.200 mm. The contour of the tip of the cone shall lie within a band defined by two concentric arcs parallel to the nominal tip contour radius (0.200 mm) but displaced from it by 0.002 mm as shown in Fig. 5. In Fig. 5(*a*) the tolerance band is shown at a magnification of 500×. In Fig. 5(*b*) possible minimum and maximum variations are shown schematically (not to scale). The surface of the cone shall blend in a tangential manner with the surface of the spherical tip. The penetrator shall be polished to such an extent that no unpolished part of its surface makes contact with the test specimen when it penetrates to a depth of 0.012 in. (0.3 mm). Since the hardness values given by a testing machine do not depend upon these dimensions alone, but also on the surface roughness, the position of the crystallographic axes of the diamond, and the seating of the diamond in the holder, a performance test is required. In this test the sphero-conical penetrator shall be used in a machine in which the load application and the depth-measuring device have been verified. Five impressions shall be made on a standardized hardness test block calibrated with a verified penetrator as prescribed in Part D. The mean of these five hardness readings shall not differ from the average of the standard test block by more than the amount shown in Table 16.

19.1.3 *Ball Penetrator*—The mean of three diameters measured on a new steel ball selected at random from a lot shall not differ from the nominal by more than the amounts shown in Table 17. The permissible difference between the largest diameter and the smallest diameter measurable on one ball shall be not more than 0.00004 in. (0.0010 mm). The steel ball penetrator shall have a Vickers hardness (HV) of at least 850 using a 10-kgf (98-N) load. Therefore, the maximum mean diagonal of the Vickers impression made on a steel ball penetrator shall not exceed the value shown in Table 18.

19.1.4 *Depth-Measuring Device* — The depth-measuring device shall be verified over not less than three ranges, including the ranges corresponding to the lowest and highest hardness for which the scale is normally used, by making known incremental movements of the penetrator or dial gage plunger. The depth-measuring device shall correctly indicate the Rockwell hardness within ±0.5 of a scale unit over each range, that is, within ±0.001 mm depth reading. The depth-measuring device shall correctly indicate the Rockwell superficial hardness within ±0.5 of a scale unit over each range, that is, within ±0.0005 mm depth reading.

19.2 *Verification by Standardized Test Block Method:*

19.2.1 A machine for Rockwell hardness testing or Rockwell superficial hardness testing used for referee, laboratory, or routine testing or machines which are verified in service may be checked by making a series of impressions on standardized hardness test blocks (Part D).

19.2.2 A minimum of five hardness readings shall be taken on the test surface of at least three blocks having different levels of hardness, as shown in Table 19, using the following test loads:

Rockwell C scale	150 kgf (1472 N) load
Rockwell B scale	100 kgf (981 N) load
Rockwell 30N scale	30 kgf (294 N) load
Rockwell 30T scale	30 kgf (294 N) load

19.2.3 When tests are made in several ranges of a scale, it is permissible to check that scale at the lower, middle, and upper levels. Examples are 20 to 30, 35 to 55, and 59 to 65 when using the C scale; and 40 to 59, 60 to 79, and 80 to 100 when using the B scale.

19.2.4 Machines for Rockwell hardness and Rockwell superficial hardness tests shall be considered verified if the results meet the requirements of Sections 21 and 22.

20. Procedure for Periodic Checks by the User

20.1 Verification by the standardized test block method (19.2) is too lengthy for daily use. Instead, the following is recommended:

20.1.1 Make at least one routine check each day that the testing machine is used.

20.1.2 Before making the check, make at least two preliminary indentations to ensure that the hardness testing machine is working freely and that the test block, penetrator, and anvil are reading correctly. The results of these preliminary indentations should be ignored.

20.1.3 Make at least five hardness readings on a standardized hardness test block on the scale and at the hardness level at which the machine is being used. If the values fall within the range of the standardized hardness test block the machine may be regarded as satisfactory. If not the machine should be verified as described in 19.2.

21. Repeatability and Error

21.1 *Repeatability:*

21.1.1 For each standardized block let R_1, $R_2 \cdots R_5$ be the hardness readings of the five indentations arranged in an increasing order of magnitude.

21.1.2 The repeatability of the hardness testing machine under the particular verification conditions is expressed by the quantity $R_5 - R_1$.

21.2 *Error:*

21.2.1 The error of the hardness testing machine under the particular verification conditions is expressed by the quantity $\bar{R} - R$, where

$$\bar{R} = (R_1 + R_2 \ldots R_5)/5$$

R_1, $R_2 \cdots R_5$ are the individual hardness values, and R is the stated hardness of the standardized hardness test block.

22. Assessment of Verification

22.1 *Repeatability*—The repeatability of the hardness testing machine is considered satisfactory if it satisfies the conditions given in Table 20.

22.2 *Error*—The mean hardness value for five impressions should not differ from the mean corresponding to the hardness of the standardized test block by more than the tolerance of the latter.

D. CALIBRATION OF STANDARDIZED HARDNESS TEST BLOCKS FOR MACHINES USED FOR ROCKWELL AND ROCKWELL SUPERFICIAL HARDNESS TESTING

23. Scope

23.1 Part D covers the calibration of standardized hardness test blocks for the verification of machines used for Rockwell and Rockwell superficial hardness testing as described in Part C.

24. Manufacture

24.1 Each metal block to be standardized shall be not less than ¼ in. (6 mm) in thickness.

24.2 Each block shall be specially prepared and heat treated to give the necessary homogeneity and stability of structure.

24.3 Each block, if of steel, shall be demagnetized by the manufacturer and maintained demagnetized by the user.

24.4 The lower surface of the test block shall have a fine ground finish.

24.5 The test (upper) surface shall be polished or fine ground and free from scratches which would influence the depth of the impression.

24.5.1 The mean surface roughness height rating of the test surfaces shall not exceed 12 μin. (0.0003 mm) center line average.

24.6 To ensure that no material is subsequently removed from the test surface of the standardized test block, an official mark or the thickness to an accuracy of ±0.005 in. (±0.1 mm) or at the time of calibration shall be marked on the test surface.

25. Standardizing Procedure

25.1 The standardizing hardness test blocks shall be calibrated on a hardness testing machine verified in accordance with the requirements of 17.1.1.

25.2 The major load shall be removed by bringing the operating lever back to its latched position according to one of the alternative methods given in 7.2.8.1 or 7.2.8.2 and 14.8.1 or 14.8.2.

26. Number of Indentations

26.1 At least five randomly distributed indentations shall be made on each test block.

26.2 The dial indicator of the depth measuring device shall be read to ±0.1 unit, that is, to the nearest 0.1 Rockwell hardness or Rockwell superficial hardness number.

27. Repeatability

27.1 Let R_1, $R_2 \cdots R_n$ be the observed Rockwell hardness or Rockwell superficial hardness number as determined by one observer, arranged in increasing order of magnitude.

27.2 The repeatability of the hardness readings on the block is defined as $R_n - R_1$.

28. Uniformity of Hardness

28.1 Unless the repeatability of the hardness readings is within the limits given in Table 21, the block cannot be regarded as sufficiently uniform for standardization purposes.

29. Marking

29.1 Each block shall be marked with the following:

29.1.1 Arithmetic mean of the hardness values found in the standardization test prefixed by the scale designation and followed by the tolerance range.

29.1.2 The name or mark of the supplier,

29.1.3 The serial number of the block, and

29.1.4 The thickness of the test block or an official mark on the test surface (see 24.6).

NOTE 10 All of the markings except the official mark or thickness shall be placed on the side of the block, the markings being upright when the test surface is the upper face.

TABLE 1 Symbols and Designations Associated with Fig. 1

Number	Symbol	Designation
1	...	Angle at the top of the diamond penetrator (120 deg)
2	...	Radius curvature at the tip of the cone (0.200 mm)
3	P_0	Minor load = 10 kgf (98 N)
4	P_1	Additional load = 90 or 140 kgf (883 or 1373 N)
5	P	Major load = $P_0 + P_1$ = 100 or 150 kgf (981 or 1472 N)
6	...	Depth of impression under minor load before application of additional load
7	...	Increase in depth of impression under additional load
8	e	Permanent increase in depth of impression under minor load after removal of additional load, the increase being expressed in units of 0.002 mm
9	xx HRA	Rockwell A hardness = 100 − e
10	xx HRC	Rockwell C hardness = 100 − e

TABLE 2 Symbols and Designations Associated with Fig. 2

Number	Symbol	Designation
1	D	Diameter of ball = 1/16 in. (1.588 mm)
3	P_0	Minor load = 10 kgf (98 N)
4	P_1	Additional load = 90 kgf (883 N)
5	P	Major load = $P_0 + P_1$ = 10 + 50, 90, or 140 = 60, 100, or 150 kgf (589, 981, or 1472 N)
6	...	Depth of impression under minor load before application of additional load
7	...	Increase in depth of impression under additional load
8	e	Permanent increase in depth of impression under minor load after removal of the additional load, the increase being expressed in units of 0.002 mm
9	xx HRF	Rockwell F hardness = 130 − e
10	xx HRB	Rockwell B hardness = 130 − e
11	xx HRG	Rockwell G hardness = 130 − e

TABLE 3 Rockwell Hardness Scales

Scale Symbol	Penetrator	Major Load, kgf	Dial Figures	Typical Applications of Scales
B	1/16-in. (1.588-mm) ball	100	red	Copper alloys, soft steels, aluminum alloys, malleable iron, etc.
C	diamond	150	black	Steel, hard cast irons, pearlitic malleable iron, titanium, deep case hardened steel, and other materials harder than B 100.
A	diamond	60	black	Cemented carbides, thin steel, and shallow case-hardened steel.
D	diamond	100	black	Thin steel and medium case hardened steel, and pearlitic malleable iron.
E	1/8-in. (3.175-mm) ball	100	red	Cast iron, aluminum and magnesium alloys, bearing metals.
F	1/16-in. (1.588-mm) ball	60	red	Annealed copper alloys, thin soft sheet metals.
G	1/16-in. (1.588-mm) ball	150	red	Malleable irons, copper-nickel-zinc and cupro-nickel alloys. Upper limit G 92 to avoid possible flattening of ball.
H	1/8-in. (3.175-mm) ball	60	red	Aluminum, zinc, lead.
K	1/8-in. (3.175-mm) ball	150	red	Bearing metals and other very soft or thin materials. Use smallest ball and heaviest load that does not give anvil effect.
L	1/4-in. (6.350-mm) ball	60	red	
M	1/4-in. (6.350-mm) ball	100	red	
P	1/4-in. (6.350-mm) ball	150	red	
R	1/2-in. (12.70-mm) ball	60	red	
S	1/2-in. (12.70-mm) ball	100	red	
V	1/2-in. (12.70-mm) ball	150	red	

TABLE 4 A Guide for Selection of Scales Using the Diamond Penetrator

NOTE—For a given thickness, any hardness greater than that corresponding to that thickness can be tested. For a given hardness, material of any greater thickness than that corresponding to that hardness can be tested on the indicated scale.

Thickness		Rockwell Scale		
		A		C
in.	mm	Dial Reading	Approximate Hardness C-Scale[a]	Dial Reading
0.014	0.36	...	...	...
0.016	0.41	86	69	...
0.018	0.46	84	65	...
0.020	0.51	82	61.5	...
0.022	0.56	79	56	69
0.024	0.61	76	50	67
0.026	0.66	71	41	65
0.028	0.71	67	32	62
0.030	0.76	60	19	57
0.032	0.81	...	...	52
0.034	0.86	...	...	45
0.036	0.91	...	...	37
0.038	0.96	...	...	28
0.040	1.02	...	...	20

[a] These approximate hardness numbers are for use in selecting a suitable scale and should not be used as hardness conversions. If necessary to convert test readings to another scale, refer to the ASTM Standard Hardness Conversion Tables E 140, for Metals (Relationship Between Brinell Hardness, Vickers Hardness, Rockwell Hardness, Rockwell Superficial Hardness, and Knoop Hardness).[3]

TABLE 5 A Guide for Selection of Scales Using the 1/16-in. (1.588-mm) Diameter Ball Penetrator

NOTE—For a given thickness, any hardness greater than that corresponding to that thickness can be tested. For a given hardness, material of any greater thickness than that corresponding to that hardness can be tested on the indicated scale.

Thickness		Rockwell Scale		
		F		B
in.	mm	Dial Reading	Approximate Hardness B-Scale[a]	Dial Reading
0.022	0.56	...	...	...
0.024	0.61	98	72	94
0.026	0.66	91	60	87
0.028	0.71	85	49	80
0.030	0.76	77	35	71
0.032	0.81	69	21	62
0.034	0.86	...	...	52
0.036	0.91	...	...	40
0.038	0.96	...	...	28
0.040	1.02	...	...	...

[a] These approximate hardness numbers are for use in selecting a suitable scale and should not be used as hardness conversions. If necessary to convert test readings to another scale refer to the ASTM Standard Hardness Conversion Tables E 140, for Metals (Relationship Between Brinell Hardness, Vickers Hardness, Rockwell Hardness, Rockwell Superficial Hardness and Knoop Hardness).[3]

Current edition approved April 29, 1974. Published June 1974. Originally published as A 356 – 52 T. Last previous edition A 356 – 68.

TABLE 6 Corrections to Be Added to Rockwell C, A, and D Values Obtained on Cylindrical Specimens[a] of Various Diameters

Dial Reading	Diameters of Cylindrical Specimens								
	¼ in. (6.4 mm)	⅜ in. (10 mm)	½ in. (13 mm)	⅝ in. (16 mm)	¾ in. (19 mm)	⅞ in. (22 mm)	1 in. (25 mm)	1¼ in. (32 mm)	1½ in. (38 mm)
	Corrections to be Added to Rockwell C, A, and D Values[b]								
20	6.0	4.5	3.5	2.5	2.0	1.5	1.5	1.0	1.0
25	5.5	4.0	3.0	2.5	2.0	1.5	1.0	1.0	1.0
30	5.0	3.5	2.5	2.0	1.5	1.5	1.0	1.0	0.5
35	4.0	3.0	2.0	1.5	1.5	1.0	1.0	0.5	0.5
40	3.5	2.5	2.0	1.5	1.0	1.0	1.0	0.5	0.5
45	3.0	2.0	1.5	1.0	1.0	1.0	0.5	0.5	0.5
50	2.5	2.0	1.5	1.0	1.0	0.5	0.5	0.5	0.5
55	2.0	1.5	1.0	1.0	0.5	0.5	0.5	0.5	0
60	1.5	1.0	1.0	0.5	0.5	0.5	0.5	0	0
65	1.5	1.0	1.0	0.5	0.5	0.5	0.5	0	0
70	1.0	1.0	0.5	0.5	0.5	0.5	0.5	0	0
75	1.0	0.5	0.5	0.5	0.5	0.5	0	0	0
80	0.5	0.5	0.5	0.5	0.5	0	0	0	0
85	0.5	0.5	0.5	0	0	0	0	0	0
90	0.5	0	0	0	0	0	0	0	0

[a] When testing cylindrical specimens, the accuracy of the test will be seriously affected by alignment of elevating screw, V-anvil, penetrators, surface finish, and the straightness of the cylinder.
[b] These corrections are approximate only and represent the averages to the nearest 0.5 Rockwell number, of numerous actual observations.

TABLE 7 Corrections to Be Added to Rockwell B, F, and G Values Obtained on Cylindrical Specimens[a] of Various Diameters

Dial Reading	Diameters of Cylindrical Specimens						
	¼ in. (6.4 mm)	⅜ in. (10 mm)	½ in. (13 mm)	⅝ in. (16 mm)	¾ in. (19 mm)	⅞ in. (22 mm)	1 in. (25 mm)
	Corrections to be Added to Rockwell B, F, and G Values[b]						
0	12.5	8.5	6.5	5.5	4.5	3.5	3.0
10	12.0	8.0	6.0	5.0	4.0	3.5	3.0
20	11.0	7.5	5.5	4.5	4.0	3.5	3.0
30	10.0	6.5	5.0	4.5	3.5	3.0	2.5
40	9.0	6.0	4.5	4.0	3.0	2.5	2.5
50	8.0	5.5	4.0	3.5	3.0	2.5	2.0
60	7.0	5.0	3.5	3.0	2.5	2.0	2.0
70	6.0	4.0	3.0	2.5	2.0	2.0	1.5
80	5.0	3.5	2.5	2.0	1.5	1.5	1.5
90	4.0	3.0	2.0	1.5	1.5	1.5	1.0
100	3.5	2.5	1.5	1.5	1.0	1.0	0.5

[a] When testing cylindrical specimens, the accuracy of the test will be seriously affected by alignment of elevating screw, V-anvil, penetrators, surface finish, and the straightness of the cylinder.
[b] These corrections are approximate only and represent the averages, to the nearest 0.5 Rockwell number, of numerous actual observations.

TABLE 8 Symbols and Designations Associated with Fig. 3

Number	Symbol	Designation
1	...	Angle at the tip of the diamond penetrator (120 deg)
2	...	Radius of curvature at the tip of the cone (0.200 mm)
3	P_0	Minor load = 3 kgf (29 N)
4	P_1	Additional load = 12, 27, or 42 kgf (118, 265, or 412 N)
5	P	Major load = $P_0 + P_1$ = 3 + 12, 27, or 42 = 15, 30, or 45 kgf (147, 294, or 441 N)
6	...	Depth of impression under minor load before application of additional load
7	...	Increase in depth of impression under additional load
8	e	Permanent increase in depth of impression under minor load after removal of additional load, the increase being expressed in units of 0.001 mm
9	xx HR15N xx HR30N xx HR45N	Rockwell 15N hardness = 100 − e Rockwell 30N hardness = 100 − e Rockwell 45N hardness = 100 − e

TABLE 9 Symbols and Designations Associated with Fig. 4

Number	Symbol	Designation
1	D	Diameter of ball = 1/16 in. (1.588 mm)
3	P_0	Minor load = 3 kgf (29 N)
4	P_1	Additional load = 12, 27, or 42 kgf (118, 265, or 412 N)
5	P	Major load = $P_0 + P_1$ = 3 + 12, 27, or 42 = 15, 30, or 45 kgf (147, 294, or 441 N)
6	...	Depth of impression under minor load before application of additional load
7	...	Increase in depth of impression under additional load
8	e	Permanent increase in depth of impression under minor load after removal of the additional load, the increase being expressed in units of 0.001 mm
9	xx HR15T xx HR30T xx HR45T	Rockwell 15N hardness = 100 − e Rockwell 30T hardness = 100 − e Rockwell 45N hardness = 100 − e

TABLE 10 Rockwell Superficial Hardness Scales

Major Load, kgf (N)	Scale Symbols				
	N Scale, Diamond Penetrator	T Scale, 1/16-in. (1.588-mm) Ball	W Scale, 1/8-in. (3.175-mm) Ball	X Scale, 1/4-in. (6.350-mm) Ball	Y Scale, 1/2-in. (12.70-mm) Ball
15 (147)	15N	15T	15W	15X	15Y
30 (294)	30N	30T	30W	30X	30Y
45 (441)	45N	45T	45W	45X	45Y

TABLE 11 A Guide for Selection of Scales Using the Diamond Penetrator

NOTE—For a given thickness, any hardness greater than that corresponding to that thickness can be tested. For a given hardness, material of any greater thickness than that corresponding to that hardness can be tested on the indicated scale.

Thickness		Rockwell Superficial Scale					
		15N		30N		45N	
in.	mm	Dial Reading	Approximate Hardness C-Scale[a]	Dial Reading	Approximate Hardness C-Scale[a]	Dial Reading	Approximate Hardness C-Scale[a]
0.006	0.15	92	65	...	...	...	...
0.008	0.20	90	60	...	...	...	...
0.010	0.25	88	55	...	...	...	...
0.012	0.30	83	45	82	65	77	69.5
0.014	0.36	76	32	78.5	61	74	67
0.016	0.41	68	18	74	56	72	65
0.018	0.46	...	...	66	47	68	61
0.020	0.51	...	...	57	37	63	57
0.022	0.56	...	...	47	26	58	52.5
0.024	0.61	...	...	...	...	51	47
0.026	0.66	...	...	...	...	37	35
0.028	0.71	...	...	...	...	20	20.5
0.030	0.76	...	...	...	...	...	...

[a] These approximate hardness numbers are for use in selecting a suitable scale, and should not be used as hardness conversions. If necessary to convert test readings to another scale, refer to the ASTM Standard Hardness Conversion Tables E 140, for Metals (Relationship Between Brinell Hardness, Vickers Hardness, Rockwell Hardness, Rockwell Superficial Hardness and Knoop Hardness).[3]

TABLE 12 A Guide for Selection of Scales Using the 1/16 in. (1.588 mm) Diameter Ball Penetrator

NOTE—For a given thickness, any hardness greater than that corresponding to that thickness can be tested. For a given hardness, material of any greater thickness than that corresponding to that hardness can be tested on the indicated scale.

Thickness		Rockwell Superficial Scale					
		15T		30T		45T	
in.	mm	Dial Reading	Approximate Hardness B-Scale[a]	Dial Reading	Approximate Hardness B-Scale[a]	Dial Reading	Approximate Hardness B-Scale[a]
0.010	0.25	91	93	...	...	...	...
0.012	0.30	86	78	...	...	...	...
0.014	0.36	81	62	79	96	...	...
0.016	0.41	75	44	73	74	71	99
0.018	0.46	68	24	64	71	62	90
0.020	0.51	...	...	55	58	53	80
0.022	0.56	...	...	45	43	43	70
0.024	0.61	...	...	34	28	31	58
0.026	0.66	...	...	...	...	18	45
0.028	0.71	...	...	...	...	4	32
0.030	0.76	...	...	...	...	...	...

[a] These approximate hardness numbers are for use in selecting a suitable scale, and should not be used as hardness conversions. If necessary to convert test readings to another scale refer to the ASTM Standard Hardness Conversion Tables E 140, for Metals (Relationship Between Brinell Hardness, Vickers Hardness, Rockwell Hardness, Rockwell Superficial Hardness and Knoop Hardness).[3]

TABLE 13 Corrections to Be Added to Rockwell Superficial 15N, 30N, and 45N Values Obtained on Cylindrical Specimens of Various Diameters

Dial Reading	Diameters of Cylindrical Specimens					
	1/8 in. (3.2 mm)	1/4 in. (6.4 mm)	3/8 in. (10 mm)	1/2 in. (13 mm)	3/4 in. (19 mm)	1 in. (25 mm)
	Corrections to be Added to Rockwell Superficial 15N, 30N, and 45N Values[b]					
20	6.0	3.0	2.0	1.5	1.5	1.5
25	5.5	3.0	2.0	1.5	1.5	1.0
30	5.5	3.0	2.0	1.5	1.0	1.0
35	5.0	2.5	2.0	1.5	1.0	1.0
40	4.5	2.5	1.5	1.5	1.0	1.0
45	4.0	2.0	1.5	1.0	1.0	1.0
50	3.5	2.0	1.5	1.0	1.0	0.5
55	3.5	2.0	1.5	1.0	0.5	0.5
60	3.0	1.5	1.0	1.0	0.5	0.5
65	2.5	1.5	1.0	0.5	0.5	0.5
70	2.0	1.0	1.0	0.5	0.5	0.5
75	1.5	1.0	0.5	0.5	0.5	0
80	1.0	0.5	0.5	0.5	0	0
85	0.5	0.5	0.5	0.5	0	0
90	0	0	0	0	0	0

[a] When testing cylindrical specimens the accuracy of the test will be seriously affected by alignment of elevating screw, V-anvil, penetrators, surface finish, and the straightness of the cylinder.

[b] These corrections are approximate only and represent the averages, to the nearest 0.5 Rockwell superficial number, of numerous actual observations.

TABLE 14 Corrections to Be Added to Rockwell Superficial 15T, 30T, and 45T Values Obtained on Cylindrical Specimens[a] of Various Diameters

Dial Reading	Diameters of Cylindrical Specimens						
	1/8 in. (3.2 mm)	1/4 in. (6.4 mm)	3/8 in. (10 mm)	1/2 in. (13 mm)	5/8 in. (16 mm)	3/4 in. (19 mm)	1 in. (25 mm)
	Corrections to be Added to Rockwell Superficial 15T, 30T, and 45T Values[b]						
20	13.0	9.0	6.0	4.5	4.5	3.0	2.0
30	11.5	7.5	5.0	3.5	3.5	2.5	2.0
40	10.0	6.5	4.5	3.5	3.0	2.5	2.0
50	8.5	5.5	4.0	3.0	2.5	2.0	1.5
60	6.5	4.5	3.0	2.5	2.0	1.5	1.5
70	5.0	3.5	2.5	2.0	1.5	1.0	1.0
80	3.0	2.0	1.5	1.5	1.0	1.0	0.5
90	1.5	1.0	1.0	0.5	0.5	0.5	0.5

[a] When testing cylindrical specimens, the accuracy of the test will be seriously affected by alignment of elevating screw, V-anvil, penetrators, surface finish, and the straightness of the cylinder.

[b] These corrections are approximate only and represent the averages, to the nearest 0.5 Rockwell superficial number, of numerous actual observations.

TABLE 15 Tolerances on Applied Loads

Load, kgf (N)	Tolerance, kgf (N)
10 (98)	±0.20 (±1.96)
60 (589)	±0.45 (±4.41)
100 (981)	±0.65 (±4.57)
150 (1472)	±0.90 (±8.83)
3 (29)	±0.060 (±0.589)
15 (147)	±0.100 (±0.981)
30 (294)	±0.200 (±1.961)
45 (441)	±0.300 (±2.943)

TABLE 17 Tolerances for Rockwell Hardness Ball Penetrators

Diameter of Ball		Tolerance[a]	
in.	mm	in.	mm
1/16	1.588	±0.0001	±0.0025
1/8	3.175	±0.0001	±0.0025
1/4	6.350	±0.0001	±0.0025
1/2	12.700	±0.0001	±0.0025

[a] For balls in the range of diameters specified, these tolerances and the permissible variation in the diameter of any one ball, as specified in 19.1.3, are met by Grade 25 steel balls of the Anti-Friction Bearing Manufacturers' Association (AFMBA).

TABLE 19 Hardness Ranges Used in Verification by Standardized Test Block Method

Rockwell Scale	Hardness Ranges
C	20 to 30 35 to 55 59 to 65
B	40 to 59 60 to 79 80 to 100
30N	40 to 50 55 to 73 75 to 80
30T	43 to 56 57 to 70 incl over 70 to 82

TABLE 21 Repeatability of Hardness Readings

Nominal Hardness of Standardized Test Block	The Repeatability of the Test Block Readings Shall Be Not Greater Than:
C Scale:	
60 and greater	0.5
Below 60	1.0
B Scale:	
60 to 100, incl	1.0
Below 60 to 40, incl	1.5
30N Scale:	
41.5 and greater	1.0
30T Scale:	
43 to 82	1.0

TABLE 16 Allowable Deviation in Hardness Readings for Verified Diamond Penetrators

For Hardness Readings in the Range of:	Allowable Deviation, Rockwell Units
C 63	±0.5
C 25	±1.0
30N 80	±0.5
30N 45	±1.0

TABLE 18 Maximum Mean Diagonal of Vickers Hardness Impression on Rockwell Hardness Balls

Ball Diameters		Maximum Mean Diagonal of Impression on the Ball Made with Vickers Indenter Under 10-kgf (98-N) Load, mm
in.	mm	
1/16	1.588	0.141
1/8	3.175	0.144
1/4	6.350	0.145
1/2	12.700	0.147

TABLE 20 Repeatability of Machines

Range of Standardized Hardness Test Blocks	The Repeatability[a] of the Machine Shall Be Not Greater Than:
Rockwell C Scale:	
20 to 30	2.0
35 to 55	1.5
59 to 65	1.0
Rockwell B Scale:	
49 to 59	2.5
60 to 79	2.0
80 to 100	2.0
Rockwell 30N Scale:	
40 to 50	2.0
55 to 73	1.5
75 to 80	1.0
Rockwell 30T Scale:	
43 to 56	2.5
57 to 70, incl	2.0
Over 70 to 82	2.0

[a] The repeatability of machines on Rockwell or Rockwell superficial hardness scales other than those given in Table 20 shall be the equivalent converted difference in hardness for those scales, except for the 15N and 15T scales. In the case of the 15N and 15T scales, the repeatability shall be no greater than 1.0 for all ranges.

Example—At C 60, typical readings of a series of impressions might range from 59 to 60, 59.5 to 60.5, 60 to 61, etc. Thus, converted A-scale values corresponding to C 59 to 60 (see Table II of Standard Tables E 140) would be A 80.7 to 81.2 and the repeatability for the A-scale would be 0.5.

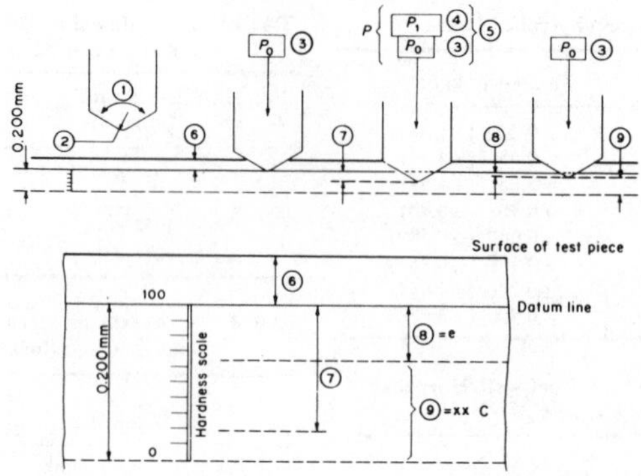

FIG. 1 Rockwell Hardness Test with Diamond Penetrator (Rockwell C) (Table 1).

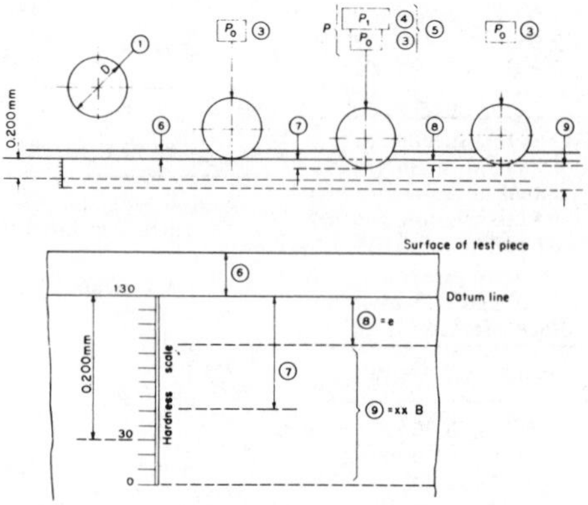

FIG. 2 Rockwell Hardness Test with Steel Ball Penetrator (Rockwell B) (Table 2).

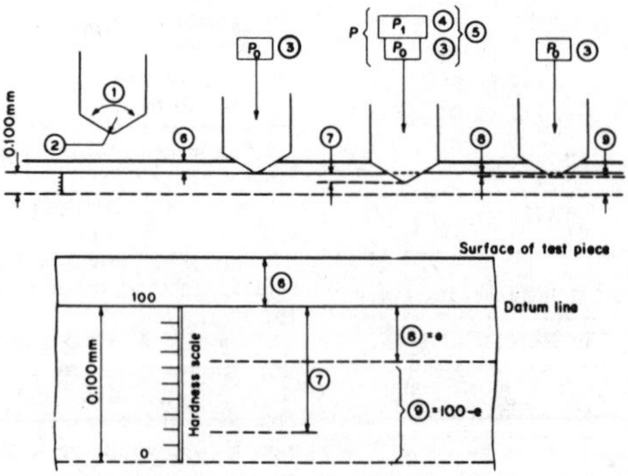

FIG. 3 Rockwell Superficial Hardness Test with Diamond Penetrator (Rockwell 30N) (Table 8).

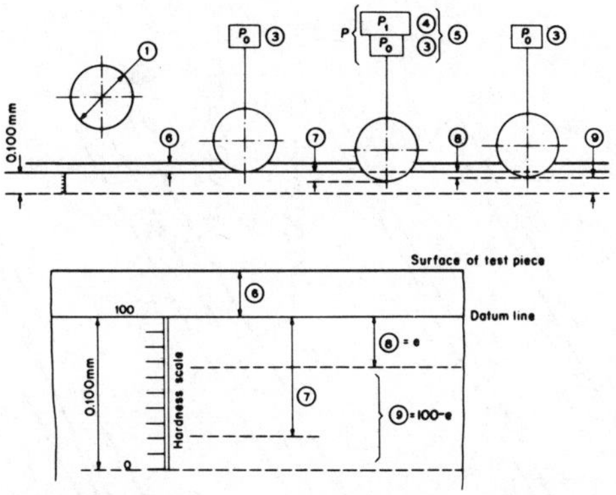

FIG. 4 Rockwell Superficial Hardness Test with Steel Ball Penetrator (Rockwell 30T) (Table 9).

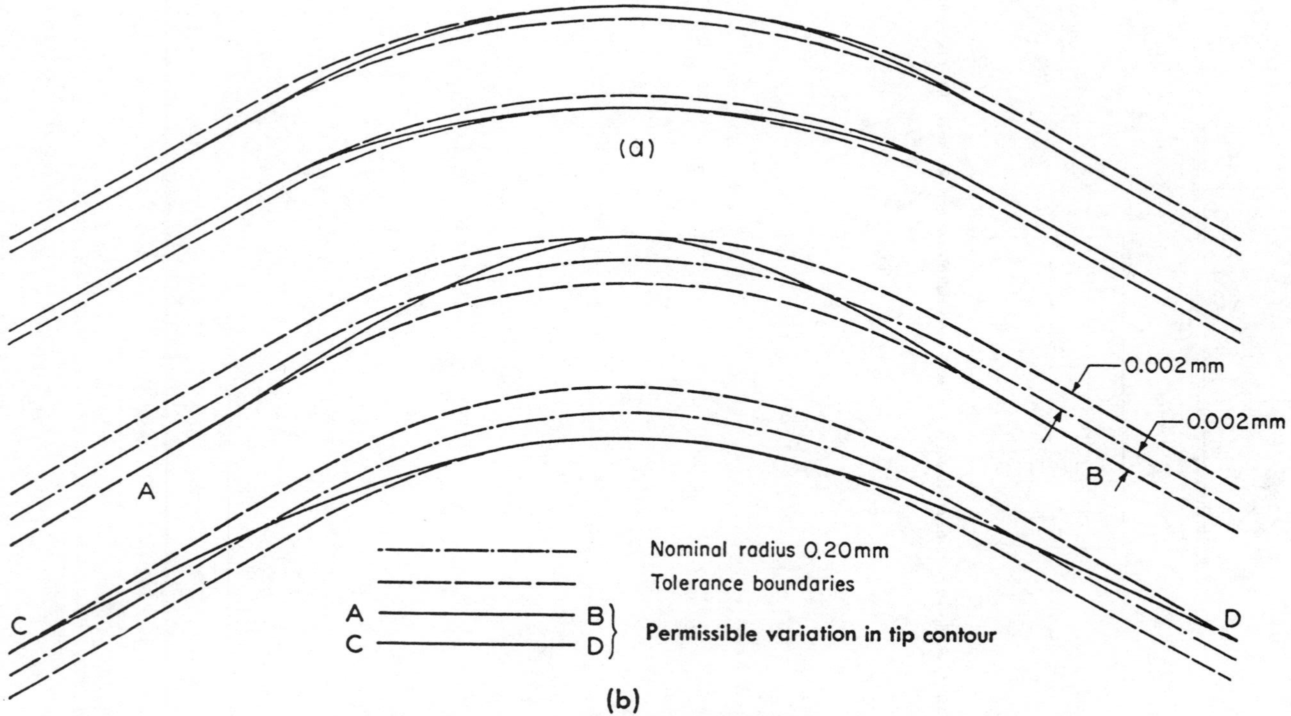

FIG. 5 Tolerance on Contour of Diamond Sphero-Conical Penetrator.
(*a*) 500× (*b*) Not to scale

APPENDIX

A1. LIST OF ASTM SPECIFICATIONS GIVING HARDNESS VALUES CORRESPONDING TO TENSILE STRENGTH

A1.1 The following ASTM specifications[3] give approximate Rockwell hardness or Rockwell superficial hardness values corresponding to the tensile strength values specified for the materials covered:

B 19, Cartridge Brass Sheet, Strip, Plate Bar, and Disks
B 36, Brass Plate, Sheet, Strip, and Rolled Bar
B 96, Copper-Silicon Alloy Plate and Sheet for Pressure Vessels
B 97, Copper-Silicon Alloy Plate, Sheet, Strip, and Rolled Bar for General Purposes
B 103, Phosphor Bronze Plate, Sheet, Strip, and Rolled Bar
B 121, Leaded Brass Plate, Sheet, Strip, and Rolled Bar
B 122, Copper-Nickel-Zinc Alloy (Nickel Silver) and Copper-Nickel Alloy Plate, Sheet, Strip, and Rolled Bar
B 130, Commercial Bronze Strip
B 134, Brass Wire
B 152, Copper Sheet, Strip, Plate, and Rolled Bar
B 291, Copper-Zinc-Manganese Alloy (Manganese Brass) Sheet and Strip
B 370, Copper Sheet and Strip for Building Construction

[3] *1974 Annual Book of ASTM Standards*, Part 6.

By publication of this standard no position is taken with respect to the validity of any patent rights in connection therewith, and the American Society for Testing and Materials does not undertake to insure anyone utilizing the standard against liability for infringement of any Letters Patent nor assume any such liability.

Designation: E 23 - 72ε

Standard Methods for
NOTCHED BAR IMPACT TESTING OF METALLIC MATERIALS[1]

This Standard is issued under the fixed designation E 23; the number immediately following the designation indicates the year of original adoption or, in the case of revision, the year of last revision. A number in parentheses indicates the year of last reapproval.

These methods have been approved by the Department of Defense to replace method 221.1 of Federal Test Method Standard No. 151 b and for listing in DoD Index of Specifications and Standards. Future proposed revisions should be coordinated with the Federal Government through the Army Materials and Mechanics Research Center, Watertown, Mass. 02172.

ε NOTE—Figure 2 was editorially changed in October 1973.

1. Scope

1.1 These methods describe notched-bar impact testing of metallic materials by the Charpy (simple-beam) apparatus and the Izod (cantilever-beam) apparatus. They give (*a*) a description of apparatus, (*b*) requirements for inspection and calibration, (*c*) dimensions and preparation of specimens, (*d*) the testing procedures, and (*e*) appended notes on the significance of notched-bar impact testing. These methods will in most cases also apply to tests on unnotched specimens.

NOTE 1—The values stated in U.S. customary units are to be regarded as the standard.

2. Summary of Methods

2.1 The essential features of an impact test are: (*a*) a suitable specimen (specimens of several different types are recognized), (*b*) an anvil or support on which the test specimen is placed to receive the blow of the moving mass, (*c*) a moving mass of known kinetic energy which must be great enough to break the test specimen placed in its path, and (*d*) a device for measuring the residual energy in the moving mass after the specimen has been broken.

3. Significance

3.1 These methods of impact testing relate specifically to the behavior of metal when subjected to a single application of a load resulting in multiaxial stresses associated with a notch, coupled with high rates of loading and in some cases with high or low temperatures. Impact tests are particularly useful in obtaining data on the mechanical behavior of certain types of metals under conditions favorable to brittle failure for which the results of the tension test are not significant. Further information on significance appears in the Appendix.

4. Apparatus

4.1 *General Requirements:*

4.1.1 The testing machine shall be a pendulum type of rigid construction and of capacity more than sufficient to break the specimen in one blow. The impact machine will be inaccurate to the extent that some energy is used in deformation or movement of its component parts or of the machine as a whole; this energy will be registered as used in fracturing the specimen. The machine shall not be used for values above 80 percent of the scale range. Looseness of parts must be avoided and deflections must be kept low; the vise and anvil region is particularly critical in this respect. The machine shall be rigidly mounted on an adequate foundation. The machine frame shall be equipped with a bubble level or a machined surface suitable for establishing levelness to within 3:1000.

4.1.2 Tests may be made at various velocities, but these shall be not less than 10 nor more than 20 ft/s (not less than 3 nor more than 6 m/s). Velocity shall always be stated as the maximum tangential velocity of the striking member at the center of strike (see Fig. 1). The impact value shall be taken as the energy absorbed in breaking the specimen

[1] These methods are under the jurisdiction of the ASTM Committee E-28 on Mechanical Testing.

Current edition approved April 10, 1972. Published June 1972. Originally published as E 23 – 33 T. Last previous edition E 23 – 66.

and is equal to the difference between the energy in the striking member at the instant of impact with the specimen and the energy remaining after breaking the specimen. The machine shall be furnished with scales graduated either in degrees or directly in foot-pounds on which readings can be estimated in increments of 0.25 percent of energy range or less. The scales may be compensated for windage and pendulum friction. The error in scale at any point shall not exceed 0.2 percent of range or 0.4 percent of reading, whichever is larger. The error in energy of blow caused by error in the weight of the pendulum shall not exceed 0.4 percent. The actual height of the pendulum in the release position shall not differ from the nominal height by more than 0.4 percent unless windage and friction are compensated for by increasing the height of drop, in which case the height may exceed the nominal value by not over 1.0 percent. Total friction and windage losses of the machine during the swing in the striking direction shall not exceed $^3/_4$ percent of scale range capacity, and pendulum energy loss from friction in the indicating mechanism shall not exceed $^1/_4$ percent of scale range capacity. Total friction and windage losses for the pendulum and indicating mechanism shall not exceed 0.4 percent of the total energy of the pendulum during the complete swing to and fro.

4.1.3 The dimensions of the pendulum shall be such that the center of percussion of the pendulum is at the center of strike within 1 percent of the distance from the axis of rotation to the center of strike. When hanging free the pendulum shall hang so that the striking edge is within 0.10 in. (2.5 mm) of the position where it would just touch the test specimen. When the indicator has been positioned to read zero energy in a free swing, it shall read within 0.2 percent of scale range when the striking edge of the pendulum is held against the test specimen. The plane of swing of the pendulum shall be perpendicular to the transverse axis of the Charpy specimen anvils or Izod vise within 3:1000. The mechanism for releasing the pendulum from its initial position shall operate freely and permit release of the pendulum in a start without initial impulse, retardation, or side vibration. Transverse play of the pendulum at the striker shall not exceed 0.030 in. (0.75 mm) under a transverse force of 4 percent of the effective weight of the pendulum applied at the center of strike. Radial play of the pendulum bearings shall not exceed 0.003 in. (0.075 mm).

4.2 *Charpy Apparatus:*

4.2.1 Means shall be provided (Fig. 1) to locate and support the test specimen against two anvil blocks in such a position that the center of the notch can be located within 0.010 in. (0.25 mm) of the midpoint between the anvils (see 8.3).

4.2.2 The supports and striking edge shall be of the forms and dimensions shown in Fig. 1. Other dimensions of the pendulum and supports should be such as to minimize interference between the pendulum and broken specimens.

4.2.3 The center line of the striking edge shall advance in the plane that is within 0.016 in. (0.40 mm) of the midpoint between the supporting edges of the specimen anvils. The striking edge shall be perpendicular to the longitudinal axis of the specimen within 5:1000. The striking edge shall be parallel within 1:1000 to the face of a perfectly square test specimen held against the anvil.

4.2.4 Specimen supports shall be square with anvil faces within 2.5:1000. Specimen supports shall be coplanar within 0.125 mm (0.005 in.) and parallel within 2.0:1000.

4.3 *Izod Apparatus:*

4.3.1 Means shall be provided (Fig. 2) for clamping the specimen in such a position that the face of the specimen is parallel to the striking edge within 1:1000. The edges of the clamping surfaces shall be sharp angles of 90 ± 1 deg with radii less than 0.016 in. (0.40 mm). The clamping surfaces shall be smooth with a 63-μm (2 μm) finish or better, and flat and parallel within 0.001 in. (0.025 mm) so as to clamp the specimen firmly up to the notch.

4.3.2 The dimensions of the striking edge and its position relative to the specimen clamps shall be as shown in Fig. 2.

5. Inspection

5.1 *Installation*—The machine shall be level to within 3:1000 and securely bolted to a concrete floor not less than 6 in. (150 mm) thick or, when this is not practical, the machine shall be bolted to a foundation having a mass not less than 40 times that of the pendulum.

5.2 *Wear of Critical Parts:*

5.2.1 *Specimen Anvils and Supports or Vise*—These shall conform to the dimensions shown in Fig. 1 or 2. To ensure a minimum of energy loss through absorption, bolts shall be secured as tightly as possible.

5.2.2 *Pendulum Striking Edge*—The striking edge (tup) of the pendulum shall conform to the dimensions shown in Figs. 1 or 2. To ensure a minimum of energy loss through absorption, the striking edge bolts shall be secured as tightly as possible. The pendulum striking edge (tup) shall comply with 4.2.3 (for Charpy tests) or 4.3.1 (for Izod tests) by bringing it into contact with a standard Charpy or Izod specimen.

5.3 *Pendulum Release Mechanism*—The mechanism for releasing the pendulum from its initial position shall comply with 4.1.3. If the same lever that is used to release the pendulum is also used to engage the brake, means shall be provided for preventing the brake from being accidentally engaged.

5.4 *Pendulum Operation:*

5.4.1 *Pendulum Alignment*—The pendulum shall comply with 4.1.3. If the side play in the pendulum or the radial play in the bearings exceeds the specified limits, adjust or replace the bearings.

5.4.2 *Potential Energy*—Determine the initial potential energy using the following procedure for the normal case where center of strike of the pendulum is coincident with the line from the center of rotation through the center of gravity. If the center of strike is more than 0.1 in. (2.5 mm) from this line, suitable corrections in elevation of the center of strike must be made in 5.4.2.2, 5.4.2.3, 5.4.5.1, and 5.4.6, so that elevations set or measured correspond to what they would be if the center of strike were on this line.

5.4.2.1 For Charpy machines place a half-width specimen (see Fig. 7), 0.394 by 0.197 in. (10 by 5.0 mm), in test position. With the striking edge in contact with the specimen a line scribed from the top edge of the specimen to the striking edge will indicate the center of strike on the striking edge. The top edge of the (0.197-in. (5.0 mm)) width indicates the center of strike of the striking edge. For Izod machines, place a specimen, so machined that the distance from the center of the notch to the top of the specimen is 0.879 in (22.31 mm), in test position. With the striking edge in contact with the specimen a line scribed from the top edge of the specimen to the striking edge will indicate the center of strike on the striking edge. The top edge of the specimen indicates center of strike of the striking edge.

5.4.2.2 Support the pendulum horizontally to within 15:1000 with two supports, one at the bearings (or center of rotation) and the other at the center of strike on the striking edge. Arrange the support at the striking edge to react upon some suitable weighing device such as a platform scale or balance, and determine the weight to within 0.4 percent. Take care to minimize friction at either point of support. Make contact with the striking edge through a round rod crossing the edge at a 90-deg angle. The weight of the pendulum is the scale reading minus the weights of the supporting rod and any shims that may be used to maintain the pendulum in a horizontal position.

5.4.2.3 Measure the height of pendulum drop for compliance with the requirement of 4.1.2. On Charpy machines measure the height from the top edge of a half-width (or center of a full-width) specimen to the elevated position of the center of strike to 0.1 percent. On Izod machines measure the height from a distance 0.879 in. (22.31 mm) above the vise to the elevated position of the center of strike to 0.1 percent.

5.4.2.4 The potential energy of the system is equal to the height from which the pendulum falls, as determined in 5.4.2.3, times the weight of the pendulum, as determined in 5.4.2.2.

5.4.3 *Impact Velocity*—Determine the impact velocity, v, of the machine, neglecting friction, by means of the following equation:

$$v = \sqrt{2gh}$$

where:

v = striking velocity, ft (or m)/s,

g = acceleration of gravity, ft (or m)/s^2 and

h = initial elevation of the striking edge, ft (or m).

5.4.4 *Center of Percussion*—To ensure that minimum force is transmitted to the point of rotation, the center of percussion shall be at a point within 1.0 percent of the distance from the axis of rotation to the center of strike in the specimen. Determine the location of the center of percussion as follows:

5.4.4.1 Using a stop watch or some other suitable time measuring device, capable of measuring time to within 0.2 s, swing the pendulum through a total angle not greater than 15 deg and record the time for 100 complete cycles (to and fro).

5.4.4.2 Determine the center of percussion by means of the following equation:

$l = 0.815p^2$, to determine l in feet

$l = 24.84p^2$, to determine l in centimeters

where:

l = distance, ft (or cm), from the axis to the center of percussion, and

p = time, s, of a complete cycle (to and fro) of the pendulum.

5.4.5 *Friction*—The energy loss from friction and windage of the pendulum and friction in the recording mechanism, if not corrected, will be included in the energy loss attributed to breaking the specimen and can result in erroneously high impact values. In machines recording in degrees, normal frictional losses are usually not compensated for by the machine manufacturer, whereas they are usually compensated for in machines recording directly in energy by increasing the starting height of the pendulum. Determine energy losses from friction as follows:

5.4.5.1 Without a specimen in the machine, release the pendulum from its starting position and record the energy value indicated. This value should indicate zero energy if frictional losses have been corrected by the manufacturer. Raise the pendulum so it just contacts the pointer at the value obtained in the free swing. Secure the pendulum at this height and determine the vertical distance from the center of strike to the top of a half-width specimen positioned on the specimen rests (see 5.4.2.1). Determine the weight of the pendulum as in 5.4.2.2 and multiply by this distance. The difference in this value and the initial potential energy is the total energy loss in the pendulum and indicator combined. Without resetting the pointer, repeatedly release the pendulum from its initial position until the pointer shows no further movement. The energy loss determined by the final position of the pointer is that due to the pendulum alone. The frictional loss in the indicator alone is then the difference between the combined indicator and pendulum losses and those due to the pendulum alone. The two frictional values obtained may with sufficient accuracy be assumed to be distributed uniformly over their respective ranges of action.

5.4.5.2 To ensure that friction and windage losses are within tolerances allowed (see 4.1.2), a simple weekly procedure may be adopted for direct reading machines. The following steps are recommended: (*a*) release the pendulum from its upright position without a specimen in the machine, and the energy reading should be 0 ft·lb; (*b*) without resetting the pointer, again release the pendulum and permit it to swing 11 half cycles. After the pendulum starts its 11th cycle, move the pointer to between 5 and 10 percent of scale range capacity and record the value obtained; and (*c*) this value, divided by 11, should not exceed 0.4 percent of scale range capacity. If this value does exceed 0.4 percent, the bearings should be cleaned or replaced.

5.4.6 *Indicating Mechanism*—To ensure that the scale is recording accurately over the entire range, check it at graduation marks corresponding to approximately 0, 10, 20, 30, 50, and 70 percent of each range. With the striking edge of the pendulum scribed to indicate the center of strike, lift the pendulum and set it in a position where the indicator reads, for example, 10.0 ft·lb. Determine the height of the pendulum to within 0.1 percent. The height of the pendulum multiplied by its weight, as determined in 5.4.2.2, is the residual energy. Increase this value by friction and windage losses in accordance with 5.4.5 and subtract from the potential energy determined in

5.4.2. The value recorded by the indicator shall equal this calculated value to within 0.2 percent of range or 0.4 percent of indicator reading, whichever is larger. Make similar calculations at other points of the scale. The scale pointer shall not overshoot or drop back with the pendulum. Make test swings from various heights to check visually the operation of the pointer over several portions of the scale.

5.5 *Specimen Clearance*—To ensure satisfactory results when testing materials of different strengths and compositions, the test specimen shall be free to leave the machine with a minimum of interference and shall not rebound into the pendulum before the pendulum completes its swing. Pendulums used on Charpy machines are of two basic designs, as shown in Fig. 3. When using a C-type pendulum, the broken specimen will not rebound into the pendulum and slow it down if the clearance at the end of the specimen is at least 0.5 in. (12.7 mm) or if the specimen is deflected out of the machine by some arrangement as is shown in Fig. 3. When using the U-type pendulum, means shall be provided to prevent the broken specimen from rebounding against the pendulum (Fig. 3). In most U-type pendulum machines, the shrouds should be designed and installed to the following requirements: (*a*) have a thickness of approximately 0.050 in. (1.3 mm), (*b*) have a minimum hardness of HRC 45, (*c*) have a radius of less than $\frac{1}{16}$ in. (1.6 mm) at the underside corners, and (*d*) be so positioned that the clearance between them and the pendulum overhang (both top and sides) does not exceed $\frac{1}{16}$ in. (1.6 mm).

NOTE 2—In machines where the opening within the pendulum permits clearance between the ends of a specimen (resting on the anvil supports) and the shrouds, and this clearance is at least 0.5 in. (12.7 mm), requirements (*a*) and (*d*) need not apply.

6. Proof Tests of Charpy Machines with Standardized Specimens

6.1 In addition to the requirements set forth in 4.1 and 4.2, Charpy machines shall be checked against standardized specimens which are available from the Army Materials and Mechanics Research Center for a nominal fee (Note 3). A set consists of fifteen 0.394 by 0.394-in. (10 by 10-mm) V-notch specimens of known energy values, five at each of three energy levels (Notes 4 and 5). The average value at each energy level as determined in the proof tests shall correspond to the nominal values of the standardized specimens within 1.0 ft·lb or 5.0 percent, whichever is the greater.

NOTE 3—Standardized specimens are available for Charpy machines only.

NOTE 4—A set of the standardized specimens may be obtained by addressing the request to: Director, Army Materials and Mechanics Research Center, ATTN: AMXMR-MQ, Watertown, Mass. 02172.

NOTE 5—The Watertown Arsenal Laboratories, now referred to as Army Materials and Mechanics Research Center, has for many years conducted a Charpy machine qualification program whereby standardized specimens are used to certify the machines of laboratories using the test as an inspection requirement on government contracts.[2] If the user desires, the results of tests with the standardized specimens will be evaluated. Participants desirous of the evaluation should complete the questionnaire provided with the standard specimens. The questionnaire provides for information such as testing temperature, the dimensions of certain critical parts, the cooling and testing techniques, and the results of the test. The broken standardized specimens are to be returned along with the completed questionnaire for evaluation (see Note 4 for address). Upon completion of the evaluation, the Army Materials and Mechanics Research Center will return a report. If a machine is producing values outside the standardized specimen tolerances, the report may suggest changes in machine design, repair or replacement of certain machine parts, a change in testing techniques, etc.

7. Test Specimens

7.1 The choice of specimen depends to some extent upon the characteristics of the material to be tested. A given specimen may not be equally satisfactory for soft nonferrous metals and hardened steels; therefore, a number of types of specimens are recognized. In general, sharper and deeper notches are required to distinguish differences in the more ductile materials or with lower testing velocities. The specimens shown in Figs. 4 and 5 are those most widely used and most generally satisfactory, particularly for ferrous metals. For precise results, it is necessary

[2] Driscoll, D. E., "Reproducibility of Charpy Impact Test," Symposium on Impact Testing, *ASTM STP No. 176*, ASTTA, Am. Soc. Testing Mats., 1955, p. 170.

that the specimens be held to the dimensional tolerances shown. These specimens are not considered suitable for tests of cast iron,[3] die-casting alloys, and metal powder structural parts. The specimen commonly found suitable for die-cast alloys is shown in Fig. 6. When the amount of material available does not permit making the standard impact test specimens shown in Figs. 4 and 5, smaller specimens may be used, but the results obtained on different sizes of specimens cannot be compared directly (Section A1.3). When Charpy specimens smaller than the standard are necessary or specified, it is recommended that they be selected from Fig. 7.

7.2 *Notching*—Notches shall be smoothly machined but polishing has proven generally unnecessary. However, since variations in notch dimensions will seriously affect the results of the tests, it is necessary to adhere to the tolerances given in Fig. 4 (A1.2 illustrates the effects from varying notch dimensions) on Type A specimens. In some materials, the impact test results will be affected by specimen orientation and notch direction. Therefore, unless otherwise specified the specimen shall be taken in the direction of major working, and the base of the notch shall be perpendicular to the surface of the plate. Notches shall be machined after any heat treating. In keyhole specimens, the round hole shall be carefully drilled with a slow feed. The slot may be cut by any feasible method. Care must be exercised in cutting the slot to see that the surface of the drilled hole is not damaged.

7.3 *Stamping*—When specimens are stamped for identification, such stamping shall be on the ends of the specimen or in a manner that will not interfere with proper support. All stamping shall be done in a way that avoids cold deforming of the metal at the root of the notch.

7.4 *Supplementary Specimens*—For economy in preparation of test specimens, special specimens of round or rectangular cross section are sometimes used for cantilever beam test. These are shown as Specimens X, Y, and Z in Figs. 8 and 9. Specimen Z is sometimes called the Philpot specimen after the name of the original designer. Its use requires special grips in the machine specimen clamp. The results cannot be related to those obtained with specimens of other size or shape as previously stated. In case of hard materials the machining of the flat surface struck by the pendulum is sometimes omitted. The Type Y round specimen occasionally used in the cantilever beam test has a circumferential V-notch as shown in Fig. 8.

7.5 *Metal Powder Structural Parts*—The specimens commonly found suitable for sintered metals are shown in Figs. 10 and 11. The specimen surface may be in the as-produced condition or smoothly machined, but polishing has proven generally unnecessary. Unnotched specimens are used with P/M materials. In P/M materials, the impact test results will be affected by specimen orientation. Therefore, unless otherwise specified, the position of the specimen in the machines shall be such that the pendulum will strike a surface that is parallel to the compacting direction.

8. Charpy Testing Procedure

8.1 *Daily Checking Procedure*—After the testing machine has been ascertained to comply with Sections 4 and 5, the routine daily checking procedures shall be as follows:

8.1.1 Prior to testing a group of specimens and before a specimen is placed in position to be tested, check the machine by a free swing of the pendulum. With the indicator at the initial position, a free swing of the pendulum shall indicate zero energy on machines reading directly in energy which are compensated for frictional losses. On machines recording in degrees, the indicated values when converted to energy shall compensate for frictional losses.

8.2 *Temperature of Testing*—In most materials, impact values vary with temperature. Unless otherwise specified, tests shall be made at 21 to 32 C (70 to 90 F). Accuracy of results when testing at other temperatures requires the following procedure:

8.2.1 *Low-Temperature Tests Down to −196 C (−320 F)*—For liquid cooling fill

[3] See 1933 Report of Subcommittee XV on Impact Testing, of Committee A-3 on Cast Iron, *Proceedings*, ASTE, Am. Soc. Testing Mats., Vol 33, Part 1, 1933 p. 87.

a suitable container, which has a grid raised at least 1 in. (25 mm) from the bottom, with liquid so that the specimen when immersed will be covered with at least 1 in. of the liquid. Cool the liquid to the desired temperature. The device used to measure the temperature of the bath should be placed in the center of a group of the specimens being cooled. Calibrate all temperature measuring equipment at least twice annually to ensure accurate testing temperatures. Hold specimens cooled in a liquid medium in an agitated bath at the desired temperature within +0 and −1.5 C (+0 and −3 F) for at least 5 min. Hold specimens cooled in gas at the desired temperature within +0 and −1.5 C (+0 and −3 F) for at least 60 min. Leave the mechanism used to remove the specimen from the coolant in the coolant except when handling the specimens. Remove the specimen from the coolant and break within 5 s.

8.2.2 *High-Temperature Tests*—Tests at elevated temperatures up to 260 C (+500 F) (Note 6) may be conducted by immersing the specimen in agitated oil or other suitable liquid and holding at test temperature within −0 and +1.5 C (−0 and +3 F) for at least 10 min. Above 260 C (500 F) use a furnace or heating oven, and hold the specimens at temperature within −0 and +2.5 C (−0 and +5 F) for 1 h. All other requirements of 8.2.1 shall also apply.

NOTE 6—Temperature up to +260 C (+500 F) may be obtained with certain oils, but "flash-point" temperatures must be carefully observed.

8.3 *Location of Test Specimen in Machine*—It is recommended that self-centering tongs similar to those shown in Fig. 12 be used in placing the specimen in the machine (see 4.2.1). The tongs illustrated in Fig. 12 are for centering V-notch specimens. If keyhole specimens are used, modification of the tong design may be necessary. If an end-centering device is used, caution must be taken to ensure that low-energy high-strength specimens will not rebound off this device into the pendulum and cause erroneously high recorded values. Many such devices are permanent fixtures of machines, and if the clearance between the end of a specimen in test position and the centering device is not approximately 0.5 in. (12.7 mm), the broken specimens may rebound into the pendulum.

8.4 *Operation of the Machine:*

8.4.1 Set the energy indicator at its initial position, take the test specimen from its cooling (or heating) medium, place it in proper position on the specimen rests, and release the pendulum smoothly. This entire sequence shall take less than 5 s.

8.4.2 Strike the test specimen on the side opposite the notch while supported against the supporting blocks in such a position that the center line of the striking edge strikes in line with the center line of the notch. The notch shall be midway between the points of support as defined in 4.2.1.

8.4.3 If any specimen fails to break, do not repeat the blow but record the fact, indicating whether the failure to break occurred through extreme ductility or lack of sufficient energy in the blow. Such results shall not be included in the average.

8.4.4 If any specimen jams in the machine, disregard the results and check the machine thoroughly for damage or maladjustment, which would affect its calibration.

8.4.5 To prevent recording an erroneous value caused by jarring the indicator when locking the pendulum in its upright position, read the indicated value from the indicator prior to locking the pendulum for the next test.

9. Izod Testing Procedure

9.1 *Daily Checking Procedure*—The general provisions for daily checking shall conform to 8.1 and 8.1.1.

9.2 *Temperature of Testing*—The specimen-holding fixture for Izod specimens is in most cases part of the base of the machine and cannot be readily cooled (or heated). For this reason, Izod testing is not recommended at other than room temperature.

9.3 *Operation of the Machine*—Strike the test specimen on the notched face while gripped in the support vise so that the center line of the notch is in line 0.005 in. (0.125 mm) with the line of support over which the specimen will be broken. Sections 8.4.3 to 8.4.5, inclusive, also apply when testing Izod specimens.

10. Report

10.1 A complete report shall include the following:

10.1.1 Type and model of machine,

NOTE 7—For commercial acceptance testing this information is usually sufficient, unless otherwise specified.

10.1.2 Type of specimen used (Note 7),

10.1.3 Linear velocity in feet (or meters) per second of the hammer at the instant of striking. This may be calculated from the initial height of the striking edge as given in 5.4.3.

10.1.4 Energy loss due to friction,

10.1.5 Energy of the blow with which the specimen was struck,

10.1.6 Temperature of the specimen or of the room conditions if both are the same (Note 7),

10.1.7 Energy actually absorbed by the specimen in breaking, reported in total foot-pounds as the "impact value" of the specimen (Note 7),

10.1.8 Appearance of fractured surface, and

10.1.9 Number of specimens failing to break (8.4.3).

11. Frequency of Calibration

11.1 Charpy machines shall have been proof tested within one year prior to the time of testing. It is not intended that parts not subject to wear (pendulum weight and scale linearity) need be measured at these intervals. Charpy machines shall, however, be rechecked immediately after making repairs or adjustments, after they have been moved, or whenever there is reason to doubt the accuracy of the results, without regard to the time interval.

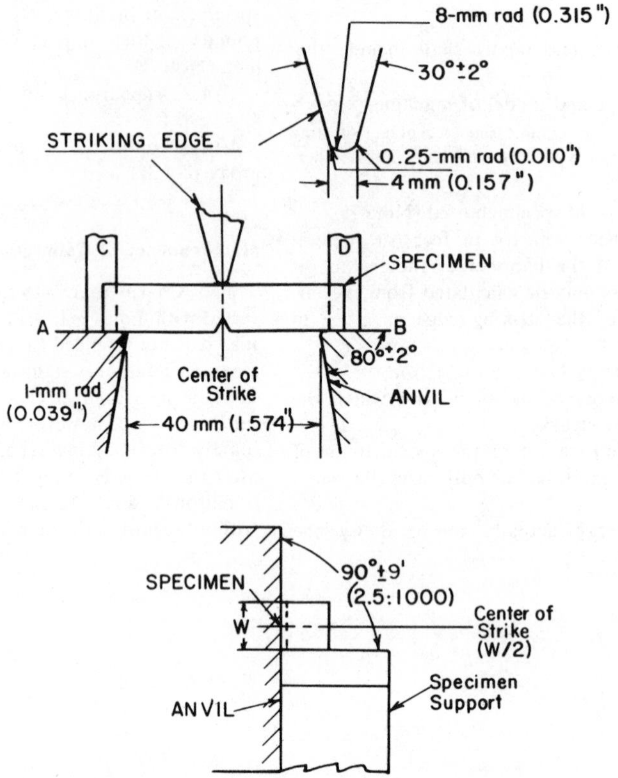

All dimensional tolerances shall be ±0.05 mm (0.002 in.) unless otherwise specified.

NOTE 1—A shall be parallel to B within 2:1000 and coplanar with B within 0.002 in. (0.05 mm).

NOTE 2—C shall be parallel to D within 2.0:1000 and coplanar with D within 0.005 in. (0.125 mm).

NOTE 3—Finish on unmarked parts shall be 125 μin. (4 μm).

FIG. 1 Charpy (Simple-Beam) Impact Test.

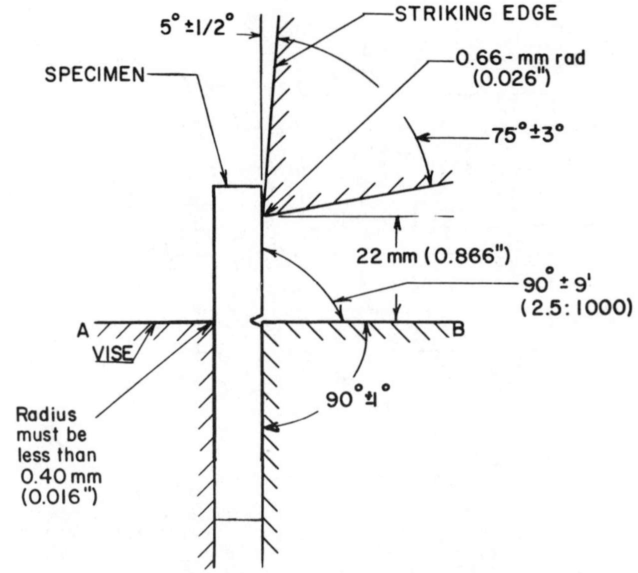

All dimensional tolerances shall be ±0.002 in. (0.05 mm) unless otherwise specified.

NOTE 1—The clamping surfaces of A and B shall be flat and parallel within 0.001 in. (0.025 mm).

NOTE 2—Finish on unmarked parts shall be 63 μin. (2 μm).

NOTE 3—Striker width must be greater than that of the specimen being tested.

FIG. 2 Izod (Cantilever-Beam) Impact Test.

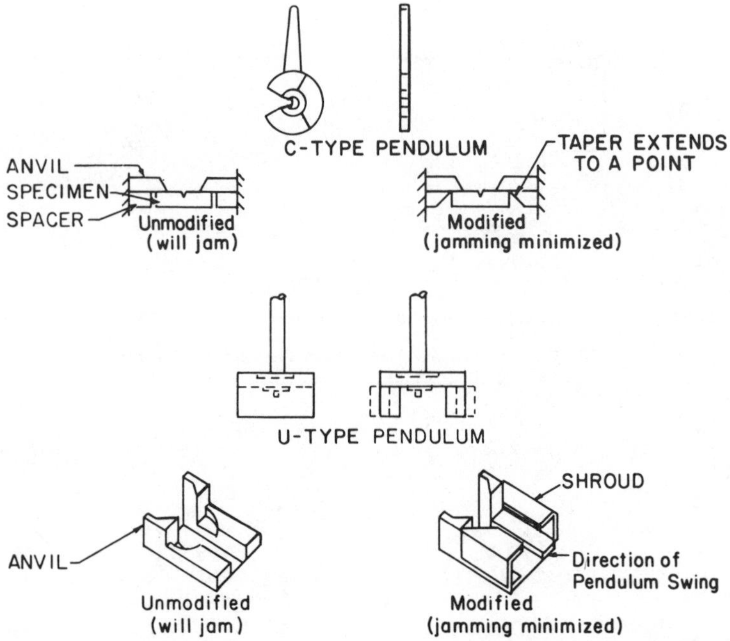

FIG. 3 Typical Pendulums and Anvils for Charpy Machines, Shown with Modifications to Minimize Jamming.

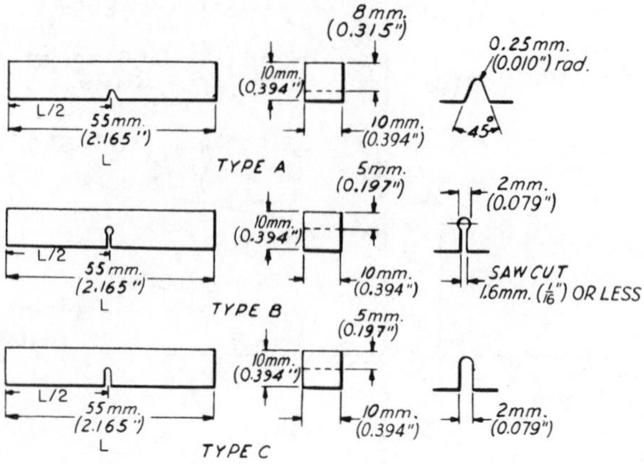

NOTE—Permissible variations shall be as follows:

Adjacent sides shall be at	90 deg ± 10 min
Cross-section dimensions	±0.001 in. (±0.025 mm)
Length of specimen (L)	+0, −0.100 in. (2.5 mm)
Centering of notch ($L/2$)	±0.039 in. (1 mm): When an end-centering device is necessary to center the specimen in the anvil, see 8.3, it is necessary that the notch be accurately centered to ensure compliance with 4.2.1.
Angle of notch	±1 deg
Radius of notch	±0.001 in. (0.025 mm)
Dimensions to bottom of notch:	
Type A specimen	0.315 ± 0.001 in. (8 ± 0.025 mm)
Types B and C specimen	0.197 ± 0.002 in. (5 ± 0.05 mm)
Finish requirements	63 μin. (2 μm) on notched surface and opposite face; 125 μin. (4 μm) on other two surfaces

FIG. 4 Charpy (Simple-Beam) Impact Test Specimens, Types A, B, and C.

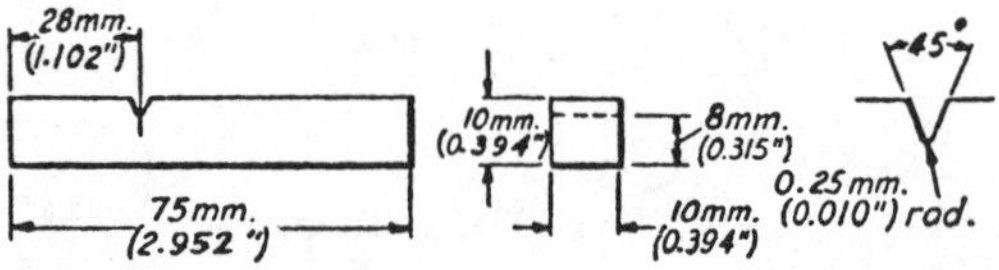

NOTE—Permissible variations shall be as follows:

Cross-section dimensions	±0.001 in. (0.025 mm)
Length of specimen	+0, −0.100 in. (2.5 mm)
Angle of notch	±1 deg
Radius of notch	±0.001 in. (0.025 mm)
Dimension to bottom of notch	0.315 ± 0.001 in. (8 ± 0.025 mm)
Adjacent sides shall be at	90 deg ± 10 min
Finish requirements	in 63 μin. (2 μm) on notched surface and opposite face; 125 μin. (4 μm) on other two surfaces.

FIG. 5 Izod (Cantilever-Beam) Impact Test Specimen, Type D.

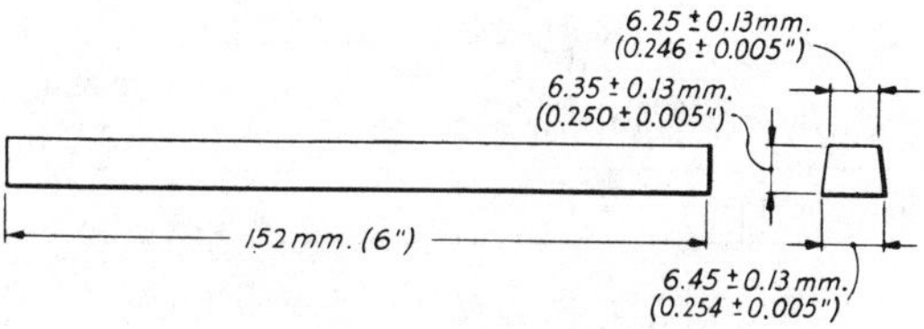

NOTE 1—Two test specimens may be cut from this bar.
NOTE 2—Blow shall be struck on narrowest face.

FIG. 6 Simple Beam Impact Test Bar for Die Castings Alloys.

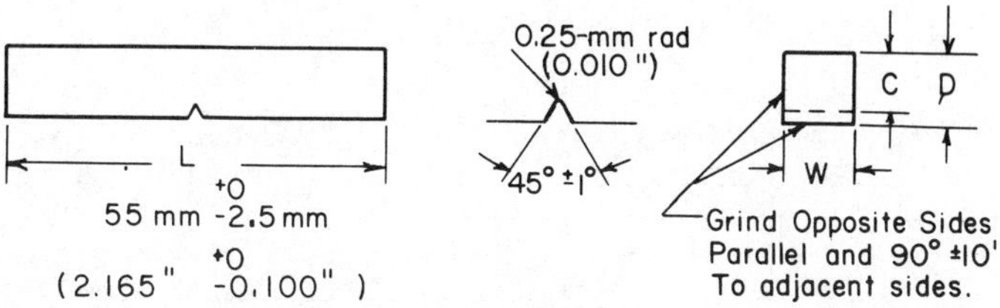

On subsize specimens the length, notch angle, and notch radius are constant (see Fig. 4); C, D, and W vary as indicated below.

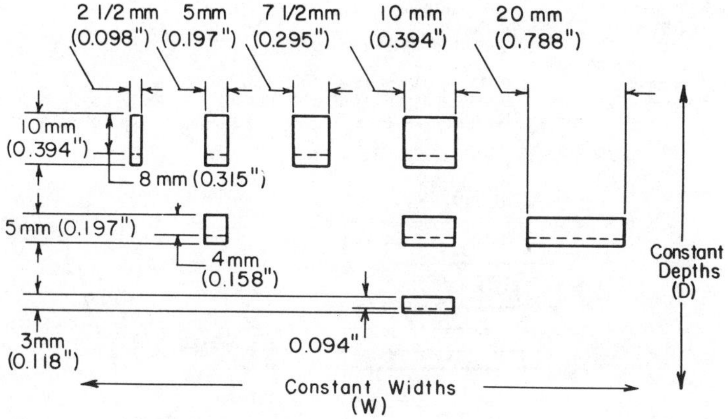

NOTE—Permissible variations shall be as follows:

Cross-section dimensions	±0.001 in. (0.025 mm)
Radius of notch	±0.001 in. (0.025 mm)
Notch dimensions	±0.001 in. (0.025 mm)
Finish requirements	63 μin. (2 μm) on notched surface and opposite face; 125 μin. (4 μm) on other two surfaces.

FIG. 7 Charpy (Simple-Beam) Subsize Impact Test Specimens.

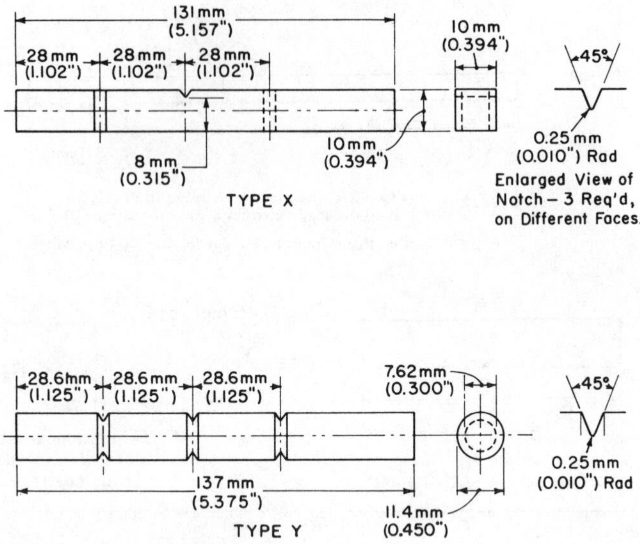

NOTE—Permissible variations shall be as follows:

Cross-section dimensions	±0.001 in. (0.025 mm)
Lengthwise dimensions	+0, −0.100 in. (2.5 mm)
Angle of notch	±1 deg
Radius of notch	±0.001 in. (0.025 mm)
Dimension to bottom of notch:	
Type X specimen	0.315 ± 0.001 in. (8 ± 0.025 mm)
Type Y specimen	0.300 ± 0.001 in. (7.62 ± 0.025 mm)

FIG. 8 Izod (Cantilever-Beam) Impact Test Specimens, Types X and Y.

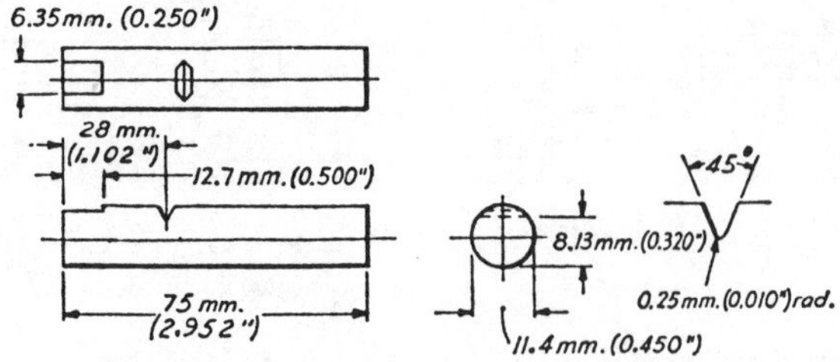

NOTE—Permissible variations shall be as follows:

Cross-section dimensions	±0.001 in. (0.025 mm)
Length of specimen	+0, −0.100 in. (2.5 mm)
Angle of notch	±1 deg
Radius of notch	±0.001 in.) (0.025 mm)
Dimension to bottom of notch	0.320 ± 0.001 in. (8.13 ± 0.025 mm)

FIG. 9 Izod (Cantilever-Beam) Impact Test Specimen (Philpot), Type Z.

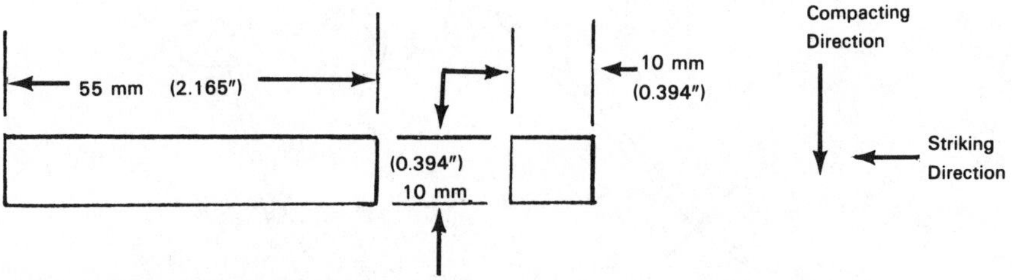

NOTE—Permissible variations shall be as follows:

Adjacent sides shall be at	90 deg. ± 10 min.
Cross section dimensions	±0.005 in. (0.125 mm)
Length of specimen	±0, −0.100 in. (2.5 mm)

FIG. 10 Charpy (Simple Beam) Impact Test Specimen for Metal Powder Structural Parts.

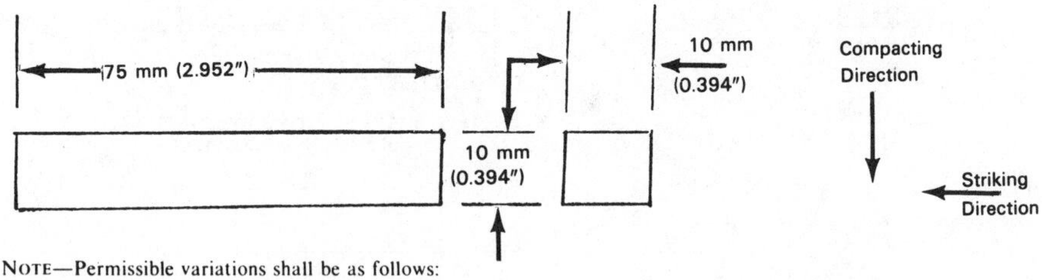

NOTE—Permissible variations shall be as follows:

Adjacent sides shall be at	90 deg. ± 10 min.
Cross section dimensions	±0.005 in. (0.125 mm)
Length of specimens	+0, −0.100 in. (2.5 mm)

FIG. 11 Izod (Cantilever-Beam) Impact Test Specimen for Metal Powder Structural Parts.

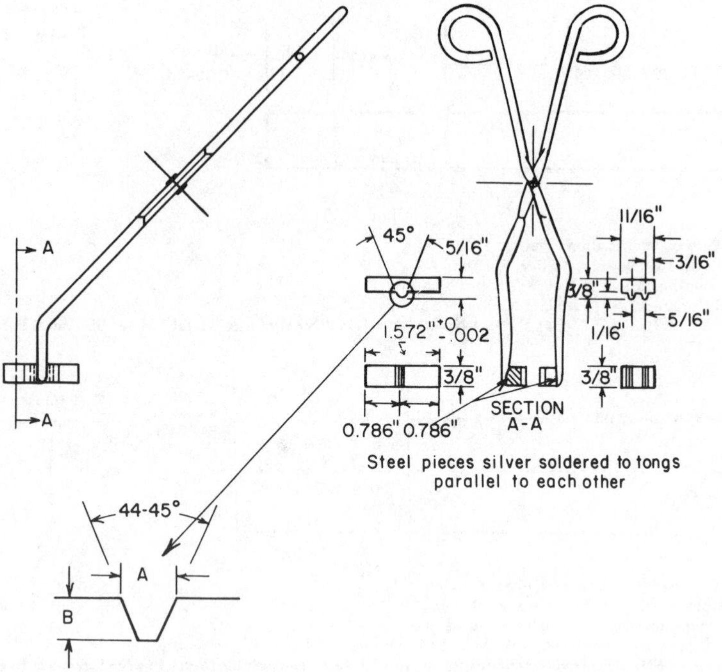

Specimen Depth, in. (mm)	Base Width (A), in. (mm)	Height (B), in. (mm)
0.394 (10)	0.063 to 0.067 (1.60 to 1.70)	0.060 to 0.065 (1.52 to 1.65)
0.197 (5)	0.029 to 0.033 (0.74 to 0.84)	0.027 to 0.032 (0.69 to 0.81)
0 118 (3)	0.016 to 0.020 (0.45 to 0.51)	0.014 to 0.019 (0.36 to 0.48)

FIG. 12 Centering Tongs for V-notch Charpy Specimens.

APPENDIX

A1. Notes on Significance of Notched-Bar Impact Testing

A1.1 Notch Behavior

A1.1.1 The Charpy and Izod type tests bring out notch behavior (brittleness versus ductility) by applying a single overload of stress. The energy values determined are quantitative comparisons on a selected specimen but cannot be converted into energy values that would serve for engineering design calculations. The notch behavior indicated in an individual test applies only to the specimen size, notch geometry, and testing conditions involved and cannot be generalized to other sizes of specimens and conditions.

A1.1.2 The notch behavior of the face-centered cubic metals and alloys, a large group of nonferrous materials and the austenitic steels can be judged from their common tensile properties. If they are brittle in tension they will be brittle when notched, while if they are ductile in tension they will be ductile when notched, except for unusually sharp or deep notches (much more severe than the standard Charpy or Izod specimens). Even low temperatures do not alter this characteristic of these materials. In contrast, the behavior of the ferritic steels under notch conditions cannot be predicted from their properties as revealed by the tension test. For the study of these materials the Charpy and Izod type tests are accordingly very useful. Some metals that display normal ductility in the tension test may nevertheless break in brittle fashion when tested or when used in the notched condition. Notched conditions include restraints to deformation in directions perpendicular to the major stress, or multiaxial stresses, and stress concentrations. It is in this field that the Charpy and Izod tests prove useful for determining the susceptibility of a steel to notch-brittle behavior though they cannot be directly used to appraise the serviceability of a structure.

A1.1.3 The testing machine itself must be sufficiently rigid or tests on high-strength low-energy materials will result in excessive elastic energy losses either upward through the pendulum shaft or downward through the base of the machine. If the anvil supports, the pendulum striking edge, or the machine foundation bolts are not securely fastened, tests on ductile materials in the range of 80 ft · lb may actually indicate values in excess of 90 to 100 ft · lb.

A1.2 Notch Effect

A1.2.1 The notch results in a combination of multiaxial stresses associated with restraints to deformation in directions perpendicular to the major stress, and a stress concentration at the base of the notch. A severely notched condition is generally not desirable, and it becomes of real concern in those cases in which it initiates a sudden and complete failure of the brittle type. Some metals can be deformed in a ductile manner even down to the low temperatures of liquid air, while others may crack. This difference in behavior can be best understood by considering the cohesive strength of a material (or the property that holds it together) and its relation to the yield point. In cases of brittle fracture, the cohesive strength is exceeded before significant plastic deformation occurs and the fracture appears crystalline. In cases of the ductile or shear type of failure, considerable deformation precedes the final fracture and the broken surface appears fibrous instead of crystalline. In intermediate cases the fracture comes after a moderate amount of deformation and is part crystalline and part fibrous in appearance.

A1.2.2 When a notched bar is loaded, there is a normal stress across the base of the notch which tends to initiate fracture. The property that keeps it from cleaving, or holds it together, is the "cohesive strength." The bar fractures when the normal stress exceeds the cohesive strength. When this occurs without the bar deforming it is the condition for brittle fracture.

A1.2.3 In testing, though not in service because of side effects, it happens more commonly that plastic deformation precedes fracture. In addition to the normal stress, the applied load also sets up shear stresses which are about 45 deg to the normal stress. The elastic behavior terminates as soon as the shear stress exceeds the shear strength of the material and deformation or plastic yielding sets in. This is the condition for ductile failure.

A1.2.4 This behavior, whether brittle or ductile, depends on whether the normal stress exceeds the cohesive strength before the shear stress exceeds the shear strength. Several important facts of notch behavior follow from this. If the notch is made sharper or more drastic, the normal stress at the root of the notch will be increased in relation to the shear stress and the bar will be more prone to brittle fracture (see Table A1). Also, as the speed of deformation increases, the shear strength increases and the likelihood of brittle fracture increases. On the other hand, by raising the temperature, leaving the notch and the speed of deformation the same, the shear strength is lowered and ductile behavior is promoted, leading to shear failure.

A1.2.5 Variations in notch dimensions will seriously affect the results of the tests. Tests on E 4340 steel specimens[4] have shown the effect of dimensional variations on Charpy results (see Table A1).

A1.3 Size Effect

A1.3.1 Increasing either the width or the depth of the specimen tends to increase the volume of metal subject to distortion, and by this factor tends to increase the energy absorption when breaking the specimen. However, any increase in size, particularly in width, also tends to increase the degree of restraint and by tending to induce brittle fracture, may decrease the amount of energy absorbed. Where a standard-size specimen is on the verge of brittle fracture, this is particularly true, and a doublewidth specimen may actually require less energy for rupture than one of standard width.

A1.3.2 In studies of such effects where the size of the material precludes the use of the standard specimen, as for example when the material is 0.25-in. (6.35 mm) plate, subsize specimens are necessarily used. Such specimens (Fig. 7) are based on the Type A specimen of Fig. 4.

A1.3.3 General correlation between the energy values obtained with specimens of different size or shape is not feasible, but limited correlations may be established for specification purposes on the basis of special studies of particular materials and particular specimens. On the other hand, in a study of the relative effect of process variations, evaluation by use of some arbitrarily selected specimen with some chosen notch will in most instances place the methods in their proper order.

A1.4 Effects of Testing Conditions

A1.4.1 The testing conditions also affect the notch behavior. So pronounced is the effect of temperature on the behavior of steel when notched that comparisons are frequently made by examining specimen fractures and by plotting energy value and fracture appearance versus temperature from tests of notched bars at a series of temperatures. When the test temperature has been carried low enough to start cleavage fracture, there may be an extremely sharp drop in impact value or there may be a relatively gradual falling off toward the lower temperatures. This drop in energy value starts when a specimen begins to exhibit some crystalline appearance in the fracture. The transition temperature at which this embrittling effect takes place varies considerably with the size of the part or test specimen and with the notch geometry.

A1.4.2 Some of the many definitions of transition temperature currently being used are: (*a*) the

[4] N. H. Fahey, "Effects of Variables in Charpy Impact Testing," *Materials Research & Standards*, MTRSA Vol. 1, No. 11, November 1961, p. 872.

lowest temperature at which the specimen exhibits 100 percent fibrous fracture, (*b*) the temperature where the fracture shows a 50 percent crystalline and a 50 percent fibrous appearance, (*c*) the temperature corresponding to the energy value 50 percent of the difference between values obtained at 100 percent and 0 percent fibrous fracture, and (*d*) the temperature corresponding to a specific energy value.

A1.4.3 A problem peculiar to Charpy-type tests occurs when high-strength, low-energy specimens are tested at low temperatures. These specimens may not leave the machine in the direction of the pendulum swing but rather in a sidewise direction. To ensure that the broken halves of the specimens do not rebound off some component of the machine and contact the pendulum before it completes its swing, modifications may be necessary in older model machines. These modifications differ with machine design. Nevertheless the basic problem is the same in that provisions must be made to prevent rebounding of the fractured specimens into any part of the swinging pendulum. Where design permits, the broken specimens may be deflected out of the sides of the machine and yet in other designs it may be necessary to contain the broken specimens within a certain area until the pendulum passes through the anvils. Some low-energy high-strength steel specimens leave impact machines at speeds in excess of 50 ft/s (15.2m/s) although they were struck by a pendulum traveling at speeds approximately 17 ft/s (5.2 m/s). If the force exerted on the pendulum by the broken specimens is sufficient, the pendulum will slow down and erroneously high energy values will be recorded. This problem accounts for many of the inconsistencies in Charpy results reported by various investigators within the 10 to 25-ft · lb range. Section 5.5 illustrates the two basic machine designs and a modification found to be satisfactory in minimizing jamming.

A1.5 Velocity of Straining

A1.5.1 Velocity of straining is likewise a variable that affects the notch behavior of steel. The impact test shows somewhat higher energy absorption values than the static tests above the transition temperature and yet, in some instances, the reverse is true below the transition temperature.

A1.6 Correlation with Service

A1.6.1 While Charpy or Izod tests may not directly predict the ductile or brittle behavior of steel as commonly used in large masses or as components of large structures, these tests can be used as acceptance tests or tests of identity for different lots of the same steel or in choosing between different steels, when correlation with reliable service behavior has been established. It may be necessary to make the tests at properly chosen temperatures other than room temperature. In this, the service temperature or the transition temperature of full-scale specimens does not give the desired transition temperatures for Charpy or Izod tests since the size and notch geometry may be so different. Chemical analysis, tension, and hardness tests may not indicate the influence of some of the important processing factors that affect susceptibility to brittle fracture nor do they comprehend the effect of low temperatures in inducing brittle behavior.

TABLE A1 Effect of Varying Notch Dimensions on Standard Specimens

	High-Energy Specimens, ft · lbf	Medium-Energy Specimens, ft · lbf	Low-Energy Specimens, ft · lbf
Specimen with standard dimensions	76.0 ± 3.8	44.5 ± 2.2	12.5 ± 1.0
Depth of notch, 0.084 in. (2.13 mm)[a]	72.2	41.3	11.4
Depth of notch, 0.0805 in. (2.04 mm)[a]	75.1	42.2	12.4
Depth of notch, 0.0775 in. (1.97 mm)[a]	76.8	45.3	12.7
Depth of notch, 0.074 in. (1.88 mm)[a]	79.6	46.0	12.8
Radius at base of notch, 0.005 in. (0.13 mm)[b]	72.3	41.7	10.8
Radius at base of notch, 0.015 in. (0.38 mm)[b]	80.0	47.4	15.8

[a] Standard 0.079 ± 0.002 in. (2.0 ± 0.05 mm).
[b] Standard 0.010 ± 0.001 in. (0.25 ± 0.025 mm).

By publication of this standard no position is taken with respect to the validity of any patent rights in connection therewith, and the American Society for Testing and Materials does not undertake to insure anyone utilizing the standard against liability for infringement of any Letters Patent nor assume any such liability.

American Association State
Highway and Transportation Officials Standard
AASHTO No.: T 19

AMERICAN NATIONAL STANDARD ANSI/ASTM C 29 - 76

Standard Test Method for
UNIT WEIGHT OF AGGREGATE[1]

This Standard is issued under the fixed designation C 29; the number immediately following the designation indicates the year of original adoption or, in the case of revision, the year of last revision. A number in parentheses indicates the year of last reapproval.

1. Scope

1.1 This method covers the determination of the unit weight of fine, coarse, or mixed aggregates.

NOTE 1—The values stated in U.S. customary units are to be regarded as the standard. The metric equivalents of U.S. customary units may be approximate.

2. Apparatus

2.1 *Balance*—A balance or scale accurate within 0.1 percent of the test load at any point within the range of use. The range of use shall be considered to extend from the weight of the measure empty to the weight of the measure plus its contents at 100 lb/ft^3 (1600 kg/m^3).

2.2 *Tamping Rod*—A round, straight steel rod, ⅝ in. (16 mm) in diameter and approximately 24 in. (600 mm) in length, having one end rounded to a hemispherical tip of the same diameter as the rod.

2.3 *Measure*—A cylindrical metal measure, preferably provided with handles. It shall be watertight, with the top and bottom true and even, preferably machined to accurate dimensions on the inside, and sufficiently rigid to retain its form under rough usage. The top rim shall be smooth and plane within 0.01 in (0.25 mm) and shall be parallel to the bottom within 0.5 deg (Note 2). Measures of the two larger sizes listed in Table 1 shall be reinforced around the top with a metal band, to provide an over-all wall thickness of not less than 0.20 in. (5 mm) in the upper 1½ in. (38 mm). The capacity and dimensions of the measure shall conform to the limits in Table 1 or 2 (Note 3).

NOTE 2—The top rim is satisfactorily plane if a 0.01-in. (0.25-mm) feeler gage cannot be inserted between the rim and a piece of ¼-in. (6-mm) or thicker plate glass laid over the measure. The top and bottom are satisfactorily parallel if the slope between pieces of plate glass in contact with the top and bottom does not exceed 1 percent in any direction.

NOTE 3—Dimensional tolerances and thicknesses of metal prescribed here are intended to be applied to measures acquired after January 1, 1968. Measures acquired before that date may conform either to this standard or to Section 2(*c*) of Method C 29 - 60.[2]

3. Sample

3.1 The sample of aggregate shall be dried to essentially constant weight, preferably in an oven, at 220 to 230 F (105 to 110 C) and thoroughly mixed.

4. Calibration of Measure

4.1 Fill the measure with water at room temperature and cover with a piece of plate glass in such a way as to eliminate bubbles and excess water.

4.2 Determine the net weight of water in the measure to an accuracy of ± 0.1 percent.

4.3 Measure the temperature of the water and determine its unit weight, from Table 3, interpolating if necessary.

4.4 Calculate the factor for the measure by dividing the unit weight of the water by the weight required to fill the measure.

COMPACT WEIGHT DETERMINATION

5. Rodding Procedure

5.1 The rodding procedure is applicable to

[1] This method is under the jurisdiction of ASTM Committee C-9 on Concrete and Concrete Aggregates and is the direct responsibility of Subcommittee C09.03.05 on Methods of Testing and Specifications for Physical Characteristics of Concrete Aggregates.

Current edition approved Sept. 24, 1976. Published November 1976. Originally published as C 29 - 20 T. Last previous edition C 29 - 71.

[2] Discontinued, see *1966 Book of ASTM Standards*, Part 10.

aggregates having a maximum size of 1½ in. (40 mm) or less.

5.1.1 Fill the measure one-third full and level the surface with the fingers. Rod the layer of aggregate with 25 strokes of the tamping rod evenly distributed over the surface. Fill the measure two-thirds full and again level and rod as above. Finally, fill the measure to overflowing and again rod as above. Level the surface of the aggregate with the fingers or a straightedge in such a way that any slight projections of the larger pieces of the coarse aggregate approximately balance the larger voids in the surface below the top of the measure.

5.1.2 In rodding the first layer, do not allow the rod to strike the bottom of the measure forcibly. In rodding the second and third layers, use only enough force to cause the tamping rod to penetrate the previous layer of aggregate.

5.1.3 Weigh the measure and its contents and record the net weight of the aggregate to the nearest 0.1 percent. Multiply this weight by the factor calculated as described in 4.4. The product is the compact unit weight of the aggregate.

6. Jigging Procedure

6.1 The jigging procedure is applicable to aggregates having a maximum size greater than 1½ in. (40 mm) and not to exceed 4 in. (100 mm).

6.1.1 Fill the measure in three approximately equal layers as described in 5.1.1, compacting each layer by placing the measure on a firm base, such as a cement-concrete floor, raising the opposite sides alternately about 2 in. (50 mm), and allowing the measure to drop in such a manner as to hit with a sharp, slapping blow. The aggregate particles, by this procedure, will arrange themselves in a densely compacted condition. Compact each layer by dropping the measure 50 times in the manner described, 25 times on each side. Level the surface of the aggregate with the fingers or a straightedge in such a way that any slight projections of the larger pieces of the coarse aggregate approximately balance the larger voids in the surface below the top of the measure.

6.1.2 Weigh the measure and its contents and record the net weight of the aggregate to the nearest 0.1 percent. Multiply this weight by the factor calculated as described in 4.4. The product is the compact unit weight of the aggregate.

Loose Weight Determination

7. Shoveling Procedure

7.1 The shoveling procedure is applicable to aggregates having a maximum size of 4 in. (100 mm) or less.

7.1.1 Fill the measure to overflowing by means of a shovel or scoop, discharging the aggregate from a height not to exceed 2 in. (50 mm) above the top of the measure. Exercise care to prevent, so far as possible, segregation of the particle sizes of which the sample is composed. Level the surface of the aggregate with the fingers or a straightedge in such a way that any slight projections of the larger pieces of the coarse aggregate approximately balance the larger voids in the surface below the top of the measure.

7.1.2 Weigh the measure and its contents and record the net weight of the aggregate to the nearest 0.1 percent. Multiply this weight by the factor calculated as described in 4.4. The product is the loose unit weight of the aggregate.

8. Precision

8.1 The multilaboratory standard deviation has been found to be 1.5 lb/ft^3 (24 kg/m^3)[3] for nominal $^3/_4$-in. (19.0-mm) maximum size, normal weight, coarse aggregate using $^1/_2$ ft^3 (15 litres) measures. Therefore, results of two properly conducted tests from two different laboratories on samples of the same coarse aggregate should not differ by more than 4.2 lb/ft^3 (67 kg/m^3).[3] The corresponding single-operator standard deviation has been found to be 0.7 lb/ft^3 (11 kg/m^3).[3] Therefore, results of two properly conducted tests by the same operator on the same coarse aggregate should not differ by more than 2.0 lb/ft^3 (32 kg/m^3).[3]

[3] These numbers represent, respectively the (1S) and (D2S) limits as described ASTM Recommended Practice C 670, for Preparing Precision Statements for Construction Materials, *Annual Book of ASTM Standards*, Part 14.

TABLE 1 Dimensions of Measures, U.S. Customary System[a]

Capacity, ft³	Inside Diameter, in.	Inside Height, in.	Thicknesses of metal, min, in.		Size of Aggregate, max, in.[b]
			Bottom	Wall	
1/10	6.0 ± 0.1	6.1 ± 0.1	0.20	0.10	½
⅓	8.0 ± 0.1	11.5 ± 0.1	0.20	0.10	1
½	10.0 ± 0.1	11.0 ± 0.1	0.20	0.12	1½
1	14.0 ± 0.1	11.2 ± 0.1	0.20	0.12	4

[a] The indicated size of container may be used to test aggregates of a maximum nominal size equal to or smaller than that listed.

[b] Based on sieves with square openings.

TABLE 2 Dimensions of Measures, Metric System[a]

Capacity, litres	Inside Diameter, mm	Inside Height, mm	Thicknesses of metal, min, mm		Size of Aggregate, max, mm[b]
			Bottom	Wall	
3	155 ± 2	160 ± 2	5.0	2.5	12.5
10	205 ± 2	305 ± 2	5.0	2.5	25
15	255 ± 2	295 ± 2	5.0	3.0	40
30	355 ± 2	305 ± 2	5.0	3.0	100

[a] The indicated size of container may be used to test aggregates of a maximum nominal size equal to or smaller than that listed.

[b] Based on sieves with square openings.

TABLE 3 Unit Weight of Water

Temperature		lb/ft³	kg/m³
deg F	deg C		
60	15.6	62.366	999.01
65	18.3	62.336	998.54
70	21.1	62.301	997.97
(73.4)	(23.0)	(62.274)	(997.54)
75	23.9	62.261	997.32
80	26.7	62.216	996.59
85	29.4	62.166	995.83

The American Society for Testing and Materials takes no position respecting the validity of any patent rights asserted in connection with any item mentioned in this standard. Users of this standard are expressly advised that determination of the validity of any such patent rights, and the risk of infringement of such rights, is entirely their own responsibility.

Designation: C 31 - 69 (Reapproved 1975)

American National Standard A37.17-1970 (R1975)
Reaffirmed Dec. 8, 1975
American National Standards Institute

Standard Method of

MAKING AND CURING CONCRETE TEST SPECIMENS IN THE FIELD[1]

This Standard is issued under the fixed designation C 31; the number immediately following the designation indicates the year of original adoption or, in the case of revision, the year of last revision. A number in parentheses indicates the year of last reapproval.

1. Scope

1.1 This method covers procedures for making and curing test specimens from concrete being used in construction.

NOTE 1—For making and curing test specimens in the laboratory, see ASTM Method C 192, Making and Curing Concrete Test Specimens in the Laboratory.[2]

NOTE 2—The values stated in U.S. customary units are to be regarded as the standard.

2. Apparatus

2.1 *Molds, General*—Molds for specimens or fastenings thereto in contact with the concrete shall be made of steel, cast iron, or other nonabsorbent material, nonreactive with concrete containing portland or other hydraulic cements. Molds shall hold their dimensions and shape under conditions of severe use. Molds shall be watertight during use as judged by their ability to hold water poured into them. A suitable sealant, such as heavy grease, modeling clay, or microcrystalline wax shall be used where necessary to prevent leakage through the joints. Positive means shall be provided to hold base plates firmly to the molds. Molds shall be lightly coated with mineral oil or a suitable nonreactive release material before use.

2.2 *Cylinder Molds:*

2.2.1 *Reusable Vertical Molds*—Reusable vertical molds shall be made of heavy gage metal or other rigid nonabsorbent material (Note 3). The plane of the rim of the mold shall be at right angles of the axis. The molds shall not vary from the prescribed diameter by more than 1/16 in. (1.6 mm) nor from the prescribed height by more than 1/4 in. (6.4 mm). Molds shall be provided with a machined metal base plate in the instance of metal molds, a smooth flat metal or integrally molded flat bottom of the same material as the sides in the instance of molds other than metal, with means for securing it to the mold at a right angle to the axis of the cylinder.

NOTE 3—Satisfactory reusable molds may be made from cold-drawn, seamless steel tubing or from steel pipe. These tubular sections shall be cut to the proper length, slotted by machine on one side parallel to the axis, and fitted with a circumferential metal band and bolt or a set of two or three clamps for closing. Split tube molds should be machined inside, if necessary, to ensure compliance with dimensional tolerances after slotting and clamping. Satisfactory reusable molds may also be made from iron or steel castings. Molds made from less rigid materials than steel tubing or iron or steel castings may require special care to ensure that they are not deformed more than the stipulated tolerances during use.

2.2.2 *Single-Use Molds*—Single-use molds shall conform to ASTM Specification C 470, for Molds for Forming Concrete Test Cylinders Vertically.[2]

NOTE 4—Special attention and supervision may be required to ensure that the specified tolerances for absorption and elongation of cardboard molds are not exceeded. Molds made from formed sheet metal or cardboard should be used with care to ensure that they are not deformed more than the

[1] This method is under the jurisdiction of ASTM Committee C-9 on Concrete and Concrete Aggregates and is the direct responsibility of Subcommittee C09.03.01 on Methods of Testing Concrete for Strength.

Current edition effective Oct. 3, 1969. Originally issued 1920. Replaces C 31 – 66.

[2] *Annual Book of ASTM Standards*, Part 14.

stipulated tolerances during use. The use of a tube of heavy metal around sheet metal molds when the specimen is being molded will preserve dimensional integrity.

2.3 *Beam Molds*—Beam molds shall be rectangular in shape and of the dimensions required to produce the specimens stipulated in 3.2. The inside surfaces of the molds shall be smooth and free from blemishes. The sides, bottom, and ends shall be at right angles to each other and shall be straight and true and free of warpage. Maximum variation from the nominal cross section shall not exceed ⅛ in. (3.2 mm) for molds with depth or breadth of 6 in. (152 mm) or more. Molds shall not be more than 1/16 in. (1.6 mm) shorter than the required length, but may exceed it by more than that amount.

2.4 *Tamping Rod*—The rod shall be a round, straight steel rod ⅝ in. (16 mm) in diameter and 24 in. (610 mm) long, with at least the tamping end rounded to a hemispherical tip of the same diameter. Both ends may be rounded, if preferred.

2.5 *Vibrators*—Internal vibrators may have rigid or flexible shafts, preferably powered by electric motors. The frequency or vibration shall be 7000 vibrations/min or greater while in use. The outside diameter or side dimension of the vibrating element shall be at least 0.75 in. (19 mm) and not greater than 1.50 in. (38 mm). The combined length of the shaft and vibrating element shall exceed the maximum depth of the section being vibrated by at least 3 in. (76 mm). External vibrators may be of two types: table or plank. The frequency for external vibrators shall be not less than 3600 vibrations/min, and preferably higher. For both table and plank vibrators, provision shall be made for clamping the mold securely to the apparatus. A vibrating-reed tachometer should be used to check the frequency of vibration.

NOTE 5—Vibratory impulses are frequently imparted to a table or plank vibrator through electromagnetic means, or by use of an eccentric weight on the shaft of an electric motor or on a separate shaft driven by a motor.

2.6 *Small Tools*—Tools and items which may be required such as shovels, pails, trowels, wood float, magnesium float, blunted trowels, straightedge, feeler gage, scoops, and rulers.

2.7 *Slump Apparatus*—The apparatus for measurement of slump shall conform to the requirements of ASTM Method C 143, Test for Slump of Portland Cement Concrete.[2]

2.8 *Sampling and Mixing Receptacle*—The receptacle shall be a suitable heavy gage metal pan, wheelbarrow, or flat, clean nonabsorbent mixing board of sufficient capacity to allow easy mixing by shovel or trowel of the entire sample.

2.9 *Air Content Apparatus*—The apparatus for measuring air content shall conform to the requirements of ASTM Method C 173, Test for Air Content of Freshly Mixed Concrete by the Volumetric Method,[2] or of ASTM Method C 231, Test for Air Content of Freshly Mixed Concrete by the Pressure Method.[2]

3. Test Specimens

3.1 *Compressive Strength Specimens*—Compressive strength specimens shall be cylinders of concrete cast and hardened in an upright position, with a length equal to twice the diameter. The standard specimen shall be the 6 by 12-in. (152 by 305 mm) cylinder when the nominal maximum size of the coarse aggregates does not exceed 2 in. (50 mm). When the nominal maximum size of the coarse aggregate exceeds 2 in. the diameter of the cylinder shall be at least three times the nominal maximum size of the coarse aggregate in the concrete (Note 6). Unless required by the project specifications, cylinders smaller than 6 by 12 in. shall not be made in the field.

NOTE 6—In general the nominal maximum size will be that size next larger than the sieve on which 15 percent of the coarse aggregate is retained. See ASTM Specification C 33, for Concrete Aggregates,[2] and ASTM Specification D 448, for Standard Sizes of Coarse Aggregates for Highway Construction.[2]

3.2 *Flexural Strength Specimens*—Flexural strength specimens shall be rectangular beams of concrete cast and hardened with long axes horizontal. The length shall be at least 2 in. (51 mm) greater than three times the depth as tested. The ratio of width to depth as molded shall not exceed 1.5. The standard beam shall be 6 by 6 in. (152 by 152 mm) in cross section, and shall be used for concrete with maximum size coarse aggregate up to 2 in. (50 mm). When the nominal maximum size of the coarse aggregate exceeds 2 in., the smaller cross-sectional dimension of the beam shall be at least three times the nominal maximum size of the coarse aggregate (Note 6). Unless required by project specifications, beams made in the field shall not have a width or depth of less than 6 in.

4. Sampling Concrete

4.1 Take samples of concrete for test specimens in accordance with ASTM Method C 172, Sampling Fresh Concrete.[2] Note the place of depositing in the structure of the sampled batch of concrete in the job records.

5. Slump and Air Content

5.1 *Slump*—Measure the slump of each batch of concrete, from which specimens are made, immediately after mixing, in accordance with the provisions of Method C 143. Discard concrete used for the slump test.

5.2 *Air Content*—Determine the air content, when required, in accordance with either Method C 173 or Method C 231. Discard concrete used for air content determination.

6. Molding Specimens

6.1 *Place of Molding*—Mold specimens promptly on a level, rigid, horizontal surface, free from vibration and other disturbances, at a place as near as practicable to the location where they are to be stored during the first 24 h. If it is not practicable to mold the specimens where they are to be stored, move them to the place of storage immediately after being struck off (Note 7). Avoid jarring, striking, tilting, or scarring of the surface of the specimens when moving the specimens to a safe place.

NOTE 7—A trowel slipped under the bottom of a cardboard mold will aid in preventing distortion of the bottom during moving.

6.2 *Placing the Concrete*—Place the concrete in the molds using a scoop, blunted trowel, or shovel. Select each scoopful, trowelful, or shovelful of concrete from the mixing pan to ensure that it is representative of the batch. It may be necessary to remix the concrete in the mixing pan with a shovel or trowel to prevent segregation during the molding of specimens. Move the scoop or trowel around the top edge of the mold as the concrete is discharged in order to ensure a symmetrical distribution of the concrete and to minimize segregation of coarse aggregate within the mold. Further distribute the concrete by use of a tamping rod prior to the start of consolidation. In placing the final layer the operator shall attempt to add an amount of concrete that will exactly fill the mold after compaction. Do not add nonrepresentative concrete to an underfilled mold.

6.2.1 *Number of Layers*—Make specimens in layers as indicated in Table 1.

6.3 *Consolidation:*

6.3.1 *Methods of Consolidation*—Preparation of satisfactory specimens requies different methods of consolidation. The methods of consolidation are rodding, and internal or external vibration. Base the selection of the method of consolidation on the slump, unless the method is stated in the specifications under which the work is being performed. Rod concretes with a slump greater than 3 in. (75 mm). Rod or vibrate concretes with slump of 1 to 3 in. (25 to 75 mm). Vibrate concretes with slump of less than 1 in. (25 mm).

NOTE 8—Concretes of such low water content that they cannot be properly consolidated by the methods described herein, or requiring other sizes and shapes of specimens to represent the product or structure, are not covered by this method. Specimens for such concretes shall be made in accordance with the requirements of Method C 192 with regard to specimen size and shape and method of consolidation.

6.3.2 *Rodding*—Place the concrete in the mold, in the required number of layers of approximately equal volume. For cylinders, rod each layer with the rounded end of the rod using the number of strokes specified in Table 2. The number of roddings per layer required for beams is one for each 2 in.^{2} (13 cm^2) top surface area of the specimen. Rod the bottom layer throughout its depth. Distribute the strokes uniformly over the cross section of the mold and for each upper layer allow the rod to penetrate about ½ in. (12 mm) into the underlying layer when the depth of the layer is less than 4 in. (100 mm), and about 1 in. (25 mm) when the depth is 4 in. or more. If voids are left by the tamping rod, tap the sides of the mold lightly to close the voids. After each layer is rodded, spade the concrete along the sides and ends of beam molds with a trowel or other suitable tool.

6.3.3 *Vibration*—Maintain a standard duration of vibration for the particular kind of concrete, vibrator, and specimen mold involved. The duration of vibration required will depend upon the workability of the concrete and the effectiveness of the vibrator. Usually sufficient vibration has been applied as soon as the surface of the concrete has become relatively smooth. Continue vibration only long enough to achieve proper consolidation of the concrete. Overvibration may cause segregation. Fill the molds and vibrate in the required number of approximately equal layers. Place all the concrete for each layer in the mold before starting vibration of that layer. Add the final layer, so as to avoid overfilling by more than ¼ in. (6 mm). Finish the surface either during or after vibration where external vibration is used. Finish the surface after vibration when internal vibration is used. When the finish is applied after vibration, add only enough concrete with a trowel to overfill the mold about ⅛ in. (3 mm), work it into the surface and then strike it off.

6.3.3.1 *Internal Vibration*—The diameter of the vibrating element, or thickness of a square vibrating element, shall not exceed one third of the width of the mold in the case of beams. For cylinders, the ratio of the diameter of the cylinder to the diameter of the vibrating element shall be 4.0 or higher. In compacting the specimen the vibrator shall not be allowed to rest on or touch the bottom or sides of the mold. Carefully withdraw the vibrator in such a manner that no air pockets are left in the specimen. After vibration of each layer tap the sides of the molds to ensure removal of large entrapped air bubbles at the surface of the mold.

6.3.3.2 *Cylinders*—Use three insertions of the vibrator at different points for each layer. Allow the vibrator to penetrate through the layer being vibrated, and into the layer below, approximately 1 in. (25 mm).

6.3.3.3 *Beams*—Insert the vibrator at intervals not exceeding 6 in. (150 mm) along the center line of the long dimension of the specimen. For specimens wider than 6 in., use alternating insertions along two lines. Allow the shaft of the vibrator to penetrate into the bottom layer approximately 1 in. (25 mm).

6.3.4 *External Vibration*—When external vibration is used, take care to ensure that the mold is rigidly attached to or securely held against the vibrating element or vibrating surface (Note 8).

6.4 *Finishing*—After consolidation by any of the methods, unless the finishing has been performed during the vibration (6.3.3), strike off the surface of the concrete and float or trowel it as required. Perform all finishing with the minimum manipulation necessary to produce a flat even surface that is level with the rim or edge of the mold and that has no depressions or projections larger than ⅛ in. (3.2 mm).

6.4.1 *Cylinders*—After consolidation finish the top surfaces by striking them off with the tamping rod where the consistency of the concrete permits, or with a wood float or trowel. If desired, cap the top surface of freshly made cylinders with a thin layer of stiff portland cement paste which is permitted to harden and cure with the specimen. See Section 4 of ASTM Method C 617, Capping Cylindrical Concrete Specimens.[2]

6.4.2 *Beams*—Beams shall be finished with a wood or magnesium float.

7. Curing

7.1 *Covering After Finishing*—To prevent evaporation of water from the unhardened concrete cover the specimens immediately after finishing, preferably with a nonabsorptive, nonreactive plate or a sheet of tough, durable, impervious plastic. Wet burlap may be used for covering, but care must be exercised to keep the burlap wet until the specimens are removed from the molds. Placing a sheet of plastic over the burlap will facilitate keeping it wet. Protect the outside surfaces of cardboard molds from all contact with wet burlap or other sources of water for the first 24 h after cylinders have been molded in them. Water may cause the molds to expand and damage specimens at this early age.

7.2 *Initial Curing*—During the first 24 h after molding, store all test specimens under conditions that maintain the temperature immediately adjacent to the specimens in the range of 60 to 80 F (16 to 27 C) and prevent loss of moisture from the specimens. Storage temperatures may be regulated by means of ventilation or by evaporation of water from sand or burlap (Note 9), or by using heating devices such as stoves, electric light bulbs, or thermostatically controlled heating cables. A temperature record of the specimens may be established by means of maximum-minimum thermometers. Store specimens in tightly constructed, firmly braced wooden boxes, damp sand pits, temporary buildings at construction sites, under wet burlap in favorable weather, or in heavyweight closed plastic bags, or use other suitable methods, provided the foregoing requirements limiting specimen temperature and moisture loss are met. Specimens formed in cardboard molds (2.2.2) shall not be stored for the first 24 h in contact with wet sand or wet burlap or under any other condition that will allow the outside surfaces of the mold to absorb water.

NOTE 9—The temperature within damp sand and under wet burlap or similar materials will always be lower than the temperature in the surrounding atmosphere if evaporation takes place.

7.3 *Curing Cylinders for Checking the Adequacy of Laboratory Mixture Proportions for Strength or as the Basis for Acceptance or for Quality Control*—Remove test specimens made for checking the adequacy of the laboratory mixture proportions for strength, or as the basis for acceptance, from the molds at the end of 20 ± 4 h and stored in a moist condition at 73.4 ± 3 F (23 ± 1.7 C) until the moment of test (Note 9). As applied to the treatment of demolded specimens, moist curing means that the test specimens shall have free water maintained on the entire surface area at all times. This condition is met by immersion in saturated lime water and may be met by storage in a moist room or cabinet meeting the requirements of ASTM Specification C 511. for Moist Cabinets and Rooms Used in the Testing of Hydraulic Cements and Concretes.[2] Specimens shall not be exposed to dripping or running water.

7.4 *Curing Cylinders for Determining Form Removal Time or When a Structure May Be Put into Service*—Store test specimens made for determining when forms may be removed or when a structure may be put in service in or on the structure as near to the point of use as possible, and shall receive, insofar as practicable, the same protection from the elements on all surfaces as is given to the portions of the structure which they represent. Test specimens in the moisture condition resulting from the specified curing treatment. To meet these conditions, specimens made for the purpose of determining when a structure may be put in service shall be removed from the molds at the time of removal of form work. Follow the provisions of 7.6, where applicable, for removal of specimens from molds.

7.5 *Curing Beams for Checking the Adequacy of Laboratory Mixture Proportions for Strength or as the Basis for Acceptance or for Quality Control*—Remove test specimens made for checking the adequacy of the laboratory mixture proportions for flexural strength, or as the basis for acceptance, or for quality control, from the mold between 20 and 48 h after molding and cure according to the provisions of 7.3 except that storage for a minimum period of 20 h immediately prior to testing shall be in saturated lime water at 73.4 ± 3 F (23 ± 1.7 C). At the end of the curing period, between the time the specimen is removed from curing until testing is completed, prevent drying of the surfaces of the specimen.

NOTE 10—Relatively small amounts of drying of the surface of flexural specimens induce tensile stresses in the extreme fibers that will markedly reduce the indicated flexural strength.

7.6 *Curing Beams for Determining when a Structure May Be Put into Service*—Cure test specimens for determining when a structure may be put into service, as nearly as practicable, in the same manner as the concrete in the structure. At the end of 48 ± 4 h after molding, take the specimens in the molds to a location preferably near a field laboratory and remove from the molds. Store specimens representing pavements or slabs on grade by placing them on the ground as molded, with their top surfaces up. Bank the sides and ends of the specimens with earth or sand that shall be kept damp, leaving the top surfaces exposed to the specified curing treatment. Store specimens representing structure concrete as near the point in the structure they represent as possible and afford them the same temperature protection and moisture environment as the structure. At the end of the curing period leave the specimens in place exposed to the weather in the same manner as the structure. Remove all beam specimens from field storage and store in lime water at 73.4 ± 3 F (23 ± 1.7 C) for 24 ± 4 h immediately before time of testing to ensure uniform moisture condition from specimen to specimen. Observe the precautions given in 7.5 to guard against drying between time of removal from curing to testing.

8. Shipment to Laboratory

8.1 Cylinders and beams shipped from the field to the laboratory for testing shall be packed in sturdy wooden boxes or other suitable containers surrounded by wet sand or wet sawdust, or other suitable packing material, and protected from freezing during shipment. Upon receipt by the laboratory they shall be placed immediately in the required curing at 73.4 ± 3 F (23 ± 1.7 C).

TABLE 1 Number of Layers Required for Specimens

Specimen Type and Size, as Depth, in. (mm)	Mode of Compaction	Number of Layers	Approximate Depth of Layer, in. (mm)
Cylinders:			
12 (305)	rodding	3 equal	4 (100)
Over 12 (305)	rodding	as required	4 (100)
12 (305) to 18 (460)	vibration	2 equal	half depth of specimen
Over 18 (460)	vibration	3 or more	8 (200) as near as practicable
Beams:			
6 (152) to 8 (200)	rodding	2 equal	half depth of specimen
Over 8 (200)	rodding	3 or more	4 (100)
6 (152) to 8 (200)	vibration	1	depth of specimen
Over 8 (200)	vibration	2 or more	8 (200) as near as practicable

TABLE 2 Number of Roddings to be Used in Molding Cylinder Specimens

Diameter of Cylinder, in. (mm)	Number of Strokes/Layer
6 (152)	25
8 (200)	50
10 (250)	75

By publication of this standard no position is taken with respect to the validity of any patent rights in connection therewith, and the American Society for Testing and Materials does not undertake to insure anyone utilizing the standard against liability for infringement of any Letters Patent nor assume any such liability.

Standard Test Method for

SPECIFIC GRAVITY AND ABSORPTION OF COARSE AGGREGATE[1]

This Standard is issued under the fixed designation C 127; the number immediately following the designation indicates the year of original adoption or, in the case of revision, the year of last revision. A number in parentheses indicates the year of last reapproval.

1. Scope

1.1 This method covers the determination of bulk and apparent specific gravity, 73.4/73.4 F (23/23 C), and absorption of coarse aggregate. Bulk specific gravity is the characteristic generally used for calculations of the volume occupied by the aggregate in portland-cement concrete.

1.2 This method determines (after 24 h in water) the bulk specific gravity and the apparent specific gravity as defined in Definitions E 12, the bulk specific gravity on the basis of weight of saturated surface-dry aggregate, and the absorption as defined in Definitions C 125.

2. Applicable Documents

2.1 *ASTM Standards:*

C 125 Definitions of Terms Relating to Concrete and Concrete Aggregates[2]

C 670 Recommended Practice for Preparing Precision Statements for Test Methods for Construction Materials[2]

C 702 Reducing Field Samples of Aggregates to Testing Size[3]

E 12 Definitions of Terms Relating to Density and Specific Gravity of Solids, Liquids, and Gases[3]

Manual of Concrete Testing[3]

3. Apparatus

3.1 *Balance*—A weighing device having a capacity of 5 kg or more, as required for the sample size selected; sensitive and readable to 0.5 g or 0.0001 times the sample weight, whichever is greater; and accurate within 0.1 percent of the test load at any point within the range used for this test. Within any 500-g range of test load, a difference between readings shall be accurate within 0.5 g or 0.0001 times the sample weight, whichever is greater.

3.2 *Sample Container*—A wire basket of No. 6 (3-mm), or finer mesh, or a bucket, of approximately equal breadth and height, with a capacity of 4000 to 7000 cm^3 for 1½ in. (38.1 mm) nominal maximum size aggregate or smaller, and a larger capacity container in the range from 8000 to 16000 cm^3 for the testing of large maximum size aggregate.

3.3 Suitable apparatus for suspending the sample container in water from the center of the scale pan or balance.

4. Test Specimen

4.1 Thoroughly mix the sample of aggregate to be tested and reduce it to the approximate quantity needed by use of a sample splitter or by quartering (Note 1). Reject all material passing a No. 4 (4.75-mm) sieve. In many instances it may be desirable to test a coarse aggregate in several separate size fractions; and if the sample contains more than 15 percent retained on the 1½-in. (38.1-mm) sieve, test the plus 1½ in. fraction or fractions separately from the smaller size fractions. The minimum weight of sample to be used is given below; when an aggregate is tested in separate size fractions, use the sample size corresponding to the nominal maximum size of each fraction:

---

[1] This method is under the jurisdiction of ASTM Committee C-9 on Concrete and Concrete Aggregates and is the direct responsibility of Subcommittee C09.03.05 on Methods of Testing and Specifications for Physical Characteristics of Concrete.

Current edition approved Jan. 14, 1977. Published March 1977. Originally published as C 127 - 36 T. Last previous edition C 127 - 73.

[2] *Annual Book of ASTM Standards*, Parts 14 and 15.

[3] *Annual Book of ASTM Standards*, Part 14.

Nominal Maximum Size, in. (mm)	Minimum Weight of Sample, kg
½ (12.5) or less	2
¾ (19.0)	3
1 (25.0)	4
1½ (37.5)	5
2 (50)	8
2½ (63)	12
3 (75)	18
3½ (90)	25

NOTE 1—The process of quartering and the correct use of a sample splitter are discussed in the Manual of Concrete Testing and in Method C 702.

5. Procedure

5.1 After thoroughly washing to remove dust or other coatings from the surface of the particles, dry the sample to constant weight at a temperature of 212 to 230 F (100 to 110 C), cool in air at room temperature for 1 to 3 h, and then immerse in water at room temperature for a period of 24 ± 4h.

NOTE 2—Where the absorption and specific gravity values are to be used in proportioning concrete mixtures in which the aggregates will be in their naturally moist condition, the requirement for initial drying to constant weight may be eliminated, and, if the surfaces of the particles in the sample have been kept continuously wet until test, the 24-h soaking may also be eliminated. Values for absorption and for specific gravity in the saturated-surface-dry condition may be significantly higher for aggregate not oven dried before soaking than for the same aggregate treated in accordance with 5.1. Therefore, any exceptions to the procedure of 5.1 should be noted on reporting the results.

5.2 Remove the specimen from the water and roll it in a large absorbent cloth until all visible films of water are removed. Wipe the larger particles individually. Take care to avoid evaporation of water from aggregate pores during the operation of surface-drying. Weigh the specimen in the saturated surface-dry condition. Record this and all subsequent weights to the nearest 0.5 g or 0.0001 times the sample weight, whichever is greater.

5.3 After weighing, immediately place the saturated-surface-dry specimen in the sample container and determine its weight in water at 73.4 ± 3 F (23 ± 1.7 C), having a density of 0.997 ± 0.002 g/cm³. Take care to remove all entrapped air before weighing by shaking the container while immersed.

NOTE 3—The container should be immersed to a depth sufficient to cover it and the test specimen during weighing. Wire suspending the container should be of the smallest practical size to minimize any possible effects of a variable immersed length.

5.4 Dry the specimen to constant weight at a temperature of 212 to 230 F (100 to 110 C), cool in air at room temperature 1 to 3 h, and weigh.

6. Bulk Specific Gravity

6.1 Calculate the bulk specific gravity, 73.4/73.4 F (23/23 C), as defined in Definitions E 12, as follows:

$$\text{Bulk sp gr} = A/(B - C)$$

where:

A = weight of oven-dry specimen in air, g,
B = weight of saturated-surface-dry specimen in air, g, and
C = weight of saturated specimen in water, g.

7. Bulk Specific Gravity (Saturated-Surface-Dry Basis)

7.1 Calculate the bulk specific gravity, 73.4/73.4 F (23/23 C), on the basis of weight of saturated-surface-dry aggregate as follows:

$$\text{Bulk sp gr (saturated-surface-dry basis)} = B/(B - C)$$

8. Apparent Specific Gravity

8.1 Calculate the apparent specific gravity, 73.4/73.4 F (23/23 C), as defined in Definitions E 12, as follows:

$$\text{Apparent sp gr} = A/(A - C)$$

9. Absorption

9.1 Calculate the percentage of absorption, as defined in Definitions C 125, as follows:

$$\text{Absorption, percent} = [(B - A)/A] \times 100$$

10. Calculation of Average Values

10.1 When the sample is tested in separate size fractions the average value for bulk specific gravity, bulk specific gravity (saturated-surface-dry basis), or apparent specific gravity can be computed as the weighted average of the values as computed in accordance with Section 6, 7, or 8 using the following equation:

$$G = \frac{1}{\frac{P_1}{100\,G_1} + \frac{P_2}{100\,G_2} + \ldots \frac{P_n}{100\,G_n}}$$

(see Appendix X1)

where:

G = average specific gravity. Either bulk-dry, saturated-surface-dry, or apparent specific gravities can be averaged in this manner.

$G_1, G_2, \ldots G_n$ = appropriate specific gravity values for each size fraction depending on the type of specific gravity being averaged.

$P_1, P_2, \ldots P_n$ = weight percentages of each size fraction present in the original sample.

10.2 For absorption the average value is the weighted average of the values as computed in Section 9, weighted in proportion to the weight percentages of the size fractions in the original fraction as follows:

$$A = (P_1A_1/100) + (P_2A_2/100) + \ldots (P_nA_n/100)$$

where:

A = average absorption, percent,

$A_1, A_2, \ldots A_n$ = absorption percentages for each size fraction, and

$P_1, P_2, \ldots P_n$ = weight percentages of each size fraction present in the original sample.

11. Precision

11.1 For normal weight aggregate with less than 3 percent absorption, the precision indexes are as follows:

	Standard Deviation (1S)[A]	Difference Between Two Tests (D2S)[A]
Multilaboratory:		
Bulk specific gravity	0.014	0.040
SSD specific gravity	0.010	0.028
Approximate specific gravity	0.011	0.031
Absorption, %	0.18	0.51
Single-operator:		
Bulk specific gravity	0.011	0.031
SSD specific gravity	0.008	0.023
Approximate specific gravity	0.007	0.020
Absorption, %	0.15	0.42

[A] These numbers represent, respectively, the (1S) and (D2S) limits as described in Recommended Practice C 670.

APPENDIX

X1. DEVELOPMENT OF EQUATIONS

X1.1 The derivation of the equation is apparent from the following simplified cases using two solids. Solid 1 has a weight W_1 in grams and a volume V_1 in millilitres; its specific gravity (G_1) is therefore W_1/V_1.
Solid 2 has a weight W_2 and volume V_2, and $G_2 = W_2/V_2$. If the two solids are considered together, the specific gravity of the combination is the total weight in grams divided by the total volume in millilitres:

$$G = (W_1 + W_2)/(V_1 + V_2)$$

Manipulation of this equation yields the following:

$$G = \frac{1}{\dfrac{V_1 + V_2}{W_1 + W_2}} = \frac{1}{\dfrac{V_1}{W_1 + W_2} + \dfrac{V_2}{W_1 + W_2}}$$

$$G = \frac{1}{\dfrac{W_1}{W_1 + W_2}\left(\dfrac{V_1}{W_1}\right) + \dfrac{W_2}{W_1 + W_2}\left(\dfrac{V_2}{W_2}\right)}$$

However, the weight fractions of the two solids are:

$$W_1/(W_1 + W_2) = P_1/100 \text{ and } W_2/(W_1 + W_2) = P_2/100$$

and,

$$1/G_1 = V_1/W_1 \text{ and } 1/G_2 = V_2/W_2$$

Therefore,

$$G = 1/[(P_1/100)(1/G_1) + (P_2/100)(1/G_2)]$$

An example of the computation is given in Table X1.

TABLE X1 Example of Calculation of Average Values of Specific Gravity and Absorption for a Coarse Aggregate Tested in Separate Sizes

Size Fraction, in. (mm)	Percent in Original Sample	Sample Weight Used in Test, g	Bulk Specific Gravity (SSD)	Absorption, percent
No. 4 to ½ (12.5)	44	2213.0	2.72	0.4
½ to 1½ (37.5)	35	5462.5	2.56	2.5
1½ to 2½ (63)	21	12593.0	2.54	3.0

Average Specific Gravity (SSD)

$$G_{SSD} = \frac{1}{\frac{0.44}{2.72} + \frac{0.35}{2.56} + \frac{0.21}{2.54}} = 2.62$$

Average Absorption

$$A = (0.44)(0.4) + (0.35)(2.5) + (0.21)(3.0) = 1.7 \text{ percent}$$

The American Society for Testing and Materials takes no position respecting the validity of any patent rights asserted in connection with any item mentioned in this standard. Users of this standard are expressly advised that determination of the validity of any such patent rights, and the risk of infringement of such rights, is entirely their own responsibility.

Designation: C 128 – 73

American National Standard A37.6-1975
Approved Oct. 7, 1975
American National Standards Institute

Standard Test Method for
SPECIFIC GRAVITY AND ABSORPTION OF FINE AGGREGATE[1]

This Standard is issued under the fixed designation C 128; the number immediately following the designation indicates the year of original adoption or, in the case of revision, the year of last revision. A number in parentheses indicates the year of last reapproval.

1. Scope

1.1 This method covers the determination of bulk and apparent specific gravity, 73.4/73.4 F (23/23 C), and absorption of fine aggregate. Bulk specific gravity is the characteristic generally used for calculations of the volume occupied by the aggregate in portland-cement concrete.

1.2 This method determines (after 24 h in water) the bulk specific gravity and the apparent specific gravity as defined in the ASTM Definitions E 12, Terms Relating to Density and Specific Gravity of Solids, Liquids and Gases,[2] the bulk specific gravity on the basis of weight of saturated surface-dry aggregate, and the absorption as defined in ASTM Definitions C 125, for Terms Relating to Concrete and Concrete Aggregates.[3]

NOTE 1—The values stated in U.S. customary units are to be regarded as the standard.

2. Apparatus

2.1 *Balance*—A balance or scale having a capacity of 1 kg or more, sensitive to 0.1 g or less, and accurate within 0.1 percent of the test load at any point within the range of use for this test. Within any 100-g range of test load, a difference between readings shall be accurate within 0.1 g.

2.2 *Pycnometer*—A flask or other suitable container into which the fine aggregate test sample can be readily introduced and in which the volume content can be reproduced within ±0.1 cm^3. The volume of the container filled to mark shall be at least 50 percent greater than the space required to accommodate the test sample. A volumetric flask of 500 cm^3 capacity or a fruit jar fitted with a pycnometer top is satisfactory for a 500-g test sample of most fine aggregates.

2.3 *Mold*—A metal mold in the form of a frustum of a cone with dimensions as follows: 40 ± 3 mm inside diameter at the top, 90 ± 3 mm inside diameter at the bottom, and 75 ± 3 mm in height, with the metal having a minimum thickness of 0.8 mm.

2.4 *Tamper*—A metal tamper weighing 12 ± ½ oz (340 ± 15 g) and having a flat circular tamping face 1 ± ⅛ in. (25 ± 3 mm) in diameter.

3. Preparation of Test Specimen

3.1 Obtain approximately 1000 g of the fine aggregate from the sample by use of a sample splitter or by quartering (Note 2). Dry it in a suitable pan or vessel to constant weight at a temperature of 212 to 230 F (100 to 110 C). Allow it to cool to comfortable handling temperature, cover with water, and permit to stand for 24 ± 4 h (Note 3). Decant excess water with care to avoid loss of fines, spread the sample on a flat surface exposed to a gently moving current of warm air, and stir

---

[1] This method is under the jurisdiction of ASTM Committee C-9 on Concrete and Concrete Aggregates and is the direct responsibility of Subcommittee C09.03.05 on Methods of Testing and Specifications for Physical Characteristics of Concrete Aggregates.

Current edition approved Oct. 29, 1973. Published December 1973. Originally published as C 128 36. Last previous edition C 128 68.

[2] *Annual Book of ASTM Standards*, Part 14.

[3] *Annual Book of ASTM Standards*, Parts 14 and 15.

frequently to secure uniform drying. Continue this operation until the test specimen approaches a free-flowing condition. Then place a portion of the partially dried fine aggregate loosely into the mold, held firmly on a smooth nonabsorbent surface with the large diameter down, lightly tamp the surface 25 times with the tamper, and lift the mold vertically. If surface moisture is still present, the fine aggregate will retain the molded shape. Continue drying with constant stirring and test at frequent intervals until the tamped fine aggregate slumps slightly upon removal of the mold. This indicates that it has reached a surface-dry condition (Note 4). If desired, mechanical aids such as tumbling or stirring may be employed to assist in achieving the saturated surface-dry condition.

NOTE 2—The process of quartering and the correct use of a sample splitter are discussed in the Manual of Concrete Testing.[2]

NOTE 3—Where the absorption and specific gravity values are to be used in proportioning concrete mixtures with aggregates used in their naturally moist condition, the requirement for initial drying to constant weight may be eliminated and, if the surfaces of the particles have been kept wet, the 24-h soaking may also be eliminated. Values for absorption and for specific gravity in the saturated-surface-dry condition may be significantly higher for aggregate not oven dried before soaking than for the same aggregate treated in accordance with 3.1.

NOTE 4—The procedure described in Section 3 is intended to ensure that the first cone test trial will be made with some surface water in the specimen. If the fine aggregate slumps on the first trial, it has been dried past the saturated and surface-dry condition. In this case thoroughly mix a few cubic centimeters of water with the fine aggregate and permit the specimen to stand in a covered container for 30 min. The process of drying and testing for the free-flowing condition shall then be resumed.

4. Procedure

4.1 Immediately introduce into the pycnometer 500.0 g (Note 5) of the fine aggregate, prepared as described in Section 3, and fill with water to approximately 90 percent of capacity. Roll, invert, and agitate the pycnometer to eliminate all air bubbles. Adjust its temperature to 73.4 ± 3 F (23 ± 1.7 C), if necessary, by immersion in circulating water and bring the water level in the pycnometer to its calibrated capacity. Determine total weight of the pycnometer, specimen, and water (Note 6). Record this and all other weights to the nearest 0.1 g.

NOTE 5—An amount other than 500 g, but not less than 50 g, may be used provided that the actual weight is inserted in place of the figure "500" wherever it appears in the formulas of 5.1, 6.1, 7.1 and 8.1. If the weight used is less than 500 g, limits on accuracy of weighing and measuring must be scaled down in proportion.

NOTE 6—As an alternative, the quantity of water necessary to fill the pycnometer may be determined volumetrically using a buret accurate to 0.15 cm³. The total weight of the pycnometer, specimen, and water is then computed as follows:

$$C = 0.9976\,V_a + 500 + W$$

where:

C = weight of pycnometer filled with the specimen plus water, g,
V_a = volume of water added to pycnometer, cm³, and
W = weight of the pycnometer empty, g.

4.2 Remove the fine aggregate from the pycnometer, dry to constant weight at a temperature of 212 to 230 F (100 to 110 C), cool in air at room temperature for ½ to 1½ h, and weigh.

4.3 Determine the weight of the pycnometer filled to its calibration capacity with water at 73.4 ± 3 F (23 ± 1.7 C).

NOTE 7—If a volumetric flask is used and is calibrated to an accuracy of 0.15 cm³ at 20 C, the weight of the flask filled with water may be calculated as follows:

$$B = 0.9976\,V + W$$

where:

B = weight of flask filled with water, g,
V = volume of flask, cm³, and
W = weight of the flask empty, g.

5. Bulk Specific Gravity

5.1 Calculate the bulk specific gravity, 73.4/73.4 F (23/23 C), as defined in ASTM Definitions E 12, as follows:

$$\text{Bulk sp gr} = A/(B + 500 - C)$$

where:

A = weight of oven-dry specimen in air, g,
B = weight of pycnometer filled with water, g, and
C = weight of pycnometer with specimen and water to calibration mark, g.

6. Bulk Specific Gravity (Saturated Surface-Dry Basis)

6.1 Calculate the bulk specific gravity, 73.4/73.4 F (23/23 C), on the basis of weight of saturated surface-dry aggregate as follows:

$$\text{Bulk sp gr (saturated surface-dry basis)} = 500/(B + 500 - C)$$

7. Apparent Specific Gravity

7.1 Calculate the apparent specific gravity, 73.4/73.4 F (23/23 C), as defined in Definitions E 12, as follows:

$$\text{Apparent sp gr} = A/(B + A - C)$$

8. Absorption

8.1 Calculate the percentage of absorption, as defined in Definitions C 125, as follows:

$$\text{Absorption, percent} = [(500 - A)/A] \times 100$$

9. Precision

9.1 Data from carefully conducted tests on normal weight aggregate at one laboratory yielded the following for tests on the same specimen. Different specimens from the same source may vary more.

9.1.1 For specific gravity, single-operator and multi-operator precision (2S limits) less than ±0.02 from the average specific gravity. Differences greater than 0.03 between duplicate tests on the same specimen by the same or different operators should occur by chance less than 5 percent of the time (D2S limit less than 0.03).

9.1.2 For absorption, single-operator precision ±0.31 from the average percent absorption 95 percent of the time (2S limits). Multi-operator tests are probably less precise. The difference between tests by the same operator on the same specimen should not exceed 0.45 more than 5 percent of the time (D2S limit).

By publication of this standard no position is taken with respect to the validity of any patent rights in connection therewith, and the American Society for Testing and Materials does not undertake to insure anyone utilizing the standard against liability for infringement of any Letters Patent nor assume any such liability.

AMERICAN NATIONAL STANDARD ANSI/ASTM C 136 - 76

American Association State Highway and Transportation Officials Standard
AASHTO No.: T 27

Standard Test Method for
SIEVE OR SCREEN ANALYSIS OF FINE AND COARSE AGGREGATES[1]

This Standard is issued under the fixed designation C 136; the number immediately following the designation indicates the year of original adoption or, in the case of revision, the year of last revision. A number in parentheses indicates the year of last reapproval.

1. Scope

1.1 This method covers the determination of the particle size distribution of fine and coarse aggregates by sieving or screening.

2. Summary of Method

2.1 A weighed sample of dry aggregate is separated through a series of sieves or screens of progressively smaller openings for determination of particle size distribution.

3. Apparatus

3.1 *Balance*—A balance or scale accurate within 0.1 percent of the test load at any point within the range of use.

3.2 *Sieves or Screens*—The sieves or screens shall be mounted on substantial frames constructed in a manner that will prevent loss of material during sieving. Suitable sieve sizes shall be selected to furnish the information required by the specifications covering the material to be tested. The sieves shall conform to ASTM Specification E 11, for Wire-Cloth Sieves for Testing Purposes.[2]

3.3 *Oven*—An oven of appropriate size capable of maintaining a uniform temperature of 230 ± 9 F (110 ± 5 C).

4. Test Sample

4.1 The sample of aggregate to be tested for sieve analysis shall be thoroughly mixed and reduced by use of a sample splitter or by quartering (Note 1) to an amount suitable for testing. Fine aggregate shall be moistened before reduction to minimize segregation and loss of dust. The sample for test shall be approximately of the weight desired when dry and shall be the end result of the reduction method. Reduction to an exact predetermined weight shall not be permitted.

NOTE 1—The process of quartering and the correct use of a sample splitter are described in the "Manual of Concrete Testing."[3]

4.2 *Fine Aggregate*—The test sample of fine aggregate shall weigh, after drying, approximately the following amount:

Aggregate with at least 95 percent passing a No. 8 (2.36-mm) sieve	100 g
Aggregate with at least 85 percent passing a No. 4 (4.75-mm) sieve and more than 5 percent retained on a No. 8 sieve	500 g

In no case, however, shall the fraction retained on any sieve at the completion of the sieving operation weigh more than 4 g/in.^{2} of sieving surface.

NOTE 2—This amounts to 200 g for the usual 8-in. (203-mm) diameter sieve. The amount of material retained on the critical sieve may be regulated by (*1*) the introduction of a larger-opening sieve immediately above the critical sieve, or (*2*) selection of a sample of proper size.

4.3 *Coarse Aggregate*—The weight of the test sample of coarse aggregate shall conform with the following:

Maximum Nominal Size, Square Openings, in. (mm)	Minimum Weight of Sample, kg
3/8 (9.5)	2
1/2 (12.5)	4
3/4 (19.0)	8

[1] This method is under the jurisdiction of ASTM Committee C-9 on Concrete and Concrete Aggregates and is the direct responsibility of Subcommittee C 09.03.05 on Methods of Testing and Specifications for Physical Characteristics of Concrete Aggregates.

Current edition approved Sept. 24, 1976. Published November 1976. Originally published as C 136 - 38 T. Last previous edition C 136 - 71.

[2] *Annual Book of ASTM Standards*, Parts 14 and 15.

[3] *Annual Book of ASTM Standards*, Part 14.

Maximum Nominal Size, Square Openings, in. (mm)	Minimum Weight of Sample, kg
1 (25.0)	12
1½ (37.5)	16
2 (50)	20
2½ (63)	25
3 (75)	45
3½ (90)	70

NOTE 3—It is recommended that sieves mounted in frames of 16-in. (406-mm) diameter or larger be used for testing coarse aggregate.

4.4 In the case of mixtures of fine and coarse aggregates, the material shall be separated into two sizes on the No. 4 (4.75-mm) sieve. The samples of fine and coarse aggregate shall be prepared in accordance with 4.2 and 4.3.

5. Procedure

5.1 Dry the sample to constant weight at a temperature of 230 ± 9 F (110 ± 5 C).

5.2 Nest the sieves in order of decreasing size of opening from top to bottom and place the sample on the top sieve. Agitate the sieves by hand or by mechanical apparatus for a sufficient period, established by trial or checked by measurement on the actual test sample, to meet the criterion for adequacy of sieving described in 5.3.

5.3 Continue sieving for a sufficient period and in such manner that, after completion, not more than 1 weight percent of the residue on any individual sieve will pass that sieve during 1 min of continuous hand sieving performed as follows: Hold the individual sieve, provided with a snug-fitting pan and cover, in a slightly inclined position in one hand. Strike the side of the sieve sharply and with an upward motion against the heel of the other hand at the rate of about 150 times/min, turn the sieve about one sixth of a revolution at intervals of about 25 strokes. In determining sufficiency of sieving for sizes larger than the No. 4 (4.75-mm) sieve, limit the material on the sieve to a single layer of particles. If the size of the mounted testing sieves makes the described sieving motion impractical, use 8-in. (203-mm) diameter sieves to verify the sufficiency of sieving.

5.4 Dry sieving alone is usually satisfactory for routine testing of normally graded aggregates. However, when accurate determination of the total amount passing the No. 200 (75-μm) sieve is desired, first test the sample in accordance with ASTM Method C 117, Test for Materials Finer than No. 200 (75-μm) Sieve in Mineral Aggregates by Washing.[2] Add the percentage finer than the No. 200 sieve determined by that method to the percentage passing the No. 200 sieve by dry sieving of the same sample. After the final drying operation in Method C 117, dry-sieve the sample in accordance with 5.2 and 5.3.

5.5 Determine the weight of each size increment by weighing on a scale or balance conforming to the requirements specified in 3.1, to the nearest 0.1 percent of the weight of the sample.

6. Calculation

6.1 Calculate percentages on the basis of the total weight of the sample, including any material finer than the No. 200 sieve determined in accordance with Method C 117.

7. Report

7.1 The report shall include the following:

7.1.1 Total percentage of material passing each sieve, or

7.1.2 Total percentage of material retained on each sieve, or

7.1.3 Percentage of material retained between consecutive sieves, depending upon the form of the specifications for use of the material under test. Report percentages to the nearest whole number, except for the percentage passing the No. 200 (75-μm) sieve, which shall be reported to the nearest 0.1 percent.

8. Precision

8.1 For coarse aggregate with a nominal maximum size of ¾-in. (19.0 mm) the precision indexes are given in Table 1; the values are given for different ranges of percentage of aggregate retained between two consecutive coarse aggregate sieves:

TABLE 1 Precision

	Percent in Size Fraction Between Two Consecutive Coarse Aggregate Sieves	Coefficient of Variation (1S%), Percent[b]	Standard Deviation (1S), Percent[a]	Difference Between Two Tests	
				Percent of Avg (D2S%)[b]	(D2S) Percent[a]
Multilaboratory	0 to 3	35	–	99	–
	3 to 10	–	1.2	–	3.4
	10 to 20	–	1.7	–	4.8
	20 to 50	–	2.2	–	6.2
Single-Operator	0 to 3	30	–	85	–
	3 to 50	–	1.4	–	4.0

[a] These numbers represent, respectively, the (1S) and (D2S) limits as described in ASTM Recommended Practice C 670 for Preparing Precision Statements for Test Methods for Construction Materials.[3]

[b] These numbers represent, respectively, the (1S%) and (D2S%) limits as described in Recommended Practice C 670.

The American Society for Testing and Materials takes no position respecting the validity of any patent rights asserted in connection with any item mentioned in this standard. Users of this standard are expressly advised that determination of the validity of any such patent rights, and the risk of infringement of such rights, is entirely their own responsibility.

Standard Specification for
PORTLAND CEMENT[1]

This Standard is issued under the fixed designation C 150; the number immediately following the designation indicates the year of original adoption or, in the case of revision, the year of last revision. A number in parentheses indicates the year of last reapproval.

1. Scope

1.1 This specification covers eight types of portland cement, as follows (see Note 1):

1.1.1 *Type I*—For use when the special properties specified for any other type are not required.

1.1.2 *Type IA*—Air-entraining cement for the same uses as Type I, where air-entrainment is desired.

1.1.3 *Type II*—For general use, more especially when moderate sulfate resistance or moderate heat of hydration is desired.

1.1.4 *Type IIA*—Air-entraining cement for the same uses as Type II, where air-entrainment is desired.

1.1.5 *Type III*—For use when high early strength is desired.

1.1.6 *Type IIIA*—Air-entraining cement for the same use as Type III, where air-entrainment is desired.

1.1.7 *Type IV*—For use when a low heat of hydration is desired.

1.1.8 *Type V*—For use when high sulfate resistance is desired.

NOTE 1—Attention is called to the fact that cements conforming to the requirements for all of these types may not be carried in stock in some areas. In advance of specifying the use of other than Type I cement, it should be determined whether the proposed type of cement is or can be made available.

NOTE 2—The values stated in U.S. customary units are to be regarded as the standard.

2. Definitions

2.1 *portland cement*—a hydraulic cement produced by pulverizing clinker consisting essentially of hydraulic calcium silicates, usually containing one or more of the forms of calcium sulfate as an interground addition.

2.2 *air-entraining portland cement*—a hydraulic cement produced by pulverizing clinker consisting essentially of hydraulic calcium silicates, usually containing one or more of the forms of calcium sulfate as an interground addition, and with which there has been interground an air-entraining addition.

3. Basis of Purchase

3.1 The purchaser should specify the type desired, and indicate which, if any, of the optional requirements apply. When the type is not specified, the requirements of Type I shall apply.

4. Additions

4.1 The cement covered by this specification shall contain no addition except as provided for below.

4.1.1 Water or calcium sulfate, or both, may be added in amounts such that the limits shown in Table 1 for sulfur trioxide and loss-on-ignition shall not be exceeded.

4.1.2 At the option of the manufacturer, processing additions may be used in the manufacture of the cement, provided such materials in the amounts used have been shown to meet the requirements of ASTM Specifications C 465, for Processing Additions for Use

[1] This specification is under the jurisdiction of ASTM Committee C-1 on Cement and is the direct responsibility of Subcommittee C01.10 on Portland Cement.

Current edition approved Feb. 25, 1977. Published April 1977. Originally published as C 150 - 40 T. Last previous edition C 150 - 76a.

in the Manufacture of Portland Cement.[2]

4.1.3 Air-entraining portland cement shall contain an interground addition conforming to the requirements of ASTM Specification C 226, for Air-Entraining Additions for Use in the Manufacture of Air-Entraining Portland Cement.[2]

5. Chemical Requirements

5.1 Portland cement of each of the eight types shown in Section 1 shall conform to the respective standard chemical requirements prescribed in Table 1. In addition, optional chemical requirements are shown in Table 1A.

6. Physical Requirements

6.1 Portland cement of each of the eight types shown in Section 1 shall conform to the respective standard physical requirements prescribed in Table 2. In addition, optional physical requirements are shown in Table 2A.

7. Sampling

7.1 When the purchaser desires that the cement be sampled and tested to verify compliance with this specification, sampling should be performed in accordance with Methods C 183, Sampling Hydraulic Cement.[2]

7.2 Methods C 183 are not designed for manufacturing quality control and are not required for manufacturer's certification.

8. Test Methods

8.1 Determine the applicable properties enumerated in this specification in accordance with the following methods:

8.1.1 *Air Content of Mortar*—Method C 185, Test for Air Content of Hydraulic Cement Mortar.[2]

8.1.2 *Chemical Analysis*—Methods C 114, for Chemical Analysis of Hydraulic Cement.[2]

8.1.3 *Strength*—Method C 109, Test for Compressive Strength of Hydraulic Cement Mortars (Using 2-In. Cube Specimens).[2]

8.1.4 *False Set*—Method C 451, Test for False Set of Portland Cement (Paste Method).[2]

8.1.5 *Fineness by Air Permeability*—Method C 204, Test for Fineness of Portland Cement by Air Permeability Apparatus.[2]

8.1.6 *Fineness by Turbidimeter*—Method C 115, Test for Fineness of Portland Cement by the Turbidimeter.[2]

8.1.7 *Heat of Hydration*—Method C 186, Test for Heat of Hydration of Portland Cement.[2]

8.1.8 *Autoclave Expansion*—Method C 151, Test for Autoclave Expansion of Portland Cement.[2]

8.1.9 *Time of Setting by Gillmore Needles*—Method C 266, Test for Time of Setting of Hydraulic Cement by Gillmore Needles.[2]

8.1.10 *Time of Setting by Vicat Needle*—Method C 191, Test for Time of Setting of Hydraulic Cement by Vicat Needle.[2]

8.1.11 *Sulfate Expansion*—Method C 452, Test for Potential Expansion of Portland Cement Mortars Exposed to Sulfate.[2]

9. Inspection

9.1 Inspection of the material shall be made as agreed upon by the purchaser and the seller as part of the purchase contract.

10. Testing Time Requirements

10.1 The following periods from time of sampling shall be allowed for completion of testing:

1-day test	6 days
3-day test	8 days
7-day test	12 days
28-day test	33 days

11. Rejection

11.1 The cement may be rejected if it fails to meet any of the requirements of this specification.

11.2 Cement remaining in bulk storage at the mill, prior to shipment, for more than 6 months, or cement in bags in local storage in the hands of a vendor for more than 3 months, after completion of tests, may be retested before use and may be rejected if it fails to conform to any of the requirements of this specification.

11.3 Packages varying more than 3 percent from the weight marked thereon may be rejected; and if the average weight of packages in any shipment, as shown by weighing 50 packages taken at random, is less than that marked on the packages, the entire shipment may be rejected.

[2] *Annual Book of ASTM Standards*, Part 13.

12. Manufacturer's Statement

12.1 At the request of the purchaser, the manufacturer shall state in writing the nature, amount, and identity of the air-entraining agent used, and of any processing addition that may have been used, and also, if requested, shall supply test data showing compliance of such air-entraining addition with the provisions of Specification C 226, and of any such processing addition with Specification C 465.

13. Packaging and Marking

13.1 When the cement is delivered in packages, the words "Portland Cement," the type of cement, the name and brand of the manufacturer, and the weight of the cement contained therein shall be plainly marked on each package. When the cement is an air-entraining type, the words "air-entraining" shall be plainly marked on each package. Similar information shall be provided in the shipping advices accompanying the shipment of packaged or bulk cement. All packages shall be in good condition at the time of inspection.

14. Storage

14.1 The cement shall be stored in such a manner as to permit easy access for proper inspection and identification of each shipment, and in a suitable weather-tight building that will protect the cement from dampness and minimize warehouse set.

15. Manufacturer's Certification

15.1 Upon request of the purchaser in the contract or order, a manufacturer's certification that the material was tested during production or transfer in accordance with this specification together with a report of the test results shall be furnished at the time of shipment.

TABLE 1 Standard Chemical Requirements

Cement Type[a]	I and IA	II and IIA	III and IIIA	IV	V
Silicon dioxide (SiO_2), min, percent	...	21.0	...	...	...
Aluminum oxide (Al_2O_3), max, percent	...	6.0	...	...	...
Ferric oxide (Fe_2O_3), max, percent	...	6.0	...	6.5	...
Magnesium oxide (MgO), max, percent	6.0	6.0	6.0	6.0	6.0
Sulfur trioxide (SO_3),[b] max, percent					
When ($3CaO \cdot Al_2O_3$)[c] is 8 percent or less	3.0	3.0	3.5	2.3	2.3
When ($3CaO \cdot Al_2O_3$)[c] is more than 8 percent	3.5	[d]	4.5	[d]	[d]
Loss on ignition, max, percent	3.0	3.0	3.0	2.5	3.0
Insoluble residue, max, percnet	0.75	0.75	0.75	0.75	0.75
Tricalcium silicate ($3CaO \cdot SiO_2$)[c] max, percent	...	...	...	35	...
Dicalcium silicate ($2CaO \cdot SiO_2$)[c] min, percent	...	...	...	40	...
Tricalcium aluminate ($3CaO \cdot Al_2O_3$)[c] max, percent	...	8	15	7	5
Tetracalcium aluminoferrite plus twice the tricalcium aluminate[c] ($4CaO \cdot Al_2O_3 \cdot Fe_2O_3$ + 2($3CaO \cdot Al_2O_3$)), or solid solution ($4CaO \cdot Al_2O_3 \cdot Fe_2O_3$ + $2CaO \cdot Fe_2O_3$), as applicable, max, percent	...	...	...	...	20

[a] See Note 1.

[b] There are cases where optimum SO_3 for a particular cement exceeds the limit in this specification. Where it has been demonstrated by Method C 563, Test for Optimum SO_3 in Portland Cement,[2] that this condition exists, an additional amount of SO_3, in no case more than 0.5 weight percent of cement, is permissible provided that, when the cement with the additional calcium sulfate is tested by Method C 265, Test for Calcium Sulfate in Hydrated Portland Cement Mortar,[2] the calcium sulfate in the hydrated mortar at 24 ± ¼ h expressed as SO_3 does not exceed 0.50 g/litre. When the manufacturer supplies cement under this provision, he will, upon request, supply supporting data to the purchaser.

[c] The expressing of chemical limitations by means of calculated assumed compounds does not necessarily mean that the oxides are actually or entirely present as such compounds.

When the ratio of percentages of aluminum oxide to ferric oxide is 0.64 or more, the percentages of tricalcium silicate, dicalcium silicate, tricalcium aluminate, and tetracalcium aluminoferrite shall be calculated from the chemical analysis as follows:

Tricalcium silicate = (4.071 × percent CaO) − (7.600 × percent SiO_2) − (6.718 × percent Al_2O_3) − (1.430 × percent Fe_2O_3) − (2.852 × percent SO_3)

Dicalcium silicate = (2.867 × percent SiO_2) − (0.7544 × percent C_3S)

Tricalcium aluminate = (2.650 × percent Al_2O_3) − (1.692 × percent Fe_2O_3)

Tetracalcium aluminoferrite = 3.043 × percent Fe_2O_3

When the alumina-ferric oxide ratio is less than 0.64, a calcium aluminoferrite solid solution (expressed as ss(C_4AF + C_2F)) is formed. Contents of this solid solution and of tricalcium silicate shall be calculated by the following formulas:

ss(C_4AF + C_2F)) = (2.100 × percent Al_2O_3) + (1.702 × percent Fe_2O_3)

Tricalcium silicate = (4.071 × percent CaO) − (7.600 × percent SiO_2) − (4.479 × percent Al_2O_3) − (2.859 × percent Fe_2O_3) − (2.852 × percent SO_3).

No tricalcium aluminate will be present in cements of this composition. Dicalcium silicate shall be calculated as previously shown.

In the calculation of C_3A, the values of Al_2O_3 and Fe_2O_3 determined to the nearest 0.01 percent shall be used. In the calculation of other compounds the oxides determined to the nearest 0.1 percent shall be used.

All values calculated as described in this note shall be reported to the nearest 1 percent.

[d] Not applicable.

TABLE 1A Optional Chemical Requirements

NOTE—These optional requirements apply only when specifically requested

Cement Type[a]	I and IA	II and IIA	III and IIIA	IV	V	Remarks
Tricalcium aluminate ($3CaO \cdot Al_2O_3$),[b] max, percent	...	...	8	...	..	for moderate sulfate resistance
Tricalcium aluminate ($3CaO \cdot Al_2O_3$),[b] max, percent	...	...	5	...	...	for high sulfate resistance
Sum of tricalcium silicate and tricalcium aluminate,[b] max, percent	...	58[c]	...	...	...	for moderate heat of hydration
Alkalies (Na_2O + $O \cdot 658K_2O$), max, percent	0.60[d]	0.60[d]	0.60[d]	0.60[d]	0.60[d]	low-alkali cement

[a] See Note 1.

[b] The expressing of chemical limitations by means of calculated assumed compounds does not necessarily mean that the oxides are actually or entirely present as such compounds.

When the ratio of percentages of aluminum oxide to ferric oxide is 0.64 or more, the percentages of tricalcium silicate, dicalcium silicate, tricalcium aluminate and tetracalcium aluminoferrite shall be calculated from the chemical analysis as follows:

Tricalcium silicate = (4.071 × percent CaO) − (7.600 × percent SiO_2) − (6.718 × percent Al_2O_3) − (1.430 × percent Fe_2O_3) − (2.852 × percent SO_3)

Dicalcium silicate = (2.867 × percent SiO_2) − (0.7544 × percent C_3S)

Tricalcium aluminate = (2.650 × percent Al_2O_3) − (1.692 × percent Fe_2O_3)

Tetracalcium aluminoferrite = 3.043 × percent Fe_2O_3

When the alumina-ferric oxide ratio is less than 0.64, a calcium aluminoferrite solid solution (expressed as ss ($C_4AF + C_2F$)) is formed. Contents of this solid solution and of tricalcium silicate shall be calculated by the following formulas:

ss($C_4AF + C_2F$)) = (2.100 × percent Al_2O_3) + (1.702 × percent Fe_2O_3)

Tricalcium silicate = (4.071 × percent CaO) − (7.600 × percent SiO_2) − (4.479 × percent Al_2O_3) − (2.859 × percent Fe_2O_3) − (2.852 × percent SO_3).

No tricalcium aluminate will be present in cements of this composition. Dicalcium silicate shall be calculated as previously shown.

In the calculation of C_3A, the values of Al_2O_3 and Fe_2O_3 determined to the nearest 0.01 percent shall be used. In the calculation of other compounds the oxides determined to the nearest 0.1 percent shall be used.

All values calculated as described in this note shall be reported to the nearest 1 percent.

[c] This limit applies when moderate heat of hydration is required and tests for heat of hydration are not requested.

[d] This limit may be specified when the cement is to be used in concrete with aggregates that may be deleteriously reactive. Reference should be made to ASTM Specifications C 33, for Concrete Aggregates, *Annual Book of ASTM Standards*, Part 14, for suitable criteria of deleterious reactivity.

TABLE 2 **Standard Physical Requirements**

Cement Type[a]	I	IA	II	IIA	III	IIIA	IV	V
Air content of mortar,[b] volume percent:								
max	12	22	12	22	12	22	12	12
min	...	16	...	16	...	16	...	...
Fineness, specific surface, cm^2/g (alternative methods):[c]								
Turbidimeter test, min	1600	1600	1600	1600	...	...	1600	1600
Air permeability test, min	2800	2800	2800	2800	...	...	2800	2800
Autoclave expansion, max, percent	0.80	0.80	0.80	0.80	0.80	0.80	0.80	0.80
Strength, not less than the values shown for the ages indicated below:[d]								
Compressive strength, psi (MPa)								
1 day	...	...	...	...	1800 (12.4)	1450 (10.0)	...	...
3 days	1800 (12.4)	1450 (10.0)	1500 (10.3) 1000[f] (6.9)[f]	1200 (8.3) 800[f] (5.5)[f]	3500 (24.1)	2800 (19.3)	...	1200 (8.3)
7 days	2800 (19.3)	2250 (15.5)	2500 (17.2) 1700[f] (11.7)[f]	2000 (13.8) 1350[f] (9.3)[f]	...	...	1000 (6.9)	2200 (15.2)
28 days	...	...	...	...	...	...	2500 (17.2)	3000 (20.7)
Time of setting (alternative methods):[e]								
Gillmore test:								
Initial set, min, not less than	60	60	60	60	60	60	60	60
Final set, h, not more than	10	10	10	10	10	10	10	10
Vicat test:								
Initial set, min, not less than	45	45	45	45	45	45	45	45
Final set, h, not more than	8	8	8	8	8	8	8	8

[a] See Note 1.

[b] Compliance with the requirements of this specification does not necessarily ensure that the desired air content will be obtained in concrete.

[c] Either of the two alternative fineness methods may be used at the option of the testing laboratory. However, in case of dispute, or when the sample fails to meet the requirements of the air-permeability test, the turbidimeter test shall be used, and the requirements in this table for the turbidimetric method shall govern.

[d] The strength at any age shall be higher than the strength at any preceding age.

[e] The purchaser should specify the type of setting-time test required. In case he does not so specify, or in case of dispute, the requirements of the Vicat test only shall govern.

[f] When the optional heat of hydration or the chemical limit on the sum of the tricalcium silicate and tricalcium aluminate is specified.

TABLE 2A Optional Physical Requirements

NOTE—These optional requirements apply only when specifically requested.

Cement Type[a]	I	IA	II	IIA	III	IIIA	IV	V
False set, final penetration, min, percent	50	50	50	50	50	50	50	50
Heat of hydration:								
7 days, max, cal/g (kJ/kg)	...	...	70 (290)[b]	70 (290)[b]	...	...	60 (250)	...
28 days, max, cal/g (kJ/kg)	...	...	80 (330)[b]	80 (330)[b]	...	...	70 (290)	...
Strength, not less than the values shown:								
Compressive strength, psi (MPa)								
7 days	...	...	...	...	[c]	[c]	...	...
28 days	4000 (27.6)	3200 (22.1)	4000 (27.6) 3200[b] (22.1)[b]	3200 (22.1) 2560[b] (17.7)[b]	...	...	...	...
Sulfate expansion,[d] 14 days, max, percent	...	...	...	...	...	...	...	0.045

[a] See Note 1.

[b] When the heat of hydration requirements are specified, the sum of the tricalcium silicate and tricalcium aluminate shall not be specified. These strength requirements apply when either heat of hydration requirements or the sum of tricalcium silicate and tricalcium aluminate are specified.

[c] The strength at any age shall be higher than the strength at any preceding age.

[d] When the sulfate expansion is specified, it shall be instead of the limits of C_3A and $C_4AF + 2\ C_3A$ listed in Table 1.

The American Society for Testing and Materials takes no position respecting the validity of any patent rights asserted in connection with any item mentioned in this standard. Users of this standard are expressly advised that determination of the validity of any such patent rights, and the risk of infringement of such rights, is entirely their own responsibility.

ANSI/ASTM D 198 - 76

AMERICAN NATIONAL STANDARD

Standard Methods of STATIC TESTS OF TIMBERS IN STRUCTURAL SIZES[1]

This Standard is issued under the fixed designation D 198; the number immediately following the designation indicates the year of original adoption or, in the case of revision, the year of last revision. A number in parentheses indicates the year of last reapproval.

INTRODUCTION

Numerous evaluations of structural members of solid sawn timber have been conducted according to ASTM Standard Designation D 198 - 27. While the importance of continued use of a satisfactory standard should not be under-estimated, the original standard (1927) was designed primarily for sawn material such as solid wood bridge stringers and joists. With the advent of laminated timbers, wood-plywood composite members, and even reinforced and prestressed timbers, a procedure adaptable to a wider variety of wood structural members is required.

The present standard expands the original standard to permit its application to wood members of all types. It provides methods of evaluation under loadings other than flexure in recognition of the increasing need for improved knowledge of properties under such loadings as tension to reflect the increasing use of dimension lumber in the lower chords of trusses. The standard establishes practices which will permit correlation of results from different sources through the use of a uniform procedure. Provision is made for varying the procedure to take account of special problems.

FLEXURE

1. Scope

1.1 This method covers the determination of the flexural properties of structural beams made of solid or laminated wood, or of composite constructions. The method is intended primarily for beams of rectangular cross section but is also applicable to beams of round and irregular shapes, such as round posts, I-beams, or other special sections.

2. Summary of Method

2.1 The structural member, usually a straight or a slightly cambered beam of rectangular cross section, is subjected to a bending moment by supporting it near its ends, at locations called reactions, and applying transverse loads symmetrically imposed between these reactions. The beam is deflected at a prescribed rate, and coordinate observations of loads and deflections are made until rupture occurs.

3. Significance

3.1 The flexural properties established by this method provide:

3.1.1 Data for use in development of grading rules and specifications.

3.1.2 Data for use in development of working stresses for structural members.

3.1.3 Data on the influence of imperfections on mechanical properties of structural members.

3.1.4 Data on strength properties of different species or grades in various structural sizes.

[1] These methods are under the jurisdiction of ASTM Committee D-7 on Wood and are under the jurisdiction of Subcommittee D07.09 on Methods of Testing.

Current edition approved Aug. 27, 1976. Published October 1976. Originally published as D 198 - 24. Last previous edition D 198 - 67 (1974).

3.1.5 Data for use in checking existing formulas or hypotheses relating to the structural behavior of beams.

3.1.6 Data on the effects of chemical or environmental conditions on mechanical properties.

3.1.7 Data on effects of fabrication variables such as depth, taper, notches, or type of end joint in laminations.

3.1.8 Data on relationships between mechanical and physical properties.

3.2 Procedures are described here in sufficient detail to permit duplication in different laboratories so that comparisons of results from different sources will be valid. Special circumstances may require deviation from some details of these procedures. Any variations shall be carefully described in the report (see Section 9).

4. Definitions

4.1 See ASTM Definitions E 6, Terms Relating to Methods of Mechanical Testing,[2] ASTM Definitions D 9, Terms Relating to Timber,[3] and ASTM Nomenclature D 1165, of Domestic Hardwoods and Softwoods.[3] A few related terms not covered in the above methods are as follows:

4.1.1 *span*—the total distance between reactions on which a beam is supported to accommodate a transverse load (Fig. 1).

4.1.2 *shear span*—two times the distance between a reaction and the nearest load point for a symmetrically loaded beam (Fig. 1).

4.1.3 *depth of beam*—that dimension of the beam which is perpendicular to the span and parallel to the direction in which the load is applied (Fig. 1).

4.1.4 *span-depth ratio*—the numerical ratio of total span divided by beam depth.

4.1.5 *shear span-depth ratio*—the numerical ratio of shear span divided by beam depth.

4.1.6 *structural wood beam*—solid wood, laminated wood, or composite structural members for which strength, stiffness, or both are primary criteria for the intended application and which usually are used in full length and in cross-sectional sizes greater than nominal 2 by 2 in. (5 by 5 cm).

4.1.7 *composite wood beam*—a laminar construction comprising a combination of wood and other simple or complex materials assembled and intimately fixed in relation to each other so as to use the properties of each to attain specific structural advantage for the whole assembly.

5. Apparatus

5.1 *Testing Machine*—A device that provides (*1*) a rigid frame to support the specimen yet permit its deflection without restraint, (*2*) a loading head through which the force is applied without high stress concentrations in the beam, and (*3*) a force-measuring device that is calibrated to ensure accuracy according to ASTM Methods E 4, Verification of Testing Machines.[2]

5.2 *Support Apparatus:*

5.2.1 *Reaction Bearing Plates*—The beam shall be supported by metal bearing plates to prevent damage to the beam at the point of contact between beam and reaction support (Fig. 1). The size of the bearing plates may vary with the size and shape of the beam. For rectangular beams as large as 12 in. (30 cm) deep by 6 in. (15 cm) wide, the recommended size of bearing plate is $^1/_2$ in. (1.3 cm) thick by 6 in. (15 cm) lengthwise and extending entirely across the width of the beam.

5.2.2 *Reaction Bearing Roller*—The bearing plates shall be supported by either rollers and a fixed knife edge reaction or a rocker type-knife edge reaction so that shortening and rotation of the beam about the reaction due to deflection will be unrestricted (Fig. 1).

5.2.3 *Reaction Bearing Alignment*—Provisions shall be made at the reaction to allow for initial twist in the length of the beam. If the bearing surfaces of the beam at its reactions are not parallel, the beam shall be shimmed or the individual bearing plates shall be rotated about an axis parallel to the span to provide full bearing across the width of the specimen (Fig. 2).

5.2.4 *Lateral Support*—Specimens that have a depth-to-width ratio of three or greater are subject to lateral instability during loading, thus requiring lateral support. Support shall be provided at least at points located about half-way between the reaction and the load point. Additional supports may be used as

[2] *Annual Book of ASTM Standards*, Part 10.

[3] *Annual Book of ASTM Standards*, Part 22.

required. Each support shall allow vertical movement without frictional restraint but shall restrict lateral deflection (Fig. 3).

5.3 *Load Apparatus:*

5.3.1 *Load Bearing Blocks*—The load shall be applied through bearing blocks (Fig. 1) across the full beam width which are of sufficient thickness to eliminate high stress concentrations at places of contact between beam and bearing blocks. The loading surface of the blocks shall have a radius of curvature equal to two to four times the beam depth for a chord length at least equal to the depth of the beam. Load shall be applied to the blocks in such a manner that the blocks may rotate about an axis perpendicular to the span (Fig. 4). Provisions such as rotatable bearings or shims shall be made to ensure full contact between the beam and both loading blocks. Metal bearing plates and rollers shall be used in conjunction with one load bearing block to permit beam deflection without restraint (Fig. 4). The size of these plates and rollers may vary with the size and shape of the beam, the same as for the reaction bearing plates. Beams having circular or irregular cross sections shall have bearing blocks which distribute the load uniformly to the bearing surface and permit, unrestrained deflections.

5.3.2 *Load Points*—The total load on the beam shall be applied equally at two points equidistant from the reactions. The two load points will normally be at a distance from their reaction equal to one third of the span, but for special purposes other distances may be specified.

NOTE 1—One of the objectives of two-point loading is to subject the portion of the beam between load points to a uniform bending moment, free of shear, and with comparatively small loads at the load points. For example, loads applied at one-third span length from reactions would be less than if applied at one-fourth span length from reaction to develop a moment of similar magnitude. When loads are applied at the one-third points the moment distribution of the beam simulates that for loads uniformly distributed across the span to develop a moment of similar magnitude. If loads are applied at the outer one-fourth points of the span, the maximum moment and shear are the same as the maximum moment and shear for the same total load uniformly distributed across the span.

5.4 *Deflection Apparatus:*

5.4.1 *General*—For either apparent or true modulus of elasticity calculations, devices shall be provided by which the deflection of the neutral axis of the beam at the center of the span is measured with respect to either the reaction or between cross sections free of shear deflections. For calculations of approximate values of modulus of rigidity two devices are necessary, one device to measure deflection at the center of span with respect to the reaction and a second device to measure deflection at the center of the span with respect to a middle span defined as a distance, l, which is smaller than the distance between load points by two beam depths [$l \approx L - 2a - 2h$]. (See flexural notations for meaning of symbols.)

5.4.2 *Wire Deflectometer*—Deflection may be read directly by means of a wire stretched taut between two nails driven into the neutral axis of the beam directly above the reactions and extending across a scale attached at the neutral axis of the beam at midspan. Deflections may be read with a telescope or reading glass to magnify the area where the wire crosses the scale. When a reading glass is used, a reflective surface placed adjacent to the scale will help to avoid parallax.

5.4.3 *Yoke Deflectometer*—A satisfactory device commonly used for short, small beams or to measure deflection of the center of the beam with respect to any point along the neutral axis consists of a lightweight U-shaped yoke suspended between nails driven into the beam at its neutral axis and a dial micrometer attached to the center of the yoke with its stem attached to a nail driven into the beam at midspan at the neutral axis. Further modification of this device may be attained by replacing the dial micrometer with a deflection transducer for automatic recording (Fig. 4).

5.4.4 *Accuracy*—The devices shall be such as to permit measurements to the nearest 0.01 in. (0.02 cm) on spans greater than 3 ft (0.9 m) and 0.001 in. (0.002 cm) on spans less than 3 ft (0.9 m), except that deflections measured over specified distance, l, shall be measured to the nearest 0.0005 in. (0.001 cm).

6. Test Specimen

6.1 *Material*—The test specimen shall consist basically of a beam which may be solid wood, laminated wood, or a composite construction of wood or of wood combined with plastics or metals in sizes that are usually used in structural applications.

6.2 *Identification*—Material or materials of the test specimen shall be identified as fully as possible by including the origin or source of supply, species, and history of drying and conditioning, chemical treatment, fabrication, and other pertinent physical or mechanical details which may affect the strength. Details of this information shall depend on the material or materials in the beam. For example, the solid wooden beams would be identified by the character of the wood, that is, species, source, etc., whereas composite wooden beams would be identified by the characteristics of the dissimilar materials and their size and location in the beam.

6.3 *Specimen Measurements*—The weight and dimensions as well as moisture content of the specimen shall be accurately determined before test. Weights and dimensions (length and cross section) shall be measured to three significant figures. Sufficient measurements of the cross section shall be made along the length of the beam to describe the width and depth of rectangular specimen and to accurately describe the critical section or sections of nonuniform beams. The physical characteristics of the specimen as described by its density and moisture content may be determined according to ASTM Methods D 2395, Determining the Specific Gravity of Wood and Wood-Base Materials[3] and Method A of Methods D 2016, Test for Moisture Content of Wood.[3]

6.4 *Specimen Description*—The inherent imperfections or intentional modifications of the composition of the beam shall be fully described by recording the size and location of such factors as knots, checks, and reinforcements. Size and location of intentional modifications such as placement of laminations, glued joints, and reinforcing steel shall be recorded during the fabrication process. The size and location of imperfections in the interior of any beam must be deduced from those on the surface, especially in the case of large sawn members. A sketch or photographic record shall be made of each face and the ends showing the size, location, and type of growth characteristics, including slope of grain, knots, distribution of sapwood and heartwood, location of pitch pockets, direction of annual rings, and such abstract factors as crook, bow, cup, or twist which might affect the strength of the beam.

6.5 *Rules for Determination of Specimen Length*—The cross-sectional dimensions of solid wood structural beams and composite wooden beams usually have established sizes, depending upon the manufacturing process and intended use, so that no modification of these dimensions is involved. The length, however, will be established by the type of data desired. The span length is determined from knowledge of beam depth, the distance between load points, as well as the type and orientation of material in the beam. The total beam length shall also include an overhang or extension beyond each reaction support so that the beam can accommodate the bearing plates and rollers and will not slip off the reactions during test.

NOTE 2—Some evaluations will require simulation of a specific design condition where nonnormal overhang is involved. In such instances the report shall include a complete description of test conditions, including overhang at each support.

6.5.1 The span length of beams intended primarily for evaluation of shear properties shall be such that the shear span is relatively short. Beams of wood of uniform rectangular cross section having the ratio of a/h less than five are in this category and provide a high percentage of shear failures.

NOTE 3—If approximate values of modulus of rupture S_R and shear strength τ_m are known, a/h values should be less than $S_R/4\tau_m$, assuming that when $a/h = S_R/4\tau_m$ the beam will fail at the same load in either shear or in extreme outer fibers.

6.5.2 The span length of beams intended primarily for evaluation of flexural properties shall be such that the shear span is relatively long. Beams of wood of uniform rectangular cross section having a/h ratios of from 5:1 to 12:1 are in this category.

NOTE 4—The a/h values should be somewhat greater than $S_R/4\tau_m$ so that the beams do not fail in shear but should not be so large that beam deflections cause sizable thrust at reactions and thrust values need to be taken into account. A suggested range of a/h values is between approximately 0.5 S_R/τ_m and 1.2 S_R/τ_m. In this category, shear distortions affect the total deflection, so that flexural properties may be corrected by formulae provided in the Appendix.

6.5.3 The span length of beams intended

primarily for evaluation of only the deflection of specimen due to bending moment shall be such that the shear span is long. Wood beams of uniform rectangular cross section in this category have a/h ratios greater than 12:1.

NOTE 5—The shear stresses and distortions are assumed to be small so that they can be neglected; hence the a/h ratio is suggested to be greater than S_R/τ_m.

7. Procedure

7.1 *Conditioning*—Unless otherwise indicated in the research program or material specification, condition the test specimen to constant weight so it is in moisture equilibrium under the desired environmental conditions. Approximate moisture contents with moisture meters or measure more accurately by weights of samples according to Method A of Methods D 2016.

7.2 *Test Setup*—Determine the size of the specimen, the span, and the shear span in accordance with 5.3.2 and 6.5. Locate the beam symmetrically on its supports with load bearing and reaction bearing blocks as described in 5.2 to 5.4. The beams shall be adequately supported laterally according to 5.2.4. Set apparatus for measuring deflections in place (see 5.4). Full contact shall be attained between support bearings, loading blocks, and the beam surface.

7.3 *Speed of Testing*—Conduct the test at a constant rate to achieve maximum load in about 10 min, but maximum load should be reached in not less than 6 min nor more than 20 min. A constant rate of outer strain, z, of 0.0010 in./in · min (0.0010 cm/cm · min) will usually permit the tests of wood members to be completed in the prescribed time. The rate of motion of the movable head of the test machine corresponding to this suggested rate of strain when two symmetrical concentrated loads are employed may be computed from the following equation:

$$N = Za(3L - 4a)/3h$$

7.4 *Load-Deflection Curves:*

7.4.1 Obtain load-deflection data with apparatus described in 5.4.1. Note the load and deflection at first failure, at the maximum load, and at points of sudden change. Continue loading until complete failure or an arbitrary terminal load has been reached.

7.4.2 If additional deflection apparatus is provided to measure deflection over a second distance, l, according to 5.4.1, such load-deflection data shall be obtained only up to the proportional limit.

7.5 *Record of Failures*—Describe failures in detail as to type, manner and order of occurrence, and position in beam. Record descriptions of the failures and relate them to drawings or photographs of the beam referred to in 6.4. Also record notations as the the order of their occurrence on such references. Hold the section of the beam containing the failure for examination and reference until analysis of the data has been completed.

8. Calculations

8.1 Compute physical and mechanical properties and their appropriate adjustments for the beam according to the relationships in the Appendix.

9. Report

9.1 The report shall include the following:

9.1.1 Complete identification of the solid wood or composite construction, including species, origin, shape and form, fabrication procedure, type and location of imperfections or reinforcements, and pertinent physical or chemical characteristics relating to the quality of the material,

9.1.2 History of seasoning and conditioning,

9.1.3 Loading conditions to portray the load, support mechanics, lateral supports, if used, and type of equipment,

9.1.4 Deflection apparatus,

9.1.5 Depth and width of the specimen or pertinent cross-sectional dimensions,

9.1.6 Span length and shear span distance,

9.1.7 Rate of load application,

9.1.8 Computed physical and mechanical properties, including specific gravity and moisture content, flexural strength, stress at proportional limit, modulus of elasticity, and a statistical measure of variability of these values,

9.1.9 Data for composite beams include shear and bending moment values and deflections,

9.1.10 Description of failure, and

9.1.11 Details of any deviations from the prescribed or recommended methods as outlined in the standard.

COMPRESSION PARALLEL TO GRAIN (SHORT COLUMN, NO LATERAL SUPPORT, $l/r < 17$)

10. Scope

10.1 This method covers the determination of the compressive properties of elements taken from structural members made of solid or laminated wood, or of composite constructions when such an element has a slenderness ratio (length to least radius of gyration) of less than 17. The method is intended primarily for members of rectangular cross section but is also applicable to irregularly shaped studs, braces, chords, round posts, or special sections.

11. Summary of Method

11.1 The structural member is subjected to a force uniformly distributed on the contact surface of the specimen in a direction generally parallel to the longitudinal axis of the wood fibers, and the force generally is uniformly distributed throughout the specimen during loading to failure without flexure along its length.

12. Significance

12.1 The compressive properties obtained by axial compression will provide information similar to that stipulated for flexural properties under Section 3.

12.2 The compressive properties parallel to grain include modulus of elasticity, stress at proportional limit, compressive strength, and strain data beyond proportional limit.

13. Definitions

13.1 See 4.1.

14. Apparatus

14.1 *Testing Machine*—Any device having the following is suitable:

14.1.1 *Drive Mechanism*—A drive mechanism for imparting to a movable loading head a uniform, controlled velocity with respect to the stationary base.

14.1.2 *Load Indicator*—A load-indicating mechanism capable of showing the total compressive force on the specimen. This force-measuring system shall be calibrated to ensure accuracy according to ASTM Methods E 4, Verification of Testing Machines.[2]

14.2 *Bearing Blocks*—Bearing blocks shall be used to apply the load uniformly over the two contact surfaces and to prevent eccentric loading on the specimen. At least one spherical bearing block shall be used to ensure uniform bearing. Spherical bearing blocks may be used on either or both ends of the specimen, depending on the degree of parallelism of bearing surfaces (Fig. 5). The radius of the sphere shall be as small as practicable, in order to facilitate adjustment of the bearing plate to the specimen, and yet large enough to provide adequate spherical bearing area. This radius is usually one to two times the greatest cross section dimension. The center of the sphere shall be on the plane of the specimen contact surface. The size of the compression plate shall be larger than the contact surface. It has been found convenient to provide an adjustment for moving the specimen on its bearing plate with respect to the center of spherical rotation to ensure axial loading.

14.3 *Compressometer:*

14.3.1 *Gage Length*—For modulus of elasticity calculations, a device shall be provided by which the deformation of the specimen is measured with respect to specific paired gage points defining the gage length. To obtain test data representative of the test material as a whole, such paired gage points shall be located symmetrically on the lengthwise surface of the specimen as far apart as feasible, yet at least one times the larger cross-sectional dimension from each of the contact surfaces. At least two pairs of such gage points on diametrically opposite sides of the specimen shall be used to measure the average deformation.

14.3.2 *Accuracy*—The device shall be able to measure changes in deformation to three significant figures. Since gage lengths vary over a wide range, the measuring instruments should conform to their oppropriate class in ASTM Methods E 83, Verification and Classification of Extensometers.[2]

15. Test Specimen

15.1 *Material*—The test specimen shall

consist basically of a structural timber which may be solid wood, laminated wood, or a composite construction of wood or of wood combined with plastics or metals in sizes that are commercially used in structural applications, that is in sizes greater than nominal 2 by 2-in. (5 by 5-cm) cross section (see 4.1.6).

15.2 *Identification*—Material or materials of the test specimen shall be as fully described as that for beams in 6.2.

15.3 *Specimen Dimensions*—The weight and dimensions, as well as moisture content of the specimen, shall be accurately measured before test. Weights and dimensions (length and cross section) shall be measured to three significant figures. Sufficient measurements of the cross section shall be made along the length of the specimen to describe shape characteristics and to determine the smallest section. The physical characteristics of the specimen, as described by its density and moisture content, may be determined in accordance with Methods D 2395 and Method A of D 2016, respectively.

15.4 *Specimen Description*—The inherent imperfections and intentional modifications shall be described as for beams in 6.4.

15.5 *Specimen Length*—The length of the specimen shall be such that the compressive force continues to be uniformly distributed throughout the specimen during loading—hence no flexure occurs. To meet this requirement, the specimen shall be a short column having a maximum length, l, less than 17 times the least radius of gyration, r, of the cross section of the specimen (see compressive notations). The minimum length of the specimen for stress and strain measurements shall be greater than three times the larger cross section dimension or about ten times the radius of gyration.

16. Procedure

16.1 *Conditioning*—Unless otherwise indicated in the research program or material specification, condition the test specimen to constant weight so it is at moisture equilibrium, under the desired environment. Approximate moisture contents with moisture meters or measure more accurately by weights of samples according to Method A of Methods D 2016.

16.2 *Test Setup:*

16.2.1 *Bearing Surfaces*—After the specimen length has been calculated according to 15.5, cut the specimen to the proper length so that the contact surfaces are plane, parallel to each other, and normal to the long axis of the specimen. Furthermore, the axis of the specimen shall be generally parallel to the fibers of the wood.

NOTE 6—A sharp fine-toothed saw of either the crosscut or "novelty" crosscut type has been used satisfactorily for obtaining the proper end surfaces. Power equipment with accurate table guides is especially recommended for this work.

NOTE 7—It is desirable to have failures occur in the body of the specimen and not adjacent to the contact surface. Therefore, the cross-sectional areas adjacent to the loaded surface may be reinforced.

16.2.2 *Centering*—First geometrically center the specimens on the bearing plates and then adjust the spherical seats so that the specimen is loaded uniformly and axially.

16.3 *Speed of Testing*—For measuring load-deformation data, apply the load at a constant rate of head motion so that the fiber strain is 0.001 in./in · min ± 25 percent (0.001 cm/cm · min). For measuring only compressive strength, the test may be conducted at a constant rate to achieve maximum load in about 10 min, but not less than 5 nor more than 20 min.

16.4 *Load-Deformation Curves*—If load-deformation data have been obtained, note the load and deflection at first failure, at changes in slope of curve, and at maximum load.

16.5 *Records*—Record the maximum load, as well as a description and sketch of the failure relating the latter to the location of imperfections in the specimen. Re-examine the section of the specimen containing the failure during analysis of the data.

17. Calculations

17.1 Physical and mechanical properties shall be computed in accordance with ASTM Definitions E 6, Terms Relating to Methods of Mechanical Testing,[2] and as follows (see compressive notations):

17.1.1 Stress at proportional limit = P'/A in pounds per square inch (kilograms per square centimeter).

17.1.2 Compressive strength = P/A in pounds per square inch (kilograms per square centimeter).

17.1.3 Modulus of elasticity = $P'/A\epsilon$ in pounds per square inch (kilograms per square centimeter).

18. Report

18.1 The report shall include the following:

18.1.1 Complete identification,

18.1.2 History of seasoning and conditioning,

18.1.3 Load apparatus,

18.1.4 Deflection apparatus,

18.1.5 Length and cross-section dimensions,

18.1.6 Gage length,

18.1.7 Rate of load application,

18.1.8 Computed physical and mechanical properties, including specific gravity and moisture content, compressive strength, stress at proportional limit, modulus of elasticity, and a statistical measure of variability of these values,

18.1.9 Description of failure, and

18.1.10 Details of any deviations from the prescribed or recommended methods as outlined in the standard.

COMPRESSION PARALLEL TO GRAIN (CRUSHING STRENGTH OF LATERALLY SUPPORTED LONG MEMBER, EFFECTIVE $l'/r < 17$)

19. Scope

19.1 This method covers the determination of the compressive properties of structural members made of solid or laminated wood, or of composite constructions when such a member has a slenderness ratio (length to least radius of gyration) of more than 17, and when such a member is to be evaluated in full size but with lateral supports which are spaced to produce an effective slenderness ratio, l'/r, of less than 17. The method is intended primarily for members of rectangular cross section but is also applicable to irregularly shaped studs, braces, chords, round posts, or special sections.

20. Summary of Method

20.1 The structural member is subjected to a force uniformly distributed on the contact surface of the specimen in a direction generally parallel to the longitudinal axis of the wood fibers, and the force generally is uniformly distributed throughout the specimen during loading to failure without flexure along its length.

21. Significance

21.1 The compressive properties obtained by axial compression will provide information similar to that stipulated for flexural properties under Section 3.

21.2 The compressive properties parallel to grain include modulus of elasticity, stress at proportional limit, compressive strength, and strain data beyond proportional limit.

22. Definitions

22.1 See 4.1.

23. Apparatus

23.1 *Testing Machine*—Any device having the following is suitable:

23.1.1 *Drive Mechanism*—A drive mechanism for imparting to a movable loading head a uniform, controlled velocity with respect to the stationary base.

23.1.2 *Load Indicator*—A load-indicating mechanism capable of showing the total compressive force on the specimen. This force-measuring system shall be calibrated to ensure accuracy according to Methods E 4.

23.2 *Bearing Blocks*—Bearing blocks shall be used to apply the load uniformly over the two contact surfaces and to prevent eccentric loading on the specimen. One spherical bearing block shall be used to ensure uniform bearing, or a rocker-type bearing block shall be used on each end of the specimen with their axes of rotation at 0 deg to each other (Fig. 6). The radius of the sphere shall be as small as practicable, in order to facilitate adjustment of the bearing plate to the specimen, and yet large enough to provide adequate spherical bearing area. This radius is usually one to two times the greatest cross section dimension. The center of the sphere shall be on the plane of the specimen contact surface. The size of the compression plate shall be larger than the contact surface.

23.3 *Lateral Support:*

23.3.1 *General*—Evaluation of the crushing strength of long structural members requires that they be supported laterally to prevent buckling during the test without undue pressure against the sides of the specimen. Furthermore, the support shall not restrain either the longitudinal compressive deformation or load during test. The support shall be either continuous or intermittent. Intermittent supports shall be spaced so that the distance, l', between supports is less than 17 times the least radius of gyration of the cross section.

23.3.2 *Rectangular Members*—The general rules for structural members apply to rectangular structural members. However, the effective column length as controlled by intermittent support spacing on flatwise face need not equal that on edgewise face. The minimum spacing of the supports on the flatwise face shall be 17 times the least radius of gyration of the cross section which is about the centroidal axis parallel to flat face. And the minimum spacing of the supports on the edgewise face shall be 17 times the other radius of gyration (Fig. 6). A satisfactory method of providing lateral support for 2-in. (5-cm) dimension stock is shown in Fig. 7. A 27-in. (69-cm) I-beam provides the frame for the test machine. Small I-beams provide reactions for longitudinal pressure. A pivoted top I-beam provides lateral support on one flatwise face, while the web of the large I-beam provides the other. In between these steel members, metal guides on 3-in. (7.6-cm) spacing (hidden from view) attached to plywood fillers provide the flatwise support and contact surface. In between the flanges of the 27-in. (69-cm) I-beam, fingers and wedges provide edgewise lateral support.

23.4 *Compressometer:*

23.4.1 *Gage Length*—For modulus of elasticity calculations, a device shall be provided by which the deformation of the specimen is measured with respect to specific paired gage points defining the gage length. To obtain data representative of the test material as a whole, such paired gage points shall be located symmetrically on the lengthwise surface of the specimen as far apart as feasible, yet at least one times the larger cross-sectional dimension from each of the contact surfaces. At least two pairs of such gage points on diametrically opposite sides of the specimen shall be used to measure the average deformation.

23.4.2 *Accuracy*—The device shall be able to measure changes in deformation to three significant figures. Since gage lengths vary over a wide range, the measuring instruments should conform to their appropriate class in Methods E 83.

24. Test Specimen

24.1 *Material*—The test specimen shall consist basically of a structural timber which may be solid wood, laminated wood, or it may be a composite construction of wood or of wood combined with plastics or metals in sizes that are commercially used in structural applications, that is, in sizes greater than nominal 2 by 2-in. (5 by 5-cm) cross section (see 4.1.6).

24.2 *Identification*—Material or materials of the test specimen shall be as fully described as that for beams in 6.2.

24.3 *Specimen Dimensions*—The weight and dimensions, as well as moisture content of the specimen, shall be accurately measured before test. Weights and dimensions (length and cross section) shall be measured to three significant figures. Sufficient measurements of the cross section shall be made along the length of the specimen to describe shape characteristics and to determine the smallest section. The physical characteristics of the specimen, as described by its density and moisture content, may be determined in accordance with Methods D 2395 and Method A of D 2016, respectively.

24.4 *Specimen Description*—The inherent imperfections and intentional modifications shall be described as for beams in 6.4.

24.5 *Specimen Length*—The cross-sectional and length dimensions of structural members usually have established sizes, depending on the manufacturing process and intended use, so that no modification of these dimensions is involved. Since the length has been approximately established, the full length of the member shall be tested, except for trimming or squaring the bearing surface (see 25.2.1).

25. Procedure

25.1 *Preliminary*—Unless otherwise indicated in the research program or material specification, the test specimen shall be conditioned to constant weight so it is at moisture equilibrium, under the desired environment. Moisture contents may be approximated with moisture meters or more accurately measured by weights of samples according to Method A of Methods D 2016.

25.2 *Test Setup:*

25.2.1 *Bearing Surfaces*—The bearing surfaces of the specimen shall be cut so that the contact surfaces are plane, parallel to each other, and normal to the long axis of the specimen.

25.2.2 *Setup Method*—After physical measurements have been taken and recorded, the specimen shall be placed in the testing machine between the bearing blocks at each end and between the lateral supports on the four sides. The contact surfaces shall be geometrically centered on the bearing plates and then the spherical seats shall be adjusted for full contact. A slight longitudinal pressure shall be applied to hold the specimen while the lateral supports are adjusted and fastened to conform to the warp, twist, or bend of the specimen.

25.3 *Speed of Testing*—For measuring load-deformation data, the load shall be applied at a constant rate of head motion so that the fiber strain is 0.001 in./in · min ± 25 percent (0.001 cm/cm · min). For measuring only compressive strength, the test may be conducted at a constant rate to achieve maximum load in about 10 min, but not less than 5 nor more than 20 min.

25.4 *Load-Deformation Curves*—If load-deformation data have been obtained, load and deflection at first failure, at changes in slope of curve, and at maximum load should be noted.

25.5 *Records*—The maximum load shall be recorded, as well as a description and sketch of the failure relating the latter to the location of imperfections in the specimen. The section of the specimen containing the failure should be re-examined during analysis of the data.

26. Calculations

26.1 Physical and mechanical properties shall be computed in accordance with Definitions E 6 and as follows (see compressive notations):

26.1.1 Stress at proportional limit = P'/A in pounds per square inch (kilograms per square centimeter).

26.1.2 Compressive strength = P/A in pounds per square inch (kilograms per square centimeter).

26.1.3 Modulus of elasticity = $P'/A\epsilon$ in pounds per square inch (kilograms per square centimeter).

27. Report

27.1 The report shall include the following:

27.1.1 Complete identification,

27.1.2 History of seasoning conditioning,

27.1.3 Load apparatus,

27.1.4 Deflection apparatus,

27.1.5 Length and cross section dimensions,

27.1.6 Gage length,

27.1.7 Rate of load application,

27.1.8 Computed physical and mechanical properties, including specific gravity and moisture content, compressive strength, stress at proportional limit, modulus of elasticity, and a statistical measure of variability of these values,

27.1.9 Description of failure, and

27.1.10 Details of any deviations from the prescribed or recommended methods as outlined in the standard.

TENSION PARALLEL TO GRAIN

28. Scope

28.1 This method covers the determination of the tensile properties of structural elements made primarily of lumber equal to and greater than nominal 1 in. (2.5 cm) thick.

29. Summary of Method

29.1 The structural member is clamped at the extremities of its length and subjected to a tensile load so that in sections between clamps the tensile forces shall be axial and generally uniformly distributed throughout the cross sections without flexure along its length.

30. Significance

30.1 The tensile properties obtained by axial tension will provide information similar to that stipulated for flexural properties in Section 3.

30.2 The tensile properties obtained include modulus of elasticity, stress at proportional limit, tensile strength, and strain data beyond proportional limit.

31. Definitions

31.1 See 4.1.

32. Apparatus

32.1 *Testing Machine*—Any device having the following is suitable:

32.1.1 *Drive Mechanism*—A drive mechanism for imparting to a movable clamp a uniform, controlled velocity with respect to a stationary clamp.

32.1.2 *Load Indicator*—A load-indicating mechanism capable of showing the total tensile force on the test section of the tension specimen. This force-measuring system shall be calibrated to ensure accuracy in accordance with Methods E 4.

32.1.3 *Grips*—Suitable grips or fastening devices shall be provided which transmit the tensile load from the movable head of the drive mechanism to one end of the test section of the tension specimen, and similar devices shall be provided to transmit the load from the stationary mechanism to the other end of the test section of the specimen. Such devices shall not apply a bending moment to the test section, allow slippage under load, inflict damage, or inflict stress concentrations to the test section. Such devices may be either plates bonded to the specimen or unbonded plates clamped to the specimen by various pressure modes.

32.1.3.1 *Grip Alignment*—The fastening device shall apply the tensile loads to the test section of the specimen without applying a bending moment. For ideal test conditions, the grips should be self-aligning, that is, they should be attached to the force mechanism of the machine in such a manner that they will move freely into axial alignment as soon as the load is applied, and thus apply uniformly distributed forces along the test section and across the test cross section (Fig. 8(*a*)). For less ideal test conditions, each grip should be gimbaled about one axis which should be perpendicular to the wider surface of the rectangular cross section of the test specimen, and the axis of rotation should be through the fastened area (Fig. 8(*b*)). When neither self-aligning grips nor single gimbaled grips are available, the specimen may be clamped in the heads of a universal-type testing machine with wedge-type jaws (Fig. 8(*c*)). A method of providing approximately full spherical alignment has three axes of rotation, not necessarily concurrent but, however, having a common axis longitudinal and through the centroid of the specimen (Figs. 8(*d*) and 9).

32.1.3.2 *Contact Surface*—The contact surface between grips and test specimen shall be such that slippage does not occur. A smooth texture on the grip surface should be avoided, as well as very rough and large projections which damage the contact surface of the wood. Grips that are surfaced with a coarse emery paper (60X aluminum oxide emery belt) have been found satisfactory for softwoods. However, for hardwoods, grips may have to be glued to the specimen to prevent slippage.

32.1.3.3 *Contact Pressure*—For unbonded grip devices, lateral pressure should be applied to the jaws of the grip so that slippage does not occur between grip and specimen. Such pressure may be applied by means of bolts or wedge-shaped jaws, or both. Wedge-shaped jaws, such as those shown on Fig. 10, which slip on the inclined plane to produce contact pressure have been found satisfactory. To eliminate stress concentration or compressive damage at the tip end of the jaw, the contact pressure should be reduced to zero. The variable thickness jaws (Fig. 10), which cause a variable contact surface and which produce a lateral pressure gradient, have been found satisfactory.

32.1.4 *Extensometer:*

32.1.4.1 *Gage Length*—For modulus of elasticity determinations, a device shall be provided by which the elongation of the test section of the specimen is measured with respect to specific paired gage points defining the gage length. To obtain data representative of the test material as a whole, such gage points shall be symmetrically located on the lengthwise surface of the specimen as far apart as feasible, yet at least two times the larger cross-sectional dimension from each jaw edge. At least two pairs of such gage points on diametrically opposite sides of the specimen shall be used to measure the average deformation.

32.1.4.2 *Accuracy*—The device shall be able to measure changes in elongation to three significant figures. Since gage lengths vary over a wide range, the measuring instruments should conform to their appropriate class in Methods E 83.

33. Test Specimen

33.1 *Material*—The test specimen shall consist basically of a structural timber which may be solid wood, laminated wood, or it may be a composite construction of wood or wood combined with plastics or metals in sizes that are commercially used in structural "tensile" applications, that is, in sizes equal to and greater than nominal 1-in. (2.5-cm) thick lumber.

33.2 *Identification*—Material or materials of the test specimen shall be fully described as beams in 6.2.

33.3 *Specimen Description*—The specimen shall be described in a manner similar to that outlined in 6.3 and 6.4.

33.4 *Specimen Length*—The tension specimen, which has its long axis parallel to grain in the wood, shall have a length between grips equal to at least eight times the larger cross-sectional dimension when tested in self-aligning grips (see 32.1.3.1). However, when tested without self-aligning grips, it is recommended that the length between grips be at least 20 times the greater cross-sectional dimension.

34. Procedure

34.1 *Conditioning*—Unless otherwise indicated, the specimen shall be conditioned as outlined in 7.1.

34.2 *Test Setup*—After physical measurements have been taken and recorded, place the specimen in the grips of the load mechanism, taking care to have the long axis of the specimen and the grips coincide. The grips should securely clamp the specimen with either bolts or wedge-shaped jaws. If the latter are employed, apply a small preload to ensure that all jaws move an equal amount and maintain axial-alignment of specimen and grips. If either bolts or wedges are employed tighten the grips evenly and firmly to the degree necessary to prevent slippage. Under load, continue the tightening if necessary, even crushing the wood perpendicular to grain, so that no slipping occurs and a tensile failure occurs outside the jaw contact area.

34.3 *Speed of Testing*—For measuring load-elongation data, apply the load at a constant rate of head motion so that the fiber strain in the test section between jaws is 0.0006 in./in · min ± 25 percent (0.006 cm/cm min). For measuring only tensile strength, the load may be applied at a constant rate of grip motion so that maximum load is achieved in about 10 min but not less than 5 nor more than 20 min.

34.4 *Load-Elongation Curves*—If load-elongation data have been obtained throughout the test, correlate changes in specimen behavior, such as appearance of cracks or splinters, with elongation data.

34.5 *Records*—Record the maximum load, as well as a description and sketch of the failure relating the latter to the location of imperfections in the test section. Re-examine the section containing the failure during analysis of data.

35. Calculations

35.1 Compute physical and mechanical properties in accordance with Definitions, E 6 and as follows (see tensile notations):

35.1.1 Stress at proportional limit = P'/A in pounds per square inch (kilograms per square centimeter).

35.1.2 Tensile strength = P/A in pounds per square inch (kilograms per square centimeter).

35.1.3 Modulus of elasticity = $P'/A\epsilon$ in pounds per square inch (kilograms per square centimeter).

36. Report

36.1 The report shall include the following:

36.1.1 Complete identification,

36.1.2 History of seasoning,

36.1.3 Load apparatus, including type of end condition,

36.1.4 Deflection apparatus,

36.1.5 Length and cross-sectional dimensions,

36.1.6 Gage length,

36.1.7 Rate of load application,

36.1.8 Computed properties, and

36.1.9 Description of failures, and

36.1.10 Details of any deviations from the prescribed or recommended methods as outlined in the standard.

TORSION

37. Scope

37.1 This method covers the determination of the torsional properties of structural elements made of solid or laminated wood, or of composite constructions. The method is intended primarily for structural element or rectangular cross section but is also applicable to beams of round or irregular shapes.

38. Summary of Method

38.1 The structural element is subjected to a torsional moment by clamping it near its ends and applying opposing couples to each clamping device. The element is deformed at a prescribed rate and coordinate observations of torque and twist are made for the duration of the test.

39. Significance

39.1 The torsional properties obtained by twisting the structural element will provide information similar to that stipulated for flexural properties under Section 3.

39.2 The torsional properties of the element include an apparent modulus of rigidity of the element as a whole, stress at proportional limit, torsional strength and twist beyond proportional limit.

40. Definitions

40.1 See 4.1.

41. Apparatus

41.1 *Testing Machine* – Any device having the following is suitable:

41.1.1 *Drive Mechanism* – A drive mechanism for imparting an angular displacement at a uniform rate between a movable clamp on one end of the element and another clamp at the other end.

41.1.2 *Torque Indicator* – A torque-indicating mechanism capable of showing the total couple on the element. This measuring system shall be calibrated to ensure accuracy in accordance with ASTM Method E 4, Verification of Testing Machines.

41.2 *Support Apparatus*:

41.2.1 *Clamps* – Each end of the element shall be securely held by metal plates of sufficient bearing area and strength to grip the element with a vise-like action without slippage, damage, or stress concentrations in the test section when the torque is applied to the assembly. The plates of the clamps shall be symmetrical about the longitudinal axis of the cross section of the element.

41.2.2 *Clamp Supports* – Each of the clamps shall be supported by roller bearings or bearing blocks that allow the structural element to rotate about its natural longitudinal axis. Such supports may be ball bearings in a rigid frame of a torque-testing machine (Figs. 11 and 12) or they may be bearing blocks (Figs. 13 and 14) on the stationary and movable frames of a universal-type test machine. Either type of support shall allow the transmission of the couple without friction to the torque measuring device, and shall allow freedom for longitudinal movement of the element during the twisting. Apparatus of Fig. 13 is not suitable for large amounts of twist unless the angles are measured at each end to enable proper torque calculation.

41.2.3 *Frame* – The frame of the torque-testing machine shall be capable of providing the reaction for the drive mechanism, the torque indicator, and the bearings. The framework necessary to provide these reactions in a universal-type test machine shall be two rigid steel beams attached to the movable and stationary heads forming an X. The extremities of the X shall bear on the lever arms attached to the test element (Fig. 13).

41.3 *Troptometer*:

41.3.1 *Gage Length* – For modulus of rigidity calculations, a device shall be provided by which the angle of twist of the element is measured with respect to specific paired gage points defining the gage length. To obtain test data representative of the element as a whole, such paired gage points shall be located symmetrically on the lengthwise surface of the element as far apart as feasible, yet at least two times the larger cross-sectional dimension from each of the clamps. A yoke (Fig. 16) or other suitable device (Fig. 12) shall be firmly attached at each gage point to permit measurement of the angle of twist. The angle of twist is measured by observing the relative rotation of the two yokes or other devices at the gage points with the aid of any suitable apparatus including a light beam (Fig. 12), dials (Fig. 14), or string and scale (Figs. 15 and 16).

41.3.2 *Accuracy*—The device shall be able to measure changes in twist to three significant figures. Since gage lengths may vary over a wide range, the measuring instruments should conform to their appropriate class in ASTM Method E 83, Verification and Classification of Extensometer.[2]

42. Test Element

42.1 *Material*—The test element shall consist basically of a structural timber, which may be solid wood, laminated wood, or a composite construction of wood or wood combined with plastics or metals in sizes that are commercially used in structural applications.

42.2 *Identification*—Material or materials of the test element shall be as fully described as for beams in 6.2.

42.3 *Element Measurements*—The weight and dimensions as well as the moisture content shall be accurately determined before test. Weights and dimensions (length and cross section) shall be measured to three significant figures. Sufficient measurements of the cross section shall be made along the length of the specimen to describe characteristics and to determine the smallest cross section. The physical characteristics of the element, as described by its density and moisture content, may be determined in accordance with Methods D 2395[3] and D 2016, Method A,[3] respectively.

42.4 *Element Description*—The inherent imperfections and intentional modifications shall be described as for beams in 6.4.

42.5 *Element Length*—The cross-sectional dimensions of solid wood structural elements and composite elements usually are established, depending upon the manufacturing process and intended use so that normally no modification of these dimensions is involved. However, the length of the specimen shall be at least eight times the larger cross-sectional dimension.

43. Procedure

43.1 *Conditioning*—Unless otherwise indicated in the research program or material specification, condition the test element to constant weight so it is at moisture equilibrium under the desired environment. Approximate moisture contents with moisture meters, or measure more accurately by weights of samples in accordance with Method D 2016, Method A.[3]

43.2 *Test Setups*—After physical measurements have been taken and recorded, place the element in the clamps of the load mechanism, taking care to have the axis of rotation of the clamps coincide with the longitudinal centroidal axis of the element. The clamps should be tightened to securely hold the element in either type of testing machine. If the tests are made in a universal-type test machine, the bearing blocks shall be equal distances from the axis of rotation of the element.

43.3 *Speed of Testing*—For measuring torque-twist data, apply the load at a constant rate of head motion so that the angular detrusion of the outer fibers in the test section between gage points is about 0.004 radian per inch of length (0.0016 radian per centimetre of length) per minute ±50 %. For measuring only shear strength, the torque may be applied at a constant rate of twist so that maximum torque is achieved in about 10 min but not less than 5 nor more than 20 min.

43.4 *Torque-Twist Curves*—If torque-twist data have been obtained, torque and twist at first failure, at changes in slope of curve, and at maximum torque should be noted.

43.5 *Record of Failures*—Describe failures in detail as to type, manner and order of occurrence, angle with the grain, and position in the test element. Record descriptions relating to imperfections in the element. The section of the element containing the failure should be reexamined during analysis of the data.

44. Calculations

44.1 Compute physical and mechanical properties in accordance with definitions in E 6 and relationships in Appendix Table X3.

45. Report

45.1 The report shall include the following:

45.1.1 Complete identification,

45.1.2 History of seasoning and conditioning,

45.1.3 Apparatus for applying and measuring torque,

45.1.4 Apparatus for measuring angle of twist,

45.1.5 Length and cross-section dimensions,

45.1.6 Gage length,

45.1.7 Rate of twist applications,

45.1.8 Computed properties, and

45.1.9 Description of failures.

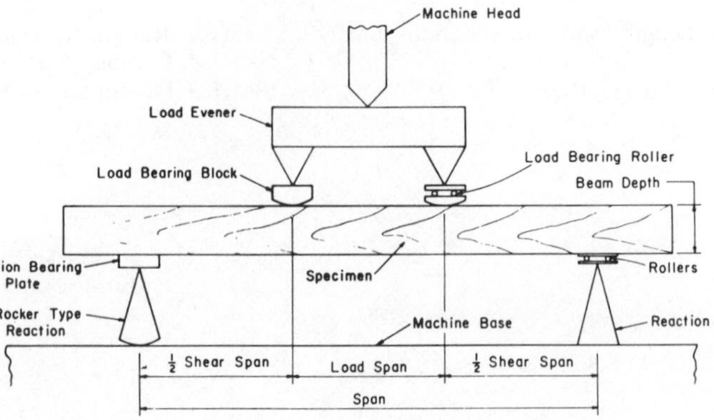

FIG. 1 Flexure Method.

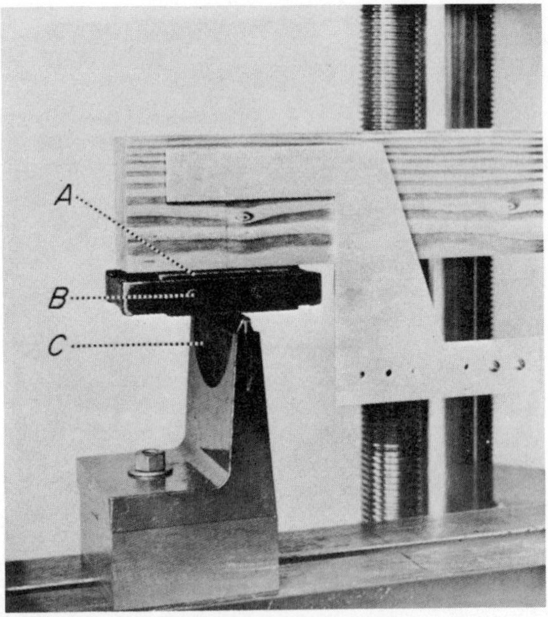

FIG. 2 Example of Bearing Plate, *A*, Rollers, *B*, and Reaction-Alignment-Rocker, *C*, for Small Beams.

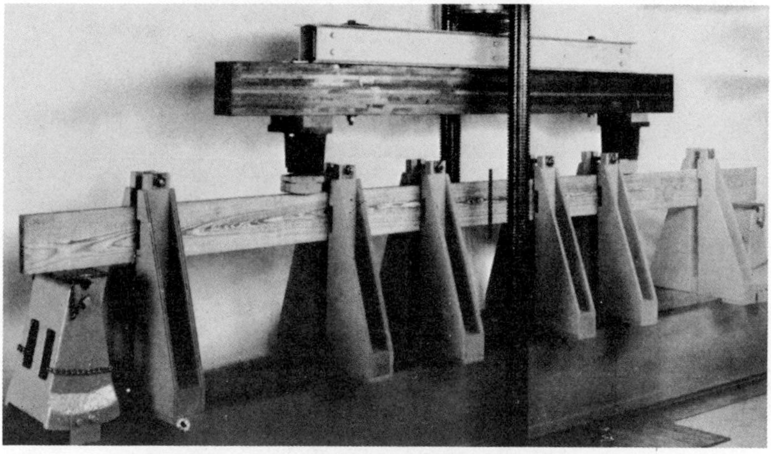

FIG. 3 Example of Lateral Support for Long, Deep Beams.

FIG. 4 Example of Curved Loading Block, *A*, Load-Alignment Rocker, *B*, Roller-Curved Loading Block, *C*, Load Evener, *D*, and Deflection-Measuring Apparatus, *E*.

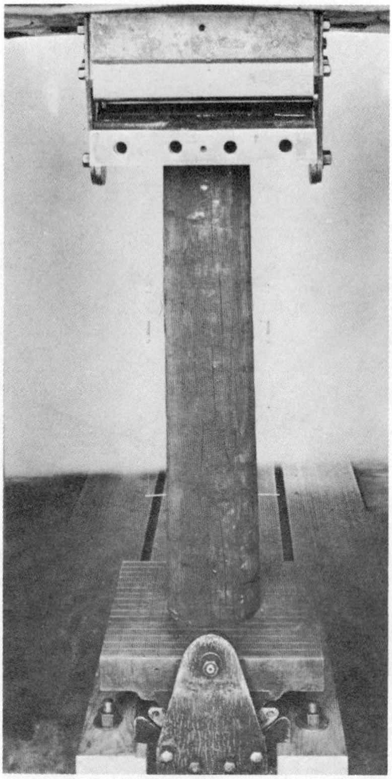

FIG. 5 Compression of a Wood Structural Element.

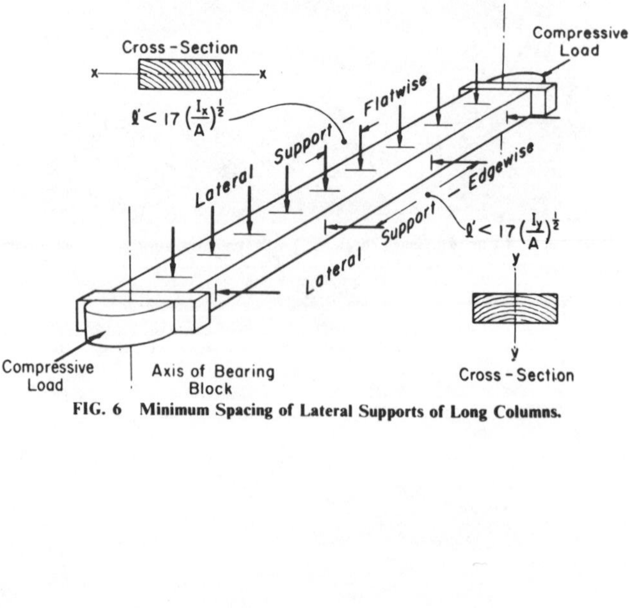

FIG. 6 Minimum Spacing of Lateral Supports of Long Columns.

FIG. 7 Compression of Long Slender Structural Member.

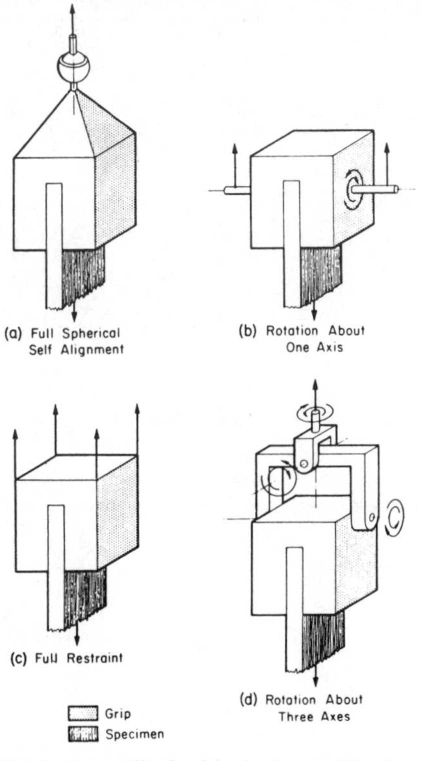

FIG. 8 Types of Tension Grips for Structural Members.

FIG. 9 Horizontal Tensile Grips for 2 by 10-in. Structural Members.

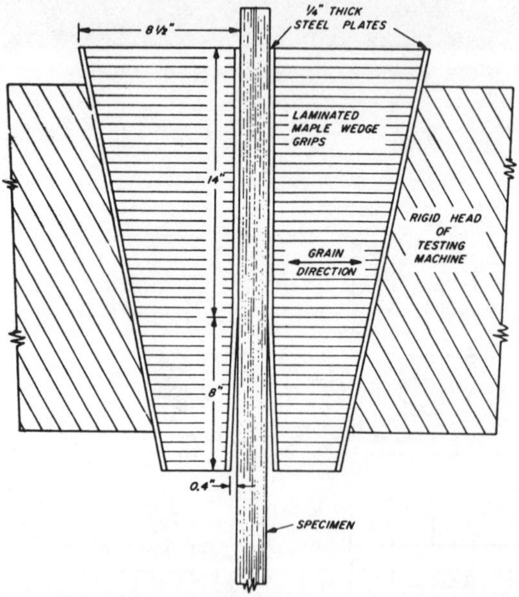

FIG. 10 Side View of Wedge Grips Used to Anchor Full-Size Structurally Graded Tension Specimens.

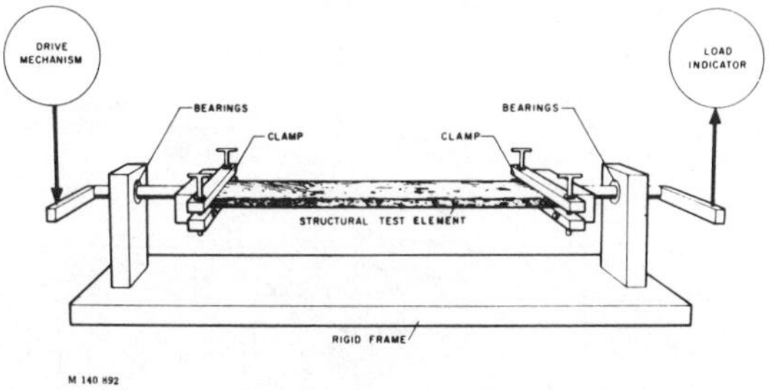

FIG. 11 Fundamentals of a Torsional Test Machine.

FIG. 12 Example of Torque-Testing Machine. (Torsion test in apparatus meeting specification requirements)

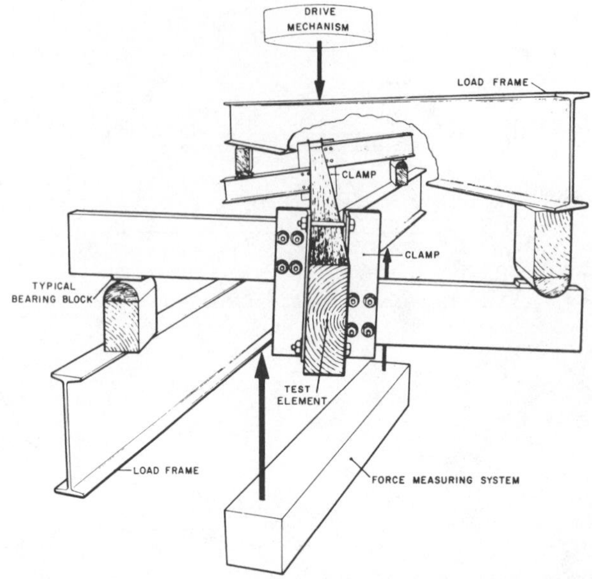

M 140 891

FIG. 13 Schematic Diagram of a Torsion Test Made in a Universal Type Test Machine.

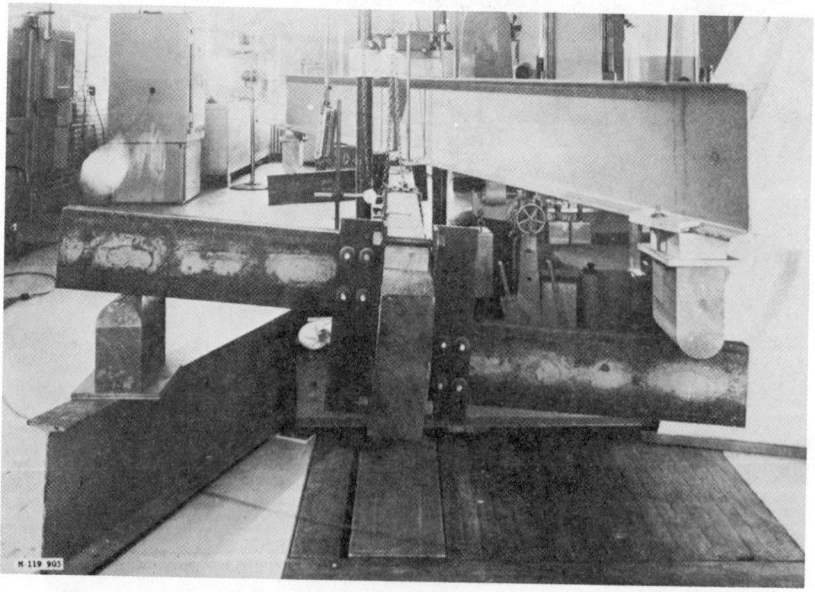

FIG. 14 Example of Torsion Test of Structural Beam in a Universal-Type Test Machine.

FIG. 15 Torsion Test with Yoke-Type Troptometer.

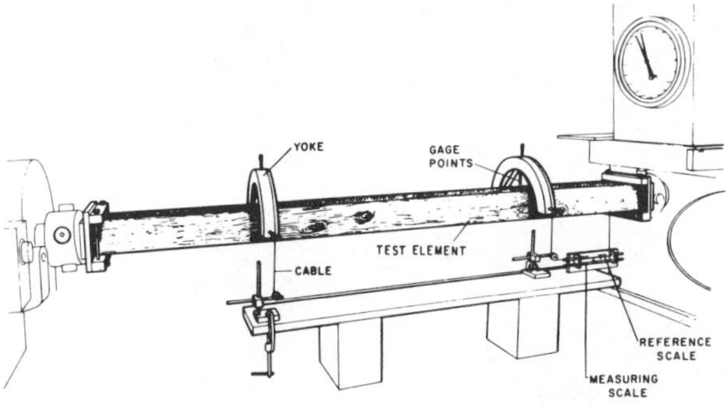

M 140 890

FIG. 16 Troptometer Measuring System.

APPENDIX

TABLE A1 Flexure Formulas[a]

Mechanical Properties	General	Two-Point Loading Rectangular Beam	Third-Point Loading Rectangular Beam
Fiber stress at proportional limit, S_f	$\frac{M'c}{I}$	$\frac{3P'a}{bh^2}$	$\frac{P'L}{bh^2}$
Modulus of rupture, S_R	$\frac{Mc}{I}$	$\frac{3Pa}{bh^2}$	$\frac{PL}{bh^2}$
Modulus of elasticity, E_f (apparent E)	$\frac{P'a}{48I\Delta}(3L^2 - 4a^2)$	$\frac{P'a}{4bh^3\Delta}(3L^2 - 4a^2)$	$\frac{P'L^3}{4.7bh^3\Delta}$
Modulus of elasticity, E_G (true E, shear corrected)	$\frac{M'l^2}{8I\Delta_l}$	$\frac{P'a(3L^2 - 4a^2)}{4bh^3\Delta\left(1 - \frac{3P'a}{5bhG\Delta}\right)}$	$\frac{P'L^3}{4.7bh^3\Delta\left(1 - \frac{P'L}{5bhG\Delta}\right)}$
		$\frac{3P'al^2}{4bh^3\Delta_l}$	$\frac{P'Ll^2}{4bh^3\Delta_l}$
Work to proportional limit per unit of volume, W_k		$\frac{P'\Delta}{2Lbh}\left[\frac{4a(3L - 4a) + \frac{24h^2E_G}{10G}}{3L^2 - 4a^2 + \frac{24h^2E_G}{10G}}\right]$	$\frac{P'\Delta}{2Lbh}\left[\frac{\frac{20}{9}L^2 + \frac{24h^2E_G}{10G}}{\frac{23}{9}L^2 + \frac{24h^2E_G}{10G}}\right]$
Approximate work to maximum load per unit of volume, W_m		$\frac{KA_m}{Lbh}\left[\frac{4a(3L - 4a) + \frac{24h^2E_G}{10G}}{3L^2 - 4a^2 + \frac{24h^2E_G}{10G}}\right]$	$\frac{KA_m}{Lbh}\left[\frac{\frac{20}{9}L^2 + \frac{24h^2E_G}{10G}}{\frac{23}{9}L^2 + \frac{24h^2E_G}{10G}}\right]$
Approximate total work per unit of volume, W_t		$\frac{KA_t}{Lbh}\left[\frac{4a(3L - 4a) + \frac{24h^2E_G}{10G}}{3L^2 - 4a^2 + \frac{24h^2E_G}{10G}}\right]$	$\frac{KA_t}{Lbh}\left[\frac{\frac{20}{9}L^2 + \frac{24h^2E_G}{10G}}{\frac{23}{9}L^2 + \frac{24h^2E_G}{10G}}\right]$
Shear stress, τ_m		$\frac{3}{4}\frac{P}{bh}$	$\frac{0.75G}{bh}$
Shear modulus, G		$\frac{6P'a}{10bh\left[\Delta - \left(\frac{L^2}{l^2} - \frac{4a^2}{3l^2}\right)\Delta_l\right]}$	$\frac{2P'L}{10bh\left[\Delta - \frac{23L^2}{27l^2}\Delta_l\right]}$

[a] For wooden beams having uniform cross section throughout their length.

TABLE A2 Physical Properties

Specific gravity (at test), $G_g = CW_g/V$ Specific gravity (ovendry), $G_d = G_g/(100 + MC)$	Method D 2395
Moisture content (percent of dry weight), $MC = 100\ (W_g/W_d - 1)$	Method A of Method D 2016

General Notations

A Cross-sectional area, in.^{2} (cm^2).

C 0.061, a constant for use when W_g is measured in grams in equation for specific gravity.
22.7, a constant for use when W_g is measured in pounds in equation for specific gravity.

ϵ Strain at proportional limit, in./in. (cm/cm).

G_d Specific gravity (ovendry).

G_g Specific gravity (at test).

I Moment of inertia of the cross section about a designated axis, in.^{4} (cm^4).

N Rate of motion of movable head, in./min (cm/min)

n Number of specimens in sample.

S Estimated standard deviation $= [(\Sigma X^2 - n\bar{X}^2)/(n-1)]^{1/2}$.

V Volume, in.^{3} (cm^3).

W_g Weight of moisture specimen (at test), lb (g).

W_d Weight of moisture specimen (ovendry), lb (g).

X Individual values.

$\bar{X}$ Average of n individual values.

Flexural Notations

a Distance from reaction to nearest load point, in. (cm) (1/2 shear span).

A_m Area of graph paper under the load-deflection curve from zero load to maximum load in.^{2} (cm^2) when deflection is measured between reaction and center of span.

A_t Area of graph paper under load-deflection curve from zero load to failing load or arbitrary terminal load, in.^{2} (cm^2), when deflection is measured between reaction and center of span.

b Width of beam, in. (cm).

c Distance from neutral axis of beam to extreme outer fiber, in. (cm).

G Modulus of rigidity in shear in psi (g/cm^2).

h Depth of beam, in. (cm).

K Graph paper scale constant for converting unit area of graph paper to load-deflection units.

l A span of the beam that is used to measure deflections caused only by the bending moment, that is, no shear distortions, in. (cm).

L Span of beam, in. (cm).

M Maximum bending moment at maximum load, in.-lb (cm-g).

M' Maximum bending moment at proportional limit load, in.-lb (cm-g).

P Maximum transverse load on beam, lb (g).

P' Load on beam at proportional limit, lb (g).

S_f Fiber stress at proportional limit, psi (g/cm^2).

S_R Modulus of rupture.

z Rate of fiber strain, in./in. (cm/cm) of outer fiber length per min.

Δ Deflection of beam, in. (cm), at neutral axis between reaction and center of beam at the proportional limit, in. (cm).

Δ_l A deflection of the beam measured at midspan over distance l, in. (cm).

τ_m Maximum shear strength, psi (g/cm^2).

Compressive Notations

l Length of compression column, in. (cm).

l' Effective length of column between supports for lateral stability, in. (cm).

P Maximum compressive load, lb (g).

P' Compressive load at proportional limit, lb (g).

r Radius of gyration $= [(I)/(A)]^{1/2}$, in. (cm).

Tensile Notations

P Maximum tensile load, lb (g).

P' Tensile load at proportional limit, lb (g).

TABLE A3 Torsion Formulas[a]

Mechanical Properties	Cross Section							
	Circle		Square		Rectangle		General[b]	
Fiber shear stress of greatest intensity at middle of long side; at proportional limit, S_s'	$2T'/\pi r^3$	(1A)	$4.808\ T'/w^3$	(1B)	$8\gamma T'/\mu w t^2$	(1C)	T'/Q	(1D)
Fiber shear strength of greatest intensity at middle of long side, S_s	$2T/\pi r^3$	(2A)	$4.808\ T/w^3$	(2B)	$8\gamma T/\mu w t^2$	(2C)	T/Q	(2D)
Fiber shear strength at middle of short side, S_s''					$8\gamma_1 T/\mu t^3$	(3C)		
Apparent modulus of rigidity, G	$2L_gT'/\pi r^4\theta$	(4A)	$7.11\ L_gT'/w^4\theta$	(4B)	$16\ L_gT'/wt^3[(16/3) - \lambda\ (t/w)]\ \theta$	(4C)	$L_gT'/\theta K$	(4D)

[a] From NACA rep. 334.

[b] Values of "Q" and "K" may be found in Roark, R. J., *Formulas for Stress and Strain*, McGraw-Hill, 1965, p. 194.

TABLE A4 Factors for Calculating Torsional Rigidity and Stress of Rectangular Prisms[a]

Ratio of Sides Column 1	λ Column 2	μ Column 3	γ Column 4	γ_1 Column 5
1.00	3.08410	2.24923	1.35063	1.35063
1.05	3.12256	2.35908	1.39651	
1.10	3.15653	2.46374	1.43956	
1.15	3.18554	2.56330	1.47990	
1.20	3.21040	2.65788	1.51753	
1.25	3.23196	2.74772	1.55268	1.13782
1.30	3.25035	2.83306	1.58544	
1.35	3.26632	2.91379	1.61594	
1.40	3.28002	2.99046	1.64430	
1.45	3.29171	3.06319	1.67265	
1.50	3.30174	3.13217	1.69512	0.97075
1.60	3.31770	3.25977	1.73889	0.91489
1.70	3.32941	3.37486	1.77649	
1.75	3.33402	3.42843	1.79325	0.84098
1.80	3.33798	3.47890	1.80877	
1.90	3.34426	3.57320	1.83643	
2.00	3.34885	3.65891	1.86012	0.73945
2.25	3.35564	3.84194	1.90543	
2.50	3.35873	3.98984	1.93614	0.59347
2.75	3.36023	4.11143	1.95687	
3.00	3.36079	4.21307	1.97087	
3.33	...	...	...	0.44545
3.50	3.36121	4.37299	1.98672	
4.00	3.36132	4.49300	1.99395	0.37121
4.50	3.36133	4.58639	1.99724	
5.00	3.36133	4.66162	1.99874	0.29700
6.00	3.36133	4.77311	1.99974	
6.67	3.36133	...	...	0.22275
7.00	3.36133	4.85314	1.99995	
8.00	3.36133	4.91317	1.99999	0.18564
9.00	3.36133	4.95985	2.00000	
10.00	3.36133	4.99720	2.00000	0.14858
20.00	3.36133	5.16527	2.00000	0.07341
50.00	3.36133	5.26611	2.00000	
100.00	3.36133	5.29972	2.00000	
∞	3.36133	5.33333	2.00000	0.00000

[a] Table I, "Factors for calculating torsional rigidity and stress of rectangular prisms," from National Advisory Committee for Aeronautics Report No. 334, "The Torsion of Members Having Sections Common in Aircraft Construction," by G. W. Trayer and H. W. March about 1929.

Torsion Notations

G Apparent modulus of rigidity, psi (gf/cm²).
K Stiffness-shape factor.
L_g Gage length of torsional element, in. (cm).
Q Stress-shape factor.
r Radius, in. (cm).
S_s' Fiber shear stress of greatest intensity at middle of long side at proportional limit, psi (g/cm²).
S_s Fiber shear strength of greatest intensity at middle of long side at maximum torque, psi (g/cm²).
S_s'' Fiber shear strength at middle of short side at maximum torque, psi (g/cm²).
T Twisting moment or torque, in.-lbf (cm-gf).
T' Torque at proportional limit, in.-lbf (cm-gf).
t Thickness, in. (cm).
w Width of element, in. (cm).
γ St. Venant constant, column 4, Table X4.
γ_1 St. Venant constant, column 5, Table X4.
θ Total angle of twist, radians (in./in. or cm/cm).
λ St. Venant constant, column 2, Table X4.
μ St. Venant constant, column 3, Table X4.

The American Society for Testing and Materials takes no position respecting the validity of any patent rights asserted in connection with any item mentioned in this standard. Users of this standard are expressly advised that determination of the validity of any such patent rights, and the risk of infringement of such rights, is entirely their own responsibility.

Designation: D 143 - 52
(Reapproved 1972)

American National Standard O4.1-1973(R-1969)
Approved April 16, 1973
By American National Standards Institute

Standard Methods of Testing
SMALL CLEAR SPECIMENS OF TIMBER[1]

This Standard is issued under the fixed designation D 143; the number immediately following the designation indicates the year of original adoption or, in the case of revision, the year of last revision. A number in parentheses indicates the year of last reapproval.

PART I. PRIMARY METHODS

Part I, Primary Methods, is the basic procedure intended for the broadest possible use in evaluating the strength and related properties of wood in the form of small clear specimens. These methods afford satisfactory results, have been widely used, and extensive data based on their use have been obtained and published. Any departure in cross-section of specimens from the 2 by 2-in. (5 by 5-cm) size employed in the Primary Methods introduces a variable that appreciably affects the results for certain properties, and thereby limits the full comparability desired for obtaining uniform results among different species. Only when relatively small trees, generally less than 12 in. (30 cm) in diameter, are available to produce the test specimens, and only when such trees because of crook, cross grain, knots, or other defects are of such quality that the longer specimens required in Part I, Primary Methods, cannot be obtained, should Part II, Secondary Methods, be employed.

INTRODUCTION

The everyday use of timber for multitudinous purposes makes manifest a continual need for data on its mechanical properties. The great variety of species, the variability of the material, the continually changing conditions of supply, the many factors affecting test results, all combine to make the technique of testing wood unique in its complexity.

In the preparation of these methods for testing small clear specimens, consideration was given both to the desirability of adopting methods that would yield results comparable to those already available and to the possibility of embodying such improvements as experience has shown desirable. In view of the many thousands of tests made under a single comprehensive plan by the U.S. Forest Service, the Forest Products Laboratories of Canada, and other similar organizations, the methods naturally conform closely to the methods used by these institutions. These methods are the outgrowth of a study of both American and European experience and methods. Their general adoption will tend toward a world-wide unification of results, permitting an interchange and correlation of data, and will establish the basis for a cumulative body of fundamental information on the timber species of the world.

[1] These methods are under the jurisdiction of ASTM Committee D-7 on Wood.
Current edition effective Sept. 30, 1952. Originally issued 1922. Replaces D 143 – 50 T.

These methods represent the entire procedure from selection of the trees to the carrying out of the tests, thus controlling factors, such as the size and proportion of test specimens and rate of loading, that may influence results. No attempt has been made to cover methods of computation and analysis, as these questions may be considered independently at any time. Such sample data and computation sheets and cards have been incorporated, however, as were thought to be of assistance to the investigator in systematizing records.

1. Scope

1.1 These methods cover tests on small clear specimens of wood that are made to afford:

1.1.1 Data for comparing the mechanical properties of various species,

1.1.2 Data for the establishment of correct strength functions which, in conjunction with results of tests of timbers in structural sizes,[2] afford the basis for fixing allowable stresses, and

1.1.3 Data upon which to determine the influence on the mechanical properties of such factors as density, locality of growth, position in cross section, height of timber in the tree, change of properties with seasoning, and change from sapwood to heartwood.

NOTE 1—The values stated in U.S. customary units are to be regarded as the standard. The metric equivalents of U.S. customary units may be approximate.

2. Summary of Method

2.1 The principal mechanical tests are static bending, compression parallel to grain, impact bending, toughness, compression perpendicular to grain, hardness, shear parallel to grain,[3] cleavage, and tension parallel to grain. The tension-perpendicular-to-grain and the nail-withdrawal tests also included are optional. These tests are made on both green and air-dry material as specified in these methods. In addition, methods for evaluating such physical properties as specific gravity, shrinkage in volume, radial shrinkage, and tangential shrinkage are presented.

2.2 The procedures for collection and preparation of the material for testing and for the various tests appear in the following order:

COLLECTION OF MATERIAL

Selection

3. Authentic Identification

3.1 The material shall be from trees selected in the forest by one qualified to identify the species and to select the trees. Whenever practicable this should be a member of the timber mechanics staff of the laboratory concerned, and where necessary, herbarium samples including leaves, fruit, twigs, and bark

[2] See ASTM Methods D 198, Static Tests of Timbers in Structural Sizes, *Annual Book of ASTM Standards*, Part 22.

[3] The test for shearing strength perpendicular to the grain (sometimes termed "vertical shear") is not included as one of the principal mechanical tests since in such a test the strength is limited by the shearing resistance parallel to the grain.

shall be obtained to ensure positive identification.

4. Selection and Number of Trees

4.1 For each species to be tested, at least five trees representative of the species shall be selected.

5. Selection and Number of Bolts

5.1 The material of each species selected for test shall be representative of the merchantable bole of the tree. One method of selection that has been found satisfactory is as follows:

5.2 From one tree of each group of five, select 8-ft (2.4-m) sections (each section representing two 4-ft (1.2-m) bolts) from various heights to afford information on the variation of properties with height in tree, as indicated in Fig. 1 and as follows:

Length of Merchantable Bolt, ft (m)	Bolts to be Selected (for explanation of letter designation of bolts, see Section 9)
16 (4.9 m)	*a, b; c, d*
20 (6.0 m)	*a, b; c, d*
24 (7.2 m)	*a, b; c, d; e*
28 (8.4 m)	*a, b; c, d; f, g*
32 (9.6 m)	*a, b; c, d; g, h*
36 (10.8 m)	*a, b; c, d; h, i*
40 (12.0 m)	*a, b; c, d; g, h; i, j*
44 (13.2 m)	*a, b; c, d; g, h; j, k*
48 (14.4 m)	*a, b; c, d; g, h; k, l*
52 (15.6 m)	*a, b; c, d; g, h; l, m*
56 (16.8 m)	*a, b; c, d; g, h; m, n*
60 (18.0 m)	*a, b; c, d; i, j; n, o*
64 (19.2 m)	*a, b; c, d; i, j; o, p*
72 (21.6 m)	*a, b; c, d; i, j; q, r*
80 (24.0 m)	*a, b; c, d; i, j; o, p; s, t*
96 (28.8 m)	*a, b; c, d; i, j; o, p; w, x*
Over 96	*a, b; c, d; k, l; s, t;* and last two bolts (each 4 ft (1.2-m) in length) at top of merchantable length

5.3 From the other trees called for in Section 4, take the 8-ft section[4] (*c-d* bolts) next above the 8-ft butt log.

6. Substitution of Flitches for Bolts

6.1 In cases where the logs or bolts are over 60 in. (1.5 m) in diameter, a single flitch 6 in. (15 cm) in thickness, taken through the pith in a north and south direction and representing the full diameter of the log, may be substituted, in the same length, for the full log or bolt specified in Section 5.

7. Selection for Important Species

7.1 For important species of wide geographical distribution, test material shall be selected from two or more localities or sites. The number of trees of a species selected from each site or locality shall conform to the requirements of Sections 4 and 5.

Field Marking

8. Tree Designation

8.1 Each tree shall be given an arabic number, the numbering in any given shipment to be consecutive for trees of a given species.

9. Bolt Designation

9.1 Each 4 ft (1.2 m) of length of a tree or log shall be considered a "bolt." Bolts shall be designated by small letters, beginning with *a* for the 4-ft section next above the stump. Bolt letters, therefore, indicate position with respect to height in tree.

10. Marking

10.1 The tree number and bolt designation shall be plainly marked upon each log selected by the collector. Thus the 16-ft (4.8-m) butt log of Tree No. 2 would be designated 2*abcd*. Steel dies are recommended for marking the butt end of the logs.

11. Indication of Cardinal Point

11.1 The north side of each log shall be indicated in some convenient manner.

12. Shipment Number

12.1 All material collected from a given locality and shipped at one time shall be given a shipment number or other designation.

Field Descriptions

13. Field Descriptions

13.1 Complete field notes describing the material shall be fully and carefully made by the collector. These notes shall, in general, supply data as outlined in Table A1.

[4] The 8-ft lengths specified are intended to provide test material in sticks of this net length. It is recommended that the bolts or sections be cut in the woods to 9-ft lengths to allow for trimming, etc.

13.2 Photographs of the standing trees selected should be taken, when practicable.

Preparation for Shipment

14. Preparation for Shipment

14.1 The bark will be left on each log, and care shall be taken to keep the bark intact. The ends of the logs shall be carefully painted to retard or prevent end drying and end checking.

14.2 Record shall be made of the shipment routing, bill of lading, kind of shipment, date of shipment, and condition of material when shipped.

14.3 Record shall also be made of date of receipt of shipment at destination, its condition, and method of storage.

DISPOSITION AT DESTINATION

Storage of Logs at Destination

15. Storage of Logs

15.1 Material shall not be kept in the bolt or log form long enough to permit damage by checks, decay, stains, or insect attack. The logs shall be piled on skids, free from contact with the soil, and shall not be stored where subjected to artificial heat. In addition they should preferably be protected from the sun and, when necessary, sprinkled regularly with water to prevent drying. As an alternative to piling on skids, the logs may be stored in water prior to testing.

Photographing, Sawing, and Final Marking

NOTE 2—In sawing, marking, and selecting test sticks, the aim should be to obtain specimens representative of the material collected. The procedure described herein is one that has been found satisfactory for most species.

16. Photographing Ends of Bolts

16.1 The top end of each *d* or *c-d* bolt shall be photographed. It is suggested that a rule be placed on the log so as to indicate the scale of the photograph and that the cardinal points be indicated on the cross-section. Figure 2 shows a photograph of this kind.

17. Sawing of Bolts

17.1 All bolts shall be marked on the top end into 2 $^1/_2$ by 2 $^1/_2$-in. (6 by 6-cm) squares as shown in Fig. 3, and sawed into nominal 2 $^1/_2$ by 2 $^1/_2$-in. sticks. The letters *N*, *E*, *S*, and *W* indicate the cardinal points.

17.2 When flitches are substituted for bolts (Section 6), the same general marking and numbering scheme of 17.1 shall be followed insofar as it is applicable.

18. Marking of Test Sticks

18.1 All test sticks shall bear the shipment number, the tree number, stick number, and bolt designation, to be known respectively as Shipment No., Piece No., Stick No., and mark. Thus, 400-1-*N*4*d* represents Stick *N*4 of Bolt *d*, Tree 1, Shipment 400.

Matching for Tests of Air-Dry Material

19. Composite Bolts

19.1 The collection of the material (Section 5) has been arranged to provide for tests of both green and air-dry specimens that are closely matched by selection from adjacent parts of the same tree. The 8-ft (2.4-m) long bolts, after being marked in accordance with 17.1, shall be sawed into 2 $^1/_2$ by 2 $^1/_2$-in. (6 by 6-cm) by 8-ft sticks, and numbered and lettered in accordance with Section 18. Each 2 $^1/_2$ by 2 $^1/_2$-in. by 8-ft stick shall then be cut into two 4-ft (1.2-m) pieces, making sure that each part carries the proper designation and bolt letter.

19.2 Part of the 2 $^1/_2$ by 2 $^1/_2$-in. by 4-ft (1.2-m) sticks from each 8-ft bolt are to provide specimens to be tested green (unseasoned) and the other part are to provide specimens to be air-dried and tested. To afford matching, the 4-ft sticks of one bolt shall be interchanged with the 4-ft sticks of the next adjacent bolt from the same tree to form two composite bolts, each being complete and being made of equal portions of the adjacent 4-ft bolts. The sticks from one of these composite bolts shall be tested green and those from the other shall be tested after air-drying. Thus, the sticks of each composite bolt shall be regarded as if they were from the same bolt.

19.3 The above procedure provides for end-to-end matching (end matching) of sticks to be tested air-dry with those to be tested green which is to be preferred when practicable. If because of the nature of the material, end matching is not practicable, side matching may be used.

20. Schedule for Forming Composite Bolts

20.1 The division of sticks into composite bolts, part to be tested green and part to be air-dried and tested, shall be made according to the following schedule, in which the numbers refer to stick numbers:

Selection of Sticks from *a* and *b* Bolts

Composite Bolt to Be Tested Green:

Bolt *a*	1	4, 5	8, 9	
Bolt *b*	2, 3	6, 7	10, etc.	

Composite Bolt to Be Air-Dried and Tested:

Bolt *a*	2, 3	6, 7	10, etc.	
Bolt *b*	1	4, 5	8, 9	

Selection of Sticks from *c* and *d* Bolts

Composite Bolt to Be Tested Green:

Bolt *c*	1	4, 5	8, 9	
Bolt *d*	2, 3	6, 7	10, etc.	

Composite Bolt to Be Air-Dried and Tested:

Bolt *c*	2, 3	6, 7	10, etc.	
Bolt *d*	1	4, 5	8, 9	

Selection of Sticks from *e* and *f* Bolts

Composite Bolt to Be Tested Green:

Bolt *e*	1	4, 5	8, 9	
Bolt *f*	2, 3	6, 7	10, etc.	

Composite Bolt to Be Air-Dried and Tested:

Bolt *e*	2, 3	6, 7	10, etc.	
Bolt *f*	1	4, 5	8, 9	

20.2 The selection of sticks from other 8-ft (2.4-m) bolts of the tree to form the composite bolts to be tested green and to be air-dried and tested shall be made in accordance with the system as indicated in 20.1.

20.3 As an example of composite bolts, assume that the cross-section, Fig. 3, represents the end of an 8-ft (2.4-m) section comprising the *c* and *d* bolts.

20.3.1 The following sticks are selected for the composite bolt to be tested green: *N1c, N2d, N3d, N4c, N5c, N6d, N7d, N8c, N9c, N10d, N11d, N12c; E3d, E4c, E5c, E6d, E7d, E8c, E9c, E10d, E11d, E12c; S1c, S2d, S3d, S4c, S5c, S6d, S7d, S8c, S9c, S10d, S11d, S12c; W3d, W4c, W5c, W6d, W7d, W8c, W9c, W10d, W11d, W12c.*

20.3.2 The following sticks are selected for the composite bolt to be air-dried and tested: *N1d, N2c, N3c, N4d, N5d, N6c, N7c, N8d, N9d, N10c, N11c, N12d; E3c, E4d, E5d, E6c, E7c, E8d, E9d, E10c, E11c, E12d; S1d, S2c, S3c, S4d, S5d, S6c, S7c, S8d, S9d, S10c, S11c, S12d; W3c, W4d, W5d, W6c, W7c, W8d, W9d, W10c, W11c, W12d.*

Disposition of Sticks

21. Green Material

21.1 The sticks (2½ by 2½ in. by 4 ft) (6 by 6 cm by 1.2 m) to be tested green shall be kept in an unseasoned condition, while awaiting preparation for test, by being stored in a framed pit or other suitable container, where they shall be close piled and covered with damp sawdust, or in some other suitable manner. As material is required for test, it shall be removed from this pit or container, surfaced on all four sides to 2 by 2 in. (5 by 5 cm) in cross section, sawed to test size, and kept covered with a damp cloth in a tightly closed container at a temperature of 68 ± 6 F (20 ± 3 C) (see Note 3, 22.5) until the time of test. Care shall be taken to avoid as much as possible the storage of green material in any form. Sticks to be tested in a green condition usually should not be sawed from the log form in quantities greater than is required to meet the testing demands for from a few days to not more than 2 weeks, depending on the prevailing conditions.

22. Air-Dry Material

22.1 The ends of the sticks to be air-dried (2½ by 2½ in. by 4 ft) (6 by 6 cm by 1.2 m) shall be dipped in melted paraffin or other substance suitable to retard checking. The material shall be piled so as to have a space of at least ½ in. (1.3 cm) on each side of each stick to permit circulation of air. The material shall be stored in a place allowing free access of air, but protected from sunshine, rain, snow, and moisture from the ground. The sticks in drying shall not be subjected to artificial heat.

22.2 All of the sticks from each composite bolt to be air-dried shall be weighed when stored and at sufficiently frequent intervals thereafter to get accurate data on the progress of seasoning. No material shall be considered thoroughly air-dried and properly conditioned for testing until practically constant weight has been reached. (Wood absorbs and gives off moisture with changing atmospheric conditions; consequently it never comes to absolutely constant weight.)

22.3 When the material has reached equilibrium, moisture sections approximately 1 in. (2.5 cm) in length shall be taken from about 10 percent of the sticks to determine the actual moisture content. These moisture specimens shall be cut not less than 1 ft (0.3 m) from the ends of the sticks, and in such a way as to prevent any appreciable loss of material for testing. When conditioned to approximately 12 percent moisture content, the material shall be surfaced on four sides to 2 by 2 in. (5 by 5 cm) in cross-section and tested.

22.4 When adequate facilities are available, the stocks may be kiln-dried instead of air-dried in order to reduce the drying time. The preparation of the sticks and procedures followed shall be similar to those used in air-drying, and the drying shall be continued until the sticks have a moisture content of approximately 12 percent. The kilns shall be operated in a manner and at such temperatures as are in accordance with best practices for drying the species in question without injury to the strength, and the drying shall be done to avoid kiln-drying defects such as "casehardening," "honeycombing," or "collapse." A record of operating conditions of the kiln shall be kept for the entire run and shall include humidity conditions and temperatures at the hottest part of the kiln. In general, the maximum kiln temperature shall not exceed 130 F (54 C), but the exact permissible limits such as are suitable for kiln-drying airplane stock without injury depend on the species.

22.5 The seasoned sticks, whether kiln-dried or air-dried, preferably should be stored in a room having controlled temperature and humidity (at 68 $\pm$ 6 F (20 $\pm$ 3 C) and 65 $\pm$ 1 percent relative humidity) before test to reduce the moisture gradient within the material, and to bring the material into equilibrium, which will be approximately 12 percent moisture content for most species.

NOTE 3—In following the recommendation that the temperature be controlled at 68 $\pm$ 6 F (20 $\pm$ 3 C), it should be understood that it is desirable to maintain the temperature as nearly constant as possible-at some temperature within this range.

ORDER, SELECTION, AND NUMBER OF TESTS

Order of Tests

23. Order of Tests

23.1 The order of tests in all cases shall be such as to eliminate as far as possible from the comparisons the effect of changes in the specimen due to such factors as storage and weather conditions.

Selection of Specimens

24. Preference in Selecting Specimens

24.1 In case the material from a given bolt should be insufficient to furnish all the test specimens hereinafter required (logs or bolts less than 24 in. (0.6 m) in diameter), additional bolts may be selected. If additional material is not available, the preferential order of mechanical tests to be used in selecting specimens shall be as follows: static bending, compression parallel to grain, impact bending, toughness, compression perpendicular to grain, hardness, shear parallel to grain, cleavage, tension parallel to grain, tension perpendicular to grain, and nail withdrawal.

25. Test Specimens from Bending Specimens After Failure

25.1 In some instances where the sticks do not provide sufficient material for all the tests, certain test specimens may be taken from the uninjured portion of the static and impact bending specimens remaining after test, provided proper care is used in the selection.

26. Quality of Test Material

26.1 Only clear straight-grained material, free of decay and other defects, shall be used for the tests. However, small knots and other similar defects may be permitted in such specimens as static bending when their location is such that it is certain they will not in any way influence the failure or otherwise affect the strength of the specimen.

Number of Tests for Each Bolt

27. Static Bending

27.1 One static bending specimen shall be taken from each pair of sticks. A pair consists of two adjacent sticks equi-distant from the pith, as *W*3 and *W*4, Fig. 3. In the composite bolts tested to afford a comparison of the strength of green and air-dry material, the pair of sticks shall be constituted as above, except that the sticks in this case will be from different bolts. Thus, *W*3*d* and *W*4*c* constitute one pair of sticks to be tested green, and *W*3*c* and *W*4*d* the corresponding pair to be tested air-dry (Section 20).

28. Compression Parallel to Grain

28.1 One compression-parallel-to-grain specimen shall be taken from each stick. Load-compression curves shall preferably be taken on all of the specimens.

29. Impact Bending

29.1 Eight impact-bending specimens shall be taken from each bolt, selection being made from the sticks remaining after obtaining the static bending tests. Two of the specimens shall be selected from near the pith, two from near the periphery, and four that are representative of the cross-section.

30. Toughness

30.1 Two toughness specimens shall be selected from the uninjured portion or end of each impact bending specimen or companion static bending specimen, making a total of 32 toughness specimens for each bolt. One from each group of two specimens from the same stick shall be tested with the load applied radially and the other tested with the load applied tangentially.

31. Compression Perpendicular to Grain

31.1 One compression-perpendicular-to-grain specimen shall be taken from each of 50 percent of the sticks selected for static bending.

32. Hardness

32.1 One hardness specimen shall be taken from each of the other 50 percent of the static-bending sticks.

33. Shear Parallel to Grain

33.1 Twelve shear-parallel-to-grain specimens shall be selected from the unused portion or ends of six sticks from which bending specimens have been selected. Two specimens shall be taken from near the pith, two from near the periphery, and eight that are representative of the average growth of the cross section of the bolt. These twelve specimens shall be selected in pairs from the six sticks. One of each pair of specimens from the same stick shall be tested in radial shear (surface of failure radial) and the other in tangential shear (surface of failure tangential).

34. Cleavage Perpendicular to Grain

34.1 Twelve cleavage specimens shall be selected from six sticks in a manner similar to that for shear (Section 33). One of each pair of specimens from the same stick shall be tested in radial cleavage (surface of failure radial) and the other in tangential cleavage (surface of failure tangential).

35. Tension Parallel to Grain

35.1 Six tension-parallel-to-grain specimens shall be chosen of which one shall be selected from near the pith, one from near the periphery, and four that are representative of the cross-section.

36. Tension Perpendicular to Grain

36.1 Twelve tension-perpendicular-to-grain specimens shall be selected from six sticks in a manner similar to that for shear (Section 33). One of each pair of specimens from the same stick shall be tested in radial tension (surface of failure radial) and the other in tangential tension (surface of failure tangential).

37. Nail Withdrawal

37.1 Twelve nail withdrawal specimens shall be selected from the unused portion or ends of twelve of the sticks from which bending or tension-parallel-to-grain specimens have been selected or, if necessary, from the uninjured portion of specimens from other tests. Six nail withdrawal specimens shall be tested in the green and six in the air-dry condition. The specimens for testing in both the green and the air-dry conditions shall be selected so as to give one from near the pith, one from near the periphery, and four that are representative of the average growth of the cross-section of the bolt.

38. Specific Gravity and Shrinkage in Volume

38.1 Six specific gravity and shrinkage-in-volume specimens shall be selected from the unused portion of bending or tension-parallel-to-grain sticks, selected so as to give one from near the pith, one from near the periphery, and four that are representative of the average growth of the cross-section of the bolt. These specimens shall be selected only from the sticks to be tested in a green condition.

39. Radial Shrinkage

39.1 Four radial shrinkage specimens shall be obtained from each *d* bolt, and where possible from the upper bolt of each pair of bolts selected at other heights in the tree. They shall be cut from the "sectors" or "quadrants" remaining after sawing (Fig. 3) or from disks cut from near the end of the bolt. When a disk is used, care must be taken to see that it is green and has not been affected by shrinking and checking, which is common near the end of the bolt. The specimens shall not be surfaced. Radial shrinkage specimens shall be cut with their greatest dimension in the radial direction. Two shall be taken from the heartwood and the other two from near the periphery. When possible, two specimens shall consist entirely of sapwood.

40. Tangential Shrinkage

40.1 Four tangential shrinkage specimens shall be obtained from each *d* bolt, and where possible from the upper bolt of each pair of bolts selected at other heights in the tree. They shall be selected at the same time and in a manner similar to radial-shrinkage specimens (Section 39), except that the greatest dimension shall be in a tangential direction. The specimens shall not be surfaced. Two shall be taken from the heartwood; the other two shall be taken from near the periphery and when possible shall consist entirely of sapwood. The heartwood and the sapwood specimens shall be taken adjacent to the respective specimens selected for radial shrinkage.

Photographs of Sticks

41. Sticks to be Photographed

41.1 Four of the static bending sticks from each species shall be selected for photographing, as follows: two average growth, one fast growth, and one slow growth. These sticks shall be photographed in cross-section and on the radial and tangential surfaces. Figure 4 is a typical photograph of a cross-section of 2 by 2-in. (5 by 5-cm) test specimens and Fig. 5 of the tangential surface of such specimens.

Control of Moisture Content and Temperature

NOTE 4—In recognition of the significant influence of temperature and humidity on the strength of wood, it is highly desirable that these factors be controlled to ensure comparable test results.

42. Control of Moisture Content

42.1 As prescribed in Section 22, sticks for test in the air-dry condition shall be brought practically to constant weight before test. Should any changes in moisture content occur during final preparation of specimens, the specimens shall be reconditioned before test to constant weight under conditions as prescribed in 22.5. Tests shall then be carried out in such manner that large changes in moisture content will not occur. To prevent such changes, it is desirable that the testing room and rooms for preparation of test specimens have some means of humidity control.

43. Control of Temperature

43.1 Independent of the effect on strength of the moisture content of the test specimens as influenced by temperature, is the significant effect of temperature itself on the mechanical properties. The specimens when tested shall be at a temperature of 68 ± 6 F (20 ± 3 C) (see Note 3, 22.5). The temperature at time of test shall in all instances be recorded as a specific part of the test record.

Record of Heartwood and Sapwood

44. Proportion of Sapwood

44.1 The estimated proportion of sapwood present shall be recorded for each test specimen.

Static Bending

45. Size of Specimens

45.1 The static bending tests shall be made on 2 by 2 by 30-in. (5 by 5 by 76-cm) specimens. The actual height and width at the center and the length shall be measured (Section 127).

46. Loading Span and Supports

46.1 Center loading and a span length of 28 in. (70 cm) shall be used. Both supporting knife edges shall be provided with bearing plates and rollers of such thickness that the distance from the point of support to the central plane is not greater than the depth of the specimen (Fig. 6). The knife edges shall be adjustable laterally to permit adjustment for slight twist or warp in the specimen.[5] Alternatively, the method of supporting the specimen in trunnion-type supports that are free to move in a horizontal direction may be employed.

47. Bearing Block

47.1 A bearing block of the form and size of that shown in Fig. 7 shall be used for applying the load.

48. Placement of Growth Rings

48.1 The specimen shall be placed so that the load will be applied through the bearing block to the tangential (flat-sawed) surface nearest the pith.

49. Speed of Testing

49.1 The load shall be applied continuously throughout the test at a rate of motion of the movable crosshead of 0.10 in. (2.5 mm)/min (Section 128).

50. Load-Deflection Curves

50.1 Load-deflection curves shall be taken to or beyond the maximum load for all static bending tests. In at least one third of the tests, the curves shall be continued to a 6-in. (15-cm) deflection, or until the specimen fails to support a load of 200 lb (90 kg).

50.2 Deflections of the neutral plane at the center of the length shall be taken with respect to points in the neutral plane above the supports.

50.3 Within the proportional limit, deflection readings shall be taken to 0.001 in. (0.02 mm). After the proportional limit is reached, less refinement is necessary in observing deflections, but it is convenient to read them by means of the dial gage (Fig. 6) until it reaches the limit of its capacity, normally approximately 1 in. (2.5 cm). Where deflections beyond 1 in. are encountered, the deflections may be measured by means of the scale mounted on the loading head (Fig. 6) and a wire mounted at the neutral axis of the specimen on the side opposite the yoke. Deflections are read to the nearest 0.01 in. (0.2 mm) at 0.10-in. (2.5-mm) intervals and also after abrupt changes in load.

50.4 The load and deflection of first failure, the maximum load, and points of sudden change shall be read and shown on the curve sheet[6] although they may not occur at one of the regular load or deflection increments.

51. Description of Static Bending Failures

51.1 Static bending (flexural) failures shall be classified according to the appearance of the fractured surface and the manner in which the failure develops (Fig. 8). The fractured surfaces may be roughly divided into "brash" and "fibrous," the term "brash" indicating abrupt failure and the term "fibrous" indicating a fracture showing splinters.

52. Weight and Moisture Content

52.1 The specimen shall be weighed immediately before test, and after test a moisture section approximately 1 in. (2.5 cm) in length shall be cut near the failure (Section 126).

Compression Parallel to Grain

53. Size of Specimens

53.1 The compression-parallel-to-grain tests shall be made on 2 by 2 by 8-in. (5 by 5 by 20-cm) specimens. The actual cross-section dimensions and the length shall be measured (Section 127).

54. End Surfaces Parallel

54.1 Special care shall be used in preparing the compression-parallel-to-grain test specimens to ensure that the end grain surfaces will be parallel to each other and at right angles to the longitudinal axis. If deemed necessary, at least one platen of the testing machine shall

---

[5] Details of laterally adjustable supports may be found in Fig. 4 of ASTM Methods D 805, Testing Veneer, Plywood, and Other Glued Veneer Constructions, *Annual Book of ASTM Standards*, Part 22.

[6] See Fig. A1 for a sample static bending data sheet form.

be equipped with a spherical bearing to obtain uniform distribution of load over the ends of the specimen.

55. Speed of Testing

55.1 The load shall be applied continuously throughout the test at a rate of motion of the movable crosshead of 0.003 in./in. (cm/cm) of specimen length/min (Section 128).

56. Load-Compression Curves

56.1 Load-compression curves shall be taken over a central gage length not exceeding 6 in. (15 cm) and preferably on all of the specimens. Load-compression readings shall be continued until the proportional limit is well passed, as indicated by the curve.[7]

56.2 Deformations shall be read to 0.0001 in. (0.002 mm).

56.3 Figures 9 and 10 illustrate two types of compressometers that have been found satisfactory for wood testing.

57. Position of Test Failures

57.1 In order to obtain satisfactory and uniform results, it is necessary that the failures be made to develop in the body of the specimen. With specimens of uniform cross-section, this result can best be obtained when the ends are at a very slightly lower moisture content than the body. With green material it will usually suffice to close-pile the specimens, cover the body with a damp or wet cloth, and expose the ends for a short time. For air-dry material, it may sometimes be advisable to pile the specimens in a similar manner and place them in a desiccator should the failures in test indicate that a slight end-drying is necessary.

58. Description of Compression Failures

58.1 Compression failures shall be classified according to the appearance of the fractured surface (Fig. 11). In case two or more kinds of failures develop, all shall be described in the order of their occurrence; thus, shearing followed by brooming. The failure shall also be sketched in its proper position on the data sheet.

59. Weight and Moisture Content

59.1 The specimen shall be weighed immediately before test, and after test a moisture section approximately 1 in. (2.5 cm) in length shall be cut from the body near the failure (Section 126).

60. Ring and Summer Wood Measurement

60.1 When practicable, the number of rings per inch (centimeter) and the proportion of summer wood shall be measured over a representative inch (centimeter) of cross section of the test specimen. In determining the proportion of summer wood, it is essential that the end surface be prepared so as to permit accurate summer wood measurement. When the fibers are broomed over at the ends from sawing, a light sanding, planing, or similar treatment of the ends is recommended.

Impact Bending

61. Size of Specimens

61.1 The impact bending tests shall be made on 2 by 2 by 30-in. (5 by 5 by 76-cm) specimens. The actual height and width at the center and the length shall be measured (Section 127).

62. Loading and Span

62.1 Center loading and a span length of 28 in. (70 cm) shall be used.

63. Bearing Block

63.1 A metal tup of curvature corresponding to the bearing block shown in Fig. 7 shall be used in applying the load.

64. Placement of Growth Rings

64.1 The specimen shall be placed so that the load will be applied through the bearing block to the tangential or flat-sawed surface nearest the pith.

65. Procedure

65.1 Make the tests by increment drops in a Hatt-Turner or similar impact machine (see Fig. 12). The first drop shall be 1 in. (2.5 cm), after which increase the drops by 1-in. (2.5-cm) increments until a height of 10 in. (25 cm) is reached. Then use a 2-in. (5-cm) increment until complete failure occurs or until a 6-in.

[7] See Fig. A2 for a sample compression-parallel-to-grain data sheet form.

(15-cm) deflection is reached.

66. Weight of Hammer

66.1 A 50-lb (22.5-kg) hammer shall be used when, with drops up to the capacity of the machine (about 68 in. (1.7 m) for the small Hatt-Turner impact machine), it is practically certain that complete failure or a 6-in. (15-cm) deflection will result for all specimens of a species. For all other cases a 100-lb (45-kg) hammer shall be used.

67. Deflection Records

67.1 When desired, graphical drum records[8] giving the deflection for each drop and the set, if any, shall be made until the first failure occurs. This record will also afford data from which the exact height of drop can be scaled for at least the first four falls.

68. Drop Causing Failure

68.1 The height of drop causing complete failure or a 6-in. (15-cm) deflection shall be observed for each specimen.

69. Description of Failure

69.1 The failure shall be sketched on the data sheet[9] and described in accordance with the directions for static bending under Section 51.

70. Weight and Moisture Content

70.1 The specimen shall be weighed immediately before test, and after test a moisture section approximately 1 in. (2.5 cm) in length shall be cut near the failure (Section 126).

TOUGHNESS

NOTE 5—A single-blow impact test on a small specimen is recognized as a valuable and desirable test. Several types of machines such as the Toughness, Izod, and Amsler have been used, but insufficient information is available to decide whether one procedure is superior to another, or whether the results by the different methods can be directly correlated. If the Toughness machine is used, the following procedure has been found satisfactory. To aid in standardization and to facilitate comparisons, the size of the toughness specimen has been made equal to that accepted internationally.

71. Size of Specimen

71.1 The toughness tests shall be made on).79 by 0.79 by 11-in. (2 by 2 by 28-cm) specinens. The actual height and width at the center and the length shall be measured (Section 127).

72. Loading and Span

72.1 Center loading and a span length of 9.47 in. (24 cm) shall be used. The load shall be applied to a radial or tangential surface on alternate specimens.

73. Bearing Block

73.1 An aluminum tup (Fig. 13) having a radius of $^3/_4$ in. (18 mm) shall be used in applying the load.

74. Apparatus and Procedure

74.1 Make the tests in a Forest Products Laboratory type toughness machine (Fig. 13). Adjust the machine before test so that the pendulum hangs truly vertical and adjust it to compensate for friction. Adjust the cable so that the load is applied to the specimen when the pendulum swings to 15 deg from the vertical so as to produce complete failure by the time the downward swing is completed. Choose the weight position and initial angle (30, 45, or 60 deg) of the pendulum so that complete failure of the specimen is obtained on one drop. Most satisfactory results are obtained when the difference between the initial and final angle is at least 10 deg.

75. Calculation

75.1 The initial and final angle shall be read to the nearest 0.1 deg by means of the vernier (Fig. 13) attached to the machine.[10] The toughness shall then be calculated as follows:

$$T = wL(\cos A_2 - \cos A_1)$$

where:

T = toughness (work per specimen), in. · lb (cm · kg),
w = weight of pendulum, lb (kg),
L = distance from center of the supporting axis to center of gravity of the pendulum, in. (cm),
A_1 = initial angle (Note 5), deg, and

[8] See Fig. A3 for a sample drum record.
[9] See Fig. A5 for a sample impact bending data sheet form. Figure A4 shows a sample data and computation card.
[10] See Fig. A6 for a sample data and computation sheet for the toughness test.

A_2 = final angle the pendulum makes with the vertical after failure of the test specimen, deg.

NOTE 6—Since friction is compensated for in the machine adjustment, the initial angle may be regarded as exactly 30, 45, or 60 deg, as the case may be.

76. Weight and Moisture Content

76.1 The specimen shall be weighed immediately before test, and after test a moisture section approximately 2 in. (5 cm) in length shall be cut from the body near the failure (Section 126).

COMPRESSION PERPENDICULAR TO GRAIN

77. Size of Specimens

77.1 The compression-perpendicular-to-grain tests shall be made on 2 by 2 by 6-in. (5 by 5 by 15-cm) specimens. The actual height, width, and length shall be measured (Section 127).

78. Loading

78.1 The load shall be applied through a metal bearing plate 2 in. (5 cm) in width, placed across the upper surface of the specimen at equal distances from the ends and at right angles to the length (Fig. 14).

79. Placement of Growth Rings

79.1 The specimens shall be placed so that the load will be applied through the bearing plate to a radial (quarter-sawed) surface.

80. Speed of Testing

80.1 The load shall be applied continuously throughout the test at a rate of motion of the movable crosshead of 0.012 in. (0.3 mm)/min (Section 128).

81. Load-Compression Curves

81.1 Load-compression curves[11] shall be taken for all specimens up to 0.1-in. (2.5-mm) compression, after which the test shall be discontinued. Compression shall be measured between the loading surfaces.

81.2 Deflection readings shall be taken to 0.0001 in. (0.002 mm).

82. Weight and Moisture Content

82.1 The specimen shall be weighed immediately before test, and after test a moisture section approximately 1 in. (2.5 cm) in length shall be cut adjacent to the part under load (Section 126).

HARDNESS

83. Size of Specimens

83.1 The hardness tests shall be made on 2 by 2 by 6-in. (5 by 5 by 15-cm) specimens. The acutal cross-section dimensions and length shall be measured (Section 127).

84. Procedure

84.1 Use the modified ball test with a "ball" 0.444 in. (1.13 cm) in diameter for determining hardness (Fig. 15). Record the load at which the "ball" has penetrated to one half its diameter, as determined by an electric circuit indicator or by the tightening of the collar against the specimen.

85. Number of Penetrations

85.1 Two penetrations shall be made on a tangential surface, two on a radial surface, and one on each end. The choice between the two radial and between the two tangential surfaces shall be such as to give a fair average of the piece. The penetrations shall be far enough from the edge to prevent splitting or chipping.[12]

86. Speed of Testing

86.1 The load shall be applied continuously throughout the test at a rate of motion of the movable crosshead of 0.25 in. (6 mm)/min (Section 128).

87. Weight and Moisture Content

87.1 The specimen shall be weighed immediately before test, and after test a moisture section approximately 1 in. (2.5 cm) in length shall be cut (Section 126).

SHEAR PARALLEL TO GRAIN

NOTE 7—The following describes one method of making the shear-parallel-to-grain test that has been extensively used and found satisfactory.

[11] See Fig. A7 for a sample compression-perpendicular to-grain data sheet form.

[12] See Fig. A8 for a sample data and computation sheet for the hardness test.

88. Size of Specimens

88.1 The shear-parallel-to-grain tests shall be made on 2 by 2 by 2$^1/_2$-in. (5 by 5 by 6.3-cm) specimens notched as illustrated in Fig. 16 to produce failure on a 2 by 2-in. (5 by 5-cm) surface. The actual dimensions of the shearing surface shall be measured (Section 127).

89. Procedure

89.1 Use a shear tool similar to that illustrated in Fig. 17, providing a $^1/_8$-in. (3-mm) offset between the inner edge of the supporting surface and the plane along which the failure occurs. Apply the load to, and support the specimen on, end-grain surfaces. Take care in placing the specimen in the shear tool to see that the crossbar is adjusted so that the edges of the specimen are vertical and the end rests evenly on the support over the contact area. Observe the maximum load only.

90. Speed of Testing

90.1 The load shall be applied continuously throughout the test at a rate of motion of the movable crosshead of 0.024 in. (0.6 mm)/min (Section 128).

91. Test Failures

91.1 The failure shall be sketched on the data sheet.[13] In all cases where the failure at the base of the specimen extends back onto the supporting surface, the test shall be culled.

92. Moisture Content

92.1 The portion of the test piece that is sheared off shall be used as a moisture specimen (Section 126).

CLEAVAGE

93. Size of Specimens

93.1 The cleavage tests shall be made on specimens of the form and size shown in Fig. 18. The actual width and length at minimum section shall be measured (Section 127).

94. Procedure

94.1 The specimens shall be held during est in grips as shown in Figs. 19 and 20. Oberve the maximum load only.

95. Speed of Testing

95.1 The load shall be applied continuously throughout the test at a rate of motion of the movable crosshead of 0.10 in. (2.5 mm)/min (Section 128).

96. Sketch of Failure

96.1 The failure shall be sketched on the data sheet.[14]

97. Moisture Content

97.1 One of the pieces remaining after failure, or a section split along the surface of failure, shall be used as a moisture specimen (Section 126).

TENSION PARALLEL TO GRAIN

NOTE 8—One method of determining the tension-parallel-to-grain strength of wood is given in the following procedure.

98. Size of Specimens

98.1 The tension-parallel-to-grain tests shall be made on specimens of the size and shape shown in Fig. 21. The specimen shall be so oriented that the direction of the annual rings at the critical section, as shown on the ends of the specimens, shall be perpendicular to the greater cross-sectional dimension. The actual cross-sectional dimensions at minimum section shall be measured (Section 127).

99. Procedure

99.1 Fasten the specimen in special grips (Fig. 22). Take load-extension curves for a 2-in. (5-cm) central gage length on all specimens. Continue the load-extension readings until the proportional limit is passed.

99.2 Read deformations to 0.0001 in. (0.002 mm).

99.3 Figure 22 illustrates gripping devices and a type of extensometer that have been found satisfactory.

100. Speed of Testing

100.1 The load shall be applied continuously throughout the test at a rate of motion

[13] See Fig. A9 for a sample data and computation sheet for the tangential-shear-parallel-to-grain test.

[14] See Fig. A10 for a sample data and computation sheet for the cleavage test.

of the movable crosshead of 0.05 in. (1 mm)/min (Section 128).

101. Sketch of Failure

101.1 The failure shall be sketched on the data sheet.[15]

102. Moisture Content

102.1 A moisture section about 3 in. (7.5 cm) in length shall be cut from the reduced section near the failure (Section 126).

TENSION PERPENDICULAR TO GRAIN

(Optional Test)

103. Size of Specimens

103.1 The tension-perpendicular-to-grain tests shall be made on specimens of the size and shape shown in Fig. 23. The actual width and length at minimum sections shall be measured (Section 127).

104. Procedure

104.1 Fasten the specimens during test in grips as shown in Figs. 24 and 25. Observe the maximum load only.

105. Speed of Testing

105.1 The load shall be applied continuously throughout the test at a rate of motion of the movable crosshead of 0.10 in. (2.5 mm)/min (Section 128).

106. Sketch of Failure

106.1 The failure shall be sketched on the data sheet.[16]

107 Moisture Content

107.1 One of the pieces remaining after failure, or a section split along the surface of failure, shall be used as a moisture specimen (Section 126).

NAIL WITHDRAWAL

(Optional Test)

NOTE 9—Presented herewith is a recommended procedure for making nail withdrawal tests. When necessary, alternative procedures that give comparable results may be used.

108. Size of Specimens

108.1 The nail withdrawal tests shall be made on 2 by 2 by 6-in. (5 by 5 by 15-cm) specimens. The actual cross-sectional dimensions and length shall be measured (Section 127).

109. Nails

109.1 Nails used for withdrawal tests shall be nominally 0.0985 in. (2.5 mm) in diameter.[17] Bright diamond-point nails shall be used, and all nails shall be cleaned before use to remove any coating or surface film that may be present as a result of manufacturing operations. Each nail shall be used but once.

110. Preparation of Specimens

110.1 Nails shall be driven at right angles to the face of the specimen to a total penetration of 1$^1/_4$ in. (3.2 cm). Two nails shall be driven on a tangential surface, two on a radial surface, and one on each end. The choice between the two radial and between the two tangential surfaces shall be such as to give a fair average of the piece. On radial and tangential faces, the nails shall be driven a sufficient distance from the edges and ends of the specimen to avoid splitting. In general, nails should not be driven closer than $^3/_4$ in. (18 mm) from the edge or 1$^1/_2$ in. (37 mm) from the end of a piece, and the two nails on a radial or tangential face should not be driven in line with each other or less than 2 in. (5 cm) apart.

111. Procedure

111.1 Withdraw all six nails in a single specimen immediately after driving. Fasten the specimens during the test in grips as shown in Figs. 26 and 27. Observe the maximum load only.[18]

112. Speed of Testing

112.1 The load shall be applied continuously throughout the test at a rate of motion of the movable crosshead of 0.075 in. (2 mm)/min (Section 128).

[15] See Fig. A11 for a sample tension-parallel-to-grain data and computation sheet.

[16] See Fig. A12 for a sample data and computation sheet for the tension-perpendicular-to-grain test.

[17] A fivepenny common nail meets this requirement. If difficulty is experienced in woods of higher density in pulling the nails without breaking the heads, a sevenpanny cement-coated sinker nail, with coating removed by use o a suitable solvent, may be used.

[18] See Fig. A13 for a sample nail-withdrawal test dat sheet form.

113. Weight and Moisture Content

113.1 The specimen shall be weighed immediately before driving the nails and after test a moisture section approximately 1 in. (2.5 cm) in length shall be cut from the body of the specimen (Section 126).

SPECIFIC GRAVITY AND SHRINKAGE IN VOLUME

114. Size of Specimens

114.1 The specific gravity and shrinkage in volume tests shall be made on 2 by 2 by 6-in. (5 by 5 by 15-cm) specimens. The actual cross-sectional dimensions and length shall be measured (Section 127).

115. Procedure

115.1 Obtain both specific gravity and shrinkage-in-volume determinations on the same specimen. These determinations should be made at approximately 12 percent moisture content and in the oven-dry condition.

115.2 A carbon impression of the end of the green specimen may be made on the back of the data sheet.[19] In like manner, a carbon impression of the same end may be made after the specimen has been conditioned (115.4 and 115.5).

115.3 Weigh the specimen when green (Section 126) and determine the volume by the immersion method.

115.4 Open-pile the green specimens after immersion and allow them to air-season under room conditions to a uniform moisture content of approximately 12 percent. The specimens should then be weighed and the volume determined by the immersion method.

115.5 Then, open-pile the specimens used for specific gravity and shrinkage determinations at 12 percent moisture content, or duplicate specimens on which green weight and volume measurements have been made prior to conditioning to approximately 12 percent moisture content, in an oven and dry at 103 ± 2 C until approximately constant weight is reached.

115.6 After over-drying, weigh the specimens (Section 126) and while still warm immerse them in a hot paraffin bath, taking care to remove them quickly to ensure a thin coating.

115.7 Determine the volume of the paraffin-coated specimen by immersion as before.

115.8 Figure 28 illustrates the apparatus used in determining the specific gravity and shrinkage in volume. The use of an automatic balance will facilitate increased rapidity and accuracy of measurements.

RADIAL AND TANGENTIAL SHRINKAGE

116. Size of Specimen

116.1 The radial- and tangential-shrinkage determinations shall be made on 1 by 4 by 1-in. (2.5 by 10 by 2.5-cm) specimens.

117. Initial Measurement

117.1 The specimen shall be measured across the 4-in. (10-cm) dimension in which the shrinkage is to be determined (Section 127).

118. Weight

118.1 The specimen shall be weighed when green and after subsequent oven-drying (Section 126).

119. Drying

119.1 The green specimens shall be open-piled and allowed to air-season under room conditions to a uniform moisture content of approximately 12 percent.

119.2 The specimens shall then be open-piled in an oven and dried at 103 ± 2 C until approximately constant weight in attained.

120. Final Measurement

120.1 Measurements shall be made on the air-dry and on the oven-dry specimens.[20]

121. Method of Measurement

121.1 Figure 29 illustrates the method of making the radial- and tangential-shrinkage measurements. An ordinary micrometer of required accuracy is suitable for this work (Section 127).

---

[19] See Fig. A14 for a sample data and computation sheet for the specific gravity and shrinkage-in-volume test.

[20] See Fig. A15 for a sample data and computation sheet for the radial- and tangential-shrinkage test.

MOISTURE DETERMINATION

122. Selection

122.1 The sample for moisture determinations of each test specimen shall be selected as hereinbefore described for each test.

123. Weighing

123.1 Immediately after obtaining the moisture sample, all loose splinters shall be removed and the sample shall be weighed (Section 126).

124. Drying

124.1 The moisture samples shall be open-piled in an oven and dried at a temperature of 103 ± 2 C until approximately constant weight is attained, after which the oven-dry weight shall be determined (Section 125).

125. Moisture Content

125.1 The loss in weight, expressed in percentage of the over-dry weight as above determined, shall be considered the moisture content of the specimen.

PERMISSIBLE VARIATIONS

126. Weights

126.1 The weight of test specimens and of moisture samples shall be determined to an accuracy of not less than ±0.2 percent.

127. Measurements

127.1 Measurements of test specimens shall be made to an accuracy of not less than ±0.3 percent, except that in no case shall the measurements be made to less than 0.01 in. (0.2 mm), except measurements of the radial and tangential shrinkage specimens that shall be made to the nearest 0.001 in. (0.02 mm).

128. Testing Machine Speeds

128.1 The testing machine speed used should not vary by more than ±25 percent from that specified for a given test. If the specified speed cannot be obtained, the speed used shall be recorded on the data sheet. The crosshead speed shall mean the free-running or no-load speed of crosshead for testing machines of the mechanical drive type and the loaded crosshead speed for testing machines of the hydraulic loading type.

CALIBRATION

129. Calibration

129.1 All apparatus used in obtaining data shall be calibrated at sufficiently frequent intervals to ensure accuracy.[21]

[21] ASTM Methods E 4, Verification of Testing Machines, *Annual Book of ASTM Standards*, Part 10.

PART II. SECONDARY METHODS

Part II, Secondary Methods, is intended for use in evaluating the properties of wood only when relatively small trees, generally less than 12 in. (30 cm) in diameter, are available to provide the test specimens and only when such trees because of crook, cross grain, knots or other defects are of such quality that the longer clear, straight-grained specimens required by Part I, Primary Methods, cannot reasonably be obtained. Whenever possible, the procedure for Part I, Primary Methods, shall be used regardless of size of trees.

INTRODUCTION

The standard methods of testing small clear specimens of timber, Part I, Primary Methods, provide for cutting the bolts (log sections) systematically into sticks of nominal $2^1/_2$ by $2^1/_2$ in. (6 by 6 cm) in cross section, that are later surfaced to provide the test specimens 2 by 2 in. (5 by 5 cm) in cross section, on which the system is based. These methods have served as an excellent basis for the evaluation of the various mechanical and related physical properties of different species of wood. They have beer extensively used, and a large amount of data based on these methods have been obtained and published.

The 2 by 2-in. test specimen has the advantage that it embraces a number of growth rings, is less influenced by springwood and summerwood differences than smaller specimens, and is large enough to represent a considerable proportion of the material. Because of the cross-sectional size and the length of specimen required for some of the tests (30 in. (76 cm) for static and impact bending) it is, however, sometimes difficult to obtain test specimens in adequate number and entirely free of defects from bolts representing smaller trees, particularly trees under 12 to 15 in. (30 to 38 cm) in diameter. With increasing need for evaluating the properties of species involving smaller trees, and the increasing importance of second-growth timber that is expected to be harvested much before it reaches the sizes attained in virgin stands, there has developed a need for secondary methods of test in which at least the longer test specimens are smaller than 2 by 2 in. in cross section. It is axiomatic that test results are intimately related to and dependent upon the test methods employed. The problem has hence been to develop secondary methods that give test results directly comparable to those obtained by the present primary methods employing 2 by 2-in. cross section for all test specimens, and thus ensure a continuing accumulation of data directly comparable to the extensive data already available. Such methods are provided by Part II, Secondary Methods, for testing small clear specimens of timber.

The exceedingly rapid rate of growth and corresponding wide annual rings in much second-growth material, together with the desirability of incorporating more than a single year's growth increment in a test specimen, has necessitated limiting the minimum cross section of test piece in these secondary methods to 1 by 1 in. (2.5 by 2.5 cm). This cross section is established for the compression-parallel-to-grain and static-bending tests. The 2 by 2-in. cross section is retained for impact bending, and for the other tests not requiring specimens longer than 6 in. (15 cm) namely, compression perpendicular to grain, hardness, shear parallel to grain, cleavage, and tension perpendicular to grain. Toughness and tension parallel to grain are special tests based on specimens of smaller cross section.

Investigations have shown that for the more important properties obtained from static bending and compression-parallel-to-grain tests (modulus of rupture and modulus of elasticity in static bending, and maximum crushing strength and modulus of elasticity in compression parallel to grain) results from specimens of 1 by 1 in. in cross section can be substituted directly for those obtained from specimens of 2 by 2 in. in cross section with but little error. Present data indicate that values of fiber stress at proportional limit in static bending and in compression parallel to grain may be slightly higher for the 1 by 1-in. than for the 2 by 2-in. specimens, based on standardized testing procedures appropriate to the two sizes, but no special reason why this should be expected is apparent. The work values in static bending, however, are related somewhat to size of specimen, and total work also to the arbitrary load and deflection limit established for terminating the test. The relationship of work values for the two sizes also varies among different species, hence work values in static bending should not be regarded as directly comparable as between the Primary Methods and the Secondary Methods. In reporting results of tests it is recommended that the size of specimen be given, or that the data be referenced to the Primary or Secondary Methods.

Since exactly the same size and form of test specimens for compression perpendicular to grain, hardness, impact bending, shear parallel to grain, cleavage, tension parallel to grain, tension perpendicular to grain, and toughness are used for the Primary and the Secondary Methods, identical test results are obviously obtained.

Since the procedure for the Secondary Methods for many features, such as in selection and care of material, and in conducting certain tests, is identical with the Primary Methods, the Secondary Methods presented herewith are referenced to the Primary Methods, and procedure is given only where it differs therefrom. For convenience the section numbers in the Secondary Methods correspond in the last two digits with the numbering of the Primary Methods but the numbering begins with 201 in the former. Thus Section 201 in the Secondary Methods corresponds in subject matter to Section 1 of the Primary Methods, etc.

201. Scope

201.1 See Section 1.

202. Summary of Method

202.1 See 2.1.

202.2 The procedures for collection and preparation of the material for testing and for the various tests appear in the following order:

	Section
Collection of Material:	
Selection	203 to 207
Field Marking	208 to 212
Field Descriptions	213
Preparation for Shipment	214
Disposition at Destination:	
Storage of Logs at Destination	215
Photographing, Sawing, and Final Marking	216 to 218
Matching for Tests of Air-Dry Material	219 and 220
Disposition of Sticks	221 and 222
Order, Selection, and Number of Tests:	
Order of Tests	223
Selection of Specimens	224 to 226
Number of Tests for Each Bolt	227 to 240
Photographs of Sticks	241
Control of Moisture Content and Temperature	242 and 243
Record of Heartwood and Sapwood	244
Static Bending	245 to 252
Compression Parallel to Grain	253 to 260
Impact Bending	261 to 270
Toughness	271 to 276
Compression Perpendicular to Grain	277 to 282
Hardness	283 to 287
Shear Parallel to Grain	288 to 292
Cleavage	293 to 297
Tension Parallel to Grain	298 to 302
Tension Perpendicular to Grain	303 to 307
Nail Withdrawal	308 to 313
Specific Gravity and Shrinkage in Volume	314 and 315
Radial and Tangential Shrinkage	316 to 321
Moisture Determination	322 to 325
Permissible Variations	326 to 328
Calibration	329

COLLECTION OF MATERIAL

Selection

203. Authentic Identification

203.1 See Section 3.

204. Selection and Number of Trees

204.1 For each species to be tested, at least ten trees representative of the species shall be selected.

205. Selection and Number of Bolts

205.1 See 5.2.

205.2 From the other trees called for in Section 204, take the 8-ft (2.4-m) section[4] (*c-d* bolts) next above the 8-ft butt log if this section falls within the merchantable length of the tree, otherwise taken the 8-ft section comprising the two highest standard bolts (multiples of 4-ft (1.2-m)) within the merchantable length.

206. Substitution of Flitches for Bolts

206.1 For the small trees to which this secondary method is applicable, bolts representing the full diameter of the log are required, and flitches should not be substituted for bolts.

207. Selection for Important Species

207.1 For important species of wide geographical distribution, test material shall be selected from two or more localities or sites. The number of trees of a species selected from each site or locality shall conform to the requirements of Sections 204 and 205.

Field Marking

208. Tree Designation

208.1 See Section 8.

209. Bolt Designation

209.1 See Section 9.

210. Marking

210.1 See Section 10.

211. Indication of Cardinal Point

211.1 See Section 11.

212. Shipment Number

212.1 See Section 12.

Field Descriptions

213. Field Descriptions

213.1 See Section 13.

Preparation for Shipment

214. Preparation for Shipment

214.1 See Section 14.

DISPOSITION AT DESTINATION

Storage of Logs at Destination

215. Storage of Logs

215.1 See Section 15.

Photographing, Sawing, and Final Marking

NOTE 10—In sawing, marking, and selecting test sticks, the aim should be to obtain specimens representative of the material collected. The procedure described herein is one that has been found satisfactory for most species.

216. Photographing Ends of Bolts

216.1 See Section 16.

217. Sawing of Bolts

217.1 All bolts shall be marked on the top end into $2^1/_2$-in. (6 by 6-cm) or $1^1/_4$ by $1^1/_4$-in. (3 by 3-cm) squares as shown in Fig. 30, and sawed into nominal $2^1/_2$ or $1^1/_4$-in. sticks. The letters *N*, *E*, *S*, and *W* indicate the cardinal points. For trees that are not circular or for intermediate sizes where more than the required distance is available, sticks may be moved outward toward the periphery to whatever position between the pith and bark would products sticks with a minimum of defects. For example, in trees 9 to 12 in. (23 to 30 cm) in diameter, sticks *N*1-4 and *S*1-4 could be moved outward as much as $1^1/_2$ in. (3.8 cm) if desired.

218. Marking of Test Sticks

218.1 All test sticks shall bear the shipment number, the tree number, stick number, and bolt designation, to be known respectively as Shipment No., Piece No., Stick No., and mark. Thus, 800-1-*N*1-4*d* represents the $2^1/_2$ by $2^1/_2$-in. (6 by 6 cm) Stick *N*1-4 of Bolt *d*, Tree 1, Shipment 800, and 800-1-*W*4*c* represents the $1^1/_4$ by $1^1/_4$-in. (3 by 3-cm) Stick *W*4 of Bolt *c*, Tree 1, Shipment 800.

Matching for Tests of Air-Dry Material

219. Composite Bolts

219.1 The collection of material (Section 205) has been arranged to provide for tests of both green and air-dry specimens that are closely matched by selection from adjacent parts of the same tree. The 8-ft (2.4-m) long bolts, after being marked in accordance with Section 217, shall be sawed into $2^1/_2$ by $2^1/_2$-in. (6 by 6-cm) or $1^1/_4$ by $1^1/_4$-in. (3 by 3-cm) by 8-ft (2.4-m) sticks, and numbered and lettered in accordance with Section 218. Each 8-ft stick shall then be cut into two 4-ft (1.2-m) pieces, making sure that each part carries the proper designation and bolt letter. If the 8-ft bolt is not straight it may be found more desirable to cut it into 4-ft lengths before cutting the $2^1/_2$- or $1^1/_4$-in. square pieces. If this is done care shall be taken to secure end-matched sticks in the two 4-ft bolts, and to ensure that each part carries the proper designation and bolt letter.

219.2 Part of the $2^1/_2$ by $2^1/_2$-in. and $1^1/_4$ by $1^1/_4$-in. by 4-ft sticks from each 8-ft bolt is to provide specimens to be tested green (unseasoned) and the other part is to provide specimens to be air-dried and tested. To afford matching, the 4-ft sticks of one bolt shall be interchanged with the 4-ft sticks of the next adjacent bolt from the same tree to form two composite bolts, each being complete and being made of equal portions of the adjacent 4-ft bolts. The sticks from one of these composite bolts shall be tested green and those from the other shall be tested after air-drying. Thus, the sticks of each composite bolt shall be regarded as if they were from the same bolt.

219.3 The above procedure provides for end-to-end matching (end matching) of sticks to be tested air-dry with those to be tested green, which is to be preferred when practicable. If, because of the nature of the material, end matching is not practicable, side matching of $1^1/_4$ by $1^1/_4$-in. sticks may be used.

220. Schedule for Forming Composite Bolts

220.1 The division of sticks into composite

bolts, part to be tested green and part to be air-dried and tested, shall be made according to the following schedule, in which the numbers refer to stick numbers:

Selection of Sticks from Bolts

Composite Bolt (ab, cd, etc.) to Be Tested Green:

Lower bolt	*N*1–4, *S*5–8, *E*4, *E*5, *E*8, *W*4, *W*5, *W*8
Upper bolt	*S*1–4, *N*5–8, *E*3, *E*6, *E*7, *W*3, *W*6, *W*7

Composite Bolt (ab, cd, etc.) to Be Air-Dried and Tested:

Lower bolt	*S*1–4, *N*5–8, *E*3, *E*6, *E*7, *W*3, *W*6, *W*7
Upper bolt	*N*1–4, *S*5–8, *E*4, *E*5, *E*8, *W*4, *W*5, *W*8

220.2 As an example of composite bolts, assume that the full cross-section, Fig. 30, represents the end of an 8-ft (2.4-m) section comprising the *c* and *d* bolts.

220.2 The following sticks are selected for the composite bolt to be tested green:

2 1/2 by 2 1/2-in. (6 by 6-cm) sticks:

*N*1–4*c*, *N*5–8*d*, *S*1–4*d*, *S*5–8*c*

1 1/4 by 1 1/4-in. (3 by 3-cm) sticks:

*E*3*d*, *E*4*c*, *E*5*c*, *E*6*d*, *E*7*d*, *E*8*c*, *W*3*d*, *W*4*c*, *W*5*c*, *W*6*d*, *W*7*d*, *W*8*c*

220.2.2 The following sticks are selected for the composite bolt to be air-dried and tested:

2 1/2 by 2 1/2-in. (6 by 6-cm) sticks:

*N*1–4*d*, *N*5–8*c*, *S*1–4*c*, *S*5–8*d*

1 1/4 by 1 1/4-in. (3 by 3-cm) sticks:

*E*3*c*, *E*4*d*, *E*5*d*, *E*6*c*, *E*7*c*, *E*8*d*, *W*3*c*, *W*4*d*, *W*5*d*, *W*6*c*, *W*7*c*, *W*8*d*

Disposition of Sticks

221. Green Material

221.1 The sticks (2 1/2 by 2 1/2 in. by 4 ft (6 by 6 cm by 1.2 m), or 1 1/4 by 1 1/4 in. by 4 ft (3 by 3 cm by 1.2 m)) to be tested green, shall be kept in an unseasoned condition, while awaiting preparation for test, by being stored in a framed pit or other suitable container where they shall be close piled and covered with damp sawdust, or in some other suitable manner. As material is required for test, it shall be removed from this pit or container, surfaced on all four sides to 2 by 2 in. (5 by 5 cm) or 1 by 1 in. (2.5 by 2.5 cm) cross section, sawed to test size, and kept covered with a damp cloth in a tightly closed container at a temperature of about 68 ± 6 F (20 ± 3 C) (see Note 3, 22.5) until the time of test. Care shall be taken to avoid as much as possible the storage of green material in any form. Sticks to be tested in a green condition usually should not be sawed from the log form in quantities greater than is required to meet the testing demands for from a few days to not more than 2 weeks, depending on the prevailing conditions.

222. Air-Dry Material

222.1 The ends of the sticks to be air-dried (2 1/2 by 2 1/2 in. by 4 ft (6 by 6 cm by 1.2 m), or 1 1/4 by 1 1/4 in. by 4 ft (3 by 3 cm by 1.2 m)) shall be dipped in melted paraffin or other substance suitable to retard checking. The material shall be piled so as to have a space of at least 1/2 in. (1.3 cm) on each side of each stick to permit circulation of air. The material shall be stored in a place allowing free access of air, but protected from sunshine, rain, snow, and moisture from the ground. The sticks in drying shall not be subjected to artificial heat.

222.2 See 22.2.

222.3 When the material has reached equilibrium, moisture sections approximately 1 in. (2.5 cm) in length shall be taken from about 10 percent of the sticks to determine the actual moisture content. These moisture specimens shall be cut not less than 1 ft (0.3 m) from the ends of the sticks, and in such a way as to prevent any appreciable loss of material for testing. When conditioned to approximately 12 percent moisture content the sticks shall be surfaced on four sides to 2 by 2 in. (5 by 5 cm) or 1 by 1 in. (2.5 by 2.5 cm) in cross-section, sawed to test size, and tested.

222.4 See 22.4.

222.5 See 22.5.

Order, Selection, and Number of Tests

Order of Tests

223. Order of Tests

223.1 See Section 23.

Selection of Specimens

224. Preference in Selecting Specimens

224.1 In case the material from a given bolt should be insufficient to furnish all the test specimens hereinafter required, additional bolts may be selected. If additional material is not available, the preferential order of mechanical tests to be used in selecting specimens shall be as follows: static bending, compression parallel to grain, impact bending, toughness, compression perpendicular to grain, hardness, shear parallel to grain, cleavage, tension parallel to grain, tension perpendicular to grain, and nail withdrawal.

225. Test Specimens from Bending Specimens After Failure

225.1 See Section 25.

226. Quality of Test Material

226.1 See Section 26.

Number of Test for Each Bolt

227. Static Bending

227.1 One static bending specimen shall be taken from each pair of $1\,^1/_4$ by $1\,^1/_4$-in. (3 by 3-cm) sticks. A pair consists of two adjacent sticks equidistant from the pith, as *W7* and *W8*, Fig. 30. In the composite bolts tested to afford a comparison of the strength of green and air-dry material, the pair of sticks shall be constituted as above, except that the sticks in this case will be from different bolts. Thus, *W3d* and *W4c* constitute one pair of sticks to be tested green, and *W3c* and *W4d* the corresponding pair to be tested air-dry (Section 220). Effort shall be made to select sticks so that both bolts of the composite bolt are represented in the green and the dry tests.

228. Compression Parallel to Grain

228.1 One compression-parallel-to-grain specimen shall be taken from each pair of $1\,^1/_4$ by $1\,^1/_4$-in. (3 by 3-cm) sticks, as for static bending. Effort shall be made to select sticks so that both bolts of a composite bolt are represented in the green and the dry tests. Load-compression curves shall preferably be taken on all of the specimens.

229. Impact Bending

229.1 One impact bending specimen shall be taken from each of 50 percent of the $2\,^1/_2$ by $2\,^1/_2$-in. (6 by 6-cm) sticks from each bolt. Effort shall be made to secure the best possible representation of the cross section.

230. Toughness

230.1 Two toughness specimens shall be selected from each stick used for static bending.

231. Compression Perpendicular to Grain

231.1 One compression-perpendicular-to-grain specimen shall be taken from each $2\,^1/_2$ by $2\,^1/_2$-in. (6 by 6-cm) stick selected for impact-bending specimens.

232. Hardness

232.1 One hardness specimen shall be taken from each $2\,^1/_2$ by $2\,^1/_2$-in. (6 by 6-cm) stick used for impact bending and compression-perpendicular-to-grain tests.

233. Shear Parallel to Grain

233.1 Two shear-parallel-to-grain specimens shall be selected from each of the $2\,^1/_2$ by $2\,^1/_2$-in. (6 by 6-cm) sticks from each bolt remaining after selection of sticks for impact bending tests. One of each pair of specimens from the same stick shall be tested in radial shear (surface of failure radial) and the other in tangential shear (surface of failure tangential).

234. Cleavage Perpendicular to Grain

234.1 Two cleavage specimens shall be taken from each of the $2\,^1/_2$ by $2\,^1/_2$-in. (6 by 6-cm) sticks selected for shear-parallel-to-grain specimens. One of each pair of specimens from the same stick shall be tested in radial cleavage (surface of failure radial) and the other in tangential cleavage (surface of failure tangential).

235. Tension Parallel to Grain

235.1 One tension-parallel-to-grain specimen shall be taken from each pair of $1\,^1/_4$ by $1\,^1/_4$-in. (3 by 3-cm) sticks.

236. Tension Perpendicular to Grain

236.1 Two tension-perpendicular-to-grain specimens shall be taken from each of the $2\,^1/_2$ by $2\,^1/_2$-in. (6 by 6-cm) sticks selected for shear-parallel-to-grain specimens. One of each pair of specimens from the same stick shall be tested in radial tension (surface of failure radial) and the other in tangential tension (surface of failure tangential).

237. Nail Withdrawal

237.1 Four nail-withdrawal specimens shall

be selected from the uninjured portion of 2 by 2-in. (5 by 5-cm) specimens from other tests, selected to give satisfactory representation of the cross-section of the bolt. Two specimens shall be tested in the green condition and two in the air-dry condition.

238. Specific Gravity and Shrinkage in Volume

238.1 Two to four specific gravity and shrinkage-in-volume specimens shall be selected from the unused portions of $2^1/_2$ by $2^1/_2$-in. (6 by 6-cm) sticks, selected to give the best possible representation of the cross section and tree heights included in composite bolts. These specimens shall be selected only from the sticks to be tested in a green condition.

239. Radial Shrinkage

239.1 Two radial-shrinkage specimens shall be obtained from each *d* bolt, and where possible from the upper bolt of each pair of bolts selected at other heights in the tree. They shall be cut from disks cut from near the end of the bolt. Care shall be taken to see that the disks are green and have not been affected by shrinking and checking, which is common near the end of the bolt. The specimens shall not be surfaced. Radial shrinkage specimens shall be cut with their greatest dimension in the radial direction. One specimen shall be taken outward from near the pith and the other inward from near the bark.

240. Tangential Shrinkage

240.1 Two tangential-shrinkage specimens shall be obtained from each *d* bolt, and where possible from the upper bolt of each pair of bolts selected at other heights in the tree. They shall be selected at the same time and in a manner similar to radial-shrinkage specimens (Section 239), except that the greatest dimension shall be in a tangential direction. The specimens shall not be surfaced. One shall be taken from the near periphery and the other nearer the pith to represent an earlier period of growth. The tangential-shrinkage specimens shall be adjacent to the specimens selected for radial shrinkage.

PHOTOGRAPHS OF STICKS

241. Sticks to be Photographed

241.1 See Section 41.

CONTROL OF MOISTURE CONTENT AND TEMPERATURE

NOTE 11—In recognition of the significant influence of temperature and humidity on the strength of wood, it is highly desirable that these factors be controlled to ensure comparable test results.

242. Control of Moisture Content

242.1 See Section 42.

243. Control of Temperature

243.1 See Section 43.

RECORD OF HEARTWOOD AND SAPWOOD

244. Proportion of Sapwood

244.1 See Section 44.

STATIC BENDING

245. Size of Specimens

245.1 The static bending tests shall be made on nominal 1 by 1 by 16-in. (2.5 by 2.5 by 41-cm) specimens. The actual height and width at the center, and the length shall be measured (Section 127).

246. Loading Span and Supports

246.1 Center loading and span length of 14 in. (35 cm) shall be used. Both supporting knife edges shall be provided with bearing plates and rollers of such thickness that the distance from the point of support to the central plane is not greater than the depth of the specimen (Fig. 6). The knife edges shall be adjustable laterally to permit adjustment for slight twist or warp in the specimen.[5] Alternatively, the method of supporting the specimen in trunnion-type supports that are free to move in a horizontal direction may be employed.

247. Bearing Block

247.1 A bearing block having a radius of curvature of $1^1/_2$ in. (3.7 cm) for a chord length of not less than 2 in. (5 cm) shall be used.

248. Placement of Growth Rings

248.1 See Section 48.

249. Speed of Testing

249.1 The load shall be applied continuously throughout the test at a rate of the movable crosshead of 0.05 in. (1.3 mm)/min (Section 128).

250. Load-Deflection Curves

250.1 Load-deflection curves shall be taken to or beyond the maximum load for all static bending tests. In at least one-third of the tests, the curves shall be continued to a 3-in. (7.5-cm) deflection, or until the specimen fails to support a load of 50 lb (22.5 kg).

250.2 See 50.2.

250.3 See 50.3.

250.4 See 50.4.

251. Description of Static Bending Failures

251.1 See Section 51.

252. Weight and Moisture Content

252.1 See Section 52.

COMPRESSION PARALLEL TO GRAIN

253. Size of Specimens

253.1 The compression-parallel-to-grain tests shall be made on nominal 1 by 1 by 4-in. (2.5 by 2.5 by 10-cm) specimens. The actual cross-section dimensions and the length shall be measured (Section 127).

254. End Surfaces Parallel

254.1 See Section 54.

255. Speed of Testing

255.1 See Section 55.

256. Load-Compression Curves

256.1 Load-compression curves shall be taken over a central gage length of 2 in. (5 cm), and preferably on all of the specimens. Load-compression readings shall be continued until the proportional limit is well passed, as indicated by the curve.[7]

256.2 Deformations shall be read to 0.0001 n. (0.002 mm).

256.3 Figures 9 and 10 illustrate two types of compressometers of 6-in. (15-cm) gage length that have been found satisfactory for wood testing. Similar apparatus is available for measurements of compression over a 2-in. (5-cm) gage length.

257. Position of Test Failures

257.1 See Section 57.

258. Description of Compression Failures

258.1 See Section 58.

259. Weight and Moisture Content

259.1 See Section 59.

260. Ring and Summer Wood Measurement

260.1 See Section 60.

IMPACT BENDING

261 to 270. See Sections 61 to 70, inclusive.

TOUGHNESS

271 to 276. See Sections 71 to 76, inclusive.

NOTE 12—A single-blow impact test on a small specimen is recognized as a valuable and desirable test. Several types of machines such as the Toughness, Izod, and Amsler have been used, but insufficient information is available to decide whether one procedure is superior to another, or whether the results by the different methods can be directly correlated. If the Toughness machine is used, the procedure described in Sections 71 to 76, inclusive, has been found satisfactory. To aid in standardization and to facilitate comparisons, the size of the toughness specimen has been made equal to that accepted internationally.

COMPRESSION PERPENDICULAR TO GRAIN

277 to 282. See Sections 77 to 82, inclusive.

HARDNESS

283 to 287. See Sections 83 to 87, inclusive.

SHEAR PARALLEL TO GRAIN

288 to 292. See Sections 88 to 92, inclusive.

NOTE 13—Sections 88 to 92, inclusive, describe one method of making the shear-prallel-to-grain test that has been extensively used and found satisfactory.

CLEAVAGE

293 to 298. See Sections 93 to 97, inclusive.

TENSION PARALLEL TO GRAIN

298 to 302. See Sections 98 to 102, inclu-

sive.

NOTE 14—One method of determining the tension-parallel-to-grain strength of wood is given in Sections 98 to 102, inclusive.

TENSION PERPENDICULAR TO GRAIN

(*Optional Test*)

303 to 307. See Sections 103 to 107, inclusive.

NAIL WITHDRAWAL

(*Optional Test*)

308 to 313. See Sections 108 to 113, inclusive.

NOTE 15—A recommended procedure for nail-withdrawal test is given in Sections 108 to 113, inclusive. When necessary, alternate procedures that give comparable results may be used.

SPECIFIC GRAVITY AND SHRINKAGE IN VOLUME

314 and 315. See Sections 114 and 115.

RADIAL AND TANGENTIAL SHRINKAGE

316 to 321. See Sections 116 to 121, inclusive.

MOISTURE DETERMINATION

322 to 325. See Sections 122 to 125, inclusive.

PERMISSIBLE VARIATIONS

326 to 328. See Sections 126 to 128, inclusive.

CALIBRATION

329. See Section 129.

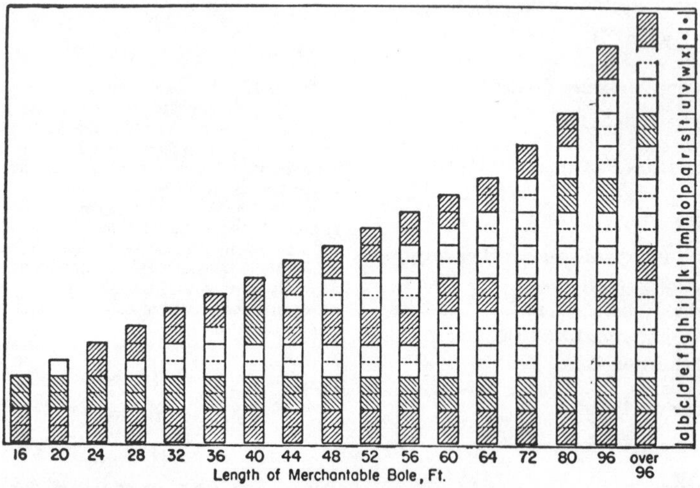

Letters *a*, *b*, *c*, etc. indicate 4-ft (1.2-m) units of the merchantable length, called bolts, and designate height in the tree.
* Indicates bolts to be taken from the top of the merchantable length, to be lettered appropriately according to their actual height in the tree.

FIG. 1 Diagram Indicating Number and Position of Bolts to be Collected from Trees Having Boles of Various Merchantable Lengths, to Evaluate Effect of Height in Tree on Properties of the Wood.

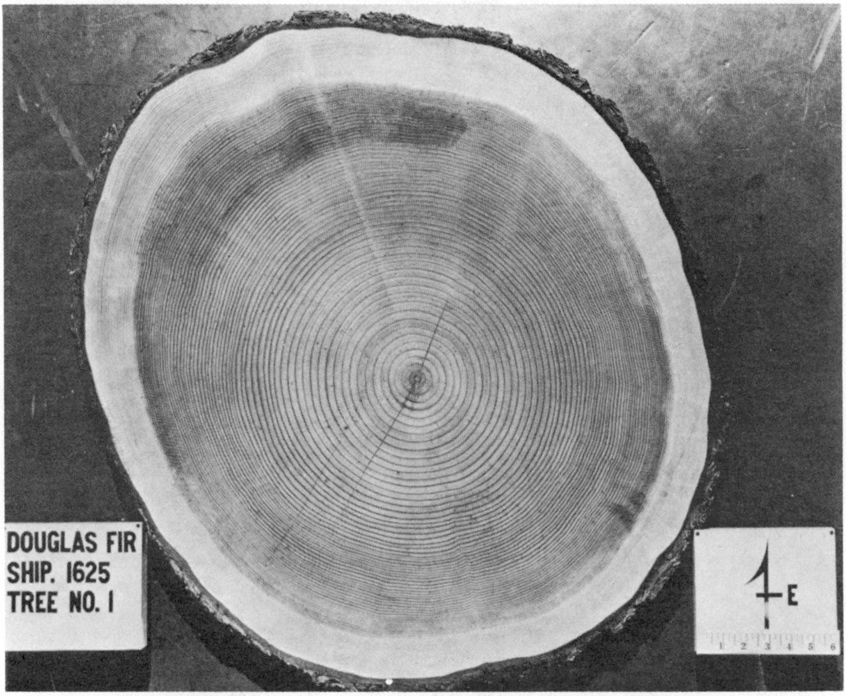

FIG. 2 Section of Log Selected for Test Material.

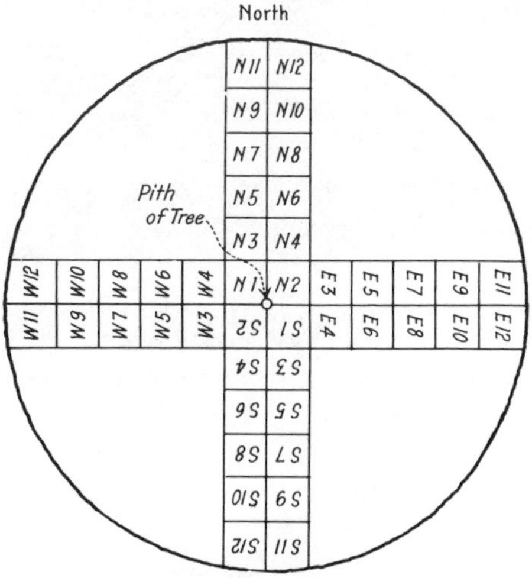

FIG. 3 Sketch Showing Method of Cutting Up the Bolt and Marking the Sticks.

FIG. 4 Cross-Sections of Bending Specimens Showing Different Rates of Growth of Longleaf Pine (2 by 2-in. (5 by 5-cm) Specimens).

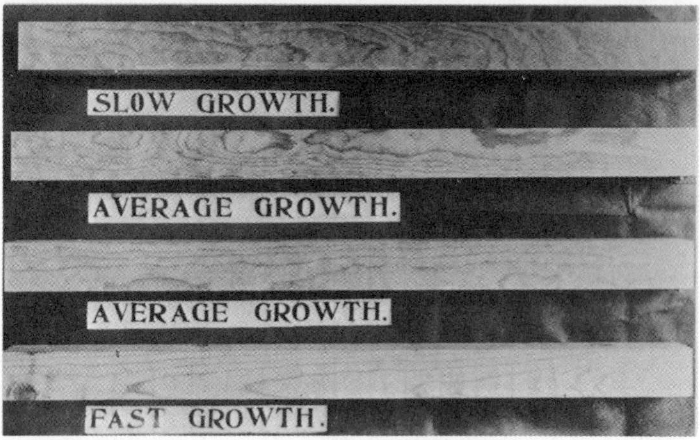

FIG. 5 Tangential Surfaces of Bending Specimens of Different Rates of Growth of Jeffrey Pine (2 by 2 by 30-in. (5 by 5 by 76-cm) Specimens).

FIG. 6 Static Bending Test Assembly Showing Method of Load Application, Specimen Supported on Rollers and Laterally Adjustable Knife Edges, and Method of Measuring Deflection at Neutral Axis by Means of Yoke and Dial Attachment. (Adjustable scale mounted on loading head is used to measure increments of deformation beyond the dial capacity.)

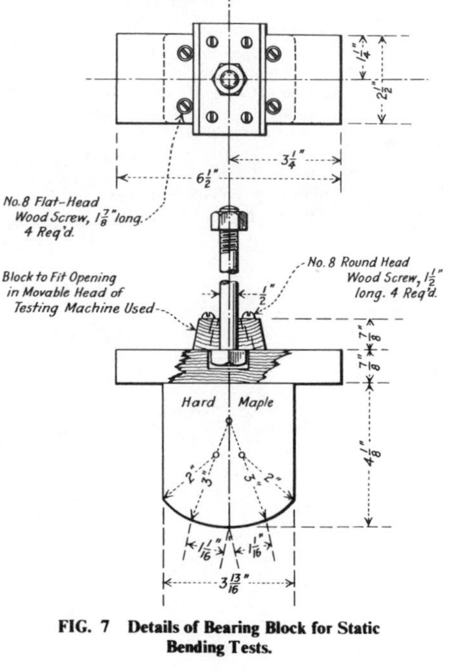

FIG. 7 Details of Bearing Block for Static Bending Tests.

(a) Simple Tension.
(Side View)

(b) Cross-Grain Tension.*
(Side View)

(c) Splintering Tension.
(View of Tension Surface)

(d) Brash Tension.
(View of Tension Surface)

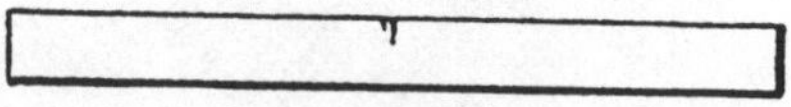

(e) Compression.
(Side View)

(f) Horizontal Shear.
(Side View)

* The term "cross grain" shall be considered to include all deviations of grain from the direction of the longitudinal axis or longitudinal edges of the specimen. It should be noted that spiral grain may be present even to a serious extent without being evident from a casual observation.

The presence of cross grain having a slope that deviates more than 1 in 20 from the longitudinal edges of the specimen shall be cause for culling the test.

FIG. 8 Types of Failures in Static Bending.

FIG. 9 Compression-Parallel-to-Grain Test Assembly Showing Method of Measuring Deformations by Means of Roller-Type Compressometer.

FIG. 10 Compression-Parallel-to-Grain Test Assembly Using an Automatic Autographic Type of Compressometer to Measure Deformations. (The wire in the lower right-hand corner connects the compressometer with the recording unit.)

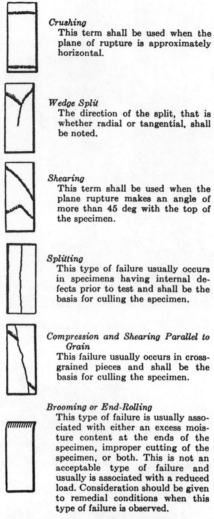

Crushing
This term shall be used when the plane of rupture is approximately horizontal.

Wedge Split
The direction of the split, that is whether radial or tangential, shall be noted.

Shearing
This term shall be used when the plane rupture makes an angle of more than 45 deg with the top of the specimen.

Splitting
This type of failure usually occurs in specimens having internal defects prior to test and shall be the basis for culling the specimen.

Compression and Shearing Parallel to Grain
This failure usually occurs in cross-grained pieces and shall be the basis for culling the specimen.

Brooming or End-Rolling
This type of failure is usually associated with either an excess moisture content at the ends of the specimen, improper cutting of the specimen, or both. This is not an acceptable type of failure and usually is associated with a reduced load. Consideration should be given to remedial conditions when this type of failure is observed.

FIG. 11 Types of Failures in Compression.

FIG. 12 Hatt-Turner Impact Machine, Illustrating Method of Conducting Impact Bending Test.

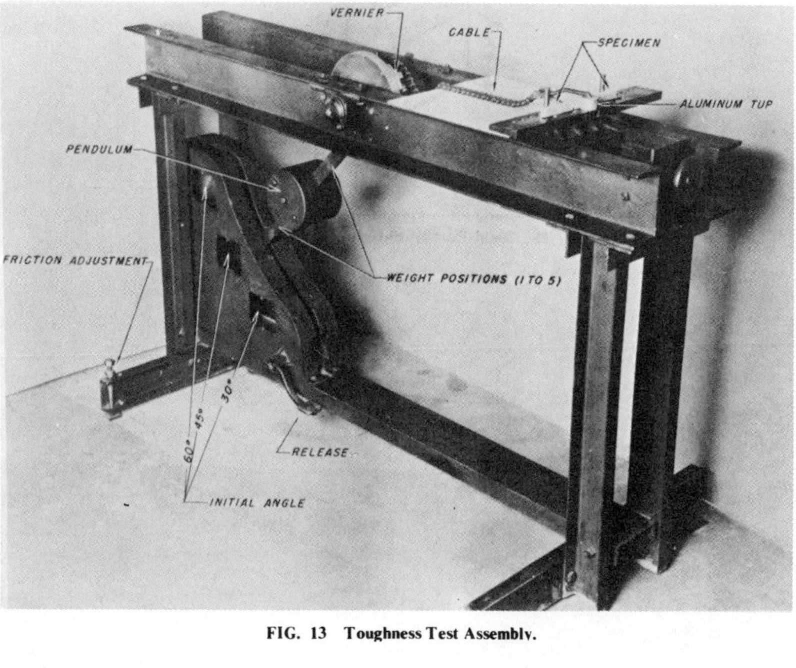

FIG. 13 Toughness Test Assembly.

FIG. 14 Compression-Perpendicular-to-Grain Test Assembly Showing Method of Load Application and Measurement of Deformation by Means of Averaging-Type Compressometer.

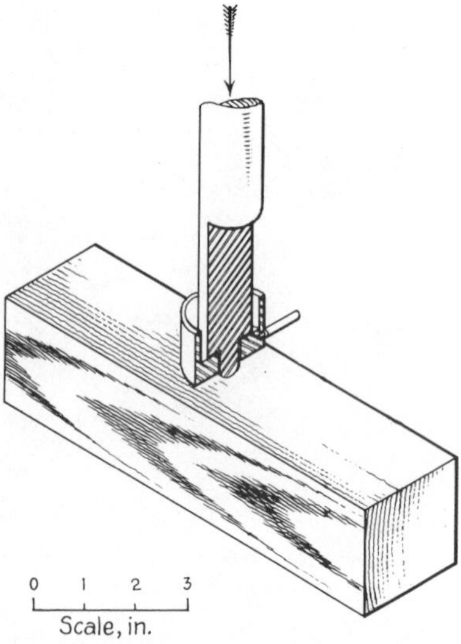

FIG. 15 Diagrammatic Sketch of Method of Conducting Hardness Test.

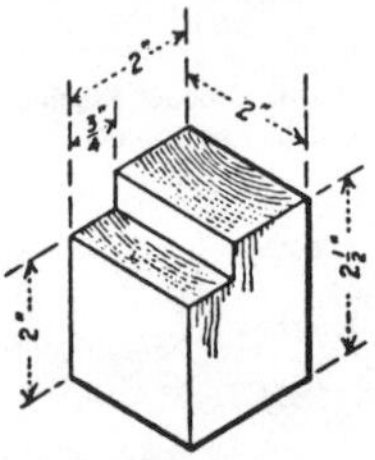

Metric Equivalents

in.	$^3/_4$	2	$2^1/_2$
cm	2	5	6

FIG. 16 Shear-Parallel-to-Grain Test Specimen.

FIG. 17 Shear-Parallel-to-Grain Test Assembly Showing Method of Load Application Through Adjustable Seat to Provide Uniform Lateral Distribution of Load.

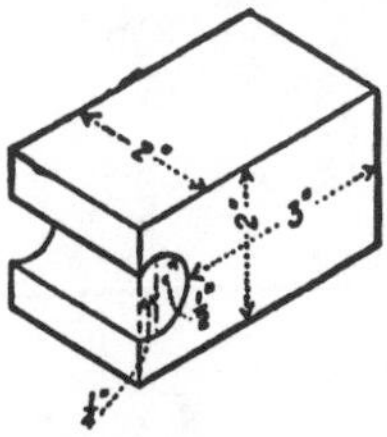

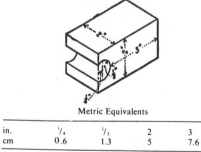

Metric Equivalents

in.	1/4	1/2	2	3
cm	0.6	1.3	5	7.6

FIG. 18 Cleavage Test Specimen.

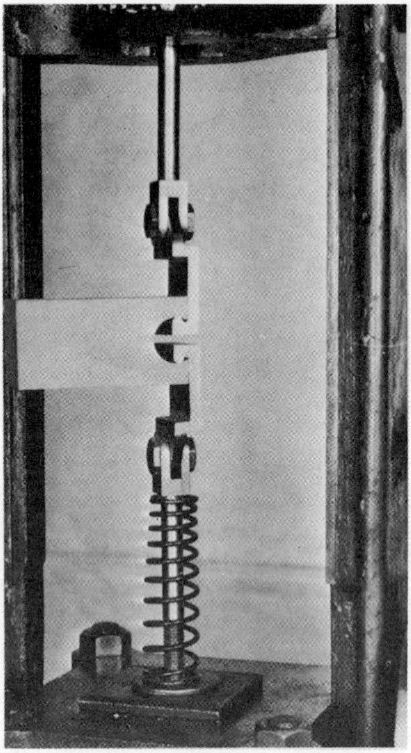

FIG. 19 Cleavage Test Assembly.

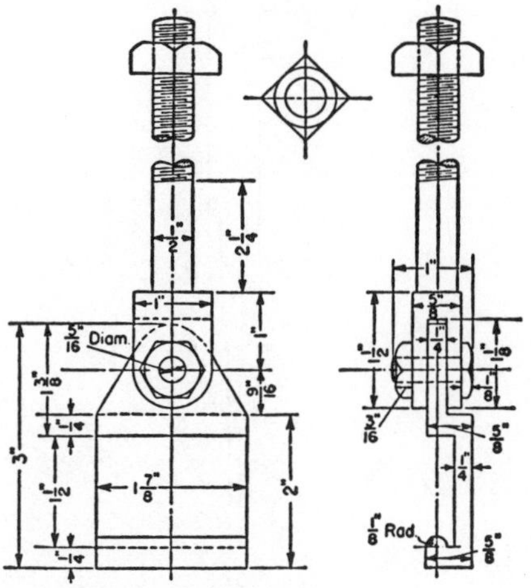

Note—Two peices included in one set:
One piece with shank 8″ long.
One piece with shank 5 1/2″ long.

Metric Equivalents

in.	mm	in.	mm
1/8	3	1 3/8	34
3/16	4.5	1 1/2	38
1/4	6	1 7/8	46
5/16	7.5	2	51
1/2	12.7	2 1/4	57
9/16	13.5	3	76
5/8	15	5 1/2	140
1	25.4	8	203
1 1/8	28		

Note—Two pieces included in one set:
One piece with shank 8″ long.
One piece with shank 5 1/2″ long.

FIG. 20 Design Details of Grips for Cleavage Test.

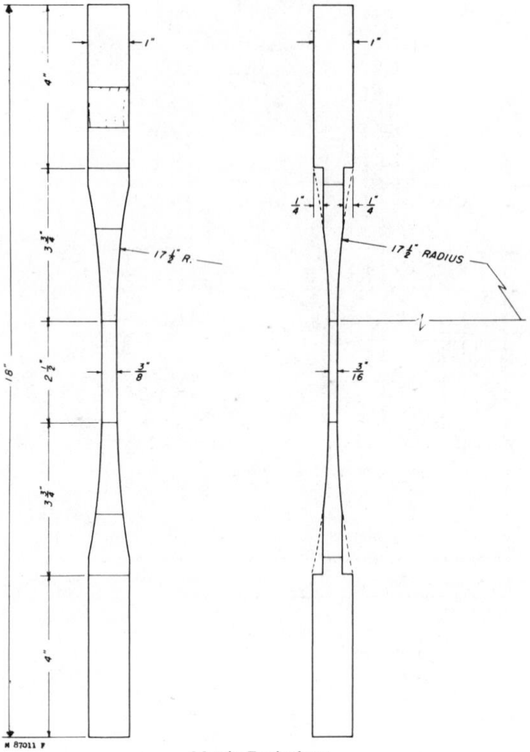

Metric Equivalents

in.	3/16	1/4	3/8	1	2 1/2	3 3/4	4	17 1/2	18
cm	0.45	0.6	0.9	2.5	6	9.5	10	44	45.7

FIG. 21 Tension-Parallel-to-Grain Test Specimen.

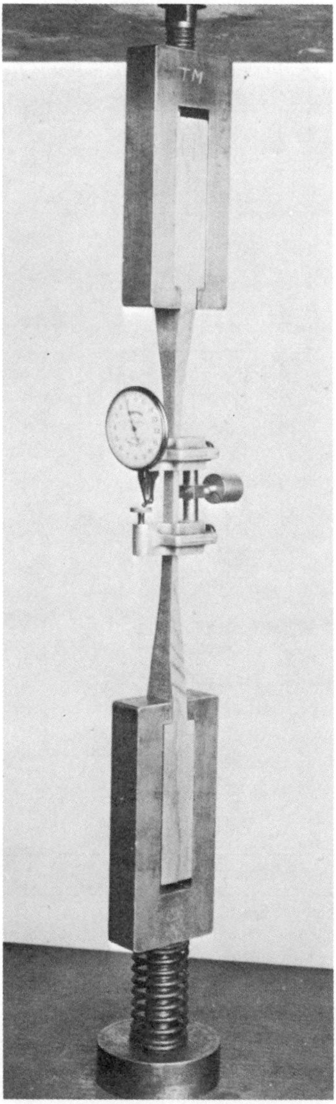

FIG. 22 Tension-Parallel-to-Grain Test Assembly Showing Grips and Use of 2-in. (5-cm) Gage Length Extensometer for Measuring Deformation.

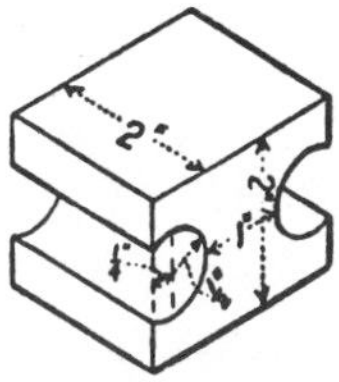

Metric Equivalents

in	1/4	1/2	1	2
cm	0.6	1.3	2.5	5

FIG. 23 Tension-Perpendicular-to-Grain Test Specimen.

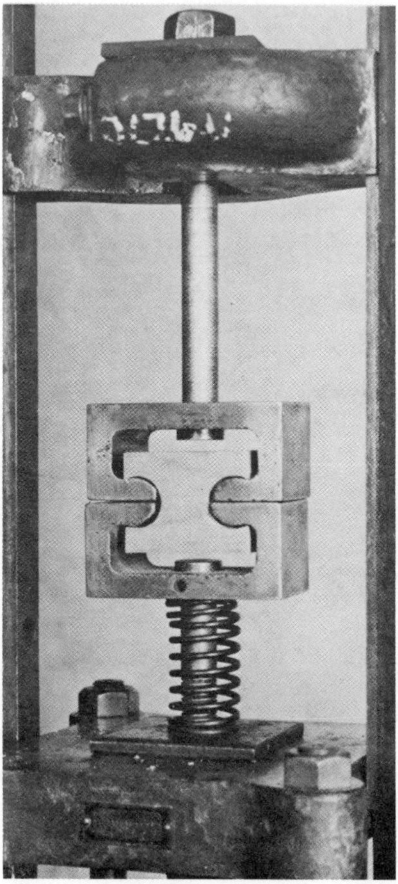

FIG. 24 Tension-Perpendicular-to-Grain Test Assembly.

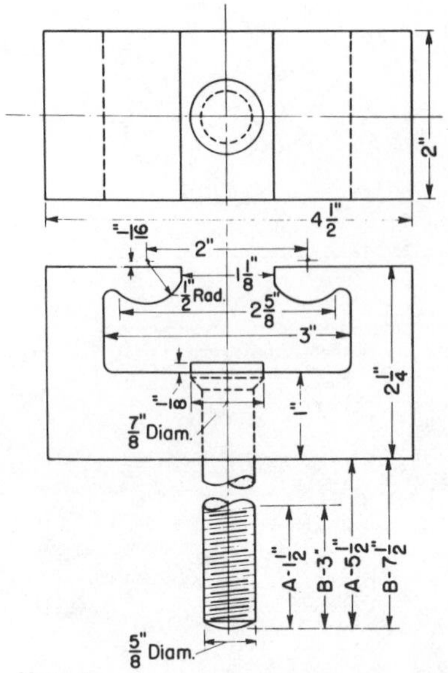

Note—Two pieces included in one set:
One marked A.
One marked B.
Scale-Full Size

Metric Equivalents

in.	cm	in.	cm
$^1/_{16}$	0.15	2	5
$^1/_8$	0.3	$2^1/_4$	5.7
$^1/_2$	1.3	$2^5/_8$	6.6
$^5/_8$	1.5	3	7.6
$^7/_8$	2	$4^1/_2$	11
1	2.5	$5^1/_2$	14
$1^1/_8$	3	$7^1/_2$	19
$1^1/_2$	3.8		

FIG. 25 Design Details of Grips for Tension-Perpendicular-to-Grain Test.

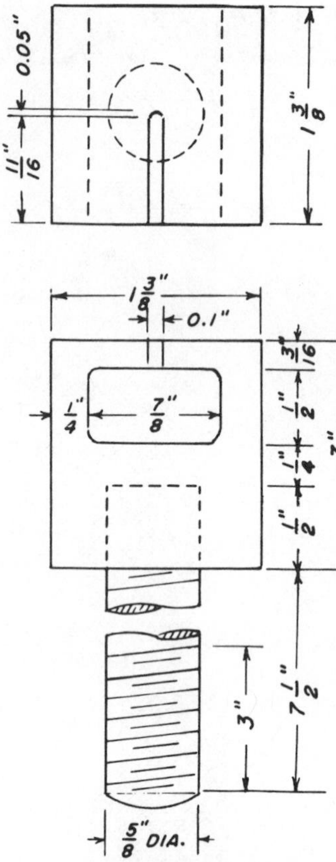

Metric Equivalents

in.	cm	in.	cm
0.05	0.12	$^{11}/_{16}$	1.7
0.1	0.25	$^7/_8$	2.
$^3/_{16}$	0.45	$1^3/_8$	3.4
$^1/_4$	0.6	$1^7/_{16}$	3.6
$^1/_2$	1.3	3	7.6
$^5/_8$	1.5	$7^1/_2$	19

FIG. 26 Design Details of Grip for Nail Withdrawal Test.

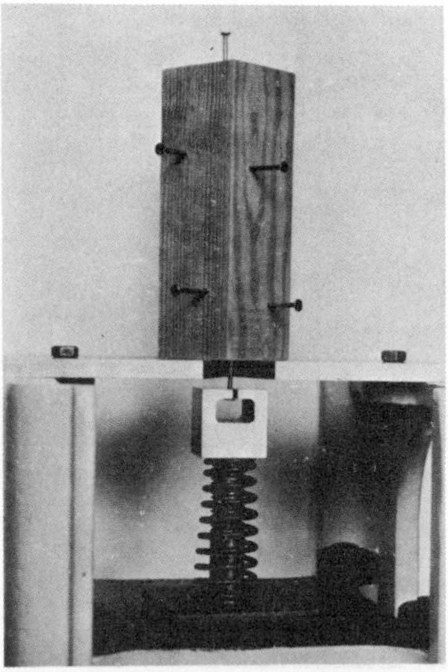

FIG. 27 Nail Withdrawal Test Assembly Showing Specimen in Position for Withdrawal of Nail Driven in One End of the Specimen.

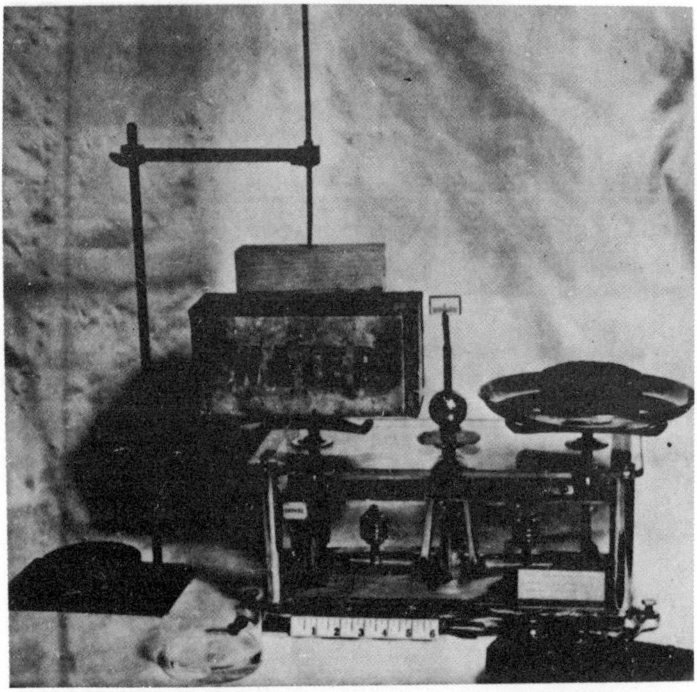

FIG. 28 Specific Gravity and Shrinkage-in-Volume Test Set-Up

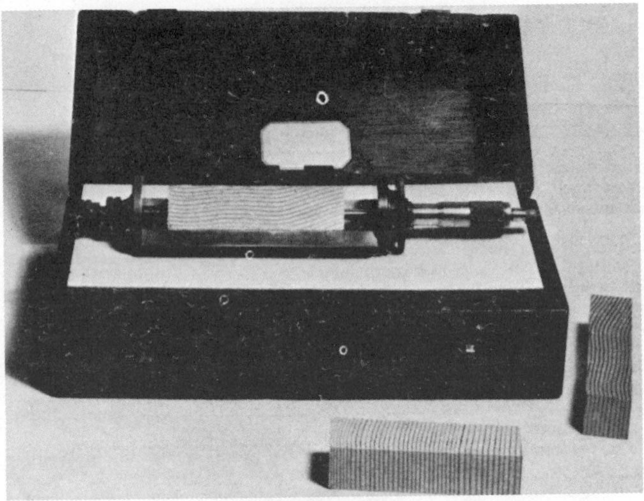

FIG. 29 Radial- and Tangential-Shrinkage Test Assembly.

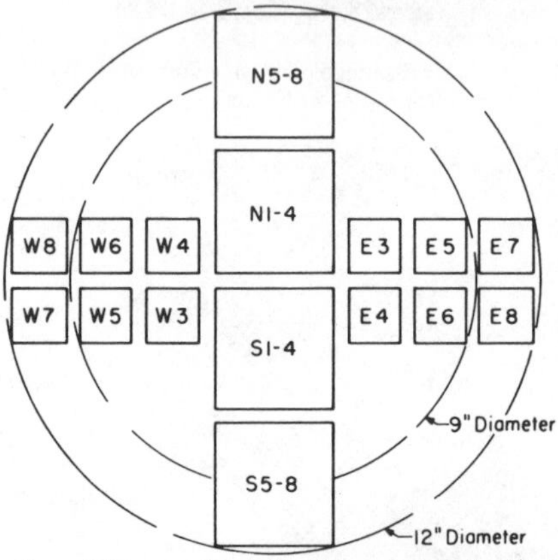

NOTE—Sticks cut from the *N-S* axis shall be 2$^1/_2$ by 2$^1/_2$ in. (6 by 6 cm) when green. Sticks cut from the *E-W* axis shall be 1$^1/_4$ by 1$^1/_4$ in. (3 by 3 cm) when green.

FIG. 30 Sketch Showing Method of Cutting Up the Bolt and Marking the Sticks for the Secondary Methods.

Ship Descr. 1

TABLE A1 Shipment Description—Field Notes

Date *Sept. 1946*

Shipment No. *1625* Species *Douglas-fir* Project No. *259*

State *Oregon* County *Lane*

Local contact: National Forest *Willamette* Private owner ———

Name ——————— Address *Eugene, Oregon*

A. General ecological description of locality.*

1. Climatic factors:
   a. Precipitation in inches: average annual amount *38.4*; seasonal distribution—spring *9.4*, summer *2.5*, autumn *10.5*, winter *16.0*
   b. Temperature, degrees Fahrenheit: mean annual *52*, mean summer *64* mean winter *41*, maximum summer *99*, minimum winter −3
   Season between killing frosts: (dates) *Oct. 31* to *April 17*
   c. Relative humidity: high, medium, or low in spring *high* summer, *low* autumn *high*, winter *high*
   d. Prevailing wind direction: *Westerly;* summer ______ winter ______
2. Topographic factors:
   a. Character of topography: level, rolling, *mountainous*, to *precipitous*
   b. Elevations: absolute *3,000 ft.*, relative *2,000 to 4,000 ft.*
   c. Presence of streams, lakes, swamps: numerous, *few*, *permanent*, intermittent *(Salt Creek Valley)*
   d. Geological history—original rock formations: (1) *igneous*, (2) sedimentary—shales, sand stones, limestones——————
   Secondary formations: *local*, glacial, alluvial, loessal *from basalt or tuffaceous conglomerate. The soils are mainly derived from weathered consolidated rocks. Stones and boulders are abundant over the surface and embedded in soil material.*

* Prepare page 1 of this form for each shipment. Underline descriptive words and fill in all blank spaces that apply. Use reverse of this form for additional notes.

Ship. Descr. 2

TABLE A1 (*Continued*)

Date *Sept. 23-25, 1946*

Shipment No. *1625* Species *Douglas-fir* Project No. *259*
State *Oregon* Co. *Lane* Twp. *22S* Range *5E* Sec. *22*
Locate on *map* or sketch; *4* miles to *McCredie Springs, Oregon*

B. Site description*

1. Site quality class: I, *II*, III, IV, V; or age *100 years*, height *170 ft.* of dominant trees
2. Physiography
   a. Elevation: absolute *3,000 ft.*, range over area *±50 ft.*
   b. Surface: level, undulating; *slope* ___, upper ___, *lower* *cove*, *degree* *15°*; ridge, valley bottom
   c. Aspect: N, NE, E, SE, S, *SW*, W, NW
   d. Distance to streams, lakes, swamps *Salt Creek—⅛ mile*
   e. Soil type: sand, sandy loam, *loam*, clay loam, clay
      U. S. Soil Survey name *Olympic Series*
      (1) Color *brown,* (2) *acid* or alkaline, (3) depth *3 to 6 ft.*
      (4) *boulders*, *stones*, or gravel *numerous*
   f. Soil moisture: dry, *moist*, wet, flooded.
      (1) Natural drainage: *good*, medium, poor
      (2) Depth of water table, ___ feet

C. Forest description:

1. Origin: *natural*, artificial; *seed*, sprout, planted ___
2. History: fires, *lumbering;* wind, snow, sleet, tapping for exudates
   a. Year: *1941 to 1945*, ___, ___, *Some pilling removed*
3. Pathological condition: *good*, medium, poor.
   a. Species attacked ——— Injury to trunk, branch, leaf, root ___
4. Type name *Douglas-fir*; all aged, *even aged*———
   a.

Species	Percent	age Av. or range	D.B.H. Av. or *range*
Douglas-fir	*100*	*100 years*	*12 to 48 in.*

   b. No. of trees over *10* in. d.b.h. per acre *100* (¼ acre, 57 ft. radius; ⅛ acre, 41.7 ft. radius; $^1/_{10}$ acre, 37.3 ft. radius).
   c. Density of crown cover: 0.1, 0.2, 0.3, 0.4, 0.5, 0.6, 0.7, *0.8*, 0.9, 1.0.
   d. Underbrush: species *Vine maple*
      amount ———
5. Forest floor
   a. Ground cover: weeds, grass, *bracken*, briars, vines. *Sword fern, Salal brush.*
   b. Litter: kind *Dead needles and branches* amount *moderate*
   c. Raw humus or duff: depth *1 in.*
   d. Humus: depth *½ in.*
6. Photographs: *yes*, no. Taken by *A. K.*
7. Trees collected: *Nos. 1 to 10.*

* **Prepare page 2 of this form for each site or forest. Underline descriptive words and fill in all blank spaces that apply. Use reverse of this form for additional notes.**

U. S. DEPARTMENT OF AGRICULTURE
FOREST SERVICE

FILE NO. 124
COST CHARGE ______
LABORATORY NO. 100,831

STATION - MADISON   DATE - AUG. 24, 1914

MEASURED ______
TESTED ______
WEIGHED ______
SHIP. NO. L-315   STICK NO. S-8
PIECE NO. 4   MARK d

SPECIES DOUGLAS-FIR
KIND OF TEST STATIC BENDING
LOADING CENTER
SPAN 28 IN.
GAGE LENGTH ______
WIDTH OF PLATE ______
MACHINE M 1037
SPEED OF MACH. 0.105 IN. PER MIN.
WEIGHT OF HAMMER ______
HEIGHT 2.02 IN.
WIDTH 2.00 IN.
LENGTH 30.10 IN.
CROSS-SECTION ______
WEIGHT 1,251 GRAMS
RINGS PER INCH 9
SAP 0%
SUMMERWOOD 40%
SEASONING GREEN
MOISTURE 31.4%
TESTED AT 75°F. TEMP. 64 % R.H.
KIND OF FAILURE COMPRESSION FOLLOWED BY SPLINTERING TENSION
REMARKS ______

SKETCH

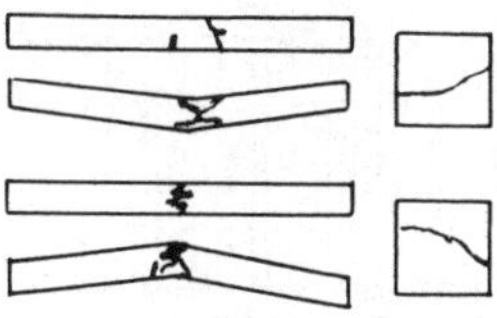

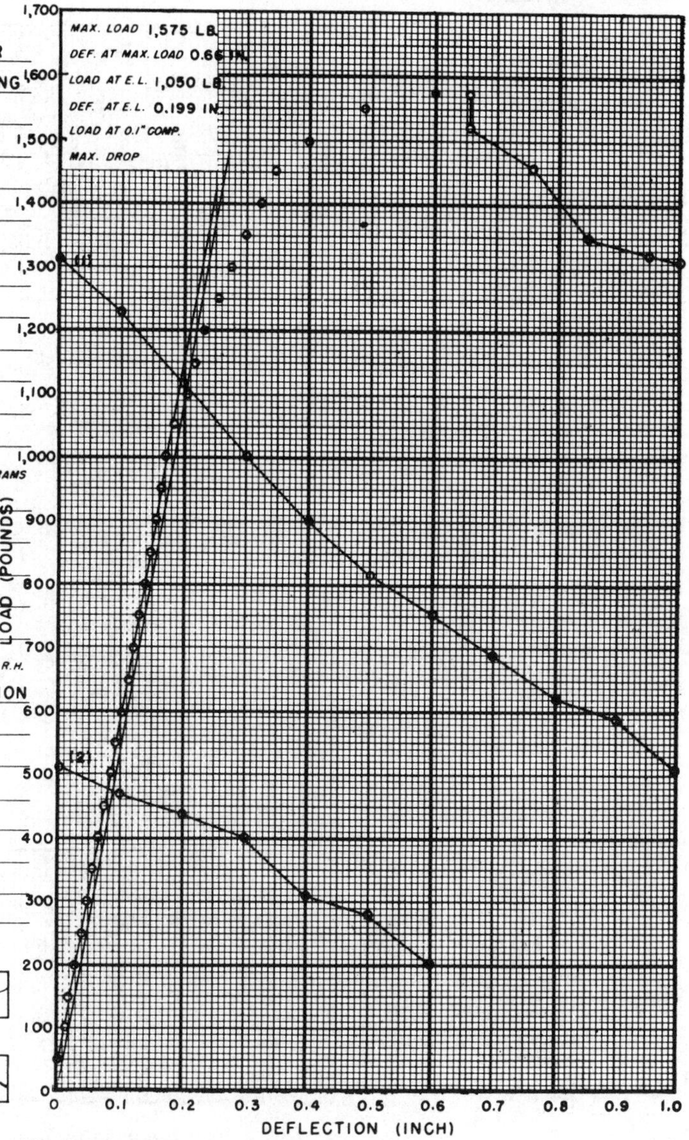

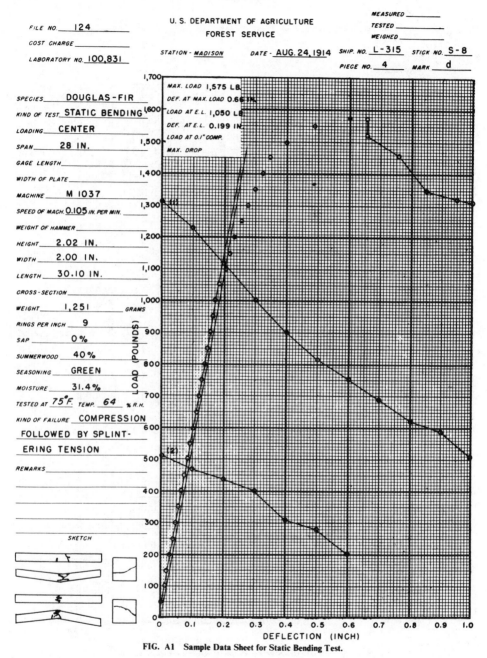

FIG. A1 Sample Data Sheet for Static Bending Test.

U. S. DEPARTMENT OF AGRICULTURE

FOREST SERVICE

MEASURED ____
TESTED ____
WEIGHED ____

FILE NO. 124
COST CHARGE ____
LABORATORY NO. 871,902

STATION - MADISON
DATE - NOV. 5, 1943
SHIP. NO. 1590
STICK NO. S-5
PIECE NO. 5
MARK d

SPECIES RED FIR
KIND OF TEST COMPRESSION PARALLEL TO GRAIN
LOADING ____
SPAN ____
GAGE LENGTH 6 IN
WIDTH OF PLATE ____
MACHINE 4272
SPEED OF MACH. 0.0223 IN. PER MIN.
WEIGHT OF HAMMER ____
HEIGHT ____
WIDTH ____
LENGTH 8.02 IN
CROSS-SECTION 2.002 X 2.014 IN.
WEIGHT 198.9 GRAMS
RINGS PER INCH 7
SAP 0%
SUMMERWOOD ____
SEASONING KILN DRY
MOISTURE 11.5%
TESTED AT 75°F. TEMP. 64 % R.H.
KIND OF FAILURE
CRUSHING NEAR END

REMARKS ____

SKETCH

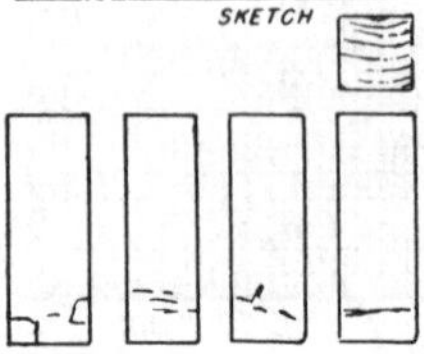

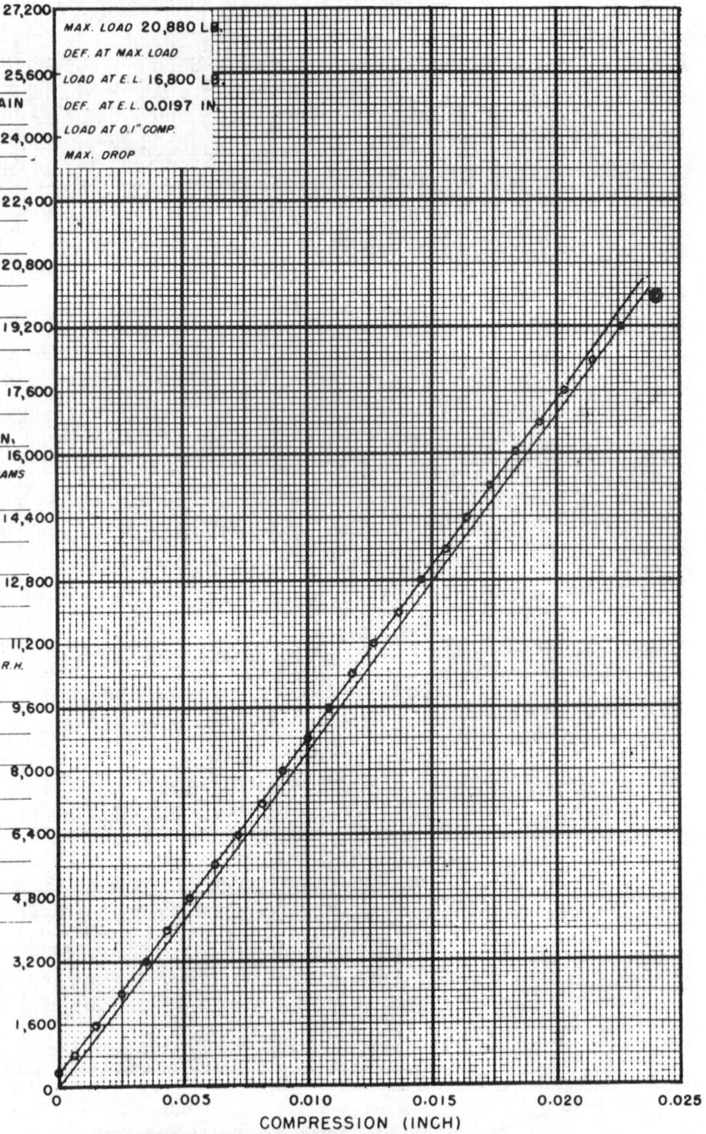

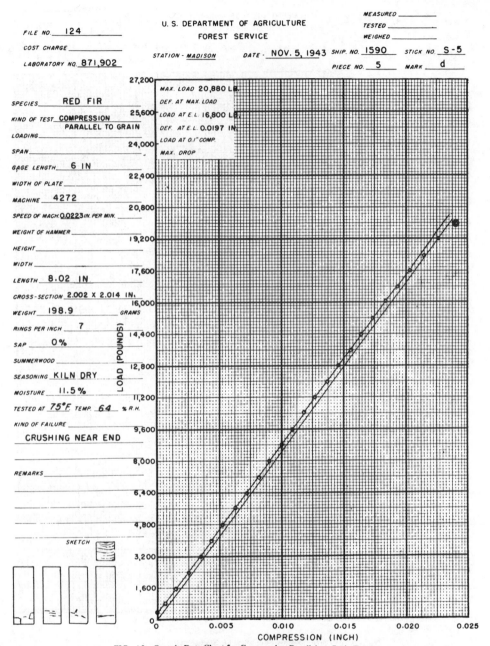

FIG. A2 Sample Data Sheet for Compression-Parallel-to-Grain Test.

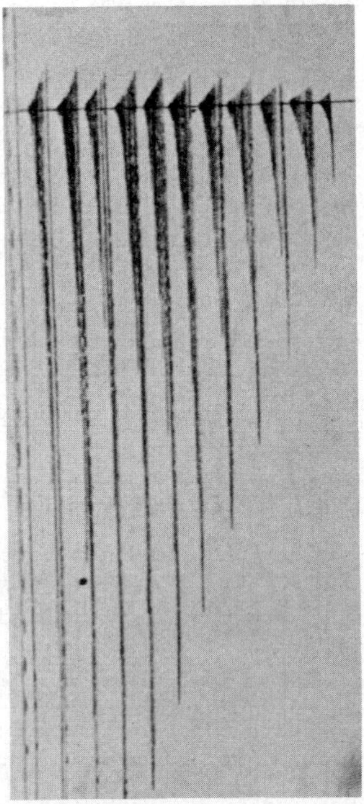

FIG. A3 Sample Drum Record of Impact Bending Test.

IMPACT BENDING

L-315 (Ship. No.) E-12 (Stick No.) 101 151 (Lab. No.)

1 (Piece No.) C (Mark) Station MADISON Date Aug. 20, 1914 124 (Project No)

Species Douglas Fir Grade Clear Seasoning Green

Rings 8 Sap 100 % Summerwood 30 % Moisture 61.4 %

Hammer 50 lbs. Span 28 in. Length 29.94 in. Height 2.00 in. Width 2.00 in. Weight 1370 g.

Drop No.	Head.	Def.	Def.²	Set.	Drop No.	Head.	Def.	Def.²	Set.		
										Sp. Gr. (at test),	0.698
1	1.0	0.13	0.017		11	12.0	0.50	0.250		Sp. Gr. (oven dry),	0.432
2	2.0	0.18	0.032		12	14.0	0.55	0.302			
3	3.0	0.22	0.048		13	16.0	0.62	0.384		F. S. at E. L.,	10 610
4	4.0	0.26	0.068		14	18.0	0.67	0.593		M. of E.,	1776
5	5.0	0.30	0.090		15					E. Resil.,	3.51
6	6.0	0.34	0.116		16						
7	7.0	0.36	0.130		17					Max. Drop,	22 in.
8	8.0	0.38	0.144		18					d,	0.010
9	9.0	0.43	0.185		19					H	7.88
10	10.0	0.46	0.212		20					Δ	0.39

Failure: Compression Followed by Splintering Tension.

FIG. A4 Sample Data and Computation Card for Impact Bending Test.

U. S. DEPARTMENT OF AGRICULTURE
FOREST SERVICE

FILE NO. 124
COST CHARGE
LABORATORY NO. 101,151

STATION - MADISON DATE - AUG. 26, 1914

MEASURED
TESTED
WEIGHED
SHIP. NO. L-315 STICK NO. E-12
PIECE NO. 1 MARK C

SPECIES DOUGLAS-FIR
KIND OF TEST IMPACT BENDING
LOADING CENTER
SPAN 28 IN.
GAGE LENGTH
WIDTH OF PLATE
MACHINE HATT-TURNER
SPEED OF MACH. IN. PER MIN.
WEIGHT OF HAMMER 50 LB.
HEIGHT 2.00 IN.
WIDTH 2.00 IN.
LENGTH 29.94 IN.
CROSS-SECTION
WEIGHT 1,370 GRAMS
RINGS PER INCH 8
SAP 100 %
SUMMERWOOD 30 %
SEASONING GREEN
MOISTURE 61.4 %
TESTED AT 75°F. TEMP. 64 % R.H.
KIND OF FAILURE COMPRESSION FOLLOWED BY SPLINTERING TENSION.
REMARKS

SKETCH

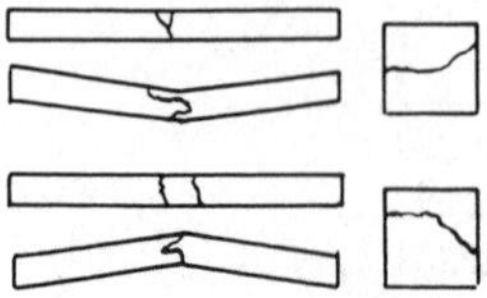

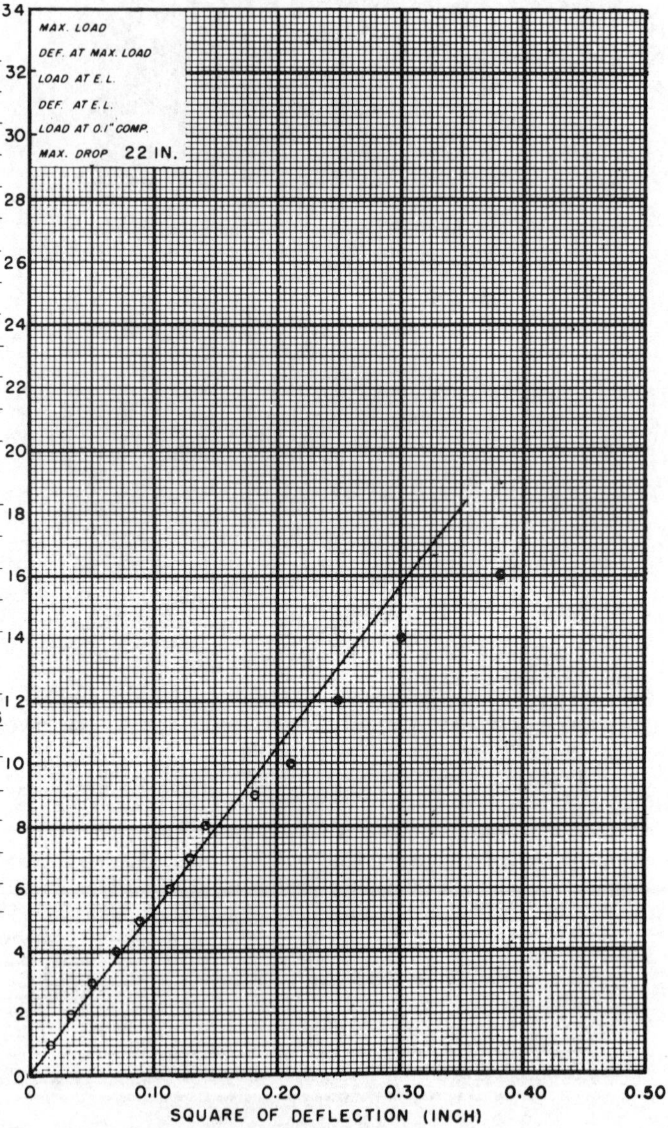

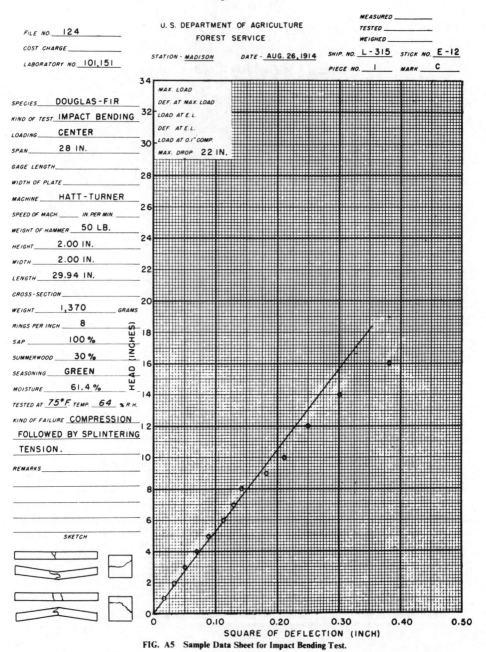

FIG. A5 Sample Data Sheet for Impact Bending Test.

TOUGHNESS

STATION - *Madison*

SPECIES *PACIFIC SILVER FIR* SHIPMENT NO. *1,651*

PROJECT *Str. IL* SEASONING *GREEN* MEASURED BY

COST CHARGE *01-3-005* SPAN *9.47 IN.* WEIGHED BY

LABORATORY NOS. *268,779A-806A* MACHINE NO. *4,715* TESTED BY

DATE *FEB. 1, 1950* TEMP. *75* °F. REL. HUMIDITY *64* %

STICK NO.	LAB. NO.	DIMENSIONS L" x H" x W"	WEIGHT GM.	MOIST. %	SP. GR.	POSITION OF RINGS * RAD.	TANG.	WEIGHT	INITIAL ANGLE °	FINAL ANGLE °	'	TOUGHNESS INCH - POUNDS	REMARKS
								3	45				
22E-3-d-1	785A	11.02X.794X.797	53.80	32.0	.357	V				32	30	143.8	
2	786A	11.02X.789X.790	52.54	31.8	.354		V			31	56	149.7	
22E-5-C-1	787A	11.02X.792X.795	53.56	35.7	.347	V				33	10	136.8	
2	788A	11.02X.794X.795	53.00	39.6	.333		V			34	4	127.6	

* "RAD." LOAD APPLIED TO RADIAL FACE; "TANG." LOAD APPLIED TO TANGENTIAL FACE.

FIG. A6 Sample Data and Computation Sheet for Toughness Test.

U. S. DEPARTMENT OF AGRICULTURE
FOREST SERVICE

MEASURED ______
TESTED ______
WEIGHED ______

FILE NO. 124
COST CHARGE ______
LABORATORY NO. 871,620

STATION - MADISON   DATE - OCT. 25, 1943   SHIP. NO. 1590   STICK NO. N-6
PIECE NO. 4   MARK C

SPECIES RED FIR
KIND OF TEST COMPRESSION
LOADING PERPENDICULAR TO GRAIN
SPAN ______
GAGE LENGTH ______
WIDTH OF PLATE 2 IN.
MACHINE 4270
SPEED OF MACH. 0.0121 IN. PER MIN.
WEIGHT OF HAMMER ______
HEIGHT 2.015 IN.
WIDTH 2.012 IN.
LENGTH 6.07 IN.
CROSS-SECTION ______
WEIGHT 175.3 GRAMS
RINGS PER INCH 25
SAP ______
SUMMERWOOD ______
SEASONING KILN DRY
MOISTURE 10.8 %
TESTED AT 75°F. TEMP. 64 % R.H.
KIND OF FAILURE ______

REMARKS ______

SKETCH

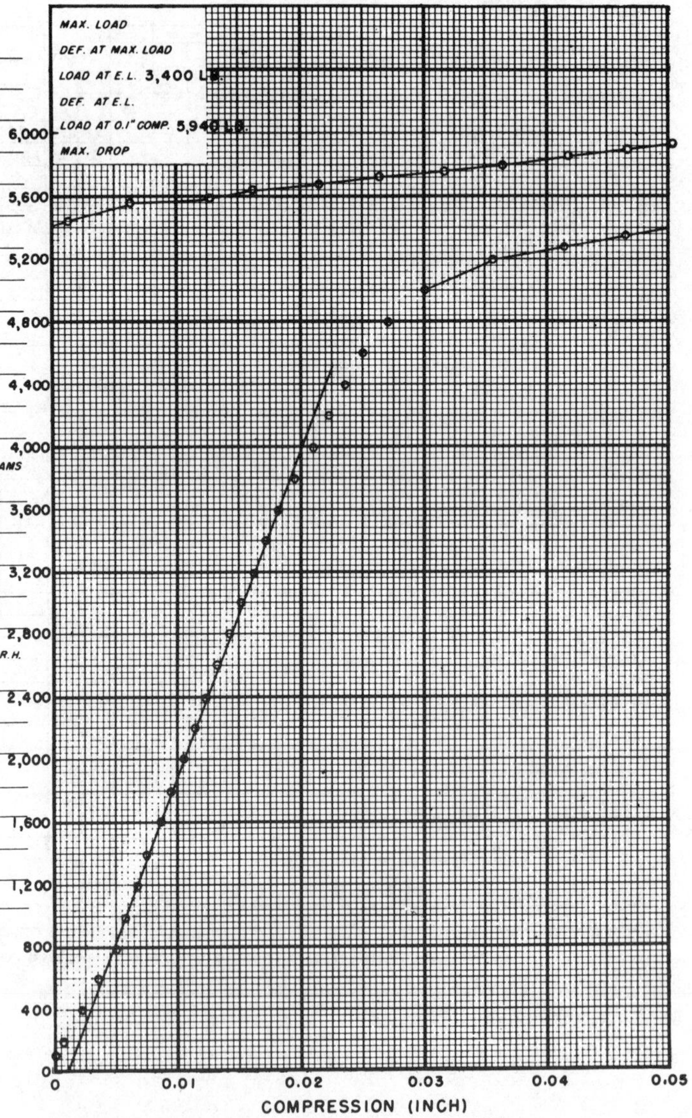

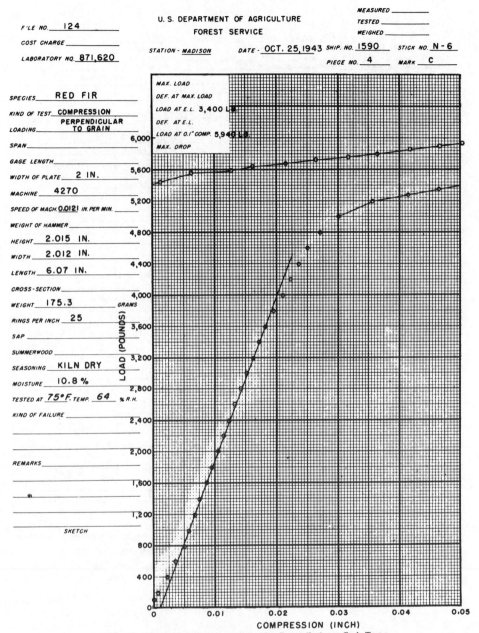

FIG. A7 Sample Data Sheet for Compression-Perpendicular-to-Grain Test.

HARDNESS

STATION -- *Madison*

SPECIES *PACIFIC SILVER FIR* SHIPMENT NO. *1,651*

PROJECT *Str. IL* SEASONING *GREEN* MEASURED BY

COST CHARGE *01-3-005* MACHINE SPEED *0.244* WEIGHED BY

LABORATORY NOS. *268,281A-290A* MACHINE NO. *4,271* TESTED BY

DATE *JAN. 31, 1951* TEMP. *75* °F. REL. HUMIDITY *64* %

STICK NO.	DIMENSIONS L" X H" X W"	WEIGHT GM.	MOIST. %	SP. GR.	HARDNESS RADIAL SURFACE LB.	HARDNESS TANGENTIAL SURFACE LB.	HARDNESS END SURFACE LB.	REMARKS	SKETCH
23-N-7-d	6.02X1.996X1.994	241.3	47.5	.416	530	470	465		
					500	515	530		
				AVERAGE	515	492	498		
AVERAGE RADIAL AND TANGENTIAL					504				
				AVERAGE					
AVERAGE RADIAL AND TANGENTIAL									
23-E-8-C	6.04X1.992X1.992	273.3	71.1	.406	370	455	510		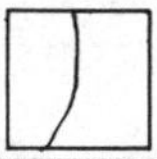
					415	435	555		
				AVERAGE	392	.445	532		
AVERAGE RADIAL AND TANGENTIAL					418				
				AVERAGE					
AVERAGE RADIAL AND TANGENTIAL									
				AVERAGE					
AVERAGE RADIAL AND TANGENTIAL									
				AVERAGE					
AVERAGE RADIAL AND TANGENTIAL									
				AVERAGE					
AVERAGE RADIAL AND TANGENTIAL									
				AVERAGE					
AVERAGE RADIAL AND TANGENTIAL									
				AVERAGE					
AVERAGE RADIAL AND TANGENTIAL									
				AVERAGE					
AVERAGE RADIAL AND TANGENTIAL									

FIG. A8 Sample Data and Computation Sheet for Hardness Test.

SHEAR

STATION - *Madison*

SPECIES *PACIFIC SILVER FIR* SHIPMENT NO. *1,651*

PROJECT *Str. IL* SEASONING *GREEN* MEASURED BY

COST CHARGE *01-3-005* MACHINE SPEED *0.0215* WEIGHED BY

LABORATORY NOS. *267,024A-029A* MACHINE NO. *4,271* TESTED BY

DATE *JAN. 16, 1951* TEMP. *75* °F. REL. HUMIDITY *64* %

STICK NO.	SHEARING SURFACE	SHEARING AREA L" X W"	MAXIMUM LOAD LB.	SHEARING STRENGTH P.S.I.	MOISTURE CONTENT %	REMARKS	SKETCH
22-N-2-d	R.	2.016 X 2.000	2770	687	40.1		
22-N-6-d	T.	2.020 X 1.998	2775	688	41.1		

FIG. A9 Sample Data and Computation Sheet for Shear-Parallel-to-Grain Test.

CLEAVAGE

STATION - *Madison*

SPECIES *PACIFIC SILVER FIR* SHIPMENT NO. *1,651*

PROJECT *Str. IL* SEASONING *GREEN* MEASURED BY ______

COST CHARGE *01-3-005* MACHINE SPEED *0.1110* WEIGHED BY ______

LABORATORY NOS. *267,036A-041A* MACHINE NO. *4269* TESTED BY ______

DATE *JAN. 17, 1951* TEMP. *75* °F. REL. HUMIDITY *64* %

STICK NO.	CLEAVAGE SURFACE	CLEAVAGE AREA L" x W"	MAXIMUM LOAD LB.	LOAD PER INCH OF WIDTH LB.	MOISTURE CONTENT %	REMARKS	SKETCH
22-N-6-d	R.	3.03 X 2.005	315	157	36.9		
22-N-6-d	T.	3.03 X 2.007	330	165	38.5		

FIG. A10 Sample Data and Computation Sheet for Cleavage Test.

U. S. DEPARTMENT OF AGRICULTURE
FOREST SERVICE

FILE NO. Str. - 1L
COST CHARGE 01-3-005
LABORATORY NO. 266,895 A

STATION - MADISON DATE - DEC. 28, 1950

MEASURED ____
TESTED ____
WEIGHED ____
SHIP. NO. 1651 STICK NO. N-1
PIECE NO. 22 MARK C

SPECIES PACIFIC SILVER FIR
KIND OF TEST TENSION PARALLEL TO GRAIN
LOADING ____
SPAN 2 IN.
GAGE LENGTH ____
WIDTH OF PLATE ____
MACHINE 4713
SPEED OF MACH. 0.036 IN. PER MIN.
WEIGHT OF HAMMER ____
HEIGHT ____
WIDTH ____
LENGTH ____
CROSS-SECTION 0.187 X 0.379
WEIGHT ____ GRAMS
RINGS PER INCH ____
SAP ____
SUMMERWOOD ____
SEASONING GREEN
MOISTURE ____
TESTED AT 75°F. TEMP. 64 % R.H.
KIND OF FAILURE SPLINTERING TENSION

REMARKS ____

SKETCH

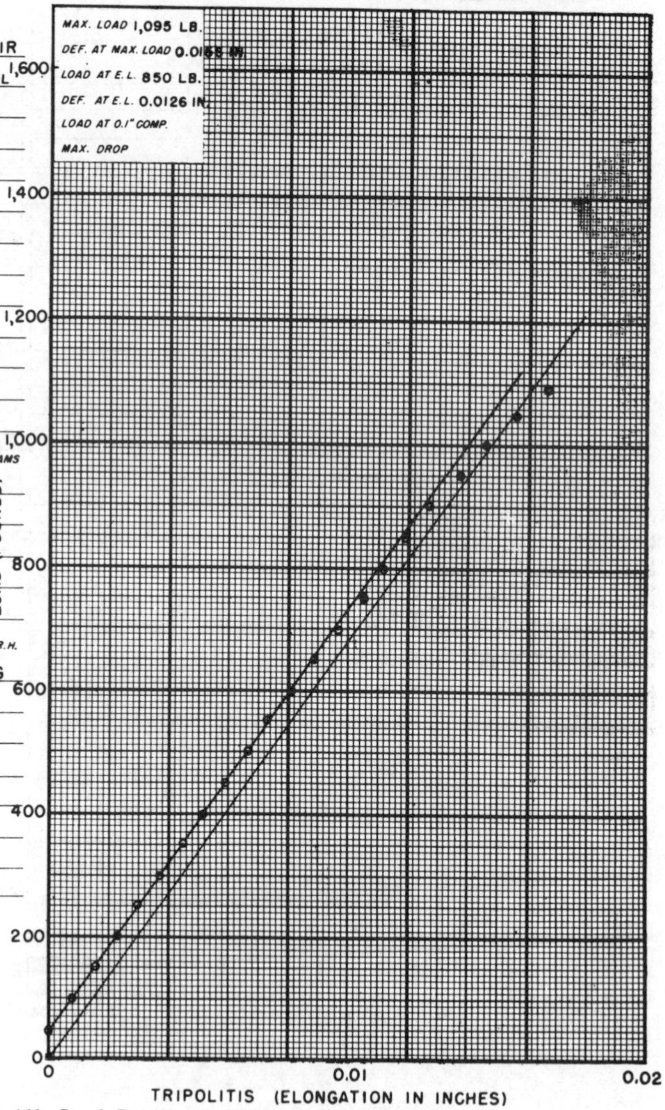

FIG. A11 Sample Data Sheet for Tension-Parallel-to-Grain Test.

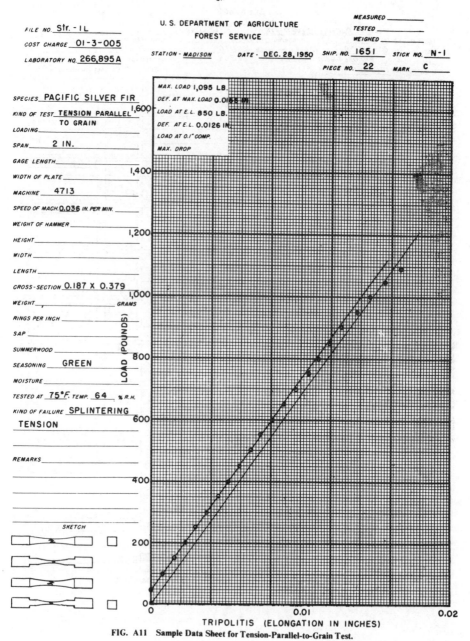

TENSION PERPENDICULAR TO GRAIN

STATION - *Madison*

SPECIES *PACIFIC SILVER FIR* SHIPMENT NO. *1,651*

PROJECT *Str. IL* SEASONING *GREEN* MEASURED BY ______

COST CHARGE *01-3-005* MACHINE SPEED *0.1080* WEIGHED BY ______

LABORATORY NOS. *267,048A-053A* MACHINE NO. *4,713* TESTED BY ______

DATE *JAN. 16, 1951* TEMP. *75* °F. REL. HUMIDITY *64* %

STICK NO.	TENSION SURFACE	TENSION AREA L" X W"	MAXIMUM LOAD LB.	TENSILE STRENGTH P.S.I.	MOISTURE CONTENT %	REMARKS	SKETCH
22-N-6-d	R.	0.98 X 2.011	575	292	33.0		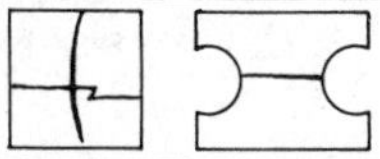
22-N-6-d	T.	1.00 X 2.001	635	317	32.4		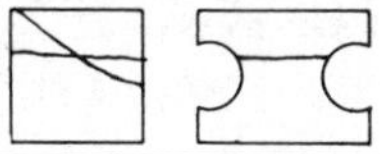

FIG. A12 Sample Data and Computation Sheet for Tension-Perpendicular-to-Grain Test.

NAIL WITHDRAWAL

STATION - *Madison*

SPECIES *PACIFIC SILVER FIR* SHIPMENT NO. *1,651*

PROJECT *Str. 1L* SEASONING *GREEN* MEASURED BY ____

COST CHARGE *01-3-005* MACHINE SPEED *0.071* WEIGHED BY ____

LABORATORY NOS. *270,270A-278A* MACHINE NO. *4269* TESTED BY ____

DATE *FEB. 2, 1951* NAILS, TYPE *7d PLAIN (SINKER) DRIVEN 1 1/4"* TEMP. *75* °F. REL. HUMIDITY *64* %

STICK NO.	DIMENSIONS L" x H" x W"	WEIGHT GM.	MOIST. %	SP. GR.	WITHDRAWAL LOADS: RADIAL SURFACE LB.	TANGENTIAL SURFACE LB.	END SURFACE LB.	REMARKS	SKETCH
23-N-5-C	6.05X1.990X1.989	326.4	77.7	.468	180	205	105		
					175	200	110		
				AVERAGE	178	202	108		
23-N-7-d	6.02X1.996X1.994	241.3	47.5	.416	180	175	110		
					185	155	75		
				AVERAGE	182	165	92		
				AVERAGE					
				AVERAGE					
				AVERAGE					
				AVERAGE					
				AVERAGE					
				AVERAGE					
				AVERAGE					
				AVERAGE					

FIG. A13 Sample Data and Computation Sheet for Nail Withdrawal Test.

SPECIFIC GRAVITY AND VOLUMETRIC SHRINKAGE

STATION - *Madison*

SPECIES *PACIFIC SILVER FIR* SHIPMENT NO. *1,651*

PROJECT *Str. IL* MEASURED BY ______

COST CHARGE *01-3-005* WEIGHED BY ______

LABORATORY NOS. *267,060A-065A* VOLUME BY ______

DATE ______

STICK NO.	DIMENSIONS L" x H" x W"	SEASONING	DATE	RINGS PER INCH	SAP %	SUMMER-WOOD %	WEIGHT GM.	MOISTURE %	VOLUME C.C.	1. SPECIFIC GRAVITY	WEIGHT POUNDS PER CUBIC FOOT	VOLUMETRIC SHRINKAGE* %
22-N-4-C	6.05 X 2.001 X	GREEN	1-9-'51	18	0		201.3	34.3	393.8	.381		
	2.002	OVEN-DRY	6-19-'51				149.9	0	332.1	.451	28.1	15.7
REMARKS		AIR-DRY	6-13-'51				168.0	12.07	360.3	.416		
REMARKS												
22-S-5-C	6.03 X 2.004	GREEN	1-9-'51	17	0		223.1	55.5	392.0	.366		
	X 2.001	OVEN-DRY	6-19-'51				143.5	0	334.2	.429	26.8	14.7
REMARKS		AIR-DRY	6-13-'51				160.9	12.13	360.9	.398		
REMARKS												
REMARKS												
REMARKS												
REMARKS												
REMARKS												
REMARKS												
REMARKS												

* BASED ON ORIGINAL VOLUME (GREEN, AIR-DRY OR KILN-DRY). NOTE: USE BACK OF SHEET FOR CARBON IMPRESSIONS.

1. BASED ON WEIGHT WHEN OVEN-DRY

FIG. A14 Sample Data and Computation Sheet for Specific Gravity and Shrinkage-in-Volume Test.

SHRINKAGE - RADIAL AND TANGENTIAL

STATION - *Madison*

SPECIES *PACIFIC SILVER FIR* SHIPMENT NO. *1,651*

PROJECT *Str. IL* MEASURED BY ____________

COST CHARGE *01-3-005*

LABORATORY NOS. *266,857A-864A* WEIGHED BY ____________

DATE ____________

STICK NO.	NOMINAL SIZE L" X H" X W"	SHRINKAGE DIRECTION	SEASONING	DATE	RINGS PER INCH	SAP %	SUMMERWOOD %	WIDTH IN.	WEIGHT GM.	MOISTURE %	SHRINKAGE* %
22-2-cd	1 X 1 X 4	R.	GREEN	12/26/50	17	15		3.997	35.50	52.5	
			AIR-DRY								
			OVEN-DRY	4/6/51				3.784	23.28		5.3
REMARKS											
22-2-cd	1 X 1 X 4	T.	GREEN	12/26/50	12	10		3.995	40.00	77.8	
			AIR-DRY								
			OVEN-DRY	4/6/51				3.602	22.50		9.8
REMARKS											
			GREEN								
			AIR-DRY								
			OVEN-DRY								
REMARKS											
			GREEN								
			AIR-DRY								
			OVEN-DRY								
REMARKS											
			GREEN								
			AIR-DRY								
			OVEN-DRY								
REMARKS											
			GREEN								
			AIR-DRY								
			OVEN-DRY								
REMARKS											
			GREEN								
			AIR-DRY								
			OVEN-DRY								
REMARKS											
			GREEN								
			AIR-DRY								
			OVEN-DRY								
REMARKS											
			GREEN								
			AIR-DRY								
			OVEN-DRY								
REMARKS											
			GREEN								
			AIR-DRY								
			OVEN-DRY								
REMARKS											

* BASED ON GREEN WIDTH.

FIG. A15 Sample Data and Computation Sheet for Radial- and Tangential-Shrinkage Tests.

By publication of this standard no position is taken with respect to the validity of any patent rights in connection therewith, and the American Society for Testing and Materials does not undertake to insure anyone utilizing the standard against liability for infringement of any Letters Patent nor assume any such liability.

AMERICAN NATIONAL STANDARD ANSI/ASTM D 1559 - 76

Standard Test Method for RESISTANCE TO PLASTIC FLOW OF BITUMINOUS MIXTURES USING MARSHALL APPARATUS[1]

This Standard is issued under the fixed designation D 1559; the number immediately following the designation indicates the year of original adoption or, in the case of revision, the year of last revision. A number in parentheses indicates the year of last reapproval.

1. Scope

1.1 This method covers the measurement of the resistance to plastic flow of cylindrical specimens of bituminous paving mixture loaded on the lateral surface by means of the Marshall apparatus. This method is for use with mixtures containing asphalt cement, asphalt cut-back or tar, and aggregate up to 1-in. (25.4-mm) maximum size.

2. Apparatus

2.1 *Specimen Mold Assembly*—Mold cylinders 4 in. (101.6 mm) in diameter by 3 in. (76.2 mm) in height, base plates, and extension collars shall conform to the details shown in Fig. 1. Three mold cylinders are recommended.

2.2 *Specimen Extractor*, steel, in the form of a disk with a diameter not less than 3.95 in. (100 mm) and ½ in. (13 mm) thick for extracting the compacted specimen from the specimen mold with the use of the mold collar. A suitable bar is required to transfer the load from the ring dynamometer adapter to the extension collar while extracting the specimen.

2.3 *Compaction Hammer*—The compaction hammer (Fig. 2) shall have a flat, circular tamping face and a 10-lb (4536-g) sliding weight with a free fall of 18 in. (457.2 mm). Two compaction hammers are recommended.

NOTE 1—The compaction hammer may be equipped with a finger safety guard as shown in Fig. 2.

2.4 *Compaction Pedestal*—The compaction pedestal shall consist of an 8 by 8 by 18-in. (203.2 by 203.2 by 457.2-mm) wooden post capped with a 12 by 12 by 1-in. (304.8 by 304.8 by 25.4-mm) steel plate. The wooden post shall be oak, pine, or other wood having an average dry weight of 42 to 48 lb/ft^3 (0.67 to 0.77 g/cm^3). The wooden post shall be secured by four angle brackets to a solid concrete slab. The steel cap shall be firmly fastened to the post. The pedestal assembly shall be installed so that the post is plumb and the cap is level.

2.5 *Specimen Mold Holder*, mounted on the compaction pedestal so as to center the compaction mold over the center of the post. It shall hold the compaction mold, collar, and base plate securely in position during compaction of the specimen.

2.6 *Breaking Head*—The breaking head (Fig. 3) shall consist of upper and lower cylindrical segments or test heads having an inside radius of curvature of 2 in. (50.8 mm) accurately machined. The lower segment shall be mounted on a base having two perpendicular guide rods or posts extending upward. Guide sleeves in the upper segment shall be in such a position as to direct the two segments together without appreciable binding or loose motion on the guide rods.

2.7 *Loading Jack*—The loading jack (Fig. 4) shall consist of a screw jack mounted in a testing frame and shall produce a uniform vertical movement of 2 in. (50.8 mm)/min. An electric motor may be attached to the jacking mechanism.

NOTE 2—Instead of the loading jack, a mechanical or hydraulic testing machine may be used

---

[1] This method is under the jurisdiction of ASTM Committee D-4 on Road and Paving Materials and is the direct responsibility of Subcommittee D04.20 on Mechanical Tests of Bituminous Mixes.

Current edition approved April 30, 1976. Published September 1976. Originally published as D 1559 - 58. Last previous edition D 1559 - 75.

provided the rate of movement can be maintained at 2 in. (50.8 mm)/min while the load is applied.

2.8 *Ring Dynamometer Assembly*—One ring dynamometer (Fig. 4) of 5000-lb (2267-kg) capacity and sensitivity of 10 lb (4.536 kg) up to 1000 lb (453.6 kg) and 25 lb (11.340 kg) between 1000 and 5000 lb (453.6 and 2267 kg) shall be equipped with a micrometer dial. The micrometer dial shall be graduated in 0.0001 in. (0.0025 mm). Upper and lower ring dynamometer attachments are required for fastening the ring dynamometer to the testing frame and transmitting the load to the breaking head.

NOTE 3—Instead of the ring dynamometer assembly, any suitable load-measuring device may be used provided the capacity and sensitivity meet the above requirements.

2.9 *Flowmeter*—The flowmeter shall consist of a guide sleeve and a gage. The activating pin of the gage shall slide inside the guide sleeve with a slight amount of frictional resistance. The guide sleeve shall slide freely over the guide rod of the breaking head. The flowmeter gage shall be adjusted to zero when placed in position on the breaking head when each individual test specimen is inserted between the breaking head segments. Graduations of the flowmeter gage shall be in 0.01-in. (0.25-mm) divisions.

NOTE 4—Instead of the flowmeter, a micrometer dial or stress-strain recorder graduated in 0.001 in. (0.025 mm) may be used to measure flow.

2.10 *Ovens or Hot Plates*—Ovens or hot plates shall be provided for heating aggregates, bituminous material, specimen molds, compaction hammers, and other equipment to the required mixing and molding temperatures. It is recommended that the heating units be thermostatically controlled so as to maintain the required temperature within 5 F (2.8 C). Suitable shields, baffle plates or sand baths shall be used on the surfaces of the hot plates to minimize localized overheating.

2.11 *Mixing Apparatus*—Mechanical mixing is recommended. Any type of mechanical mixer may be used provided it can be maintained at the required mixing temperature and will produce a well-coated, homogeneous mixture of the required amount in the allowable time, and further provided that essentially all of the batch can be recovered. A metal pan or bowl of sufficient capacity and hand mixing may also be used.

2.12 *Water Bath*—The water bath shall be at least 6 in. (152.4 mm) deep and shall be thermostatically controlled so as to maintain the bath at 140 ± 1.8 F (60 ± 1.0 C) or 100 ± 1.8 F (37.8 ± 1 C). The tank shall have a perforated false bottom or be equipped with a shelf for supporting specimens 2 in. (50.8 mm) above the bottom of the bath.

2.13 *Air Bath*—The air bath for asphalt cut-back mixtures shall be thermostatically controlled and shall maintain the air temperature at 77 F ± 1.8 F (25 ± 1.0 C).

2.14 *Miscellaneous Equipment:*

2.14.1 *Containers* for heating aggregates, flat-bottom metal pans or other suitable containers.

2.14.2 *Containers* for heating bituminous material, either gill-type tins, beakers, pouring pots, or saucepans may be used.

2.14.3 *Mixing Tool*, either a steel trowel (garden type) or spatula, for spading and hand mixing.

2.14.4 *Thermometers* for determining temperatures of aggregates, bitumen, and bituminous mixtures. Armored-glass or dial-type thermometers with metal stems are recommended. A range from 50 to 400 F (9.9 to 204 C), with sensitivity of 5 F (2.8 C) is required.

2.14.5 *Thermometers* for water and air baths with a range from 68 to 158 F (20 to 70 C) sensitive to 0.4 F (0.2 C).

2.14.6 *Balance*, 2-kg capacity, sensitive to 0.1 g, for weighing molded specimens.

2.14.7 *Balance*, 5-kg capacity, sensitive to 1.0 g, for batching mixtures.

2.14.8 *Gloves* for handling hot equipment.

2.14.9 *Rubber Gloves* for removing specimens from water bath.

2.14.10 *Marking Crayons* for identifying specimens.

2.14.11 *Scoop*, flat bottom, for batching aggregates.

2.14.12 *Spoon*, large, for placing the mixture in the specimen molds.

3. Test Specimens

3.1 *Number of Specimens*—Prepare at least three specimens for each combination of aggregates and bitumen content.

3.2 *Preparation of Aggregates*—Dry aggregates to constant weight at 221 to 230 F (105 to

110 C) and separate the aggregates by dry-sieving into the desired size fractions.[2] The following size fractions are recommended:

1 to ¾ in. (25.0 to 19.0 mm)
¾ to ⅜ in. (19.0 to 9.5 mm)
⅜ in. to No. 4 (9.5 mm to 4.75 mm)
No. 4 to No. 8 (4.75 mm to 2.36 mm)
Passing No. 8 (2.36 mm)

3.3 *Determination of Mixing and Compacting Temperatures:*

3.3.1 The temperatures to which the asphalt cement and asphalt cut-back must be heated to produce a viscosity of 170 ± 20 cSt shall be the mixing temperature.

3.3.2 The temperature to which asphalt cement must be heated to produce a viscosity of 280 ± 30 cSt shall be the compacting temperature.

3.3.3 From a composition chart for the asphalt cut-back used, determine from its viscosity at 140 F (60 C) the percentage of solvent by weight. Also determine from the chart the viscosity at 140 F (60 C) of the asphalt cut-back after it has lost 50 percent of its solvent. The temperature determined from the viscosity temperature chart to which the asphalt cut-back must be heated to produce a viscosity of 280 ± 30 cSt after a loss of 50 percent of the original solvent content shall be the compacting temperature.

3.3.4 The temperature to which tar must be heated to produce Engler specific viscosities of 25 ± 3 and 40 ± 5 shall be respectively the mixing and compacting temperature.

3.4 *Preparation of Mixtures:*

3.4.1 Weigh into separate pans for each test specimen the amount of each size fraction required to produce a batch that will result in a compacted specimen 2.5 ± 0.05 in. (63.5 ± 1.27 mm) in height (about 1200 g). Place the pans on the hot plate or in the oven and heat to a temperature not exceeding the mixing temperature established in 3.3 by more than approximately 50 F (28 C) for asphalt cement and tar mixes and 25 F (14 C) for cut-back asphalt mixes. Charge the mixing bowl with the heated aggregate and dry mix thoroughly. Form a crater in the dry blended aggregate and weigh the preheated required amount of bituminous material into the mixture. For mixes prepared with cutback asphalt introduce the mixing blade in the mixing bowl and determine the total weight of the mix components plus bowl and blade before proceeding with mixing. Care must be exercised to prevent loss of the mix during mixing and subsequent handling. At this point, the temperature of the aggregate and bituminous material shall be within the limits of the mixing temperature established in 3.3. Mix the aggregate and bituminous material rapidly until thoroughly coated.

3.4.2 Following mixing, cure asphalt cut-back mixtures in a ventilated oven maintained at approximately 20 F (11.1 C) above the compaction temperature. Curing is to be continued in the mixing bowl until the precalculated weight of 50 percent solvent loss or more has been obtained. The mix may be stirred in a mixing bowl during curing to accelerate the solvent loss. However, care should be exercised to prevent loss of the mix. Weigh the mix during curing in successive intervals of 15 min initially and less than 10 min intervals as the weight of the mix at 50 percent solvent loss is approached.

3.5 *Compaction of Specimens:*

3.5.1 Thoroughly clean the specimen mold assembly and the face of the compaction hammer and heat them either in boiling water or on the hot plate to a temperature between 200 and 300 F (93.3 and 148.9 C). Place a piece of filter paper or paper toweling cut to size in the bottom of the mold before the mixture is introduced. Place the entire batch in the mold, spade the mixture vigorously with a heated spatula or trowel 15 times around the perimeter and 10 times over the interior. Remove the collar and smooth the surface of the mix with a trowel to a slightly rounded shape. Temperatures of the mixtures immediately prior to compaction shall be within the limits of the compacting temperature established in 3.3.

3.5.2 Replace the collar, place the mold assembly on the compaction pedestal in the mold holder, and unless otherwise specified, apply 50 blows with the compaction hammer with a free fall in 18 in. (457.2 mm). Hold the axis of the compaction hammer perpendicular to the base of the mold assembly during compaction. Remove the base plate and collar,

[2] Detailed requirements for these sieves are given in ASTM Specification E 11, for Wire-Cloth Sieves for Testing Purposes see *Annual Book of ASTM Standards*, Parts 15 and 41.

and reverse and reassemble the mold. Apply the same number of compaction blows to the face of the reversed specimen. After compaction, remove the base plate and place the sample extractor on that end of the specimen. Place the assembly with the extension collar up in the testing machine, apply pressure to the collar by means of the load transfer bar, and force the specimen into the extension collar. Lift the collar from the specimen. Carefully transfer the specimen to a smooth, flat surface and allow it to stand overnight at room temperature. Weigh, measure, and test the specimen.

NOTE 5—In general, specimens shall be cooled as specified in 3.5.2. When more rapid cooling is desired, table fans may be used. Mixtures that lack sufficient cohesion to result in the required cylindrical shape on removal from the mold immediately after compaction may be cooled in the mold in air until sufficient cohesion has developed to result in the proper cylindrical shape.

4. Procedure

4.1 Bring the specimens prepared with asphalt cement or tar to the specified temperature by immersing in the water bath 30 to 40 min or placing in the oven for 2 h. Maintain the bath or oven temperature at 140 ± 1.8 F (60 ± 1.0 C) for the asphalt cement specimens and 100 ± 1.8 F (37.8 ± 1.0 C) for tar specimens. Bring the specimens prepared with asphalt cut-back to the specified temperature by placing them in the air bath for a minimum of 2 h. Maintain the air bath temperature at 77 ± 1.8 F (25 ± 1.0 C). Thoroughly clean the guide rods and the inside surfaces of the test heads prior to making the test, and lubricate the guide rods so that the upper test head slides freely over them. The testing-head temperature shall be maintained between 70 to 100 F (21.1 to 37.8 C) using a water bath when required. Remove the specimen from the water bath, oven, or air bath, and place in the lower segment of the breaking head. Place the upper segment of the breaking head on the specimen, and place the complete assembly in position on the testing machine. Place the flowmeter, where used, in position over one of the guide rods and adjust the flowmeter to zero while holding the sleeve firmly against the upper segment of the breaking head. Hold the flowmeter sleeve firmly against the upper segment of the breaking head while the test load is being applied.

4.2 Apply the load to the specimen by means of the constant rate of movement of the load jack or testing-machine head of 2 in. (50.8 mm)/min until the maximum load is reached and the load decreases as indicated by the dial. Record the maximum load noted on the testing machine or converted from the maximum micrometer dial reading. Release the flowmeter sleeve or note the micrometer dial reading, where used, the instant the maximum load begins to decrease. Note and record the indicated flow value or equivalent units in hundredths of an inch (twenty-five hundredths of a millimetre) if a micrometer dial is used to measure the flow. The elapsed time for the test from removal of the test specimen from the water bath to the maximum load determination shall not exceed 30 s.

NOTE 6—For core specimens, correct the load when thickness is other than 2½ in. (63.5 mm) by using the proper multiplying factor from Table 1.

5. Report

5.1 The report shall include the following information:

5.1.1 Type of sample tested (laboratory sample or pavement core specimen).

NOTE 6—For core specimens, the height of each test specimen in inches (or millimetres) shall be reported.

5.1.2 Average maximum load in pounds-force (or newtons) of at least three specimens, corrected when required,

5.1.3 Average flow value, in hundredths of an inch, twenty-five hundredths of a millimetre, of three specimens, and

5.1.4 Test temperature.

TABLE 1 Stability Correlation Ratios[a]

Volume of Specimen, cm^3	Approximate Thickness of Specimen, in.[b]	mm	Correlation Ratio
200 to 213	1	25.4	5.56
214 to 225	1$\frac{1}{16}$	27.0	5.00
226 to 237	1$\frac{1}{8}$	28.6	4.55
238 to 250	1$\frac{3}{16}$	30.2	4.17
251 to 264	1$\frac{1}{4}$	31.8	3.85
265 to 276	1$\frac{5}{16}$	33.3	3.57
277 to 289	1$\frac{3}{8}$	34.9	3.33
290 to 301	1$\frac{7}{16}$	36.5	3.03
302 to 316	1$\frac{1}{2}$	38.1	2.78
317 to 328	1$\frac{9}{16}$	39.7	2.50
329 to 340	1$\frac{5}{8}$	41.3	2.27
341 to 353	1$\frac{11}{16}$	42.9	2.08
354 to 367	1$\frac{3}{4}$	44.4	1.92
368 to 379	1$\frac{13}{16}$	46.0	1.79
380 to 392	1$\frac{7}{8}$	47.6	1.67
393 to 405	1$\frac{15}{16}$	49.2	1.56
406 to 420	2	50.8	1.47
421 to 431	2$\frac{1}{16}$	52.4	1.39
432 to 443	2$\frac{1}{8}$	54.0	1.32
444 to 456	2$\frac{3}{16}$	55.6	1.25
457 to 470	2$\frac{1}{4}$	57.2	1.19
471 to 482	2$\frac{5}{16}$	58.7	1.14
483 to 495	2$\frac{3}{8}$	60.3	1.09
496 to 508	2$\frac{7}{16}$	61.9	1.04
509 to 522	2$\frac{1}{2}$	63.5	1.00
523 to 535	2$\frac{9}{16}$	64.0	0.96
536 to 546	2$\frac{5}{8}$	65.1	0.93
547 to 559	2$\frac{11}{16}$	66.7	0.89
560 to 573	2$\frac{3}{4}$	68.3	0.86
574 to 585	2$\frac{13}{16}$	71.4	0.83
586 to 598	2$\frac{7}{8}$	73.0	0.81
599 to 610	2$\frac{15}{16}$	74.6	0.78
611 to 625	3	76.2	0.76

[a] The measured stability of a specimen multiplied by the ratio for the thickness of the specimen equals the corrected stability for a 2$\frac{1}{2}$-in. (63.5 mm) specimen.

[b] Volume-thickness relationship is based on a specimen diameter of 4 in. (101.6 mm).

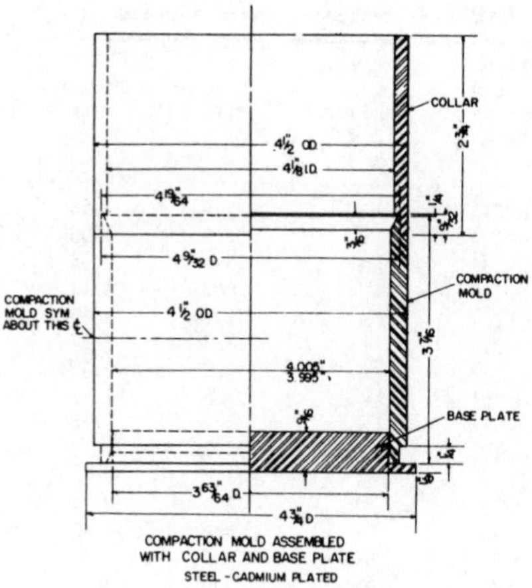

FIG. 1 Compaction Mold.

Table of Equivalents for Figs. 1 and 3

U. S. Customary Units, in.	Metric Equivalents, mm	U. S. Customary Units, in.	Metric Equivalents, mm	U. S. Customary Units, in.	Metric Equivalents, mm	U. S. Customary Units, in.	Metric Equivalents, mm
0.005	0.11	11/16	17.5	2 5/16	58.7	4 1/8	104.8
1/32	0.8	3/4	19.0	2 1/2	63.5	4 9/32	108.7
1/16	1.6	7/8	22.2	2 3/4	69.8	4 19/64	109.1
1/8	3.2	15/16	23.8	2 7/8	73.0	4 1/2	114.3
3/16	4.8	1	25.4	3	76.2	4 5/8	117.5
1/4	6.4	1 1/8	28.6	3 1/4	82.6	4 3/4	120.6
9/32	7.1	1 1/4	31.8	3 7/16	87.3	5 1/16	128.6
3/8	9.5	1 3/8	34.9	3 7/8	98.4	5 1/8	130.2
0.496	12.6	1 1/2	38.1	3 63/64	101.2	5 3/4	146.0
0.499	12.67	1 5/8	41.3	3.990	101.35	6	152.4
1/2	12.7	1 3/4	44.4	3.995	101.47	6 1/4	158.8
9/16	14.3	2	50.8	4	101.6	7 5/8	193.7
5/8	15.9	2 1/4	57.2	4.005	101.73	27	685.8

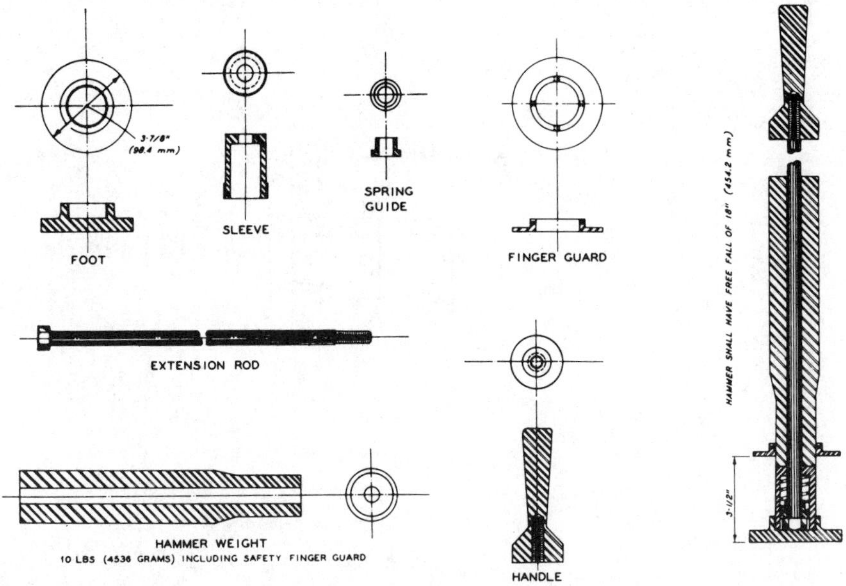

FIG. 2 Compaction Hammer.

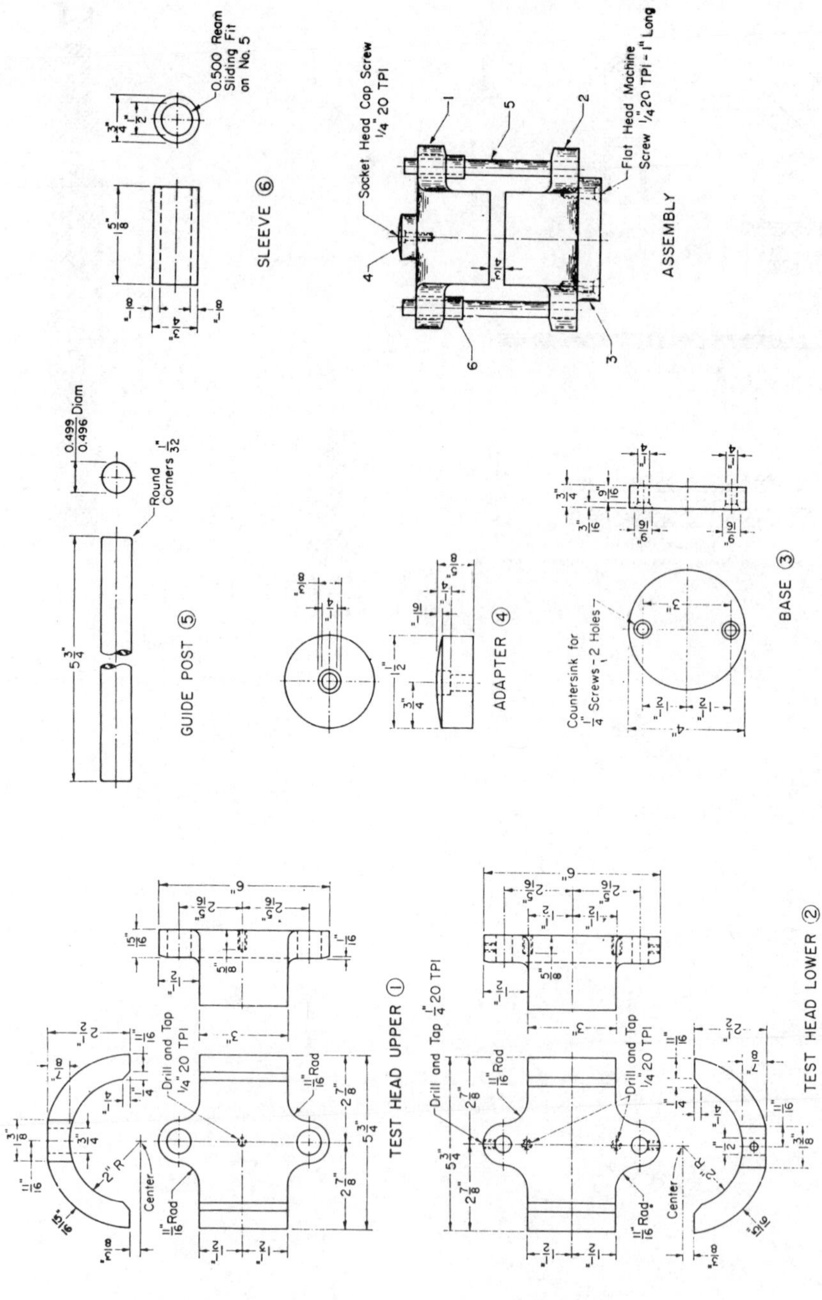

(Table of Equivalents same as for Fig. 1.)

FIG. 3 Breaking Head.

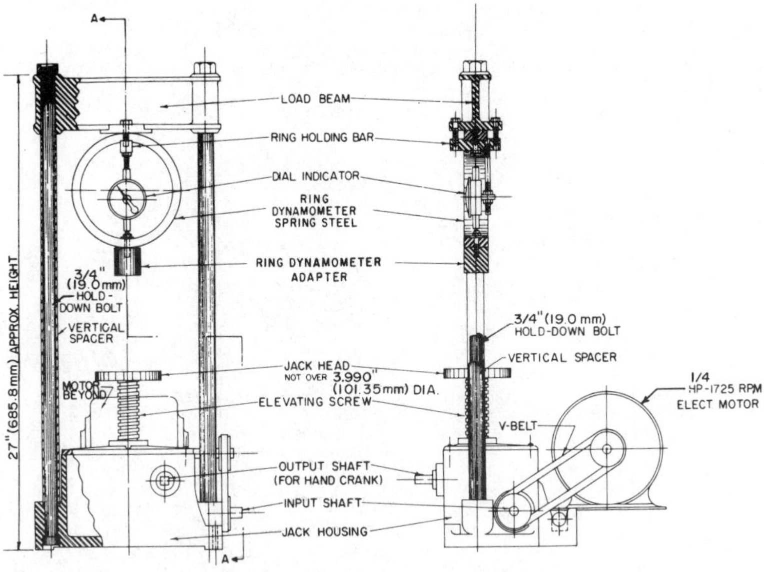

FIG. 4 Compression Testing Machine.

The American Society for Testing and Materials takes no position respecting the validity of any patent rights asserted in connection with any item mentioned in this standard. Users of this standard are expressly advised that determination of the validity of any such patent rights, and the risk of infringement of such rights, is entirely their own responsibility.

This standard is subject to revision at any time by the responsible technical committee and must be reviewed every five years and if not revised, either reapproved or withdrawn. Your comments are invited either for revision of this standard or for additional standards and should be addressed to ASTM Headquarters. Your comments will receive careful consideration at a meeting of the responsible technical committee, which you may attend. If you feel that your comments have not received a fair hearing you should make your views known to the ASTM Committee on Standards, 1916 Race St., Philadelphia, Pa. 19103, which will schedule a further hearing regarding your comments. Failing satisfaction there, you may appeal to the ASTM Board of Directors.

INDEX

D

P

Q

R

S